Lecture Notes in Computer Science 10122

Commenced Publication in 1973
Founding and Former Series Editors:
Gerhard Goos, Juris Hartmanis, and Jan van Leeuwen

More information about this series at http://www.springer.com/series/7409

Panos M. Pardalos · Piero Conca
Giovanni Giuffrida · Giuseppe Nicosia (Eds.)

Machine Learning, Optimization, and Big Data

Second International Workshop, MOD 2016
Volterra, Italy, August 26–29, 2016
Revised Selected Papers

Springer

Editors
Panos M. Pardalos
Department of Industrial and Systems Engineering
University of Florida
Gainesville, FL
USA

Piero Conca
Semantic Technology Laboratory
National Research Council (CNR)
Catania
Italy

Giovanni Giuffrida
Dipartimento di Sociologia e Metodi della Ricerca Sociale
Università di Catania
Catania
Italy

Giuseppe Nicosia
Department of Mathematics and Computer Science
University of Catania
Catania
Italy

ISSN 0302-9743 ISSN 1611-3349 (electronic)
Lecture Notes in Computer Science
ISBN 978-3-319-51468-0 ISBN 978-3-319-51469-7 (eBook)
DOI 10.1007/978-3-319-51469-7

Library of Congress Control Number: 2016961276

LNCS Sublibrary: SL3 – Information Systems and Applications, incl. Internet/Web, and HCI

Printed on acid-free paper

This Springer imprint is published by Springer Nature
The registered company is Springer International Publishing AG
The registered company address is: Gewerbestrasse 11, 6330 Cham, Switzerland

Preface

MOD is an international workshop embracing the fields of machine learning, optimization, and big data. The second edition, MOD 2016, was organized during August 26–29, 2016, in Volterra (Pisa, Italy), a stunning medieval town dominating the picturesque countryside of Tuscany.

The key role of machine learning, optimization, and big data in developing solutions to some of the greatest challenges we are facing is undeniable. MOD 2016 attracted leading experts from the academic world and industry with the aim of strengthening the connection between these institutions. The 2016 edition of MOD represented a great opportunity for professors, scientists, industry experts, and postgraduate students to learn about recent developments in their own research areas and to learn about research in contiguous research areas, with the aim of creating an environment to share ideas and trigger new collaborations.

As program chairs, it was an honor to organize a premiere workshop in these areas and to have received a large variety of innovative and original scientific contributions.

During this edition, four plenary lectures were presented:

Nello Cristianini, Bristol University, UK
George Michailidis, University of Florida, USA
Stephen H. Muggleton, Imperial College London, UK
Panos Pardalos, University of Florida, USA

There were also two tutorial speakers:

Luigi Malagó, Shinshu University, Nagano, Japan
Luca Oneto and Davide Anguita, Polytechnic School, University of Genova, Italy

Furthermore, an industrial panel on "Machine Learning, Optimization and Data Science for Real-World Applications" was also offered:

Amr Awadallah, Founder and CTO at Cloudera, San Francisco, USA
Giovanni Giuffrida, CEO and co-founder at Neodata Group, Italy
Andy Petrella, Data Scientist and co-founder at Data Fellas, Liege, Belgium
Daniele Quercia, Head of Social Dynamics group at Bell Labs, Cambridge, UK
Fabrizio Silvestri, Facebook Inc., USA
Moderator: Donato Malerba, University of Bari, Italy and Consorzio Interuniversitario Nazionale per l'Informatica (CINI)

MOD 2016 received 97 submissions, and each manuscript was independently reviewed via a blind review process by a committee formed by at least five members. These proceedings contain 40 research articles written by leading scientists in the fields of machine learning, computational optimization, and data science presenting a substantial array of ideas, technologies, algorithms, methods and applications.

This conference could not have been organized without the contributions of these researchers, and we thank them all for participating. A sincere thank goes also to all the Program Committee, formed by more than 300 scientists from academia and industry, for their valuable work of selecting the scientific contributions.

Finally, we would like to express our appreciation to the keynote speakers, tutorial speakers, and the industrial panel who accepted our invitation, and to all the authors who submitted their research papers to MOD 2016.

August 2016

Panos M. Pardalos
Piero Conca
Giovanni Giuffrida
Giuseppe Nicosia

Organization

MOD 2016 Committees

General Chair

Giuseppe Nicosia	University of Catania, Italy

Conference and Technical Program Committee Co-chairs

Panos Pardalos	University of Florida, USA
Piero Conca	University of Catania, Italy
Giovanni Giuffrida	University of Catania, Italy
Giuseppe Nicosia	University of Catania, Italy

Tutorial Chair

Giuseppe Narzisi	New York Genome Center, NY, USA

Industrial Session Chairs

Ilaria Bordino	UniCredit R&D, Italy
Marco Firrincieli	UniCredit R&D, Italy
Fabio Fumarola	UniCredit R&D, Italy
Francesco Gullo	UniCredit R&D, Italy

Organizing Committee

Piero Conca	CNR and University of Catania, Italy
Jole Costanza	Italian Institute of Technology, Milan, Italy
Giuseppe Narzisi	New York Genome Center, USA
Andrea Patane'	University of Catania, Italy
Andrea Santoro	University of Catania, Italy
Renato Umeton	Harvard University, USA

Publicity Chair

Giovanni Luca Murabito	DiGi Apps, Italy

Technical Program Committee

Ajith Abraham	Machine Intelligence Research Labs, USA
Andy Adamatzky	University of the West of England, UK
Agostinho Agra	University of Aveiro, Portugal
Hernán Aguirre	Shinshu University, Japan
Nesreen Ahmed	Intel Research Labs, USA

Youhei Akimbo	Shinshu University, Japan
Leman Akoglu	Stony Brook University, USA
Richard Allmendinger	University College London, UK
Paula Amaral	University Nova de Lisboa, Portugal
Ekart Aniko	Aston University, UK
Paolo Arena	University of Catania, Italy
Ashwin Arulselvan	University of Strathclyde, UK
Jason Atkin	University of Nottingham, UK
Martin Atzmueller	University of Kassel, Germany
Chloé-Agathe Azencott	Mines ParisTech Institut Curie, France
Jaume Bacardit	Newcastle University, UK
James Bailey	University of Melbourne, Australia
Baski Balasundaram	Oklahoma State University, USA
Wolfgang Banzhaf	Memorial University, Canada
Helio Barbosa	Laboratório Nacional Computação Científica, Brazil
Thomas Bartz-Beielstein	Cologne University of Applied Sciences, Germany
Simone Bassis	University of Milan, Italy
Christian Bauckhage	Fraunhofer IAIS, Germany
Aurélien Bellet	Télécom ParisTech, France
Gerardo Beni	University of California at Riverside, USA
Tanya Berger-Wolf	University of Illinois at Chicago, USA
Heder Bernardino	Universidade Federal de Juiz de Fora, Brazil
Daniel Berrar	Shibaura Institute of Technology, Japan
Martin Berzins	University of Utah, USA
Rajdeep Bhowmik	Cisco Systems, Inc., USA
Albert Bifet	University of Waikato, New Zealand
Mauro Birattari	Université Libre de Bruxelles, Belgium
J. Blachut	University of Liverpool, UK
Konstantinos Blekas	University of Ioannina, Greece
Maria J. Blesa	Universitat Politècnica de Catalunya, Spain
Christian Blum	Basque Foundation for Science, Spain
Flavia Bonomo	Universidad de Buenos Aires, Argentina
Gianluca Bontempi	Université Libre de Bruxelles, Belgium
Pascal Bouvry	University of Luxembourg, Luxembourg
Larry Bull	University of the West of England, UK
Tadeusz Burczynski	Polish Academy of Sciences, Poland
Róbert Busa-Fekete	University of Paderborn, Germany
Sergio Butenko	Texas A&M University, USA
Stefano Cagnoni	University of Parma, Italy
Mustafa Canim	IBM T.J. Watson Research Center, USA
Luigia Carlucci Aiello	Sapienza Università di Roma, Italy
Tania Cerquitelli	Politecnico di Torino, Italy
Uday Chakraborty	University of Missouri St. Louis, USA
Lijun Chang	University of New South Wales, Australia
W. Art Chaovalitwongse	University of Washington, USA
Ying-Ping Chen	National Chiao Tung University, Taiwan

Koke Chen	Wright State University, USA
Kaifeng Chen	NEC Labs America, USA
Silvia Chiusano	Politecnico di Torino, Italy
Miroslav Chlebik	University of Sussex, UK
Sung-Baa Cho	Yonsei University, South Korea
Siang Yew Chong	University of Nottingham, Malaysia Campus, Malaysia
Philippe Codognet	University of Tokyo, Japan
Pietro Colombo	Università dell'Insubria, Italy
Ernesto Costa	University of Coimbra, Portugal
Jole Costanza	Fondazione Istituto Italiano di Tecnologia, Italy
Maria Daltayanni	University of California Santa Cruz, USA
Raj Das	University of Auckland, New Zealand
Mahashweta Das	Hewlett Packard Labs, USA
Kalyanmoy Deb	Michigan State University, USA
Noel Depalma	Joseph Fourier University, France
Clarisse Dhaenens	University of Lille 1, France
Luigi Di Caro	University of Turin, Italy
Gianni Di Caro	IDSIA, Switzerland
Tom Diethe	University of Bristol, UK
Federico Divina	Pablo de Olavide University, Spain
Stephan Doerfel	University of Kassel, Germany
Karl Doerner	Johannes Kepler University Linz, Austria
Rafal Drezewski	AGH University of Science and Technology, Poland
Ding-Zhou Du	University of Texas at Dallas, USA
George S. Dulikravich	Florida International University, USA
Talbi El-Ghazali	University of Lille, France
Michael Emmerich	Leiden University, The Netherlands
Andries Engelbrecht	University of Pretoria, South Africa
Roberto Esposito	University of Turin, Italy
Cesar Ferri	Universitat Politècnica de València, Spain
Steffen Finck	Vorarlberg University of Applied Sciences, Austria
Jordi Fonollosa	Institute for Bioengineering of Catalonia, Spain
Carlos M. Fonseca	University of Coimbra, Portugal
Giuditta Franco	University of Verona, Italy
Piero Fraternali	Politecnico di Milano, Italy
Valerio Freschi	University of Urbino, Italy
Enrique Frias-Martinez	Telefonica Research, Spain
Marcus Gallagher	University of Queensland, Australia
Patrick Gallinari	Pierre et Marie Curie University, France
Xavier Gandibleux	University of Nantes, France
Amir Hossein Gandomi	The University of Akron, USA
Inmaculada Garcia Fernandez	University of Almeria, Spain
Deon Garrett	Icelandic Institute Intelligent Machine, Iceland
Paolo Garza	Politecnico di Torino, Italy
Martin Josef Geiger	Helmut Schmidt University, Germany

Michel Gendreau	École Polytechnique de Montréal, Canada
Kyriakos Giannakoglou	National Technical University of Athens, Greece
Giovanni Giuffrida	University of Catania, Italy and Neodata Intelligence Inc., Italy
Aris Gkoulalas Divanis	IBM Dublin Research Lab, Ireland
Christian Gogu	Université Toulouse III, France
Michael Granitzer	University of Passau, Germany
Mario Guarracino	ICAR-CNR, Italy
Heiko Hamann	University of Paderborn, Germany
Jin-Kao Hao	University of Angers, France
William Hart	Sandia Labs, USA
Richard F. Hartl	University of Vienna, Austria
Mohammad Hasan	Indiana University Purdue University, USA
Geir Hasle	SINTEF, Norway
Verena Heidrich-Meisner	Extraterrestrial Physics CAU Kiel, Germany
Eligius Hendrix	Universidad de Málaga, Spain
Carlos Henggeler Antunes	University of Coimbra, Portugal
Alfredo G. Hernández-Díaz	Pablo de Olvide University, Spain
Francisco Herrera	University of Granada, Spain
J. Michael Herrmann	University of Edinburgh, UK
Jaakko Hollmén	Aalto University, Finland
Vasant Honavar	Penn State University, USA
Hongxuan Huang	Tsinghua University, China
Fabric Huet	University of Nice, France
Sam Idicula	Oracle, USA
Yoshiharu Ishikawa	Nagoya University, Japan
Christian Jacob	University of Calgary, Canada
Hasan Jamil	University of Idaho, USA
Gareth Jones	Dublin City University, Ireland
Laetitia Jourdan	Inria/LIFL/CNRS, France
Narendra Jussien	Ecole des Mines de Nantes/LINA, France
Valeriy Kalyagin	Higher School of Economics, Russia
George Karakostas	McMaster University, Canada
George Karypis	University of Minnesota, USA
Ioannis Katakis	University of Athens, Greece
Saurabh Kataria	Xerox Research, USA
Graham Kendall	University of Nottingham, UK
Kristian Kersting	TU Dortmund University, Germany
Zeynep Kiziltan	University of Bologna, Italy
Joshua Knowles	University of Manchester, UK
Andrzej Kochut	IBM T.J. Watson Research Center, USA
Yun Sing Koh	University of Auckland, New Zealand
Igor Konnov	Kazan University, Russia
Petros Koumoutsakos	ETHZ, Switzerland
Georg Krempl	University of Magdeburg, Germany
Erhun Kundakcioglu	Ozyegin University, Turkey

Halina Kwasnicka	Wroclaw University of Technology, Poland
Joerg Laessig	University of Applied Sciences Zittau/Görlitz, Germany
Albert Y.S. Lam	The University of Hong Kong, Hong Kong, SAR China
Niklas Lavesson	Blekinge Institute of Technology, Sweden
Kang Li	Groupon Inc., USA
Edo Liberty	Yahoo Labs, USA
Arnaud Liefooghe	University of Lille, France
Weifeng Liu	China University of Petroleum, China
Giosue' Lo Bosco	Università di Palermo, Italy
Fernando Lobo	University of Algarve, Portugal
Marco Locatelli	University of Parma, Italy
Manuel Lopez-Ibanez	University of Manchester, UK
Jose A. Lozano	The University of the Basque Country, Spain
Paul Lu	University of Alberta, Canada
Angelo Lucia	University of Rhode Island, USA
Luigi Malagò	Shinshu University, Japan
Lina Mallozzi	University of Naples Federico II, Italy
Vittorio Maniezzo	University of Bologna, Italy
Yannis Manolopoulos	Aristotle University of Thessaloniki, Greece
Marco Maratea	University of Genova, Italy
Elena Marchiori	Radboud University, The Netherlands
Tiziana Margaria	Lero, Ireland
Juan Enrique Martinez-Legaz	Universitat Autònoma de Barcelona, Spain
Basseur Matthieu	LERIA Angers, France
Giancarlo Mauri	University of Milano-Bicocca, Italy
Suzanne McIntosh	NYU Courant Institute and Cloudera Inc., USA
Gabor Melli	VigLink, USA
Silja Meyer-Nieberg	Universität der Bundeswehr München, Germany
Alessio Micheli	University of Pisa, Italy
Martin Middendorf	University of Leipzig, Germany
Taneli Mielikäinen	Nokia, Finland
Kaisa Miettinen	University of Jyväskylä, Finland
Marco A. Montes De Oca	Clypd, Inc., USA
Antonio Mora	University of Granada, Spain
Christian L. Müller	Simons Center for Data Analysis, USA
Mohamed Nadif	University of Paris Descartes, France
Hidemoto Nakada	National Institute of Advanced Industrial, Japan
Amir Nakib	Université Paris Est Créteil, France
Mirco Nanni	ISTI-CNR Pisa, Italy
Giuseppe Nicosia	University of Catania, Italy
Jian-Yun Nie	Université de Montréal, Canada
Xia Ning	Indiana University Purdue, USA
Eirini Ntoutsi	Ludwig-Maximilians-Universitüt München, Germany

Salvatore Orlando	Università Ca' Foscari Venezia, Italy
Sinno Jialin Pan	Nanyang Technological University, Singapore
Pan Pan	Alibaba Inc., China
George Papastefanatos	IMIS/RC Athena, Greece
Luis Paquete	University of Coimbra, Portugal
Andrew J. Parkes	University of Nottingham, USA
Ioannis Partalas	Viseo R&D, France
Jun Pei	University of Florida, USA
Nikos Pelekis	University of Piraeus, Greece
David Pelta	University of Granada, Spain
Diego Perez	University of Essex, UK
Vincenzo Piuri	University of Milan, Italy
Silvia Poles	Noesis Solutions NV, Belgium
George Potamias	FORTH-ICS, Greece
Adam Prugel-Bennett	University of Southampton, UK
Buyue Qian	IBM T.J. Watson Research Center, USA
Chao Qian	Nanjing University, China
Günther Rail	Technische Universität Wien, Austria
Helena Ramalhinho	Universitat Pompeu Fabra, Spain
Jan Ramon	Inria Lille, France
Vitorino Ramos	Technical University of Lisbon, Portugal
Zbigniew Ras	University of North Carolina, USA
Khaled Rasheed	University of Georgia, USA
Jan Rauch	University of Economics Prague, Czech Republic
Steffen Rebennack	University of Florida, USA
Celso Ribeiro	Universidade Federal Fluminense, Brazil
Florian Richoux	Université de Nantes, France
Juan J. Rodriguez	University of Burgos, Spain
Andrea Roli	University of Bologna, Italy
Samuel Rota Bulò	Fondazione Bruno Kessler, Italy
Arnab Roy	Fujitsu Laboratories of America, USA
Alessandro Rozza	Università di Napoli-Parthenope, Italy
Thomas Runarsson	University of Iceland, Iceland
Berc Rustem	Imperial College London, UK
Florin Rusu	University of California, Merced, USA
Nick Sahinidis	Carnegie Mellon University, USA
Lorenza Saitta	Università del Piemonte Orientale, Italy
Horst Samulowitz	IBM Research, USA
Ganesh Ram Santhanam	Iowa State University, USA
Vítor Santos Costa	Universidade do Porto, Portugal
Claudio Sartori	University of Bologna, Italy
Frédéric Saubion	University of Angers, France
Andrea Schaerf	University of Udine, Italy
Robert Schaefer	AGH University of Science and Technology, Poland
Fabio Schoen	University of Florence, Italy
Christoph Schommer	University of Luxembourg, Luxembourg

Oliver Schuetze	CINVESTAV-IPN, Mexico
Michèle Sebag	University of Paris-Sud, France
Giovanni Semeraro	University of Bari, Italy
Roberto Serra	University of Modena Reggio Emilia, Italy
Marc Sevaux	Université de Bretagne-Sud, France
Junming Shao	University of Electronic Science and Technology, China
Ruey-Lin Sheu	National Cheng-Kung University, Taiwan
Patrick Siarry	Université de Paris 12, France
Dan Simovici	University of Massachusetts Boston, USA
Karthik Sindhya	University of Jyväskylä, Finland
Anthony Man-ChoSo	The Chinese University of Hong Kong, Hong Kong, SAR China
Christine Solnon	LIRIS — CNRS, France
Oliver Stein	Karlsruhe Institute of Technology, Germany
Catalin Stoean	University of Craiova, Romania
Thomas Stützle	Université Libre de Bruxelles, Belgium
Ponnuthurai Suganthan	Nanyang Technological University, Singapore
Johan Suykens	K.U. Leuven, Belgium
El-Ghazali Talbi	University of Lille, France
Domenico Talia	University of Calabria, Italy
Wei Tan	IBM, USA
Letizia Tanca	Politecnico di Milano, Italy
Ke Tang	University of Science and Technology of China, China
Andrea Tettamanzi	University Nice Sophia Antipolis, France
Jerzy Tiuryn	Warsaw University, Poland
Heike Trautmann	TU Dortmund University, Germany
Vincent S. Tseng	National Chiao Tung University, Taiwan
Theodoros Tzouramanis	University of the Aegean, Greece
Satish Ukkusuri	Purdue University, USA
Giorgio Valentini	University of Milan, Italy
Pascal Van Hentenryck	University of Michigan, USA
Analucia Varbanescu	University of Amsterdam, The Netherlands
Carlos A. Varela	Rensselaer Polytechnic Institute, USA
Iraklis Varlamis	Harokopio University of Athens, Greece
Eleni Vasilaki	University of Sheffield, UK
Sébastien Verel	Université du Littoral Côte d'Opale, France
Vassilios Verykios	Hellenic Open University, Greece
Henna Viktor	University of Ottawa, Canada
Maksims Volkovs	University of Toronto, Canada
Dean Vucinic	Vrije Universiteit Brussel, Belgium
Jason Wang	New Jersey Institute of Technology, USA
Jianwu Wang	University of Maryland, USA
Lipo Wang	NTU, Singapore
Liqiang Wang	University of Wyoming, USA
Lin Wu	The University of Adelaide, Australia

Yinglong Xia	IBM T.J. Watson Research Center, USA
Ning Xiong	Mälardalen University, Sweden
Chang Xu	Peking University, China
Xin Xu	George Washington University, USA
Shengxiang Yang	De Montfort University, UK
Qi Yu	Rochester Institute of Technology, USA
Tina Yu	Memorial University of Newfoundland, Canada
Kunpeng Zhang	University of Illinois at Chicago, USA
Nan Zhang	The George Washington University, USA
Qingfu Zhang	City University of Hong Kong, Hong Kong, SAR China
Rui Zhang	IBM Research — Almaden, USA
Ying Zhao	Tsinghua University, China
Bin Zhou	University of Maryland, USA
Zhi-Hua Zhou	Nanjing University, China
Djamel A. Zighed	University of Lyon 2, France
Antennas Zilinskas	Vilnius University, Lithuania
Julius Zilinskas	Vilnius University, Lithuania

Contents

Machine Learning: Multi-site Evidence-Based Best Practice Discovery

Eva K. Lee[1,2,3](✉), Yuanbo Wang[1,2,3], Matthew S. Hagen[1,2,3], Xin Wei[1,2,3], Robert A. Davis[4,5], and Brent M. Egan[4,5]

[1] Center for Operations Research in Medicine and HealthCare, Atlanta, GA, USA
eva.lee@gatech.edu
[2] NSF I/UCRC Center for Health Organization Transformation, Atlanta, GA, USA
[3] Georgia Institute of Technology, Atlanta, GA, USA
[4] University of South Carolina School of Medicine, Greenville, SC, USA
[5] Care Coordination Institute, Greenville, SC, USA

Abstract. This study establishes interoperability among electronic medical records from 737 healthcare sites and performs machine learning for best practice discovery. A mapping algorithm is designed to disambiguate free text entries and to provide a unique and unified way to link content to structured medical concepts despite the extreme variations that can occur during clinical diagnosis documentation. Redundancy is reduced through concept mapping. A SNOMED-CT graph database is created to allow for rapid data access and queries. These integrated data can be accessed through a secured web-based portal. A classification model (DAMIP) is then designed to uncover discriminatory characteristics that can predict the quality of treatment outcome. We demonstrate system usability by analyzing Type II diabetic patients. DAMIP establishes a classification rule on a training set which results in greater than 80% blind predictive accuracy on an independent set of patients. By including features obtained from structured concept mapping, the predictive accuracy is improved to over 88%. The results facilitate evidence-based treatment and optimization of site performance through best practice dissemination and knowledge transfer.

1 Introduction

Individual health systems provide various services and allocate different resources for patient care. Healthcare resources including professional and staff time are often constrained. Making clinical decisions is a complicated task since it requires physicians to infer information from a given case and determine a best treatment based on their knowledge [1]. Addressing these problems is essential for delivering effective care plans to patients.

Data from electronic medical records (EMRs) can reveal critical variables that impact treatment outcomes and inform allocation of limited time and resources, allowing physicians to practice evidence-based treatment tailored to individual patient

P.M. Pardalos et al. (Eds.): MOD 2016, LNCS 10122, pp. 1–15, 2016.
DOI: 10.1007/978-3-319-51469-7_1

conditions. On a larger scale, realistically modifiable social determinants of health that will improve community health can potentially be discovered and addressed.

Although EMR adoption is spreading across the industry, many providers continue to document clinical findings, procedures and outcomes with "free text" natural language on their EMRs [2]. They have difficulty (manually) mapping concepts to standardized terminologies and struggle with application programs that use structured clinical data. This creates challenges for multi-site comparative effectiveness studies.

Standardized clinical terminologies are essential in facilitating interoperability among EMR systems. They allow seamless sharing and exchange of healthcare information for quality care delivery and coordination among multiple sites. However, the volume and number of available clinical terminologies are large and are expanding. Further, due to the increase in medical knowledge, and the continued development of more advanced computerized medical systems, the use of clinical terminologies has extended beyond diagnostic classification [3]. SNOMED-CT is a multidisciplinary terminology system that is used for clinical decision support, ICU monitoring, indexing medical records, medical research, and disease surveillance [4]. LOINC is a set of universal names for expressing laboratory tests and clinical observations [5]. RxNorm is a normalized naming system for medicines and drugs [6]. The Unified Medical Language System (UMLS) is a terminology integration system developed by the US National Library of Medicine (NLM) to promote the interoperability of systems and mapping between the multitude of available clinical terminologies [7].

Many systems have been developed to map heterogeneous terminologies to support communication and semantic interoperability between healthcare centers. STRIDE mapped RxNorm concepts to the SNOMED-CT hierarchy and used the RxNorm relationships in UMLS to link pharmacy data from two EMR sources in the Stanford University Medical Center [8]. Carlo et al. classified ICD-9 diagnoses from unstructured discharge summaries using mapping tables in UMLS [9]. Patel and Cimino used the existing linkages in UMLS to predict new potential terminological relationships [10]. While many of these studies only focus on one standardized concept, our work herein designs a mapping system that has the ability to accurately map medications, laboratory results, and diagnosis entries from multiple EMRs. Each of the entries is mapped to predefined terms in the SNOMED-CT ontology. Due to the hierarchical nature of SNOMED-CT, similarities between patient diagnoses, laboratory results, and medications can be found more easily. Hence mapped concepts can be generalized to find shared characteristics among patients. Our work thus creates a more powerful system than previous studies, because multiple types of concepts can be mapped and compared in a hierarchical manner.

This study includes 737 clinical sites and de-identified data for over 2.7 million patients with data collected from January 1990 to December 2012. To the best of our knowledge analysis of EMR data across hundreds of healthcare sites and millions of patients has not been attempted previously. Such analysis requires effective database management, data extraction, preprocessing, and integration. In addition, temporal data mining of longitudinal health data cannot currently be achieved through statistically and computationally efficient methodologies and is still under-explored [1]. This is a particularly important issue when analyzing outcome and health conditions for chronic disease patients.

In this paper, we first establish interoperability among EMRs from 737 clinical sites by developing a system that can accurately map free text to concise structured medical concepts. Multiple concepts are mapped, including patient diagnoses, laboratory results, and medications, which allows shared characterization and hierarchical comparison. We then leverage a mixed integer programming-based classification model (DAMIP) [11] to establish classification rules with relatively small subsets of discriminatory features that can be used to predict diabetes treatment outcome. Previous studies in diabetes have identified features such as demographics, obesity, hypertension, and genetics that appear to be linked to the development and progression of type II diabetes [12–14]. However, little is known about how these features interact with treatment characteristics to affect patient outcome.

2 Methods and Design

2.1 Integrating and Mapping of Data to Standardized Clinical Concepts and Terminologies

This study utilized EMR data of 2.7 million patients collected from 737 healthcare facilities. A relational database was first designed with Postgres 9.3 to store these data. Thirteen tables were created containing patient records pertaining to procedures, demographics, diagnosis codes, laboratory measurements and medications. Indexes were developed for each table to enable rapid search and table joins for data querying. The data size for indexes is an additional 11 GB, totaling 27 GB for the entire database. We label this the CCI-health database, where CCI stands for Care Coordination Institute.

In the CCI-health database, 2.46 million patients are associated with a diagnosis, 1.33 million are associated with laboratories, and 955,000 are linked to medications. Laboratory and medication records are described with free text entries without unique codes for each entry. Since clinicians may describe identical treatments with many possible variations, it is essential to map entries to structured concepts without ambiguity. Overall 803 unique lab phrases and 9,755 medication phrases were extracted from the patient records. Metamap [15–17], developed at the National Library of Medicine, is a natural language processing tool that maps biomedical text to UMLS [7] concepts. In this study, we used Metamap to recognize laboratory terms from the LOINC [5] terminology and medication terms from the RxNorm [6] terminology. This was done respectively using the UMLS MRCONSO and RXNCONSO tables. The final step was to map associated terms to concepts in the SNOMED-CT ontology. SNOMED-CT [4] is a comprehensive medical ontology with detailed hierarchical relationships containing over 521,844 concepts and 4.65 million relationships (July 2015 release). LOINC and RxNorm terms established from the CCI-health database were linked to SNOMED using the UMLS MRREL and RXNREL tables. In our implementation, for LOINC, only concepts that have the name 'procedure' were returned from the MRREL table. For RxNorm, only concepts that have "has_form", "has_ingredient", and "has_tradename" relationships were returned from the RXNREL table. When medication entries in an EMR and a SNOMED concept were named

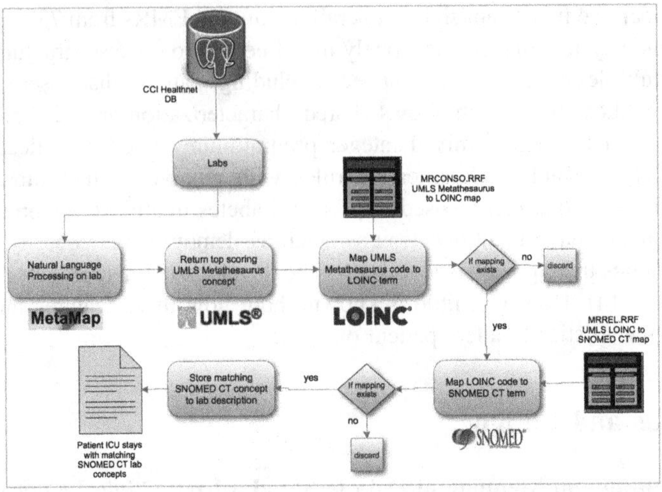

Fig. 1. The diagram shows the mapping procedure for laboratory phrases to SNOMED-CT concepts.

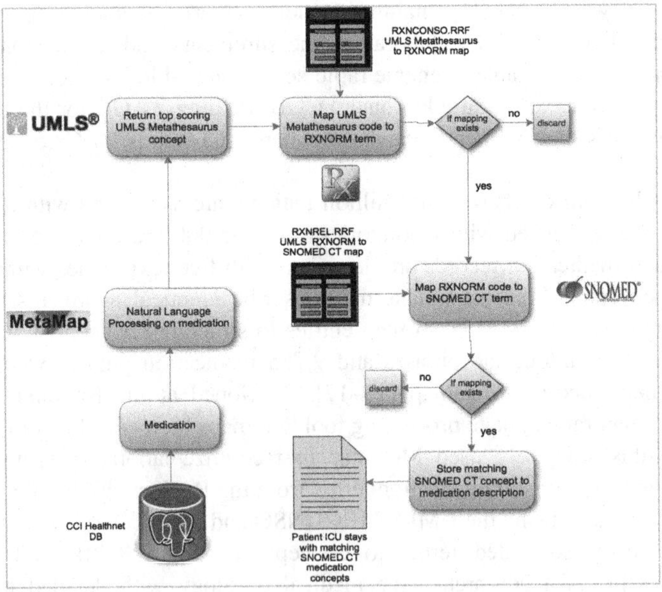

Fig. 2. This diagram shows the mapping procedure for medication phrases to SNOMED-CT concepts.

completely differently, relationships could still be found due to rules such as trade-names and ingredients. Figures 1 and 2 show the workflows for mapping laboratory and medication phrases to SNOMED-CT concepts.

The CCI-health database employs ICD-9 [18] codes for patient diagnoses. This makes the mapping procedure to SNOMED-CT concepts slightly different from those designed for laboratories and medications. The ICD9CM_SNOMED_MAP table in UMLS can be used to map ICD-9 directly to SNOMED-CT concepts. However, this does not include all ICD-9 codes that are associated with patients in the CCI-health database. Metamap was then used to analyze the descriptions of the remaining ICD codes that are not found in the ICD9CM_SNOMED_MAP table. The MRCONSO table was used to map the UMLS concepts returned by Metamap to associated SNOMED-CT concepts (Fig. 3).

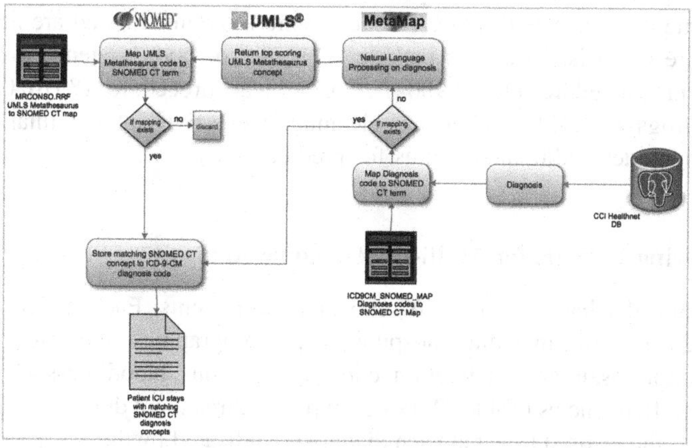

Fig. 3. This figure shows the mapping of diagnosis ICD-9 codes to SNOMED-CT concepts.

SNOMED-CT provides a rich hierarchy enabling semantic distances between concepts to be measured by path distances. We developed a Neo4j Graph Database for the CCI-health data to rapidly compute common ancestor queries between the mapped SNOMED CT terms. In our Neo4j Graph Database, tab delimited files of all SNOMED concepts and relationships are exported from SNOMED CT Postgres relational database tables. The tab delimited files are then directly loaded into Neo4j community edition 2.0.0 using their batch inserter (http://docs.neo4j.org/chunked/milestone/batchinsert.html). This results in a SNOMED Graph Database that has many cycles. The cycles greatly impede graph operations such as returning all paths between nodes. Using the Cypher query language, we can quickly identify all cycles with 2 or more nodes. Each cycle can then be removed by deleting an edge based on some criteria involving node depth and in-degree (number of incoming edges to a node).

After an acyclic SNOMED-CT Graph database was created, graph computations such as shortest paths and common ancestor queries can be performed rapidly. This is beneficial since laboratories, diagnoses, and medications are all mapped to many SNOMED-CT concepts that can be too specific for machine learning analysis. In this study, all nodes are assigned a depth level according to the minimum number of edges that must be traversed to reach the root node. The root node is at depth level 0. All the mapped SNOMED-CT concepts can then be generalized to concepts at a higher depth level. It is important to choose an appropriate depth level to accurately distinguish patient characteristics from one another. For medications and diagnosis, a depth level of 2 is chosen. A depth level of 3 is chosen for laboratories, since assigning lower depth levels returned concepts that are too general. For a given SNOMED-CT concept, Neo4j can quickly calculate all possible paths to the root node. With the Cypher query language, Neo4j returns all nodes for a given depth level that are crossed from all possible paths to the root of the hierarchy. This method was used for all mapped SNOMED-CT concepts, and they can be converted into equivalent nodes at a more general depth level. After conversion, the data was manipulated so that each row contains one patient with a set of yes/no columns for each general SNOMED-CT concept. Some mapped SNOMED-CT concepts are too common and are necessary to remove before analysis. Examples include "Disorder of body system (362965005)", "Measurement procedure (122869004)", "Chemical procedure (83762000)", and "Types of drugs (278471000)". Upon cleaning, we arrived at the final integrated dataset that includes additional features for predictive analysis.

2.2 Clustering Patients for Multi-site Outcome Analysis

The CCI-health database contains 267,666 diabetic patients. Each patient is characterized by 24 features including hospital site, demographics, laboratory tests and results, prescriptions, treatment duration, chronic conditions, blood pressure, number of visits and visit frequencies (Table 1). For each patient, treatment duration is determined by calculating the elapsed time between diagnosis (indicated by the first prescription of a diabetic medication) and the last recorded activity (i.e. procedure, lab, etc.). These variables are considered potential features that may influence treatment outcome. They are used as input for our classification analysis.

To establish the outcome group status for these diabetic patients, Glycated hemoglobin (HbA1c) lab measurement series throughout the treatment duration were used as indicators of treatment outcome. In our analysis, only patients with 7 or more HbA1c measurements recorded and no missing features were included. This resulted in 3,875 patients. On each patient's HbA1c measurement series, we performed sliding window with a size of five measurements with equal weights to reduce potential noise. Figure 4 shows the comparison of a patient's HbA1c data before and after sliding window is performed.

The 3,875 patients were clustered into two outcome groups based on their smoothed HbA1c lab measurement series. Since each patient has different numbers of records, a method for clustering time series of different lengths is required. Here we compared these measurements based on the method proposed by Caiado et al. [19]: First a periodogram of each patient's smoothed HbA1c measurements was calculated. Next, discrepancy statistics were calculated for each pair of periodograms and used as the distance between each pair of patients. In the case when their recorded measurements are not equal in length, the shorter series is extended by adding zeros and the zero-padding discrepancy statistics between the two series were used. Lastly, using the distance matrix filled with discrepancy statistics, agglomerative clustering with average

Table 1. Name and description of the 24 features for each diabetic patient.

Features used in training classifiers	Description
Treatment duration	Total time elapsed since first prescription of a diabetic medication
Visit frequency	Number of vitals measurements recorded/treatment duration
Diagnoses (6 features)	6 binary variables (true or false) indicating if patient has been diagnosed with the following conditions: hypertension, hyperlipidemia, cardiovascular disease, stroke, emphysema, asthma
Race	Patient race
Gender	Patient gender
Age	Patient age
Height	Patient height measured during first visit
Weight	Patient weight measured during first visit
Provider site	Clinical site that patient receives treatment
Systolic blood pressures (5 features)	5 systolic blood pressure measurements sampled equi-distantly throughout measurements over the entire treatment period after sliding average with window size of 5 is applied
Diastolic blood pressure (5 features)	5 diastolic blood pressure measurements sampled equi-distantly throughout measurements over the entire treatment period after sliding average with window size of 5 is applied

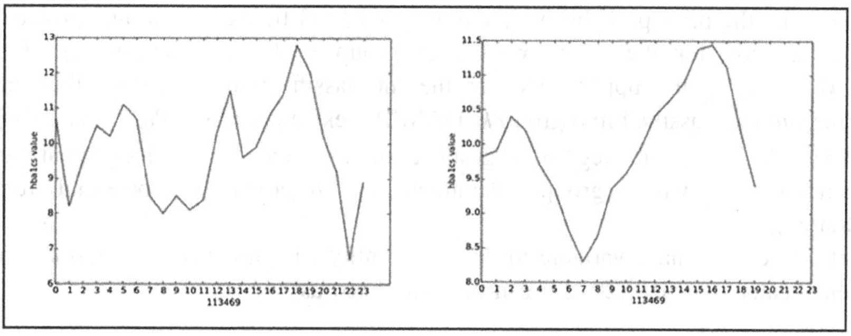

Fig. 4. Patient Glycated hemoglobin (HbA1c), a laboratory measurement (over the entire treatment duration) is used to assess glucose control in diabetic patients, before (left) and after sliding window (right) with size of five measurements and equal weight.

linkage (SciPy v0.17.0, scipy.cluster.hierarchy.linkage) was performed. A threshold of 650 and the "distance" criterion were used to form two flat patient clusters. As a result, 400 patients were clustered into the "Good outcome" group, and the remaining 3,475 patients were clustered in the "medium outcome" group. We characterized the two groups as "good" versus "medium" based on the trend of HbA1c lab measurements of patients in each group.

To establish the classification rule, the 3,875 diabetic-patient dataset were partitioned into a training set and an independent set for blind-prediction using stratified

Table 2. Partitioning of periodogram-based patient clusters for classification analysis.

	10-fold cross validation training set			Blind prediction set		
Total Patients	Total	Medium outcome	Good outcome	Total	Medium outcome	Good outcome
3,875	2,325	2,074	251	1,550	1,401	149

random sampling. The training set consisted of 2,325 patients (60% of the population), and the blind prediction set consisted of 1,550 patients (40% of the population). Table 2 summarizes the number of patients in each set.

2.3 Optimization-Based Classifier: Discriminant Analysis via Mixed Integer Program (DAMIP)

Suppose we have n entities from K groups with m features. Let $\mathcal{G} = \{1, 2, \ldots, K\}$ be the group index set, $\mathcal{O} = \{1, 2, \ldots, n\}$ be the entity index set, and $\mathcal{F} = \{1, 2, \ldots, m\}$ be the feature index set. Also, let $\mathcal{O}_k$, $k \in \mathcal{G}$ and $\mathcal{O}_k \subseteq \mathcal{O}$, be the entity set which belong to group k. Moreover, let $\mathcal{F}_j$, $j \in \mathcal{F}$, be the domain of feature j, which could be the space of real, integer, or binary values. The i th entity, $i \in \mathcal{O}$, is represented as $(y_i, \boldsymbol{x}_i) = (y_i, x_{i1}, \ldots, x_{im}) \in \mathcal{G} \times \mathcal{F}_1 \times \cdots \times \mathcal{F}_m$, where y_i is the group to which entity i belongs, and $(x_{i1}, \ldots, x_{im})$ is the feature vector of entity i. The classification model finds a function $f : (\mathcal{F}_1 \times \cdots \times \mathcal{F}_m) \rightarrow \mathcal{G}$ to classify entities into groups based on a selected set of features.

Let π_k be the prior probability of group k and $f_k(\boldsymbol{x})$ be the conditional probability density function for the entity $\boldsymbol{x} \in \mathbb{R}^m$ of group k, $k \in \mathcal{G}$. Also let $\alpha_{hk} \in (0, 1)$, $h, k \in \mathcal{G}$, $h \neq k$, be the upperbound for the misclassification percentage that group h entities are misclassified into group k. DAMIP seeks a partition $\{P_0, P_1, \ldots, P_K\}$ of $\mathbb{R}^K$, where P_k, $k \in \mathcal{G}$, is the region for group k, and P_0 is the reserved judgement region with entities for which group assignment are reserved (for potential further exploration).

Let u_{ki} be the binary variable to denote if entity i is classified to group k or not. Mathematically, DAMIP [20–22] can be formulated as

$$\max \sum_{i \in \mathcal{O}} u_{y_i i} \tag{1}$$

$$\text{s.t } L_{ki} = \pi_k f_k(\boldsymbol{x}_i) - \sum_{h \in \mathcal{G}, h \neq k} f_h(\boldsymbol{x}_i) \lambda_{hk} \quad \begin{array}{l} \forall i \in \mathcal{O}, \\ k \in \mathcal{G} \end{array} \tag{2}$$

$$u_{ki} = \begin{cases} 1 & \text{if } k = \arg\max\{0, L_{hi} : h \in \mathcal{G}\} \\ 0 & \text{otherwise} \end{cases} \quad \forall i \in \mathcal{O}, k \in \{0\} \cup \mathcal{G} \tag{3}$$

$$\sum_{k \in \{0\} \cup \mathcal{G}} u_{ki} = 1 \quad \forall i \in \mathcal{O} \tag{4}$$

$$\sum_{i:i\in\mathcal{O}_h} u_{ki} \leq \lfloor \alpha_{hk} n_h \rfloor \quad \forall h,\, k \in \mathcal{G},\, h \neq k \tag{5}$$

$$u_{ki} \in \{0,1\} \quad \forall i \in \mathcal{O},\, k \in \{0\} \cup \mathcal{G}$$

$$L_{ki} \text{ unrestricted in sign} \quad \forall i \in \mathcal{O},\, k \in \mathcal{G}$$

$$\lambda_{hk} \geq 0 \quad \forall h,\, k \in \mathcal{G},\, h \neq k$$

The objective function (1) maximizes the number of entities classified into the correct group. Constraints (2) and (3) govern the placement of an entity into each of the groups in G or the reserved-judgment region. Thus, the variables L_{ki} and λ_{hk} provide the shape of the partition of the groups in the G space. Constraint (4) ensures that an entity is assigned to exactly one group. Constraint (5) allows the users to preset the desirable misclassification levels, which can be specified as overall errors for each group, pairwise errors, or overall errors for all groups together. With the reserved judgment in place, the mathematical system ensures that a solution that satisfies the preset errors always exists.

Mathematically, we have proven that DAMIP is *NP-hard.* The model has many appealing characteristics including: (1) the resulting DAMIP-classification rule is *strongly universally consistent*, given that the Bayes optimal rule for classification is known [20, 22]; (2) the misclassification rates using the DAMIP method are consistently lower than other classification approaches in both simulated data and real-world data; (3) the classification rules from DAMIP appear to be insensitive to the specification of prior probabilities, yet capable of reducing misclassification rates when the number of training entities from each group is different; and (4) the DAMIP model generates stable classification rules on imbalance data, regardless of the proportions of training entities from each group. [20–23].

Computationally, DAMIP is the first multiple-group classification model that includes a reserved judgment and the ability to constrain the misclassification rates simultaneously within the model. Further, Constraints (2) serve to transform the attributes from their original space to the group space, serving as dimension reduction. In Brooks and Lee (2010) and Brooks and Lee (2014), we have shown that DAMIP is difficult to solve [22, 23]. We applied the hypergraphic structures that Lee, Wei, and Maheshwary (2013) derived to efficiently solve these instances.

The predictive model maximizes the number of correctly classified cases; therefore, it is robust and not skewed by errors committed by observation values. The associated optimal decision variable values (L_{ki} and λ_{hk}) form the classification rule, which consists of the discriminatory attributes; examples include demographics, treatment duration, health conditions, laboratory results, medications, etc.

The entities in our model correspond to the diabetic patients with two outcome groups: "Medium outcome" versus "Good outcome". The goal was to predict which patients have good treatment outcome. Identifying these patient and care characteristics will establish best practice and evidence for providers.

In the classification analysis, the model is first applied to the training set to establish the classification rules using 10-fold cross-validation unbiased estimate. The predictive accuracy of each classifier is further assessed by applying the rule to the blind-prediction set. DAMIP returns classifiers that include as few features as possible.

To demonstrate the effectiveness of inclusion of standardized clinical terminologies, we utilized the generalized SNOMED-CT concepts obtained from our mapping procedure, which introduced a total of 2,205 additional features including diagnosis, lab, and medication for each patient. Using the same classification approaches, we established new classifiers using these additional features.

We contrasted the DAMIP approach with logistic regression, naïve Bayes, radial basis function networks (RBFNetwork), Bayesian Network, J48 decision tree, and sequential minimal optimization (SMO) approaches that are implemented in Weka 3.6.13. In these Weka classifiers, feature selection was performed using the "InfoGainAttributeEval" as the Attribute Evaluator and "Ranker" as the Search Method to select at most 200 features.

3 Results

3.1 Data Integration and Mapping to Standardized Medical Concepts

Free text entries from laboratories and medications and ICD9 codes were all successfully mapped to SNOMED-CT concepts. Of the 803 unique lab phrases in the CCI-health database, 603 can be linked to SNOMED-CT; this covers 1.20 million of 1.33 million patients. Similar successes were found for concept mapping of medications and diagnoses: 5,899 of 9,755 medication phrases were mapped to SNOMED-CT; this covers 801,025 of 952,729 patients. 10,655 of 29,371 ICD-9 codes were linked to SNOMED-CT concepts. Specifically, 2.35 million of 2.46 million patients contain at least one ICD-9 diagnosis code that can be mapped to SNOMED-CT. Table 3 shows examples of medical entries for medication, laboratories, and diagnoses and their mapped SNOMED concepts. Medication can have many brand names and a range of ingredients. Laboratory procedures and diagnoses may also be described by physicians with a great deal of variation. Table 3 shows how variations and ambiguity among phrases are eliminated by our mapping algorithm.

After mapping all the medications, laboratories and diagnoses to standardized medical terms, tens of thousands of new features were created for each patient. Specifically, a total of 17,140 entries for laboratories, medications, and diagnosis codes were mapped to 11,208 SNOMED-CT concepts. For each patient, a feature must be created for each mapped medical concept indicating whether a patient received a medication, was given a diagnosis, or underwent a laboratory procedure. Furthermore, many of these features were closely related. It is thus beneficial to generalize them to reduce the total number of features per patient. In our implementation, these mappedSNOMED-CT concepts were generalized into level 2 nodes for medications and diagnoses and level 3 nodes for laboratories.

This results in a total of 2,385 SNOMED-CT general standardized and unambiguous concepts. Of these, 2,205 concepts were used for clustering and classification for diabetic patient analysis. Table 4 demonstrates examples for generalizing SNOMED-CT terms into higher level concepts for medication, laboratory procedures, and diagnosis.

Table 3. Mapping of medication, laboratory, and diagnosis phrases to SNOMED concepts.

Entity type	Entry	SNOMED concept	SNOMED code	Count
Medication	Mucinex D Tablets Extended Release	Pseudoephedrine (product)	91435002	4322
	Bromfed Dm Cough Syrup			2554
	Lortab Tablets	Acetaminophen (product)	90332006	94442
	Roxicet Tablets			40885
	Hydrocortisone Acetate	Hydrocortisone preparation (product)	16602005	705
	Proctofoam Hc Aerosol			533
Lab	lipid-triglycerides	Plasma lipid measurement (procedure)	314039006	716055
	lipid-vldl-cholesterol			429935
	urine-creatinine-24 h	Urine creatinine measurement (procedure)	271260009	46356
	urine-creatinine-random			24101
	glucose tolerance test (3 h)	Glucose tolerance test (procedure)	113076002	191
	glucose-tolerance-test-5-hour			79
Diagnosis	V13.64 Personal history of (corrected) congenital malformations of eye, ear, face and neck	Congenital deformity (disorder)	276655000	43
	V13.65 Personal history of (corrected) congenital malformations of heart and circulatory system			37
	770.86 Aspiration of postnatal stomach contents with respiratory symptoms	Respiratory symptom (finding)	161920001	27
	770.14 Aspiration of clear amniotic fluid with respiratory symptoms			10
	806.35 Open fracture of T7-T12 level with unspecified spinal cord injury	Open fracture (disorder)	397181002	2
	806.11 Open fracture of C1-C4 level with complete lesion of cord			1

3.2 Classification Results on Predicting Treatment Outcome

Table 4 contrasts our in-house DAMIP classification results with six Weka classifiers. Uniformly the six classifiers suffer from group imbalance and tend to place all patients into the larger "Medium outcome" group. In contrast, the DAMIP classifier selects 5 discriminatory features among the 24 and is able to achieve relatively high classification

Table 4. Generalizing mapped concepts of medication, laboratory and diagnosis into SNOMED depth-level mapping.

Entity type	SNOMED concept	SNOMED code	SNOMED generalized concept	SNOMED general code	Count
Medication	Azithromycin (product)	96034006	Antibacterial drugs (product)	346325008	66150
	Amoxicillin (product)	27658006			64842
	Valsartan (product)	108581009	Hypotensive agent (product)	1182007	117287
	Benazepril (product)	108572003			98706
Lab	Urine screening for protein (procedure)	171247004	Evaluation of urine specimen (procedure)	442564008	4267836
	Measurement of fasting glucose in urine specimen using dipstick (procedure)	442033004			2799670
	Hemoglobin variant test (procedure)	302763003	Hematology procedure (procedure)	33468001	18262531
	Red blood cell count (procedure)	14089001			16487444
	Histamine release from basophils measurement (procedure)	63987006	Immunologic procedure (procedure)	108267006	145032
	C-reactive protein measurement (procedure)	55235003			79804
Diagnosis	Tongue tie (disorder)	67787004	Congenital anomaly of gastrointestinal tract (disorder)	128347007	1717
	Idiopathic congenital megacolon (disorder)	268209004			1177
	Hemorrhage of rectum and anus (disorder)	266464001	Hemorrhage of abdominal cavity structure (disorder)	444312002	41580
	Hematoma AND contusion of liver without open wound into abdominal cavity (disorder)	24179004	Hemorrhage of abdominal cavity structure (disorder)		1031

and blind prediction accuracies for both groups. We remark that the commonly used Pap Smear diagnosis test has an accuracy of roughly 70%.

Table 5 shows classification results after including 2,205 additional features for each patient. With these added features, DAMIP improves its prediction accuracies of the "Good outcome" group while keeping a high accuracy for the "Medium outcome" group. We observe that instead of selecting the "provider site" as one of the discriminatory

Table 5. Comparison of DAMIP results against other classification methods.

Classifier	Features set	10-fold cross validation accuracy			Blind prediction accuracy		
		Overall	Medium outcome group	Good outcome group	Overall	Medium outcome group	Good outcome group
Logistic regression	treatment duration, visit frequency, hypertension, hyperlipidemia, cvd, stroke, emphysema, asthma, race, gender, age, height, weight, patient site, 5 systolic blood pressure measurements, and 5 diastolic blood pressure measurements	89.68%	96.87%	30.28%	90.77%	97.14%	30.87%
Naïve Bayes		87.05%	92.77%	39.84%	87.94%	92.72%	42.95%
RBFNetwork		88.22%	98.36%	4.38%	89.03%	97.22%	12.08%
Bayesian Network		87.66%	92.96%	43.82%	87.23%	91.86%	43.62%
J48		89.16%	97.69%	18.73%	90.65%	97.72%	24.16%
SMO		89.16%	99.42%	4.38%	90.77%	99.64	7.38%
DAMIP (5 features)	treatment duration, visit frequency, hyperlipidemia, asthma, provider site	**80.1%**	**80.6%**	**75.1%**	**81.9%**	**81.9%**	**81.3%**

features, DAMIP selects the type of procedures and medications used instead. Identifying these features facilitates dissemination of best practice and target treatment regime to specific patients among the sites. For the six Weka classifiers, the results improve only slightly on the "Good outcome" group. (Table 6).

Table 6. Comparison of prediction accuracies with additional 2,205 medical terminologies obtained from mapping.

Classifier	Features set	10-fold cross validation accuracy			Blind prediction accuracy		
		Overall	Medium outcome group	Good outcome group	Overall	Medium outcome group	Good outcome group
Logistic regression	treatment duration, patient site (practice), height, asthma, visit frequency, race, hypertension, 2 diastolic blood pressure measurements, and 191 new features from terminology mapping	87.83%	94.07%	36.25%	89.81%	94.93%	41.61%
Naïve Bayes		74.75%	76.81%	57.77%	75.29%	77.73%	52.35%
RBFNetwork		88.73%	99.42%	0.40%	90.39%	100.0%	0.00%
Bayesian Network		80.34%	83.37%	55.38%	80.84%	83.73%	53.69%
J48		89.12%	96.62%	27.09%	87.94%	95.07%	20.81%
SMO		89.25%	96.77%	27.09%	90.52%	97.29%	26.85%
DAMIP (9 features)	treatment duration, visit frequency, asthma, + 6 features from mapping: Injection of therapeutic agent (procedure), Oral form levofloxacin (product), Clopidogrel (product). Aspirin (product), Nystatin (product), and Metformin (product)	**88.7%**	**88.6%**	**89.5%**	**87.9%**	**86.7%**	**89.7%**

4 Discussion

In this study, we establish interoperability among EMRs from 737 clinical facilities using a mapping process that disambiguates free text entries. The mapping provides a unique way to link to structured medical concepts despite the extreme variations that can occur during clinical diagnosis and documentation. It enables more powerful systems to be developed for future studies where semantic distances can be calculated between patient records due to their association with hierarchical concepts. The graph database allows for rapid data access and queries.

To test the usability of such heterogeneous multi-site data, we develop a DAMIP machine learning model for uncovering patient and practice characteristics that can discriminate "Good outcome" from other outcomes on a group of diabetic patients. Compared to other classification models, the DAMIP model is able to select a small set of discriminatory features to establish good classification rules. Specially, the DAMIP-rules can predict patients with good outcome from those with medium outcome with over 80% blind predictive accuracy. By including medical concepts obtained from terminology mapping, we are able to further improve the predictive capabilities of our model, creating a more powerful evidence-based best practice discovery system. The ultimate goal in our study is to establish and identify practice characteristics that result in good outcome and to disseminate this practice among the 737 clinical sites for population health improvement. The features identified, including treatment duration, frequency, co-existing condition, and type of regimens allow for design of "best practice" clinical practice guidelines.

The next step is to use the discriminatory features identified and the criteria developed via the DAMIP-machine learning framework to design and optimize evidence-based treatment plans and to disseminate such knowledge through "rapid learning" across the multiple sites. With these, we seek to increase quality and timeliness of care and maximize outcome and service performance of the healthcare system. We will investigate and contrast our methodologies on other types of diseases. We will also perform predictive health surveillance on the healthy patients (there are 1.8 million healthy individuals in our CCI-health database). Health status will be monitored and rapid analytics will be performed based on existing data to identify the first sign of health risk. This facilitates proactive patient-centered pre-disease and behavior intervention.

Acknowledgment. This paper receives the 2016 NSF Health Organization Transformation award (second place). The work is partially supported by a grant from the National Science Foundation IIP-1361532. Findings and conclusions in this paper are those of the authors and do not necessarily reflect the views of the National Science Foundation.

References

1. Jensen, P.B., Jensen, L.J., Brunak, S.: Mining electronic health records: towards better research applications and clinical care. Nat. Rev. Genet. **13**(6), 395–405 (2012)
2. Park, H., Hardiker, N.: Clinical terminologies: a solution for semantic interoperability. J. Korean Soc. Med. Inform. **15**(1), 1–11 (2009)

3. Rosenbloom, S.T., et al.: Interface terminologies: facilitating direct entry of clinical data into electronic health record systems. J. Am. Med. Inform. Assoc. **13**(3), 277–288 (2006)
4. Donnelly, K.: SNOMED-CT: the advanced terminology and coding system for eHealth. Stud. Health Technol. Inform. **121**, 279 (2006)
5. McDonald, C.J., et al.: LOINC, a universal standard for identifying laboratory observations: a 5-year update. Clin. Chem. **49**(4), 624–633 (2003)
6. Liu, S., et al.: RxNorm: prescription for electronic drug information exchange. IT Prof. **7**(5), 17–23 (2005)
7. Bodenreider, O.: The unified medical language system (UMLS): integrating biomedical terminology. Nucleic Acids Res. **32**(suppl 1), D267–D270 (2004)
8. Hernandez, P., et al.: Automated mapping of pharmacy orders from two electronic health record systems to RxNorm within the STRIDE clinical data warehouse. American Medical Informatics Association (2009)
9. Carlo, L., Chase, H.S., Weng, C.: Aligning structured and unstructured medical problems using umls. American Medical Informatics Association (2010)
10. Patel, C.O., Cimino, J.J.: Using semantic and structural properties of the unified medical language system to discover potential terminological relationships. J. Am. Med. Inform. Assoc. **16**(3), 346–353%@ 1067–5027 (2009)
11. Gallagher, R.J., Lee, E.K.: Mixed integer programming optimization models for brachytherapy treatment planning. In: Proceedings of the AMIA Annual Fall Symposium. American Medical Informatics Association (1997)
12. Haffner, S.M.: Epidemiology of type 2 diabetes: risk factors. Diab. Care **21**(Suppl. 3), C3–C6 (1998)
13. Chan, J.M., et al.: Obesity, fat distribution, and weight gain as risk factors for clinical diabetes in men. Diab. Care **17**(9), 961–969 (1994)
14. Estacio, R.O., et al.: Effect of blood pressure control on diabetic microvascular complications in patients with hypertension and type 2 diabetes. Diab. Care **23**, B54 (2000)
15. Aronson, A.R.: Effective mapping of biomedical text to the UMLS Metathesaurus: the MetaMap program. In: Proceedings of the AMIA Symposium. American Medical Informatics Association (2001)
16. Aronson, A.R.: Metamap: Mapping Text to the UMLS Metathesaurus. NLM, NIH, DHHS, Bethesda, pp. 1–26 (2006)
17. Aronson, A.R., Lang, F.-M.: An overview of MetaMap: historical perspective and recent advances. J. Am. Med. Inform. Assoc. **17**(3), 229–236 (2010)
18. National Center for Health Statistics (U.S.). ICD-9-CM: The International Classification of Diseases, 9th Revision, Clinical Modification. 1978: Commission on Professional and Hospital Activities
19. Caiado, J., Crato, N., Peña, D.: Comparison of times series with unequal length in the frequency domain. Commun. Stat. Simul. Comput.® **38**(3), 527–540 (2009)
20. Lee, E.K.: Large-scale optimization-based classification models in medicine and biology. Ann. Biomed. Eng. **35**(6), 1095–1109 (2007)
21. Lee, E.K., et al.: A clinical decision tool for predicting patient care characteristics: patients returning within 72 hours in the emergency department. In: AMIA Annual Symposium Proceedings. American Medical Informatics Association (2012)
22. Brooks, J.P., Lee, E.K.: Solving a multigroup mixed-integer programming-based constrained discrimination model. INFORMS J. Comput. **26**(3), 567–585 (2014)
23. Brooks, J.P., Lee, E.K.: Analysis of the consistency of a mixed integer programming-based multi-category constrained discriminant model. Ann. Oper. Res. **174**(1), 147–168 (2010)

Data-Based Forest Management with Uncertainties and Multiple Objectives

Markus Hartikainen[1(✉)], Kyle Eyvindson[2], Kaisa Miettinen[1], and Annika Kangas[3]

[1] Faculty of Information Technology, University of Jyvaskyla, P.O. Box 35, 40014 University of Jyvaskyla, Finland
{markus.hartikainen, kaisa.miettinen}@jyu.fi

[2] Department of Biological and Environmental Science, University of Jyvaskyla, P.O. Box 35, 40014 University of Jyvaskyla, Finland
kyle.j.eyvindson@jyu.fi

[3] Natural Resources Institute Finland (Luke), Economics and Society Unit, Yliopistokatu 6, 80101 Joensuu, Finland
annika.kangas@luke.fi

Abstract. In this paper, we present an approach of employing multiobjective optimization to support decision making in forest management planning. The planning is based on data representing so-called stands, each consisting of homogeneous parts of the forest, and simulations of how the trees grow in the stands under different treatment options. Forest planning concerns future decisions to be made that include uncertainty. We employ as objective functions both the expected values of incomes and biodiversity as well as the value at risk for both of these objectives. In addition, we minimize the risk level for both the income value and the biodiversity value. There is a tradeoff between the expected value and the value at risk, as well as between the value at risk of the two objectives of interest and, thus, decision support is needed to find the best balance between the conflicting objectives. We employ an interactive method where a decision maker iteratively provides preference information to find the most preferred management plan and at the same time learns about the interdependencies of the objectives.

Keywords: Forest management planning · Multiobjective optimization · Interactive multiobjective optimization · Pareto optimality · Uncertainty

1 Introduction

In forest management, a forest area is typically divided to decision units, so-called stands, which are relatively homogeneous with respect to the age structure and the species composition of trees. Forest management planning means selecting optimal harvest schedules including one or more treatment option(s) and their timing for each of these stands. The treatment options may include harvesting all the trees (final felling) or a part of them (thinning) within any stand, and planting new seedlings and tending them after a harvest has been carried out. The timing of the treatment options in the

P.M. Pardalos et al. (Eds.): MOD 2016, LNCS 10122, pp. 16–29, 2016.
DOI: 10.1007/978-3-319-51469-7_2

schedules is described by dividing the planning horizon to several planning periods of one or more years.

For most decision makers, forest management planning is a multiobjective decision making problem. Harvesting implies incomes from forests but, on the other hand, it diminishes the recreational and esthetical values of the forest, and it may have adverse effects on the natural values of the forest area, for instance, the biodiversity within the area and the viability of wildlife populations living in the area.

Importantly, forest management planning involves a lot of uncertainty as it is not possible to measure all trees within a forest area. This means that there is uncertainty concerning the current state of the forests. Furthermore, measuring biodiversity is prohibitively expensive, so using proxy variables (biodiversity indices) is the only possibility. Finally, all forest decisions concern the future (typically the next 5–20 years), so that the state of the forest stands and the biodiversity as well as the consequences of the treatment options need to be predicted using statistical models. As we do not know the exact consequences of the management decisions, the decisions involve uncertainty which the decision makers may wish to manage.

In this paper, we present an application that accounts for the conflicting objectives in forestry (incomes and biodiversity) and manage the risk involved in them using the value at risk (often denoted by VaR) concept. We apply an interactive multiobjective optimization method to find the most preferred treatment option for each stand considered.

2 Background and Related Work

Selecting the optimal treatment schedule for each stand is a combinatorial optimization problem, where the size of the problem depends on the number of separate stands and the number of possible treatment schedules for them. Linear programming has been widely used to solve this problem since the 1960s (Kangas et al. 2015). Most applications have multiple objectives, and have typically been solved using the ε-constraint method (Miettinen 1999). Heuristic optimization has also been used since the 1990s. Within a heuristic framework, multiobjective optimization has typically been based on a weighted (additive) utility function. The main reason for not utilizing multiobjective optimization more has been the lack of computational tools to do so.

Multiobjective optimization problems can in general be formulated as

$$
\begin{aligned}
&\max\{f_1(x),\ldots,f_k(x)\}\\
&\text{s.t } g_j(x) \geq 0 \text{ for all } j = 1,\ldots,J\\
&h_k(x) = 0 \text{ for all } k = 1,\ldots,K\\
&a_i \leq x_i \leq b_i \text{ for all } i = 1,\ldots,n\\
&x = (x^i, x^r), \text{ where } x^i \in Z^a,\ x^i \in R^{n-z}.
\end{aligned}
$$

In the above problem, f_i are objective functions to be maximized simultaneously, g_j and h_k are inequality and equality constraints, respectively, bounds $a_i \leq x_i \leq b_i$ are called box constraints and, finally, the decision variable vector $x = (x^i, x^r)$ consists of integer-valued variables x^i and real-valued variables x^r. All the constraints form a feasible set Q which is a subset of the decision space. The image of Q mapped with the objective functions is called the feasible objective set *f(Q)* and elements of it are so-called objective vectors z.

In multiobjective optimization, there typically does not exist a single optimal solution but, instead, there exist multiple so-called Pareto optimal solutions, where none of the objective functions can be improved without impairing at least one of the others. For this reason, one needs additional preference information to decide which of the Pareto optimal solutions is the best one. The person giving this information is called a decision maker (DM).

One way of giving preferences is providing aspiration levels, which are values of objective functions that should desirably be achieved, and they constitute a reference point $z^{ref} \in R^k$. Aspiration levels are employed in so-called achievement scalarizing functions (Wierzbicki 1986). The main idea behind achievement scalarizing functions is that they measure the preferability of a solution given a reference point in a way that is theoretically justifiable. This is defined as order-consistency by Wierzbicki (1986).

There exist many achievement scalarizing functions following the characterization of Wierzbicki (1986). In this paper, we use the following achievement scalarizing function to be maximized

$$\begin{aligned} s^{asf} &: f(Q) \times R^k \to R, \\ (z, z^{ref}) &\mapsto \min_{i=1,\ldots,k}(z_i - z_i^{ref})/(z_i^{ideal} - z_i^{nadir}) \\ &+ \rho \sum\nolimits_{i=1}^{k} z_i/(z_i^{ideal} - z_i^{nadir}). \end{aligned}$$

In the above problem, z is a so-called objective vector in the image space of the feasible set, z^{ideal} is the ideal vector of the problem containing the maximum values of the individual objectives and z^{nadir} is the nadir vector containing the minimum values of the individual objectives within the set of Pareto optimal solutions (see e.g. Miettinen, 1999). The summation term at the end is called an augmentation term and the constant ρ is a small positive constant, e.g. 0.0001. The augmentation term guarantees that the solutions are indeed Pareto optimal.

Employing the achievement scalarizing function means solving the optimization problem

$$\max_{x \in Q} s^{asf}(f(x), z^{ref})$$

for any reference point z^{ref} given by the DM. Given a reference point, the optimal solution of the above problem is a Pareto optimal solution to the original multiobjective optimization problem (Wierzbicki 1986). The variant of the reference point method (Wierzbicki 1986) employed utilizes the achievement scalarizing function.

A vast majority of applications within forest management planning assumes a decision situation under certainty. A couple of applications of stochastic optimization have been published since the 1980s, and most of them in the last 10 years. To our knowledge, any kind of risk management has only been included in two published papers (Eyvindson and Kangas 2016a, Eyvindson and Chen 2016). The latter employed conditional value at risk (Rockafeller and Uryasev 2000). Applications utilizing value at risk (Duffie and Pan 1997) have not yet been published in forest management planning. To our knowledge, applications including two or more different value at risk concepts have not been published in any field. However, in forest planning, it is quite possible that the DM is willing to accept high risks for poor incomes but only a low risk for losing biodiversity or vice versa. The main reason why the uncertainties have been ignored so far is that the problem including uncertainties requires quite heavy calculations, which can be regarded too demanding for large areas with a lot of stands.

In interactive multiobjective optimization methods (e.g. Miettinen 1999, Miettinen et al. 2008), the final Pareto optimal solution, to be called the most preferred solution, is identified by iterating the steps of re-defining the preferences and producing a solution fulfilling these preferences as well as possible, until the DM is satisfied. The idea is that in this way the DM learns about what kind of preferences are attainable and what kind of solutions are achievable. As mentioned, the preferences can be expressed, for instance by giving aspiration levels. Even though the benefits of interactive methods have been acknowledged in the forest management planning field (Pykäläinen 2000), only a few applications have been published.

3 Multiobjective Optimization of Forest Inventory

One complication in forest management planning is that most decisions concern the future. Even when a decision is implemented immediately, the consequences of the actions in the future need to be predicted. This prediction is carried out using a forest growth simulator.

In the Finnish forestry, there are three different operational simulators that are used by all organizations making forest planning. In this study, we used the SIMO simulator, which includes more than 400 equations to predict, e.g., the diameter growth and the height growth of each tree, and the probability that a tree dies in the next years. The simulator also predicts the total volume of timber available from a harvest carried out at a specific time, and the income based on the harvested volume. Forest growth can be predicted fairly accurately for the next 1–5 years, but as the time period becomes longer, the uncertainties grow.

While forest growth can be simulated, biodiversity cannot really be measured in practical forestry. Biodiversity should include genetic variation within species, variation of species within each stand and the variation of habitats (types of forest stands) within the forest area. Of these, only the habitat variation is an operational objective for optimization. It is even more difficult to predict the consequences of different actions on the biodiversity. Therefore, the usual approach is to utilize a proxy variable, a so-called biodiversity (BD) index, in the optimization. The BD indices are based on the characteristics of the stands, which enables analyzing the consequences of actions also from the biodiversity point of view (Kangas and Pukkala 1996). As these forest characteristics include uncertainty, and using a proxy in itself includes uncertainty, these estimates are highly uncertain.

The stochasticity involved can be dealt with by using a set of scenarios. Parametric distributions are utilized to describe the uncertainty in any one of the input variables of a system, and a parametric distribution can also be used to describe the uncertainty in the simulated growth. The variables of interest, like incomes from harvests, are a result of several input variables and statistical models, and therefore describing the uncertainty is easiest to carry out with a set of scenarios, each of them essentially describing one possible future development of the forest area. Then, the whole set of scenarios describes the uncertainty in the variables of interest. Using a set of scenarios also enables describing the stochastic optimization problem involving uncertainties in a way where linear optimization can be used (a so-called deterministic equivalent of the stochastic problem).

It is possible to evaluate the quality of the set of scenarios used to describe the stochastic problem, through the use of evaluation techniques. The so-called Sample Average Approximation (SAA, Kleywegt et al. 2001) method compares the solution generated by a smaller number of scenarios to the solution generated by a much larger number of scenarios. This is iterated several times to generate confidence intervals of the gap in optimality and expected objective value of the solution. One application of this method has been realized in forestry, and the number of scenarios required to effectively represent the uncertainty depended on the quantity of uncertainty and risk preferences of the DM (Eyvindson and Kangas 2016b).

4 Modelling Forest Management as a Multiobjective Mixed Integer Linear Problem

We assume that we have S stands, T treatment schedules for each stand, R scenarios and P periods of time to consider. We have simulated values $I_{t,s,r,p}$ for the income and $B_{t,s,r,p}$ for the biodiversity with indices denoting

- treatment schedules t (including one or more timed treatments),
- stands s,
- scenarios of future development of the forest stand r
- and 5-year periods p.

The problem of choosing the best treatment schedules for all the stands can be formulated as a multiobjective optimization problem

$$\max\left\{\min_{p\in P, r\in R_I}\sum_{t\in T, s\in S} x_{t,s}I_{t,s,r,p},\ \min_{p\in P, r\in R_B}\sum_{t\in T, s\in S} x_{t,s}B_{t,s,r,p},\ \mathop{\mathbb{E}}_{r\in R}\sum_{t\in T, s\in S} x_{t,s}\min_{p\in P} I_{t,s,r,p},\ \mathop{\mathbb{E}}_{r\in R}\sum_{t\in T, s\in S} x_{t,s}B_{t,s,r,p}, \delta_I, \delta_B\right\}$$

$$\begin{aligned}
\text{s.t. } & \mathrm{R}_I \subset R, \#\mathrm{R}_I \geq \delta_I \#R,\\
& \mathrm{R}_B \subset R, \#\mathrm{R}_B \geq \delta_B \#R,\\
& \sum\nolimits_{t\in T} x_{t,s} = 1 \text{ for all } s \in S,\\
& x_{t,s} \in \{0,1\} \text{ for all } t \in T, s \in S.
\end{aligned}$$

In the above problem, # denotes the number of elements in a set and we have six objectives to be maximized:

1. Minimum income in the scenarios that are in the set of scenarios R_I (in euros). This objective will be denoted by VaR_I in what follows. According to the first constraint, the set R_I is a subset of the complete set of scenarios R and the number of scenarios in this subset is greater or equal to δ_I times the number of scenarios in the complete set of scenarios. This means that this is the income at risk for the risk level $1 - \delta_I$.
2. Minimum biodiversity index in the scenarios that are in the set of scenarios R_B. This objective will be denoted by VaR_B. Being an index, this variable is unitless, and only the relative differences can be interpreted. According to the second constraint, the set R_B is a subset of the complete set of scenarios R and the number of scenarios in this subset is greater or equal to δ_B times the number of scenarios in the complete set of scenarios. This means that this is the biodiversity at risk for the risk level $1 - \delta_B$.
3. Expected minimum income across the periods p in the complete set of scenarios R(in euros). This objective will be denoted by E_I.
4. Expected minimum biodiversity index across the periods in the complete set of scenarios R. This objective will be denoted by E_B.
5. Probability δ_I of the set R_I. The risk level for the value at risk for the income is, thus, $1 - \delta_I$.
6. Probability δ_B of the set R_B. The risk level for the value at risk for the biodiversity is, thus, $1 - \delta_B$.

Traditionally in forestry, the variation of income over periods has been handled by seeking a so-called even-flow solution. In an even-flow solution, the income is equal over all periods. With biodiversity, stability is even more important. We, however, maximize the minimum income and biodiversity over periods, instead of seeking the even-flow solution. The solution that maximizes the minimum income or biodiversity over the periods is sometimes the even-flow solution, if all the periodic incomes are the same as the minimum income, or it is better than the best available even-flow solution, if the income or the biodiversity in one of the periods is higher than the minimum income or biodiversity. For this reason, we believe that it makes more sense to maximize the minimum over the periods than to seek for the even-flow solution.

Our decision variables in the problem formulated are both the sets R_I and R_B and the treatment decisions $x_{t,s}$ for all treatment schedules t and stands s. The treatment decision $x_{t,s}$ is one, if the treatment schedule is chosen for the stand s and 0 if not. Because only one treatment schedule can be chosen for each stand, the third constraint allows only one of the $x_{t,s}$ values to be one, and the others must be zero for all the stands.

In order to solve the optimization problem efficiently, it can be converted into a mixed integer linear problem (MILP). Once we can re-formulate the problem as a MILP, we can use the efficient MILP solvers available e.g., IBM ILOG CPLEX Optimization Studio (see e.g., http://www-01.ibm.com/support/knowledgecenter/SSSA5P_12.6.3/ilog.odms.studio.help/Optimization_Studio/topics/COS_home.html) or Gurobi Optimization (see e.g., http://www.gurobi.com/documentation/). Our problem can be re-formulated as a multiobjective MILP

$$\max\left\{ I_{min,subset},\ \mathrm{B}_{min,subset},\ \sum_{r\in R}\sum_{t\in T,s\in S} x_{t,s}\frac{\min_{p\in P} I_{t,s,d,p}}{\#R},\ \sum_{r\in R}\sum_{t\in T,s\in S} x_{t,s}\frac{\min_{p\in P} B_{t,s,d,p}}{\#R},\ \delta_I, \delta_B \right\}$$

$$\begin{aligned}
\text{s.t.} & \sum\nolimits_{r\in R} t_r^I \geq \delta_I \#R,\\
& \sum\nolimits_{r\in R} t_r^B \geq \delta_B \#R,\\
& \sum\nolimits_{t\in T} x_{t,s} = 1 \ \text{ for all } s \in S,\\
& t_r^I \in \{0,1\} \ \text{ for all } r \in R,\\
& t_r^B \in \{0,1\} \ \text{ for all } r \in R,
\end{aligned}$$

where

$$I_{min,subset} = \min_{p\in P, r\in R}\Big(\sum_{t\in T,s\in S} x_{t,s} I_{t,s,r,p} + (1 - t_r^I)M\Big)$$

$$\mathrm{B}_{min,subset} = \min_{p\in P, r\in R}\Big(\sum_{t\in T,s\in S} x_{t,s} B_{t,s,r,p} + (1 - t_r^B)M\Big).$$

In the re-formulated problem, the constant M is a big number that allows for the minimum over the scenarios R in both the biodiversity and income to be the minimum over the scenarios, for which the variable t_r^I or t_r^B has the value one. Because of the two first new constraints, the new variables t_r^I and t_r^B must be one for the ratio of scenarios given by their respective probability variables δ_I and δ_B.

The re-formulated problem is, however, computationally extremely expensive to solve when the number of stands is high. For this reason, we approximate it with a problem where the decision variables for the treatment schedules $x_{t,s}$ are allowed to take any real value between 0 and 1, instead of being binary variables. This is a common approximation in forest management. The interpretation of treatment schedules with non-binary values is that a part of the stand is treated with a different schedule. For this reason, we can do this approximation and still get solutions which can be implemented as treatment schedules for the stands. This leaves us with a multiobjective optimization problem, with both integer $(t_r^I$ and $t_r^B)$ and real-valued $(x_{t,s})$ variables.

5 Experiment and Results

5.1 An Overview of the Experiment

We have conducted an experiment and Fig. 1 shows the data flow in it. We first need to acquire the data (1), which is estimated for a set of 0.04 ha segments of a forest area. To make the data more manageable, this data is segmented (2) into fairly homogenous stands. To include uncertainty into the model, randomization (3) is included into the forest data. With each instance of the stand level data, the forest stand is simulated (4) to predict future forest resources according to different treatment schedules. Once the future resources are predicted and different scenarios are created, we approximate (5) the ideal and nadir vectors using optimization. Then we can start the (6) solution process with an interactive method using reference points from a DM. Once the DM is satisfied, we produce a list of treatment schedules for all stands, and we can implement the plan for the forest area.

In what follows, Sect. 5.2 describes the data, segmentation of the forest area, randomization of the data and simulation. Section 5.3 outlines single-objective optimization and interactive multiobjective optimization methods applied. The output of our experiment are preferred treatment schedules for the stands that take into account the conflict between income and biodiversity and the DM's preferences and handle the uncertainties inherent to the problem.

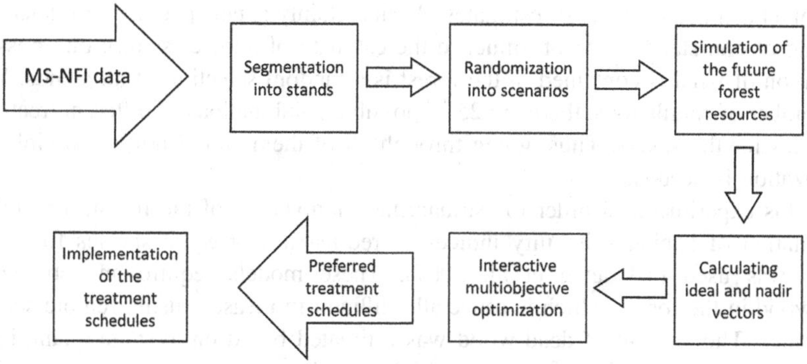

Fig. 1. Data flow in the conducted experiment

5.2 Forest Inventory Data with Uncertainties

The forest inventory data has been acquired through the combined use of remote sensing and statistical models based on a field plot measurement. The forest data, obtained from the Multi-source National Forest Inventory (MS-NFI) for 2011, was available in a raster dataset with a pixel size representing a 20 × 20 m footprint, provided by the Natural Resources Institute Finland (Luke), available from http://kartta.luke.fi/index-en.html. Our area of interest is a large forest area (8,415 ha) to the north of the city of Jyväskylä in Finland. The entire forest area has been segmented into a set of 2,842 stands where each stand represents a relatively homogenous forest area. The segmentation is based on a mainly cloud free Landsat image (LC81880162013155LGN00), utilizing an algorithm developed by the Technical Research Centre of Finland (VTT). In the experiment, we only used 300 stands, and 50 scenarios for the whole forest area. The forest-level scenarios were constructed by randomly selecting one of 25 stand-level scenarios for each stand, resulting in a problem with a size similar to a 15,000 stand problem. The selection of 50 scenarios was based on previous research (Eyvindson and Kangas 2016b).

As collected, the data are estimates of the expected value of the forest resources, and the actual forest resources may differ from the expectation. Two sources of uncertainty were included: initial inventory errors and forest stand age. We assumed that the inventory errors for the height, basal area (a measure of the forest density) and forest stand age were normally distributed, with a mean of 0 and a standard deviation equal to 10% of the mean of both variables. These estimates of errors reflect the current state-of-the-art inventory methods for Finnish forest conditions (Mäkinen et al. 2010).

The simulation of future forest resources was conducted using an open-source forest simulation and optimization software (SIMO, Rasinmäki et al. 2009). For each stand, SIMO generates a list of possible treatment scenarios which could be used in the forest during the planning period. They include silvicultural (i.e. planting, tending and fertilizing) and harvesting options (i.e. final felling or thinning). The maximum number of possible sets of treatment schedules for all stands was 34. For each stand, a total of 25 simulations were run for all treatment schedules. Each simulation was run with a random adjustment to the error estimates. As uncertainty is not spatially correlated (the estimate of one stand does not influence the estimate of another stand), one possible realization of what is contained in the forest is a random selection of all stands. With this number of simulations, there are 25^{300} possible combinations of different treatment schedules for the stands. Thus, going through all of them would not be possible and optimization is needed.

In this experiment, in order to estimate the biodiversity of the forest, a weighted combination of habitat suitability indices of red-listed saproxylic species for boreal forests was used (Tikkanen et al. 2007). These models require an estimate of dead-wood in the forest, which is especially difficult to measure using remote sensing techniques. The amount of dead-wood was estimated based on average quantities of dead-wood in Finland (Tomppo et al. 1999) and the age of the forest stands. Two functions were used to estimate the quantity of deadwood, one to represent the increase of dead-wood from a middle aged forest to an old-aged forest, while the other represented the decrease of deadwood from a recently harvested stand to a middle aged one.

5.3 Decision Making

The decision making environment has been implemented using a Jupyter Notebook, which has been made freely available at https://github.com/maeehart/MOD2016. The data used in decision making is also available at the same repository. The data of the stands is represented as a text file with the format required by the IBM® ILOG® CPLEX® Optimization Studio. The DM involved was an expert in forest management planning. The experiment was run on a sixteen core Intel Xeon E5-2670 processor with 64 GB of RAM. The computer was running Ubuntu Linux 14.04.3 LTS.

The multiobjective optimization problem of forest management planning was modeled using the Optimization Programming Language (OPL) of IBM. The problem was not, however, directly converted into a text file using OPL but, instead, the essential components (i.e., objectives, constraints and decision variables) of the problem were included in a Python dictionary. This is because OPL does not directly support multiobjective optimization, but the problems need to be scalarized (in this case with the achievement scalarizing function) in order to be completely modelled with OPL and then to be solved with IBM® ILOG® CPLEX® Optimization Studio.

Before starting the decision making process, one had to estimate the ideal and nadir vectors of the problem. This was done using the pay-off table approach (Miettinen 1999), where one first optimizes each objective individually, evaluates the values of all objectives at solutions obtained and forms the k-dimensional ideal vector from the best value of each objective and estimates the components of the nadir vector from the worst values obtained. The pay-off table for the problem considered is given in Table 1. In the pay-off table, the rows represent objective function values calculated at the solution where an objective function obtained the best value (ideal on the diagonal) and the components of the nadir vector are the worst values of each column. The biggest values in the table are written in bold face and the smallest values are underlined. The ideal vector is, thus

$$(VaR_I, VaR_B, E_I, E_B, \delta_I, \delta_B) = (5246601, 520, 4828656, 520, 1, 1)$$

and the nadir vector is all zeros.

Table 1. Pay-off table for the problem considered

VaR_I	VaR_B	E_I	E_B	δ_I	δ_B
5246601	358	4275987	358	0	1
0	**520**	0	**520**	1	0
3588782	398	**4828656**	399	0	1
0	519	0	**520**	0	**1**
0	0	0	0	**1**	**1**
0	0	0	0	**1**	**1**

The Pareto optimal solution corresponding to the reference point specified by the DM was obtained by solving the achievement scalarizing function introduced earlier. The achievement scalarizing function was implemented using a Python function, which compiles a string, which is then written into a text file. The text file was, finally, solved by calling the IBM® ILOG® CPLEX® Optimization Studio and attaching the data files to the call.

Before asking the DM for any reference point, a so-called neutral compromise solution (Wierzbicki 1999) for the problem was computed and shown to the DM together with the ideal and the nadir vectors. The neutral compromise solution was

$$(4595853, 456, 4631499, 456, 0.54, 0.68).$$

The neutral compromise solution gives information about the objective function values that are roughly in the middle of Pareto optimal solutions. This solution means that

- there is a 100%–54% = 46% chance that income is worse than 4595853€,
- there is 32% chance that the biodiversity is worse than 456; and that
- the mean income and biodiversity are 4631499€ and 456, respectively.

The DM found the solution quite good. However, she wanted to improve the security on the biodiversity and was willing to give up on the income. Thus, she gave a reference point

$$(2500000, 504, 2600000, 505, 0.7, 0.85).$$

In it, the confidence on the biodiversity is higher and both the mean and value at risk for the biodiversity are larger, and the values for income are worse than in the neutral compromise solution.

The solution of the achievement scalarizing function corresponding to the reference point was

$$(2545940, 509, 2642280, 510\ 0.8, 1).$$

This Pareto optimal solution is better in all objectives than the reference point. Thus, the DM was really happy. In the reference point method, the DM specifies a new reference point as long as she wishes to continue iterating.

As the DM wanted to still see whether it would be possible to improve the expected income at the cost of the reliabilities, she gave a new reference point

$$(250000, 508, 3000000, 509, 0.7, 0.85),$$

which has a higher value for the expected income. However, this reference point led to a Pareto optimal solution

$$(2913004, 507, 2986231, 508, 1, 0.98),$$

which means that the higher expected income came at the cost of biodiversity instead of the reliabilities.

This was not what the DM wanted. For this reason, she wanted to make one more check whether the income could be improved at the cost of reliabilities and gave one more reference point

$$(2000000, 504, 3000000, 510, 0.7, 0.85),$$

where she had improved the expected biodiversity and allowed the value at risk for the income to get worse. This resulted in a Pareto optimal solution

$$(2755038, 507, 2982154, 508, 0.98, 1).$$

The DM was satisfied with this solution, because it is better than the reference point that she gave in all of the objectives except the expected biodiversity and income, and even in these objectives, the value is very close. In addition, the DM was really happy that the probability of the biodiversity being over the value at risk is 100%. For this reason, the DM chose this solution as the final, most preferred solution.

The interpretation of this final preferred solution is that

- expected values of the minimum income and minimum biodiversity (i.e., minima over the periods) are 2982154 € and 508;
- there is only a two percent risk that the minimum income is smaller than 2755038 €; and
- the minimum biodiversity is guaranteed to be over 507 with 100% probability.

This was very much what the DM was hoping for, because the expected values are at satisfactory levels and risk levels in the value at risk are really low (i.e., 2% and 0%). Overall, the interactive solution process enabled the DM to get convinced of the goodness of the final solution and gain insight about the interdependencies of the objectives.

6 Conclusions

We have presented an approach of using interactive multiobjective optimization to handle the conflict between both

- income and biodiversity and
- risk and expected outcomes

in forest management planning. Our approach is based on

- simulating forest growth using standard models,
- modelling uncertainty using a scenario-based approach,
- converting the decision problem into a six-objective optimization problem, and
- using a reference point based interactive method to find preferable treatment schedules for the forest stands.

In the solution process, we applied the reference point method of Wierzbicki (1986) in its simplest form. Thanks to the interactive nature of the method, the DM could learn about the interdependencies between the objectives and the attainability of her preferences. Thanks to this, she could gain confidence of the final solution selected.

From the DM's point of view, an interactive method is essential. As the biodiversity is a unitless index, interpreting the results is only possible in relative terms. Relating the current solution to the previous solutions is needed in order to be able to construct a set of preferences.

Having the risk level and the value at risk at that particular risk level both as objectives for both the incomes and biodiversity further emphasizes the need for an interactive approach. It would be very difficult to give pre-defined hopes (like weights) for these variables. It may be easy enough to set e.g. weights when there is only one expected value and the value at risk at stake, but the setting would require quite an abstract analysis unless the task can be carried out in iterative fashion enabling learning.

A drawback of the solution process was that the computation times with the given data for a single given reference point were rather long, ranging from less than one day to almost five days. This is a problem when using interactive methods, when the DM may get tired in waiting for new solutions to be computed (as argued e.g., in Korhonen and Wallenius 1996). This is a major challenge to extend this method to scale, which is needed with large forests. Large forests may contain up to hundreds of thousands of stands, while our data contained only 300 stands. In many cases, such forest properties can be divided into parts that can be handled independently from each other, but still there is a need to make the method less time-consuming. One way could be the use of hierarchical planning. For instance, Pantuso, Fagerholt and Wallace (2015) have developed a method of solving hierarchical stochastic programs for a maritime fleet renewal problem. They were able to solve large problems in a hierarchical framework, which could not be solved using CPLEX alone.

Acknowledgements. We acknowledge Natural Resources Institute Finland Luke (for the MS-NFI data) and the Finish Research Institute VTT (for the segmentation). In addition, we thank IBM for allowing IBM® ILOG® CPLEX® Optimization Studio being used for academic work through the Academic initiative.

References

Eyvindson, K., Cheng, Z.: Implementing the conditional value at risk approach for even-flow forest management planning. Can. J. For. Res. **46**(5), 637–644 (2016)

Eyvindson, K., Kangas, A.: Evaluating the required scenario set size for stochastic programming in forest management planning: incorporating inventory and growth model uncertainty. Can. J. For. Sci. **46**(3), 340–347 (2016a)

Eyvindson, K., Kangas, A.: Integrating risk preferences in forest harvest scheduling. Ann. For. Sci. **73**, 321–330 (2016b)

Kangas, A., Kurttila, M., Hujala, T., Eyvindson, K., Kangas, J.: Decision Support for Forest Management. Managing forest ecosystems, 2nd edn., vol. 30. 307 p. Springer, New York (2005)

Korhonen, P., Wallenius, J.: Behavioural issues in MCDM: neglected research questions. J. Multi-Criteria Decis. Anal. **5**, 178–182 (1996)

Miettinen, K.: Nonlinear Multiobjective Optimization, 298 p. Kluwer, Boston (1999)

Miettinen, K., Ruiz, F., Wierzbicki, A.P.: Introduction to multiobjective optimization: Interactive approaches. In: Branke, J., Deb, K., Miettinen, K., Slowinski, R. (eds.) Multiobjective Optimization. LNCS, vol. 5252, pp. 27–57. Springer, Heidelberg (2008)

Mäkinen, A., Kangas, A., Mehtätalo, L.: Correlations, distributions and trends of inventory errors and their effects on forest planning. Can. J. For. Res. **40**(7), 1386–1396 (2010)

Kangas, J., Pukkala, T.: Operationalization of biological diversity as a decision objective in tactical planning. Can. J. For. Res. **26**, 103–111 (1996)

Kleywegt, A., Shapiro, A., Homem-de Mello, T.: The sample average approximation method for stochastic discrete optimization. SIAM J. Optim. **12**(2), 479–502 (2001)

Pantuso, G., Fagerholt, K., Wallace, S.W.: Solving hierarchical stochastic programs: application to the maritime fleet renewal problem. INFORMS J. Comput. **27**(1), 89–102 (2015)

Pykäläinen, J.: Interactive use of multi-criteria decision analysis in forest planning. Dissertation. Faculty of Forestry, University of Joensuu (2000)

Rasinmäki, J., Mäkinen, A., Kalliovirta, J.: SIMO: an adaptable simulation framework for multiscale forest resource data. Comput. Electron. Agric. **66**(1), 76–84 (2009)

Tikkanen, O.P., Heinonen, T., Kouki, J., Matero, J.: Habitat suitability models of saproxylic red-listed boreal forest species in long-term matrix management: cost-effective measures for multi-species conservation. Biol. Conserv. **140**(3), 359–372 (2007)

Tomppo, E.: Keski-Suomen metsäkeskuksen alueen metsävarat ja niiden kehitys 1967–96. Metsätieteen aikakauskirja 2B/1999, pp. 309–387 (1999). (in Finnish)

Wierzbicki, A.P.: On the completeness and constructiveness of parametric characterizations to vector optimization problems. OR Spectr. **8**(2), 73–87 (1986)

Wierzbicki, A.P.: Reference point approaches. In: Gal, T., Stewart, T.J., Hanne, T. (eds.) Multicriteria Decision Making: Advances in MCDM Models, Algorithms, Theory, and Applications, pp. 9-1–9-39. Kluwer Academic Publishers, Boston (1999)

Rockafellar, R.T., Uryasev, S.: Optimization of conditional value-at-risk. J. Risk **2**, 21–42 (2000)

Duffie, D., Pan, J.: An overview of value at risk. J. Deriv. **4**(3), 7–49 (1997)

Metabolic Circuit Design Automation by Multi-objective BioCAD

Andrea Patané[1], Piero Conca[1,2](✉), Giovanni Carapezza[1], Andrea Santoro[1,2,3], Jole Costanza[3], and Giuseppe Nicosia[1]

[1] Department of Mathematics and Computer Science, University of Catania, Catania, Italy
nicosia@dmi.unict.it
[2] Consiglio Nazionale delle Ricerche (CNR), Catania, Italy
pieroconca@gmail.com
[3] Istituto Italiano di Tecnologia (IIT), Milan, Italy

Abstract. We present a thorough *in silico* analysis and optimization of the genome-scale metabolic model of the mycolic acid pathway in *M. tuberculosis*. We apply and further extend MEGDMO to account for finer sensitivity analysis and post-processing analysis, thanks to the combination of statistical evaluation of strains robustness, and clustering analysis to map the phenotype-genotype relationship among *Pareto optimal* strains. In the first analysis scenario, we find 12 Pareto-optimal single gene set knockout, which completely shut down the pathway, hence critically reducing the pathogenicity of *M. tuberculosis*; as well as 34 genotypically different strains in which the production of mycolic acid is severely reduced.

Keywords: Metabolic pathways · Mycolic acid maximization · M. tuberculosis · Global sensitivity analysis · Robustness analysis · Clustering analysis · Optimization

1 Introduction

The attention given to *in silico* methods for the design of microbial strains overproducing metabolites of interest has drastically increased in the last few years [19]. The number of recent successes in the field of *synthetic biology* [23] seems indeed to shake off all but little doubt that in the near future the latter will be standard practice in the production of therapeutic drugs [24], renewable bio-materials [4] and biofuels [25]. However the intrinsic complexity of biological systems and organisms makes of paramount importance the design of mathematical and computational approaches to fully exploit the potential of this emerging discipline [26]. In fact, recent advances in *system biology* [2] have paved the way for accurate and computational efficient *in silico* analysis of biological system, in particular that of bacteria. Modelling paradigms used in the field include ordinary, partial and stochastic differential equation [1], agent-based modelling [6]

P.M. Pardalos et al. (Eds.): MOD 2016, LNCS 10122, pp. 30–44, 2016.
DOI: 10.1007/978-3-319-51469-7_3

and metabolic reconstructions [7]. Among the others, thought based on many simplifying assumptions, steady-state analysis of GEnome-scale Metabolic Models (GEMs), such as the one carried out by *Flux Balance Analysis* (FBA) [3], along with the ease of implementation of *-omics* data sets information into GEMs, has proven to be a computational efficient and reliable modelling approach for systematic *in silico* analysis of many organisms [4].

In this work we present a thorough analysis and knockout based optimization of an FBA model for the *Mycolic Acid Pathway* (MAP) of *M. tuberculosis*, focusing the analysis onto its pathogenicity. *Tuberculosis* is one of the worlds deadliest diseases, with over two million related deaths each year. The particular composition of the cell envelop of *M. tuberculosis*, which is primarily made of mycolates, is strongly associated with its pathogenicity [22]. Indeed it furnishes protection against the host immune system. In fact mycolic acid synthesis is the target of most anti-tuberculosis drugs. However the appearance of many drug resistant strains urges for the need of finding new candidate targets for therapeutic drugs. By using the metabolic reconstruction of the MAP of *M. tuberculosis* presented in [22], we investigate the *trade-off* among mycolic acid production and knock-out cost, i.e. number of targeted enzymes, in a struggle to extensively analyse possible metabolic targets in within *M. tuberculosis* for therapeutic purposes. To do this we present a *flux regulation* based sensitivity analysis, which ranks reactions of the networks accordingly to the influence that their lower and upper bound on the flux has on the production of mycolates. We hence formulate multi-objective minimization/maximization problems associated to mycolate production and knock-out cost, thorough-fully analysing more the 100 different strains, associated to different combination of targeted enzymes. Finally we analyse the robustness of the strains by statistically evaluating the robustness of their associated metabolic networks, under different values for the *perturbation strength* and the *robustness threshold*.

Related Work. A great amount of work has recently been done in the field of *in silico* analysis and optimization of metabolic networks. The authors in [27] introduce optimization in the field of metabolic networks analysed by FBA. This was achieved by directly manipulating the upper and lower bound on the reaction fluxes. The approach was further improved to account for better modelling of genetic manipulation in different directions. Examples are the use of genetic knockouts [27] and the modelling of enzymes up/down-regulation [5]. Heuristic optimization [32,33] has in many occasions proved to be an effective and efficient tool for hard optimization problems [34] in the field of system and synthetic biology, especially in metabolic networks. Genetic Design through Local Search (GDLS) [29] in which the MILP is iteratively solved in small region of the design space; Enhancing Metabolism with Iterative Linear Optimization (EMILiO) [28], that use a successive linear programming approach in order to solve efficiently the MILP obtained through the Karush-Kuhn-Tucker method. A review can be found in [19]. Sensitivity analysis and robustness are also extensively used concept in system biology. Some examples can be found in [20,21].

2 Genetic Re-programming of Genome-Scale Metabolic Models

In this section we introduce Genome-Scale Metabolic models and modelling approach for genetic manipulation that will be the used in the remaining sections. A genome-scale FBA model is a steady-state model based on the mass conservation assumption. Let $S = (s_{ij})$ be the $m \times n$ stoichiometric matrix associated with the metabolism of an organism, where m is the number of metabolites and n is the number of reactions which build up the organism metabolism, i.e. s_{ij} is the stoichiometric coefficient of the ith metabolite in the jth reaction. Let $v = (v_1, \ldots, v_n)$ be the fluxes vector, then assuming the organism is at steady-state we obtain the condition: $Sv = 0$. The linear system in Eq. ?? defines the *network capabilities* vector space, i.e. the subset of v allowed from a strictly physical point of view. Of course we cannot expect that stoichiometric information can account for the global behaviour of a cell. This is mathematically reflected by the condition $m < n$ which usually leads to a very large *network capabilities* space. Biological information are summarized into an n dimensional objective coefficient vector f experimentally tuned for the accurate modelling of each specific organism, which represents the *biological objective* of the organism. Although several alternatives are possible, the most used *biological objective* is the cell *biomass* production. By using FBA modelling the dynamic of a cell is hence retrieved by solving the linear programming problem

$$\begin{aligned} &\text{maximize } f^T v \\ &\text{subject to } Sv = 0 \\ &\qquad\qquad v^- \leq v \leq v^+ \end{aligned} \tag{1}$$

where v^- and v^+ are lower and upper bounds vector on the fluxes, whose actual values are based on empirical observations.

Of great importance for the task we are tackling is the possibility of augmenting an FBA model to a genome-scale model by using genomics information. This is accomplished by introducing the *Gene-Protein-Reaction* (GPR) mapping $G = (g_{lj})$. Firstly, the genes (or a subset of the genes) of a cell are reorganized into gene sets, i.e. group of genes linked by Boolean relations accordingly to common reactions that their associated proteins catalyse. For example, a gene set of the form "G_1 *and* G_2" implies that both G_1 and G_2 are needed for a particular reaction to be catalysed (i.e. they represent an *enzymatic complex*), whereas a gene set of the form "G_1 *or* G_2" implies that at least one among G_1 and G_2 is needed for that particular reaction to be catalysed (i.e. G_1 and G_2 code for *isoenzymes*). The GPR hence relates sets of reactions to sets of genes, which code for proteins catalysing for the former sets. Namely, g_{lj} is equal to 1 if and only if the lth gene set is related to the jth reaction; g_{lj} is equal to 0 otherwise. This allow us to perform *in silico* analysis of the effect of genetic knock-outs/knock-ins to the cell metabolism by simple manipulation of the FBA model [27]. Namely, the knockout of the lth gene set is modelled by constraining to a zero flux all the reactions j such that g_{lj} is equal to 1. Finally, an important index for the analysis of gene set knockouts is the Knock-out Cost (KC) associated with the gene-sets. This is recursively defined over the form of the gene set. Briefly, if a gene set is composed by two smaller gene sets related by an "*and*", then the KC of the composite gene set is the smallest KC of the two gene sets that compose it (knocking out either one these two will knock-out the *enzymatic complex*). If whereas the two smaller gene set are linked by means of an "*or*" then the KC of the gene set is the sum of the KCs of the smaller gene sets (since they are isoenzymes we need to knock-out both of them).

3 Analysis and Optimization of Metabolic-Models

In this section we review the basic principle of GDMO (Genetic Design through Multi-objective Optimization) bioCAD tool [8]. Then, we extend GDMO with a novel approach for the statistical analysis of the relation between the genotype and the phenotype of Pareto optimal strains, based on a clustering technique.

Pareto Optimality. GDMO is built on the concept of Pareto optimality, which has proven to be of paramount importance in the *in silico* design analysis of FBA models [9]. Pareto optimality is an extension of the concept of optimality used in single-objective optimization problems [34], in which the ordering relationship is extended along each coordinate direction. Intuitively, Pareto optimality comes into play when for a particular design it is of interest to optimize several objective functions, which are in contrast with each other. For example there usually exists a trade-off between the growth rate of a bacterium and the production rate of a particular metabolite. In order to enhance the production of the latter the bacterium has to redirect its resources from the pathways involved into growth to the pathways involved into the production of the metabolite. Pareto optimality allows a rigorous analysis of the trade-offs among these two production rates.

Formally, we define a strict ordering relationship $\prec$ for each $x, y \in \mathbb{R}^n$: $x \prec y \iff x_i \leq y_i \quad i = 1, \ldots, n \quad \text{and} \quad \exists\ j \text{ s.t. } x_j < y_j$ i.e., each component of x is less than or equal to its corresponding component of y, and at least one x component is strictly less than y component. This is used for the definition of Pareto optimality. Consider a generic multi-objective optimization problem [10] of the form: minimize $F(x_1, \ldots, x_k)$ subject to $(x_1, \ldots, x_k) \in T$ where $F : \mathbb{R}^k \to \mathbb{R}^n$ is the n-dimensional objective function and T is the feasible region of the problem. We define the *domination* relationship associated with the objective function F for domain elements $x, y \in \mathbb{R}^k$ by using the $\prec$ of the codomain: $x \prec_F y \iff F(x) \prec F(y)$ i.e. a point x of the domain dominates y if each objective function component of x is less than or equal to the corresponding component of y, and at least one component is strictly less. Finally we say that a point $x \in T$ is Pareto optimal for the problem defined by Eq. 3 if there are no points $y \in T$ which dominate x. Notice that the $\prec$ is a *partial* order hence, generally, many Pareto optimal points will exist, which we group in to the *Pareto front*. The goal of a multi-objective optimization algorithm is thus to find the *Pareto front* of the problem.

Sensitivity Analysis. Sensitivity Analysis (SA) [11] is a statistical model analysis, which we use to evaluate the *global* effect that a group of controllable parameters has on the output of a model. In a optimization problem, it can be used to identify the parameters of a model that do not affect the performances significantly (*unsensitive* parameters). We use a *qualitative* SA method (i.e. it may be used only to rank parameters according to their effect) based on the *one-factor-at-a-time* Morris method [12], and that extend *PoSA* [13] to account for a finer regulation of reaction fluxes.

In practical terms the Morris method returns a mean μ_i^* and a standard deviation σ_i for each parameter. These two values measure the effect caused by several random parameter variations on the output(s). If the mean value returned by the method is "relatively" high, then the associated parameter has an important influence on the output. Analogously, a high standard deviation returned for a specific parameter indicates that either the latter is interacting with other design parameters or that it has strongly nonlinear effects on the model output.

In our modelling framework the output of the model to be considered is a vector v of fluxes satisfying Eq. 1. For what it concerns the input parameters, we extend the *PoSA* method to consider a *flux regulation* analysis.

This analysis evaluates the effect that the up/down-regulation of any reaction included in the FBA model has on the flux distribution vector v. The controllable parameters that we consider in this analysis are thus the lower and upper bounds of the *internal* reactions. Reactions for which high values of μ^* are found are hence the reactions which strongly modify the flux distribution vector v and are, therefore, of interest for a *in silico* knockout analysis of the model.

Optimization and Trade-off Analysis. MEGDMO optimization engine builds upon NSGA-II algorithm [14]. The latter is a state-of-the-art algorithm for multi-objective optimization. It is based on the paradigm of genetic computation. It works by sampling from the domain an initial set of allowed solution to the optimization problem, i.e. a *population*, and it iteratively attempts to improve the *fitness*, i.e. a problem dependent measure of the Pareto optimality of candidate solutions, of the population by applying several *evolutionary operators* to the solutions [10] and by using the survival-of-the-fittest principle. The pseudo-code of the optimization algorithm MEGDMO in shown in Algorithm 1. The input of the procedure are: (i) *pop*, the size of the population; (ii) *maxGen*, the maximum number of *generations* (i.e. iterations of the algorithm main loop) to be performed; (iii) *dup*, the strength of the *cloning operator* [15]; and (iv) *uKC*, the maximum knock-out cost allowed to be taken into account by the algorithm. The initial population, $P^{(0)}$, is randomly initialized by the routine *InitPop*, who just randomly sample the domain of the problem (i.e. randomly apply few mutations to the wild type strain). We hence apply the FBA to each strain in $P^{(0)}$, and, accordingly to the value of the production rates of metabolite of interest, we compute *rank* and *crowding distance* of each member of the population [14]. The former ensures the *Pareto-orientation* of our procedure, redirecting the search towards the problem Pareto front. The *crowding distance*

Algorithm 1. meGDMO Optimization Algorithm

procedure MEGDMO(*pop*,*maxGen*,*dup*, *uKC*)
    $P^{(0)} \leftarrow InitPop(pop)$
    $FBA(P^{(0)})$
    $Rank\_and\_crowding\_distance(P^{(0)})$
    $gen \leftarrow 0$
    while $gen < maxGen$ **do**
        $Pool^{(gen)} \leftarrow Selection(P^{(gen)}, \lceil \frac{pop}{2} \rceil)$
        $Q_{dup}^{(gen)} \leftarrow GenOffspring(Pool^{(gen)}, dup)$
        $Q_{dup}^{(gen)} \leftarrow Force\_to\_feasible(Q^{(gen)}, uKC)$
        $FBA(Q_{dup}^{(gen)})$
        $Rank\_and\_crowding\_distance(Q_{dup}^{(gen)})$
        $(Q^{(gen)}) \leftarrow BestOutOfDup(Q_{dup}^{(gen)}, dup)$
        $P^{(gen+1)} \leftarrow Best(P^{(gen)} \cup Q^{(gen)}, pop)$
        $gen \leftarrow gen + 1$
    return $\left(\bigcup_{gen} P^{(gen)}\right)$

whereas is a rough estimation of the population density near each candidate solution. During the optimization main loop, candidate solutions in *unexplored* regions of the objective space (thus having small values of crowding distance) are preferred to those which lie in "crowded" regions of the objective space. This has the purpose of obtaining good approximations of the actual Pareto front of the problem. We then initialize the generation counter, and enter the main loop, which is performed *maxGen* times. At the beginning of each generation the *Selection* procedure generate a mating pool $Pool^{(gen)}$, by selecting individual from the current population $P^{(gen)}$. This is done following a *binary tournament selection* approach. Namely, tournaments are performed until there are $\lceil pop/2 \rceil$ individuals (*parents*) in the mating pool. Each tournament consist of randomly choosing two individuals from $P^{(gen)}$, and putting the best of the two individuals (in terms of rank and crowding distance) into $Pool^{(gen)}$. *Children* individuals are thus generated from the *parents* by using binary *mutation*. Namely we randomly generate *dup* different children from each parent, generating the $Q_{dup}^{(gen)}$. Then, we keep only the best solution of these *dup* children for each parent, hence defining the actual offspring set $Q^{(gen)}$. The reason for this lies in the fact that many of the mutations allowed in an FBA model are *lethal* mutation, i.e. they severely compromise the bacteria growth. Of course, a greater value for *dup* implies that feasible mutations are more likely to be found, whereas smaller values reduce the computational burden of the optimization. In order to achieve this we firstly ensure that each individual of $Q_{dup}^{(gen)}$ is feasible with respect to our optimization problem (i.e. it has less than *uKC* knock-outs). Namely if a child is not in the allowed region, we randomly knock-in genes, until it is forced back to the feasible region. We hence evaluate the biomass and metabolites production of each new individual, and the algorithm computes new values of rank and crowding distance, for each individual. Procedure *BestOutOfDup* select from each of the *dup* children of each parent the best one and put it in the $Q^{(gen)}$ set. Finally, procedure *Best* generates a new population of *pop* individuals, considering the current best individuals and children. Output of the optimization algorithm is the union of the populations of all the generations. We then analyse the optimization results by means of Pareto analysis, hence computing the *observed* Pareto fronts, i.e. the set of $\bigcup_{gen} P^{(gen)}$ elements which are not dominated by any other element in $\bigcup_{gen} P^{(gen)}$ (notice that $\bigcup_{gen} P^{(gen)}$ covers only a portion of the feasible region, hence we talk about *observed* Pareto optimality).

Robustness Analysis. As post-processing procedure of MEGDMO, we consider various *Robustness Analysis* (RA) routines [8]. Briefly, we use RA indexes for the *in silico* test and validation of the strains obtained in the optimization phase; in the specific we investigate how notable Pareto optimal (or approximated Pareto-optimal) individuals adapt to "small" variations that inevitably occur either in within the considered bacterium itself, or in the environment surrounding the latter. We throughout fully analyse the robustness of each notable strain assigning it two different *robustness indexes*: (i) *Local Robustness* (LR_i), associated with each reaction of the metabolic network considered; and (ii) *Global Robustness* (GR), which evaluates the robustness of a strain under a global point of view. Consider a strain and let v^* be the distribution flux vector associated with it, computed by applying FBA modelling to the strain. Let ϕ be a strain-dependent function (e.g. production rate of a particular metabolite). Let $\sigma \in [0,1]$ be a perturbation strength and $\epsilon \in [0,1]$ a threshold value. Let $\Delta_i \sim \mathcal{N}(0, \sigma v_i^*)$ be a normally distributed random variable and $V^i = (0, \dots, \Delta_i, \dots, 0)$. We define the ith local robustness index of parameter σ and ϵ for the considered strain as:

$$LR_i^{\sigma,\epsilon} = P\left(\left|\phi\left(v^*\right) - \phi\left(v^* + V^i\right)\right| \leq \epsilon \left|\phi\left(v^*\right)\right|\right)$$

which is the probability that a perturbation of percentage strength σ for the flux of the ith model reaction will produce a small perturbation on the value of ϕ. The definition for the GR index of a strain is a straight forward extension to that of local robustness. Let $V = (\Delta_1, \dots, \Delta_n)$. We then define the global robustness index of a strain as:

$$GR^{\sigma,\epsilon} = P\left(\left|\phi\left(v^*\right) - \phi\left(v^* + V\right)\right| \leq \epsilon \left|\phi\left(v^*\right)\right|\right).$$

In the implementation, we compute the LR_i and GR indexes using a Monte-Carlo approach used to estimate the probabilities in the definitions.

Clustering. Solutions when represented using, for instance, the production of a metabolite and biomass production tend to form clusters. These highlight the feasible zones of the space of solutions, assuming that an exhaustive search has been performed. Performing clustering on such solutions allows to study the characteristics of the strains that belong to a cluster and potentially identify similarities. There are several clustering techniques, however we suggest that in this context density-based clustering seems to be the preferable to centroid-based or probabilistic techniques such as k-means and Expectation Maximization, since clusters generally tend to have irregular shapes.

4 Mycolic Acid Production in *M. tuberculosis*

In this section we analyse the *Mycolic Acid Pathway* (MAP) in *M. tuberculosis* [22]. We apply meGDMO to investigate the production of mycolic acid using FBA modelling. The genome-scale metabolic model of the MAP comprises 28 genes (protein) grouped in 24 gene sets; and 197 metabolites involved in 219 different reactions plus 28 exchange reactions that allow exchange of metabolites (i.e. nutrients and output metabolites for the pathway) between the model and its environment. In this case, we normalize all the reactions fluxes to either be in $[-1, 1]$ (reversible reactions) or in $[0, 1]$ (irreversible reactions); hence we obtain *relative* flux values as analysis results. The biological objective of this pathway model is, of course, the production of *mycolic acid*. Since there are six kind of different mycolates associated with the pathogenicity of *M. tuberculosis* (namely, *alpha-mycolate*, *cis-metoxy-mycolate*, *trans-metoxy-mycolate*, *cis-keto-mycolate* and *trans-keto-mycolate*), the biological objective is a weighted sum of the production rate of these 6 mycolates [22].

Sensitivity Analysis of the MAP. We apply the *flux regulation* sensitivity analysis to the MAP FBA model. Hence we compute the effect that the regulation of each reaction (modelled as a perturbation of its lower and upper bounds) have on the distribution flux vector computed by FBA modelling. Results for this analysis are shown in Fig. 1. This analysis demonstrate that regulation of the lower bounds of the reaction rates (in red in the Figure) usually have a grater effect on the fluxes vector than a corresponding regulation of the upper bounds. In fact only one upper bound (in blue in the Figure) has a mean and standard deviation significantly different from zero. This upper bound is associated with the flux

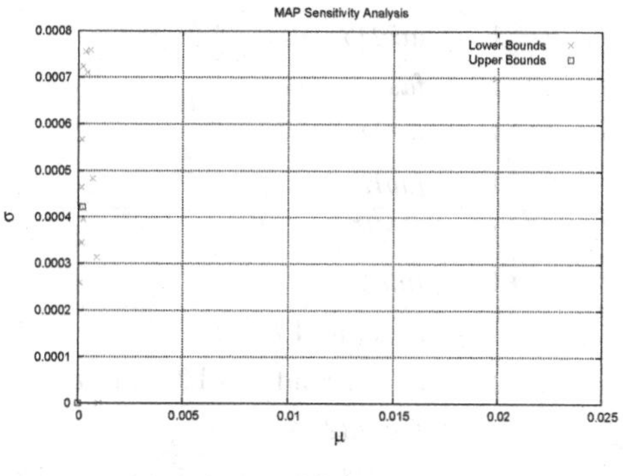

Fig. 1. *Flux regulation* sensitivity analysis of the MAP FBA model. (Color figure online)

of exchange *coenzyme-A*. The effect of the upper bound of this reaction on the flux vector is either not linear or the input interacts with other reactions. This underlines the importance of *coenzyme-A* for the virulence of *M. tuberculosis.* In fact this fatty acid is an important precursor of mycolic acid [22]. Analogous consideration may be made for all the reaction associated with all the lower bounds that are in the top left region of the figure. The reaction that however has the greatest effect on the output of the FBA of the MAP is *acetyl-CoA*, which lies in the top-right of the Figure. The latter strongly interacts with *coenzyme-A* as a precursor to mycolic acid [22].

Knock-out Analysis of the MAP. In this section we perform extensive knock-out analysis of the MAP FBA model, by applying 3 meGDMO optimization to the latter. Namely: (i) minimization of both KC and of mycolic acid production, which investigate the model for the minimum number of different enzyme targets that undermine the efficiency of the pathway; (ii) minimization of KC and maximization of mycolic acid production, which analyse the model for the minimum sets of knock-outs that optimize the efficiency of the pathway; and (iii) Maximize both $NADH$ and ATP production, in which meGDMO strives to find knock-outs that optimize the production rates of both $NADH$ and ATP.

Minimize Knock-out Cost and Mycolic Acid Production. We investigate the MAP model for minimum sets of *lethal genes*, i.e. sets of genes which if knocked-out completely compromise the biological object of the FBA model, i.e. severely reduce the virulence of *M. tuberculosis.* We find 12 different single knock-outs strains, enumerated in Table 1 as A_{1-12}. Either one of this knock-outs completely shut off the pathway, yielding a null mycolic acid production rate. Notice that in

Table 1. MOO and robustness analysis results: minimize KO cost and minimize mycolic acid production.

Strain	MA production	KC	Gene sets	GR %	LR %
A_1	0	1	*accA3*	100%	100%
A_2	0	1	*accD3*	100%	100%
A_3	0	1	*fas*	100%	100%
A_4	0	1	*fabD*	100%	100%
A_5	0	1	*fabH*	100%	100%
A_6	0	1	*UNK1*	100%	100%
A_7	0	1	*inhA*	100%	100%
A_8	0	1	*kasB* and *kasA*	100%	100%
A_9	0	1	*desA1* and *desA2* and *desA3*	100%	100%
A_{10}	0	1	*fadD32*	100%	100%
A_{11}	0	1	*accD4* and *accD5* and *accA3*	100%	100%
A_{12}	0	1	*pks13*	100%	100%

this case a single knock-out is enough for shutting off the pathway, hence Pareto points are all characterized by a KC of 1.

Minimization of KC and Maximization of Mycolic Acid Production. We optimize the MAP model for the simultaneous maximization of mycolic acid production and minimization of the cost of knock-outs performed. We set the maximum number of knock-outs allowed (i.e. uKc) to 7. Since mycolic acid production is the biological objective of the MAP FBA model, we cannot expect to improve its nominal production value (i.e. 0.0130) by just knocking-out genes. Hence this optimization boils down to an analysis of minimum required enzyme targets whose combined knock-out reduce the production of mycolates to various optimal levels, hence retrieving important information about targets which are strictly involved (however not necessary essential) to the building of the cell envelope.

In this optimization scenario, MEGDMO is able to find 10 notable strains which compose the building blocks of all the remaining Pareto optimal solutions, listed in Table 2 as B_{1-10}. Figure 2a we plot the projection on the co-domain space of all the >100 different strains explored by MEGDMO during this analysis. We find 5 strains with a $KC = 1$ with production rate identical to that of the wild type strain. These correspond to enzymes, which do not alter the production of mycolates at all. For $KC = 2$ MEGDMO finds 34 solutions with wild type mycolic acid production. Of this, 33 are obtained combining the gene sets of strains B_{1-5}; whereas the remaining one is characterized by the knock-out of the gene sets *fabG1* and *mabA* (strain B_6). Finally, for $KC = 3$, we find 45 solutions with wild type production, which are all combinations of the above strains gene sets, i.e. that of strains B_{1-6}. The remaining solutions are given by gene sets that slightly inhibit mycolic acid production, with respect to the wild type one. Building blocks strains for this solutions are strains B_{7-10}.

Table 2. MOO and robustness analysis results: minimize KO cost and maximize mycolic acid production.

Strain	MA production	KC	Gene sets	GR %	LR %
B_1	0.0130	1	*cmaA2*	0%	7.6%
B_2	0.0130	1	*mmaA3*	0%	5.2%
B_3	0.0130	1	*mmaA4*	0%	4.4%
B_4	0.0130	1	*acpS*	0%	4%
B_5	0.0130	1	*fabG2*	0%	7.2%
B_6	0.0130	2	*fabG1* or *mabA*	0%	6.4%
B_7	0.0060	1	*mmaA2*	0%	5.6%
B_8	0.0060	1	*pcaA*	0%	4.8%
B_9	0.0052	2	*pcaA* ∪ *mmaA3*	0%	8%
B_{10}	0.0007	2	*mmaA2* ∪ *mmaA3*	0%	4.8%

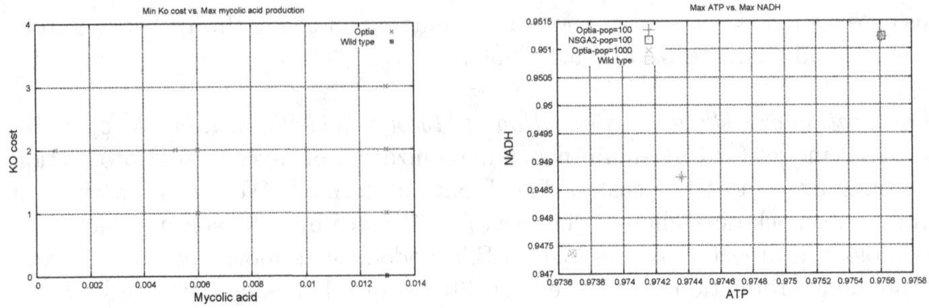

(a) KC minimization and mycolic acid production maximization.

(b) maximize ATP and NADH.

Fig. 2. Optimization results.

Notice that any strain obtained by the combined knock-out of more than 4 of any these gene sets does not produce mycolic acid. This demonstrates that none of these genes is strictly essential for the production of mycolic acid, however they are if taken together. Hence the production is characterized by various bifurcation; this is justified by the fact that more than one mycolic acid is included in the MAP FBA model.

Maximization of NADH e ATP. We here analyse the MAP metabolic network for the production of *NADH* and *ATP*. We investigate the behaviour of MEGDMO in three different settings. Namely (i) $pop = 100$, $maxGen = 300$, using the default optimization algorithm of MEGDMO; (ii) $pop = 100$, $maxGen = 300$, using Optia as optimization engine [18]; and (iii) $pop = 1000$, $maxGen = 300$, using Optia as optimization engine. We superimpose the results of the three optimization performed in Fig. 2b.

The best strains found among all these optimization (which is found in all three of them, this demonstrates the robustness of MEGDMO as an optimization tool) has values of *ATP* and *NADH* production respectively 0.2 and 0.4 times greater than these for the wild type strain. We list the two different strains, which obtain these production rate values in Table 3. Notice that these two strains, C_1 and C_2, correspond to strains B_9 and B_{10}.

Robustness Analysis of the MAP. We evaluate both the *GR* and the *LR* robustness indexes for the wild type strain. This furnishes an important index of the robustness of the set of targeted enzymes for the shut-down of the mycolates

Table 3. MOO results: maximize ATP and NADH.

Strain	*ATP* production	*NADH* production	*KC*	Gene set	GR %	LR %
C_1	0.9756	0.9512	2	*pcaA* or *mmaA3*	0%	4.4%
C_2	0.9756	0.9512	2	*mmaA2* or *mmaA3*	0%	6%

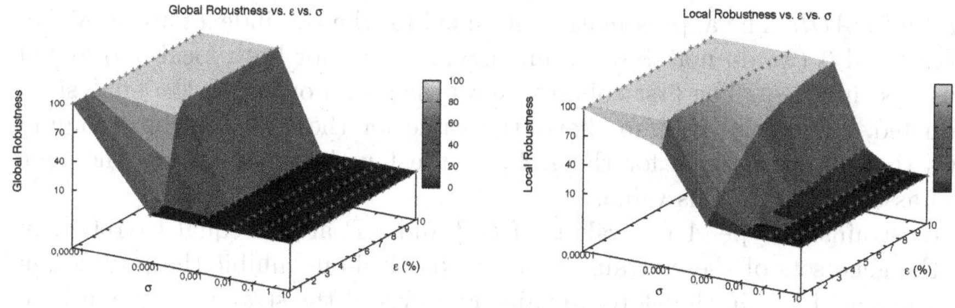

(a) Global Robustness. (b) Local Robustness.

Fig. 3. Robustness analysis for mycolic acid production.

production. An high value of robustness for a particular strain, indicates that enzyme targets associated to the latter will result in a more stable set of targets for therapeutic purposes. Hence less likely to become ineffective because of bacterium adaptations. Figures 3a and b show the results of this analysis (in percentage points) performed on the mycolic acid production rate. The *LR* index plotted is the mean value of all the local indexes associated with each reaction of the model. Values for the perturbation strength considered, σ, span in the interval $\left[10^{-6}, 10\right]$, while we consider value from the interval $[0.005, 0.01]$ for the robustness threshold ϵ. Of course, for low values of σ and for high values of ϵ both *GR* and *LR* tend to 100%. Vice versa, for high values of σ and for low values of ϵ *GR* and *LR* both tend to 0%. The results of an analogous analysis performed for the average of *ATP* and *NADH* productions are plotted in Figs. 4a and b.

We analyse the robustness indexes of the strains of type *A*, *B*, and *C*, found during the optimization phase using $\sigma = 10^{-5}$ and $\epsilon = 0.5$. For the strains of type *A* and *B* we analyse the robustness associated with mycolic acid production. Whereas for the strains of type *C* we use the average of the productions rate of

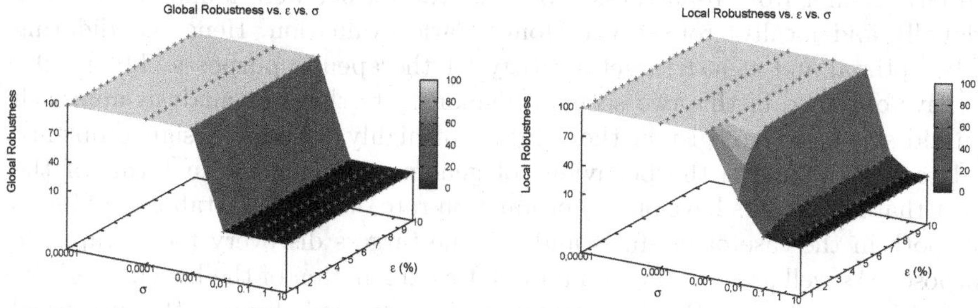

(a) Global Robustness. (b) Local Robustness.

Fig. 4. Robustness analysis for *ATP* and *NADH* production.

ATP and *NADH*. The approximation obtained for the two indexes are shown in Tables 2 and 3 (we do not show strains having 100% for both local and global robustness indexes). The first column shows the strain on which the analysis is conducted, the second column shows the value for the *GR*, the third column shows the minimum value for the *LR* and the fourth column shows the gene set(s) associated with this value.

For strains of type *A* the values of *GR* and *LR* are all equal to 100%, in fact the gene sets of these strains that are knocked-out inhibit the production of mycolic acid which, therefore, remains null even if the strains are perturbed. Instead, the type *B* strains values for the *GR* are all zero. Hence in this case small perturbations greatly affect the production of mycolic acid. Even the approximation for *LR* are small, thus type *B* strains cannot be considered robust with respect to the inhibition of mycolic acid production.

Finally, for type *C* strains values for the *GR* are also null. This does not occur in the wild type strain that has a *GR* of 100%. Same considerations can be made for *LR* values. Strains C_{1-2} have low *LR*, while the wild type has 99.6%.

5 Conclusions

In this paper we analysed two FBA models by using an extension of MEGDMO BioCAD tool, which account for finer analysis of the metabolic network models and processing of Pareto optimal strains. In particular we have performed a throughout full knock-out analysis of a metabolic reconstruction of the metabolic acid pathway of *M. tuberculosis*, by combining MEGDMO, as analysis tool, and FBA modelling, as simulation engine. The strains obtained were extensively analysed by Monte Carlo approximation of global and robustness indexes, for different values of *perturbation strength* and *robustness threshold*. In this case we found 12 candidate enzyme targets, which completely shut-down the production of mycolate in the bacterium, hence drastically reducing the pathogenicity associated to *M. tuberculosis*, as well as more than 100 different strains, in which the production of mycolic acid is severely inhibited. Thought we showed that the latter suffer from robustness problems, the former were found to be 100% (globally and locally) robust via Monte Carlo evaluation. Hence candidating to be optimal and robust target strategy for therapeutic purposes. The results we have obtained in this two scenario demonstrate that our analysis approach can aid synthetic biologist in the solution of highly complex design problems, and to better analyse the behaviour of genome-scale models in terms of the effect that knock-outs have on the production rate of several metabolite of interest, both in the case of bio-fuels and enzyme targets discovery for therapeutic purposes. As well as furnishing an automatic explanation of the knock-outs performed in a particular Pareto optimal strain, obtained through the statistical analysis of the empiric distribution of knock-outs in a particular cluster of the Pareto front.

References

1. Hasdemir, D., Hoefsloot, H.C.J., Smilde, A.K.: Validation and selection of ODE based systems biology models: how to arrive at more reliable decisions. BMC Syst. Biol. **9**, 1–9 (2015)
2. Palsson, B.: Systems Biology. Cambridge University Press, Cambridge (2015)
3. Kauffman, K.J., Prakash, P., Edwards, J.S.: Advances in flux balance analysis. Curr. Opin. Biotechnol. **14**(5), 491–496 (2003)
4. Yim, H., Haselbeck, R., Niu, W., Pujol-Baxley, C., Burgard, A., Boldt, J., Khandurina, J., et al.: Metabolic engineering of Escherichia coli for direct production of 1, 4-butanediol. Nat. Chem. Biol. **7**(7), 445–452 (2011)
5. Rockwell, G., Guido, N.J., Church, G.M.: Redirector: designing cell factories by reconstructing the metabolic objective. PLoS Comput. Biol. **9**, 1 (2013)
6. Figueredo, G.P., Siebers, P., Owen, M.R., Reps, J., Aickelin, U.: Comparing stochastic differential equations and agent-based modelling and simulation for early-stage cancer. PloS One **9**(4), e95150 (2014)
7. Hamilton, J.J., Reed, J.L.: Software platforms to facilitate reconstructing genome-scale metabolic networks. Environ. Microbiol. **16**(1), 49–59 (2014)
8. Costanza, J., Carapezza, G., Angione, C., Lió, P., Nicosia, G.: Robust design of microbial strains. Bioinformatics **28**(23), 3097–3104 (2012)
9. Patane, A., Santoro, A., Costanza, J., Carapezza, G., Nicosia, G.: Pareto optimal design for synthetic biology. IEEE Trans. Biomed. Circ. Syst. **9**(4), 555–571 (2015)
10. Deb, K.: Multi-objective Optimization Using Evolutionary Algorithms. Wiley, Hoboken (2001)
11. Saltelli, A., Ratto, M., Andres, T., Campolongo, F., Cariboni, J., Gatelli, D., Saisana, M., Tarantola, S.: Global Sensitivity Analysis: The Primer. Wiley, Hoboken (2008)
12. Morris, M.D.: Factorial sampling plans for preliminary computational experiments. Technometrics **33**(2), 161–174 (1991)
13. Angione, C., Carapezza, G., Costanza, J., Lió, P., Nicosia, G.: Computing with metabolic machines. Turing-100 **10**, 1–15 (2012)
14. Deb, K., Pratap, A., Agarwal, S., Meyarivan, T.: A fast and elitist multiobjective genetic algorithm: NSGA-II. Turing-100 **10**, 1–15 (2012)
15. Cutello, V., Narzisi, G., Nicosia, G., Pavone, M.: Clonal selection algorithms: a comparative case study using effective mutation potentials. In: Jacob, C., Pilat, M.L., Bentley, P.J., Timmis, J.I. (eds.) ICARIS 2005. LNCS, vol. 3627, pp. 13–28. Springer, Heidelberg (2005). doi:10.1007/11536444_2
16. Angione, C., Costanza, J., Carapezza, G., Lió, P., Nicosia, G.: Pareto epsilon-dominance and identifiable solutions for BioCAD modelling. In: Proceedings of the 50th Annual Design Automation Conference, pp. 43–51 (2013)
17. Carapezza, G., Umeton, R., Costanza, J., Angione, C., Stracquadanio, G., Papini, A., Liò, P., Nicosia, G.: Efficient behavior of photosynthetic organelles via Pareto optimality, identifiability, and sensitivity analysis. ACS Synth. Biol. **2**(5), 274–288 (2013)
18. Cutello, V., Narzisi, G., Nicosia, G., Pavone, M.: An immunological algorithm for global numerical optimization. In: Talbi, E.-G., Liardet, P., Collet, P., Lutton, E., Schoenauer, M. (eds.) EA 2005. LNCS, vol. 3871, pp. 284–295. Springer, Heidelberg (2006). doi:10.1007/11740698_25
19. Long, M.R., Ong, W.K., Reed, J.L.: Computational methods in metabolic engineering for strain design. Curr. Opin. Biotechnol. **34**, 135–141 (2015)

20. Marino, S., Hogue, I.B., Ray, C.J., Kirschner, D.E.: A methodology for performing global uncertainty and sensitivity analysis in systems biology. J. Theor. Biol. **254**(1), 178–196 (2008)
21. Kitano, H.: Biological robustness. Nat. Rev. Genet. **5**(11), 826–837 (2004)
22. Takayama, K., Wang, C., Besra, G.S.: Pathway to synthesis and processing of mycolic acids in Mycobacterium tuberculosis. Clin. Microbiol. Rev. **18**(1), 81–101 (2005)
23. Church, G.M., Regis, E.: Regenesis: How Synthetic Biology Will Reinvent Nature and Ourselves. Basic Books, New York (2014)
24. Church, G.M., Elowitz, M.B., Smolke, C.D., Voigt, C.A., Weiss, R.: Realizing the potential of synthetic biology. Nat. Rev. Mol. Cell Biol. **15**(3), 289–294 (2014)
25. Lee, S.K., Chou, H., Ham, T.S., Lee, T.S., Keasling, J.D.: Metabolic engineering of microorganisms for biofuels production: from bugs to synthetic biology to fuels. Curr. Opin. Biotechnol. **19**(6), 556–563 (2008)
26. Andrianantoandro, E., Basu, S., Karig, D.K., Weiss, R.: Synthetic biology: new engineering rules for an emerging discipline. Mol. Syst. Biol. **2**(1) (2006)
27. Burgard, A.P., Pharkya, P., Maranas, C.D.: Optknock: a bilevel programming framework for identifying gene knockout strategies for microbial strain optimization. Biotechnol. Bioeng. **84**(6), 647–657 (2003)
28. Yang, L., Cluett, W.R., Mahadevan, R.: EMILiO: a fast algorithm for genome-scale strain design. Metab. Eng. **13**(3), 272–281 (2011)
29. Lun, D.S., Rockwell, G., Guido, N.J., Baym, M., Kelner, J.A., Berger, B., Galagan, J.E., Church, G.M.: Large-scale identification of genetic design strategies using local search. Mol. Syst. Biol. **5**(1), 296 (2009)
30. Orth, J.D., Conrad, T.M., Na, J., Lerman, J.A., Nam, H., Feist, A.M., Palsson, B.: A comprehensive genome-scale reconstruction of Escherichia coli metabolism-2011. Mol. Syst. Biol. **7**(1), 535 (2011)
31. Ester, M., Kriegel, H., Sander, J., Xu, X., et al.: A density-based algorithm for discovering clusters in large spatial databases with noise. Kdd **96**(34), 226–231 (1996)
32. Ciccazzo, A., Conca, P., Nicosia, G., Stracquadanio, G.: An advanced clonal selection algorithm with ad-hoc network-based hypermutation operators for synthesis of topology and sizing of analog electrical circuits. In: Bentley, P.J., Lee, D., Jung, S. (eds.) ICARIS 2008. LNCS, vol. 5132, pp. 60–70. Springer, Heidelberg (2008). doi:10.1007/978-3-540-85072-4_6
33. Anile, A.M., Cutello, V., Giuseppe, N., Rascuna, R., Spinella, S.: Comparison among evolutionary algorithms and classical optimization methods for circuit design problems. In: IEEE Congress on Evolutionary Computation, CEC 2005, Edinburgh, UK, 2–5 September 2005, vol. 1, pp. 765–772. IEEE Press (2005)
34. Cutello, V., Narzisi, G., Giuseppe, N., Pavone, M.: Real coded clonal selection algorithm for global numerical optimization using a new inversely proportional hypermutation operator. In: The 21st Annual ACM Symposium on Applied Computing, SAC 2006, Dijon, France, 23–27 April 2006, vol. 2, pp. 950–954. ACM Press (2006)

A Nash Equilibrium Approach to Metabolic Network Analysis

Angelo Lucia[1(✉)] and Peter A. DiMaggio[2]

[1] Department of Chemical Engineering, University of Rhode Island, Kingston, RI 02881, USA
alucia@uri.edu
[2] Department of Chemical Engineering, Imperial College London, London SW7 2AZ, UK
p.dimaggio@imperial.ac.uk

Abstract. A novel approach to metabolic network analysis using a Nash Equilibrium formulation is proposed. Enzymes are considered to be players in a multi-player game in which each player attempts to minimize the dimensionless Gibbs free energy associated with the biochemical reaction(s) it catalyzes subject to elemental mass balances. Mathematical formulation of the metabolic network as a set of nonlinear programming (NLP) sub-problems and appropriate solution methodologies are described. A small example representing part of the production cycle for acetyl-CoA is used to demonstrate the efficacy of the proposed Nash Equilibrium framework and show that it represents a paradigm shift in metabolic network analysis.

1 Introduction

Flux balance analysis (FBA) has been the mainstay for understanding and quantifying metabolic networks for many years. See, for example, [1–6]. The basic idea behind FBA is to represent a given metabolic network at steady-state in the form of a graph with nodes that define specific biochemical reactions and fluxes that connect nodes. The constraints for the network constitute a set of under-determined steady-state linear mass balance equations. To complete the representation, a linear objective function relevant to the particular biological task at hand [e.g., maximizing the output flux [5]; minimizing the cardinality of the flux vector [4]; minimum nutrient intake; minimum ATP production minimal knockout, etc.] is selected, which together with the mass balance constraints, results in a linear programming (LP) formulation. Many variants and extensions to FBA have also been proposed over the years, including Mixed Integer Linear Programming (MILP) formulations [7], the incorporation of linearized thermodynamic constraints [8] or thermodynamic metabolic flux analysis (TMFA), dynamic FBA [9], and others [10,11].

In addition to FBA, the other main approach to metabolic network analysis involves a time-dependent or dynamic formulation that incorporates chemical reaction kinetics.

P.M. Pardalos et al. (Eds.): MOD 2016, LNCS 10122, pp. 45–58, 2016.
DOI: 10.1007/978-3-319-51469-7_4

1.1 Motivation

Both the kinetic model and FBA approaches suffer from a number of significant modeling and biological limitations. Kinetic models require a large number of parameters that are not 'directly' measurable and thus must be determined by model regression. Also, all FBA-based approaches are constrained only by reaction stoichiometry, which often results in degenerate solutions that lie on the user-specified flux bounds. Additional modeling limitations stem from the inability of either approach to accurately capture the inherent complexities within a crowded cellular environment. In particular, natural competition/cooperation among enzymes exists for the pool of continuously produced metabolites and neither approach takes into consideration the population distribution and/or heterogeneity of protein-metabolite interactions that form the basis for these reactions. From a biological perspective, these approaches also fail to model the phenotypic consequence of overproducing a given product and the subsequent regulation of overproduction in these organisms. For instance, it is well known that excess quantities of a given protein or small molecule can lead to 'higher order' interactions, which can be non-specific or the result of adaptive evolutionary pressures that activate alternative pathways to deplete these excess pools.

1.2 Nash Equilibrium Approach

This paper takes a radically different approach to metabolic network analysis by formulating the problem as a Nash Equilibrium (NE) problem using first principles (i.e., conservation of mass and rigorous reaction thermodynamics). The key idea behind our proposed NE approach is to view enzymes as 'players' in a multi-player game, in which each enzyme pursues a strategy that can be quantified using a payoff or objective function. The collection of enzymes or players has a solution called a 'NE point', which is a point that maximizes the 'payoff' of all players. Thus a NE point is the best solution for all players taken together and not necessarily a point that is best for any one player. This NE approach, in our opinion, is a more accurate representation of the evolutionary-defined competition/cooperation observed in complex metabolic pathways.

The remainder of this paper is organized in the following way. In Sect. 2, the optimization formulations, input and other data required, size of data, etc. for FBA and the Nash Equilibrium approach to metabolic pathway analysis are presented. In Sect. 3, the mathematical tools needed to solve the NE problem are described in some detail. Section 4 provides numerical results for a proof-of-concept example based on acetyl-CoA metabolism in *E. Coli* and both NE and FBA predictions of key fluxes are compared to experimental data. Finally, Sect. 5 draws conclusions from this work.

2 Mathematical Modeling

In this section, optimization formulations, input and other data required, size of data, and other aspects of FBA and the proposed NE approach are described in detail.

2.1 FBA Formulation

It is well known that conventional FBA is formulated as a linear programming (LP) problem, which in general is given by:

$$\begin{aligned} \text{optimize} \quad & c^T v \\ \text{subject to} \quad & Sv = 0 \\ & v^L \leq v \leq v^U \end{aligned} \tag{1}$$

where $c^T v$ is a linear objective function in which c are considered weighting coefficients, v is a vector of unknown metabolic fluxes, S is a stoichiometric matrix associated with the chemical reactions in the network, and the superscripts L and U denote lower and upper bounds, respectively.

Data Required for Flux Balance Analysis. The data typically required for FBA includes: (1) stoichiometric coefficients for all chemical reactions in the network, which are placed in the stoichiometric matrix, S, (2) fluxes for all inputs to the given network, (3) lower and upper bounds on all unknown metabolic fluxes.

Amount of Data Required for Flux Balance Analysis. The amount of data required for FBA is directly proportional to the number of species and chemical reactions in the network.

2.2 NE Formulation

Let the unknown variables, v, be partitioned into N subsets, $v = [v_1, v_2, ..., v_N]$. Each variable partition, v_j, has n_j unknown variables. The Nash Equilibrium (NE) formulation for an arbitrary metabolic network is quite different for that for FBA and is given by a collection of $j = 1, 2, ..., N$ nonlinear programming (NLP) sub-problems of the form:

$$\begin{aligned} \min \quad & \frac{G_j(v_j, v_{-j}^*)}{RT} \\ \text{subject to} \quad & \text{conservation of mass} \\ & v_{-j}^* \end{aligned} \tag{2}$$

where $\frac{G_j}{RT}$, the dimensionless Gibbs free energy, is the objective function associated with the appropriate enzymes involved in a particular set (or number) of metabolic reactions at a given node j in the network, R is the gas constant, and T is the temperature. The conservation of mass constraints are elemental mass balances and v_j represents the flow of metabolic material in and out of any node. Finally, the vector, v_{-j}^*, denotes the minima of all other sub-problems, $k = 1, 2, .., j-1, j+1, ..., N$. In this article the words "sub-problem" and "node" mean the same thing.

The Gibbs free energy for sub-problem j is given by:

$$\frac{G_j}{RT} = \sum_{i=1}^{C_j} x_{ij} \left[\frac{\Delta G_{ij}^0}{RT} + \ln x_{ij} + \ln \phi_{ij} \right] \tag{3}$$

where ΔG_{ij}^0 are the standard state Gibbs free energies of the components for the metabolic reactions associated with sub-problem j, x_{ij} are mole fractions, which are related to the fluxes, ϕ_{ij} are fugacity coefficients, i is a component index, and C_j and R_j are the number of components and number of reactions associated with a sub-problem j in the network.

Temperature effects can be taken into effect using the van't Hoff equation, which is given by:

$$\frac{\Delta G_{ij}^0(T)}{RT} = \frac{\Delta G_{ij}^0(T_0)}{RT_0} + \frac{\Delta H_{ij}^0(T_0)}{R} \left[\frac{T - T_0}{TT_0} \right] \tag{4}$$

where T_0 is the reference temperature (usually 25 °C), T is the temperature at which the reaction takes place (usually 37 °C), and $\Delta H_{ij}^0(T_0)$ is the standard state enthalpy of component i at node j in the network. Equations (3) and (4) are equivalent to the Gibbs free energy changes of reaction, $\Delta G_{ij}^R(T)$, which can also be computed from Gibbs free energies and enthalpies of formation, temperature effects, and reaction stoichiometry.

$$\Delta G_{ij}^{R0} = \sum_{k=1}^{n_p(ij)} s_k \Delta G_{f,ijk}^0 - \sum_{k=1}^{n_r(ij)} s_k \Delta G_{f,ijk}^0 \tag{5}$$

where the s_k's are the stoichiometric numbers and $n_p(ij)$ and $n_r(ij)$ are the number of products and number of reactants, respectively, associated with reaction i and node j.

The key attributes that distinguish the proposed NE approach from other formulations and make the problem challenging are: (a) the objective functions in all sub-problems are nonlinear and (b) chemical reaction equilibrium can be non-convex and therefore multiple solutions can exist.

Data Required for Nash Equilibrium Metabolic Pathway Analysis. The amount of data needed for the NE approach to metabolic pathway analysis is greater than that for the equivalent FBA formulation. In addition to all stoichiometric coefficients of all chemical reactions, the NE approach requires Gibbs free energies and enthalpies of formation at the reference conditions. However, numerical data for $\Delta G_{f,ijk}^0(T_0)$ and $\Delta H_{f,ijk}^0(T_0)$ are generally available in various text books [12,13] and journal articles [14,15].

Amount of Data Required for NE Metabolic Pathway Analysis. As with FBA, the amount of data required for metabolic pathway analysis using Nash Equilibrium is directly proportional to the number of species and chemical reactions in the network.

3 Solutions Methods

A summary of the solution methods for FBA and NE approaches to metabolic pathway analysis are presented in this section.

3.1 Solution Methods Used to Solve FBA Problems

As discussed in Sect. 2.1, FBA problems are formulated as LP problems and therefore linear programming methods are commonly used to solve FBA problems as well as for thermodynamic metabolic flux analysis (TMFA) problems originally proposed by [8]. LP methods are well established in the open literature and not discussed in any detail in this paper. For a well posed LP problem, it is well known that the solution must lie at a vertex of the feasible region and while the optimum value of the objective function in the LP solution is unique, the values of the fluxes need not be. Other approaches to solving FBA problems include MILP methods, in which binary variables are added to the problem formulation in order to find all bases that give vertex solutions and to determine degenerate (or multiple) solutions [7] and protein abundance design using multi-objective optimization to automatically adjust flux bounds [10].

3.2 Solution Methods for Nash Equilibrium Problems

The premise of Nash Equilibrium is to find a solution to Eq. 2 and that this solution may not represent the best solution for each individual sub-problem in Eq. 2. Many of the methodologies for solving Nash Equilibrium problems are rooted in game theory where continuity and/or continuous differentiability are difficult to guarantee and thus are not considered. However, through reformulation a number of standard techniques from complementarity problems, variational inequalities, and nonlinear programming have been adapted for solving NE problems.

Nonlinear Programming Methods. In general, NE problems are not smooth and therefore require some reformulation to apply nonlinear programming methods such as those that use the Nikaido-Isoda function and Jacobi-type trust region methods [16]. Surveys of numerical methodologies applied to NE problems can be found in Facchinei and Kanzow [17] and von Heusinger [18].

Terrain Methods. The metabolic network formulation within the framework of Nash Equilibrium (i.e., Eq. 2) does not suffer from non-differentiability. Therefore, in this work, we propose the use of the class of methods known as terrain methods [19–21] for solving the sub-problems in Eq. 2. Terrain methods are Newton-like methods that are based on the simple idea of moving up and down valleys on the surface of the objective function to find all stationary points (i.e., minima, saddles, and singular points). We only briefly summarize the terrain methodology. Additional details can be found in the references.

Given an objective function, $f(Z) : \Re^n \rightarrow \Re^n$, where $f(Z)$ is C^3, the basic steps of the terrain method are as follows:

1. Initialize Z.
2. Compute a solution (i.e., minima, saddle point or singular point) using a trust region method.
3. Perform a (partial) eigenvalue-eigenvector decomposition of the matrix $H^T H$, where H is the Hessian matrix of the function $f(Z)$.
4. Move uphill in the eigen-direction associated with the smallest eigenvalue of $H^T H$ using predictor-corrector steps to follow the valley. Predictor steps are simply uphill Newton steps that can actually drift from the valley. Therefore, corrector steps are used to return iterates to the valley by solving the nonlinear programming problem given by:

$$V = \text{opt } g^T H^T H g \quad \text{such that } g^T g = \mathcal{L} \quad \forall L \in \mathcal{L} \tag{6}$$

   where g is the gradient of $f(Z)$, L is a given level set and $\mathcal{L}$ is a set of level curves.
5. Repeat step 4 using predictor-corrector movement uphill until a saddle point on $f(Z)$ is found.
6. Perform an eigenvalue-eigenvector decomposition of $H^T H$ and choose the eigen-direction associated with the smallest eigenvalue to initiate the next downhill search. Go to step 2.

There are considerably more details to the terrain method than are summarized in the foregoing steps (e.g., perturbations from a given solution to initiate the next search, the way in which corrector steps are initiated, keeping track of solutions, overall termination criteria, etc.). See [19–21] for details.

4 Numerical Example

Numerical results are presented for a small proof-of-concept example to illustrate that the formulation of metabolic pathways as a Nash Equilibrium problem represents a paradigm shift and to allow the reader interested in the details to reproduce our results. All computations in this section were performed on a Dell Inspiron laptop computer in double precision arithmetic using the LF95 compiler.

4.1 Problem Statement

Consider the network shown in Fig. 1, which has 9 unknown fluxes as shown Table 4 in the Appendix.

4.2 Nash Equilibrium

The Nash Equilibrium formulation for the metabolic network in Fig. 1 has four players (i.e., four enzymes: pyruvate dehydrogenase, acetyl-CoA synthetase, phosphotransacetylase, and acetate kinase; see Appendix for further details) and three nonlinear programming (NLP) sub-problems that describe the competition

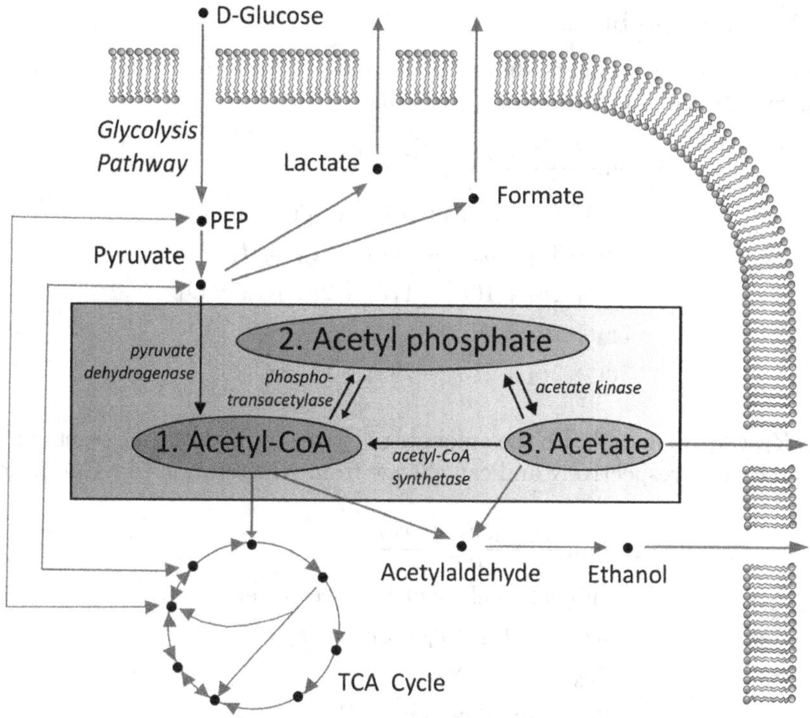

Fig. 1. Simplified metabolic network for the production of acetyl CoA based on the iJO1366 network for *E. coli* [22]. The metabolic reactions considered are represented using black arrows for acetyl-CoA, acetyl phosphate and acetate.

between those enzymes under the assumption that the goal of each enzyme is to minimize the dimensionless Gibbs free energy for the biochemical reaction it catalyzes subject to appropriate elemental mass balances (i.e., conservation of mass of carbon, hydrogen, oxygen, nitrogen, phosphorous and sulfur). Thus the overall goal of the network is find the best value of:

$$\frac{G(v)}{RT} = \sum_{j=1}^{3} \min \frac{G_j(v_j, v_{-j}^*)}{RT} \tag{7}$$

Note that the node in Fig. 1 for acetyl-CoA production is modeled by the sum of two biochemical reactions while each of the other nodes is modeled by a single biochemical reaction. The specific biochemical reactions and the enzymes (or players) used in this NE network model are illustrated in Fig. 1 and given in the Appendix, along with the flux numbers and species to which the numbers correspond.

4.3 NLP Sub-problems

All mass balance constraints shown in the sub-problem descriptions are linear and have been simplified as much as possible.

1. $$\min \frac{G_1(v_1, v_2, v_3, v_4, v_5, v_6)}{RT} \tag{8}$$
   subject to element balances given by
   $$3v_1 + 3v_2 + 32v_3 + 34v_4 + 2v_6 = H_1 \tag{9}$$
   $$3v_1 + 2v_2 + 16v_3 + 17v_4 + 2v_5 + v_6 = O_1 \tag{10}$$
   $$7v_3 + 7v_4 = N_1 \tag{11}$$
   $$3v_1 + 2v_2 + 21v_3 + 23v_4 + v_5 = C_1 \tag{12}$$

where H_1, O_1, N_1 and C_1 are the amounts of elemental hydrogen, oxygen, nitrogen, and carbon respectively and can change from NE iteration to NE iteration.

2. $$\min \frac{G_2(v_3, v_4, v_7, v_8)}{RT} \tag{13}$$
   subject to element balances given by
   $$32v_3 + 34v_4 + v_7 + 3v_8 = H_2 \tag{14}$$
   $$7v_3 + 7v_4 = N_2 \tag{15}$$
   $$3v_3 + 3v_4 + v_7 + v_8 = P_2 \tag{16}$$

where H_2, N_2, and P_2 are the amounts of elemental hydrogen, nitrogen, and phosphorous, respectively, which can change at each NE iteration.

3. $$\min \frac{G_3(v_2, v_8, v_9)}{RT} \tag{17}$$
   subject to element balances given by
   $$2v_2 + 2v_8 = C_3 \tag{18}$$
   $$v_8 + v_9 = P_3 \tag{19}$$

where C_3 and P_3 are any amount of elemental carbon and phosphorous in the acetyl phosphate pool. Note that each sub-problem has some degrees of freedom. Sub-problem 1 has six unknowns, four linear constraints, and two degrees of freedom. Sub-problems 2 and 3 each have one degree of freedom.

4.4 Numerical Solution of the Nash Equilibrium

It is important for the reader to understand that the right hand sides of the linear mass balance constraints in the NLP sub-problems will vary from NE iteration to NE iteration as one cycles through the sub-problems until the Nash Equilibrium problem is converged.

The numerical strategy for solving the Nash Equilibrium problem defined in Sects. 4.1 and 4.2 consists of the following steps:

1. Break the feedback in the network and estimate the appropriate fluxes.
2. Solve each NLP sub-problem for its associated minimum in G/RT.
3. Compare calculated feedback fluxes with estimated fluxes.
   (a) If they match within a certain tolerance, stop.
   (b) If not, replace the estimated feedback fluxes with the calculated ones and go to step 2.

The proposed approach is a direct (or successive) substitution strategy for converging the NE problem.

4.5 Numerical Results

This section provides details for the solution of the individual NLP sub-problems and the Nash Equilibrium problem. The Nash Equilibrium problem can be viewed as a master problem.

NLP Sub-problem Solution. The individual NLP sub-problem are straightforward to solve. Table 1 provides an illustration for the conversion of pyruvate and acetate to acetyl-CoA. Since the constraints are linear, four of the six unknown fluxes can be eliminated by projecting G/RT onto the mass balance constraints, leaving two unknown fluxes [e.g., v_{co2}, v_{h2o}]. As a result, the sequence of iterates generated is feasible and all other flux values can be computed using projection matrices. Table 1 shows that the NLP solution methodology is robust, generates a monotonically decreasing sequence of dimensionless Gibbs free energy values, and convergence is quite fast (10 iterations) to a tight tolerance (i.e., 10^{-12}) in the two-norm of the component chemical potentials.

Table 1. NLP solution to acetyl-CoA production

Iteration no.	Unknown variables (v_{co2}, v_{h2o})	G/RT
0	$(0.00001, 0.00001)$	-4.39621
1	$(1.41971 \times 10^{-5}, 1.37181 \times 10^{-5})$	-4.39640
2	$(2.75153 \times 10^{-5}, 2.53197 \times 10^{-5})$	-4.39695
3	$(8.23974 \times 10^{-5}, 7.13924 \times 10^{-5})$	-4.39897
4	$(3.98558 \times 10^{-4}, 3.22080 \times 10^{-4})$	-4.40867
5	$(3.02694 \times 10^{-3}, 2.25533 \times 10^{-3})$	-4.46949
6	$(2.18630 \times 10^{-2}, 1.48783 \times 10^{-2})$	-4.76591
7	$(0.111113, 0.0706099)$	-5.63168
8	$(0.337213, 0.228526)$	-6.81803
9	$(0.523871, 0.423184)$	-7.21853
10	$(0.548037, 0.459889)$	-7.22650
11	$(0.548251, 0.459857)$	-7.22650

Solution to the Nash Equilibrium Problem. The solution to the Nash Equilibrium formulation of the metabolic network shown in Fig. 1 is determined by specifying the input flux to the network (i.e., the pyruvate flux), breaking the acetate feedback to the acetyl-CoA pool, providing an initial estimate of the acetate flux, and regulating (or constraining) the fluxes of water and co-factor co-enzyme A (CoA). This allows each of the three NLP sub-problems to be solved, in turn, in the order shown in Sect. 4.3: (1) acetyl-CoA production, (2) acetyl phosphate production, and (3) acetate production. The acetate biochemical equilibrium provides a new value for the acetate flux, which is compared to the initial estimate. If the initial and calculated values do not match, the calculated value is used for the next iteration, and this process is repeated until convergence.

Table 2 gives a summary of the NE iterations and solution per nmol/h of pyruvate starting from an initial estimate of acetate flux of $v_2 = v_{ac} = 0$ nmol/h and regulated fluxes of $v_3 = v_{CoA} = 2$ nmol/h and $v_6 = v_{H2O} = 1$ nmol/h. Note that (1) our NE approach takes 33 iterations to converge, (2) convergence is linearly, and (3) the sequence of G/RT values is monotonically decreasing.

Table 2. Nash Equilibrium iterates and solution to acetyl-CoA production.

Iter.	G/RT^*	v_{actp}	v_{ac}	v_{PO_3}	v_{accoa}	v_{coa}	v_{pi}	v_{co2}	v_{h2o}	Error
1	−10.34	0.0532	0.5045	0.0661	0.4330	1.5670	0.4330	0.5143	1.0380	0.0478
2	−14.19	0.0978	0.8200	0.1249	0.7845	1.6484	0.3296	0.9032	1.0660	0.3155
3	−16.69	0.1257	1.0057	0.1625	1.0651	1.7194	0.2641	1.1833	1.0660	0.2806
4	−18.36	0.1415	1.1185	0.1833	1.2777	1.7874	0.2275	1.3800	1.0524	0.1128
5	−19.50	0.1501	1.1912	0.1943	1.4317	1.8460	0.2080	1.5163	1.0373	0.0727
6	−20.29	0.1548	1.2403	0.1999	1.5399	1.8918	0.1976	1.6099	1.0251	0.0491
7	−20.84	0.1575	1.2743	0.2030	1.6147	1.9253	0.1918	1.6739	1.0165	0.0340
8	−21.21	0.1592	1.2979	0.2047	1.6658	1.9489	0.1885	1.7175	1.0108	0.0236
9	−21.46	0.1602	1.3143	0.2056	1.7006	1.9652	0.1866	1.7471	1.0071	0.0164
10	−21.64	0.1608	1.3256	0.2062	1.7242	1.9764	0.1853	1.7673	1.0047	0.0113
⋮										
31	−22.01	0.1620	1.3501	0.2073	1.7740	1.9999	0.1830	1.8098	1.0000	3×10^{-5}
32	−22.01	0.1620	1.3501	0.2073	1.7740	2.0000	0.1830	1.8098	1.0000	2×10^{-5}
33	−22.01	0.1620	1.3501	0.2073	1.7740	2.0000	0.1830	1.8098	1.0000	1×10^{-5}

$^*G/RT = \sum_{j=1}^{3} \min G_j/RT$

Comparisons to Experiments. To assess the predictive capability of the NE approach for metabolic pathway analysis, experimental data for the metabolic fluxes for *E. coli* were examined. These fluxes have been experimentally measured using a number of technologies, such as GC-MS and ^{13}C-fluxomics, and are available in the recently curated CeCaFDB database [23] that contains data for *E. coli* strains from 31 different studies over varying conditions. However, only one study reports experimental fluxes for the metabolic sub-network containing acetyl-CoA, acetyl phosphate and acetate [24] shown in Fig. 1.

Table 3. Comparison of NE (from Table 2) and FBA predicted relative fluxes* with experimental data [24].

Component	Experimentally measured fluxes	NE predicted fluxes	FBA predicted fluxes
Acetate	0.4557	0.4109	Upper bound
Acetyl phosphate	0.0000	0.0493	0
Acetyl-CoA	0.5443	0.5398	2

*Relative fluxes are considered since the experimental data is reported per unit of cell dry weight.

Table 3 presents the relative fluxes for the *E. coli* strain ML308 using glucose as a carbon source [24]. In [24] it was found that 45.57% of acetyl-CoA (generated from the pyruvate flux) was converted to acetyl phosphate, which in turn was entirely consumed to produce acetate. Table 3 shows that normalized NE predicted fluxes are in good agreement with experiment, while those for FBA are quite poor. The NE model also predicts that acetyl phosphate constitutes 4.93% of the pool of three metabolites at equilibrium, implying that acetyl phosphate produced from acetyl-CoA was mostly converted into acetate. While the experimental data estimated that acetyl phosphate was completely converted into acetate (Table 3), recent reports have shown that intracellular acetyl phosphate does exist in concentrations up to 3mM in *E. coli* [25], which is consistent with the NE predictions. This hypothesis is further supported by the excellent agreement between experimental and the NE predicted relative flux of acetyl-CoA. While the relative flux of acetate shows the largest deviation, the predictions of the NE model are in good agreement with experimental data and far superior to those predicted by FBA for the small sub-network and modeling assumptions considered.

5 Conclusions

A paradigm shift in metabolic network analysis based on Nash Equilibrium (NE) was proposed. The key idea in this NE approach is to view enzymes as players in a multi-player game, where each player optimizes a specific objective function. Governing equations, data requirements, solution methodologies for the NE formulation and conventional FBA formulation were described and compared. A proof-of-concept example consisting of a simplified metabolic network for the production of acetyl-CoA with four players was presented.

Nomenclature

c	objective function coefficients in any LP formulation
C	number of chemical species
f	objective function

g	gradient
G	Gibbs free energy
H	enthalpy, Hessian matrix
L	level set value
n	dimension of space
n_p	number of products
n_r	number of reactants
N	number of sub-problems
R	gas constant, number of reactions
$\Re^n$	real space of dimension n
s	stoichiometric numbers
S	stoichiometric coefficients
T	temperature
x	mole fraction
Z	unknown variables

Greek Symbols

v	unknown fluxes
ϕ	fugacity coefficient

Subscripts

f	formation
i	component index
j	sub-problem or node index
$-j$	excluding j
0	reference state

Superscripts

R	reaction
L	lower bound
U	upper bound
$*$	optimal value

Appendix

Biochemical Reactions for a Simplified Metabolic Network for Acetyl CoA Production

The biochemical reactions involved at each node in the metabolic network are as follows:

1. Acetyl CoA production from pyruvate and acetate

$$C_3H_3O_3 + C_2H_3O_2 + 2C_{21}H_{32}N_7O_{16}P_3S \rightleftharpoons 2C_{23}H_{34}N_7O_{17}P_3S + CO_2 + H_2O \tag{20}$$

Pyruvate → acetyl CoA: pyruvate dehydrogenase (genes: lpd, aceE, and aceF) [26]

Acetate → acetyl CoA: acetyl CoA synthetase (genes: acs) [26]

2. Acetyl phosphate production from acetyl CoA

$$C_{23}H_{34}N_7O_{17}P_3S + HO_4P \rightleftharpoons C_2H_3O_5P + C_{21}H_{32}N_7O_{16}P_3S \tag{21}$$

Acetyl CoA ↔ acetyl phosphate: phosphotransacetylase (genes: pta or eutD) [26]

3. Acetate production from acetyl phosphate

$$C_2H_3O_5P \rightleftharpoons C_2H_3O_2^- + PO_3^{3-} \tag{22}$$

Acetyl phosphate ↔ acetate: acetate kinase (genes: ackA, tdcD, or purT) [26]

Table 4. Unknown fluxes, components and Gibbs energy of formation.

Flux no.	Species	Symbol†	Chemical formula	ΔG_f^0 (kJ/mol)
1	Pyruvate	pyr	$C_3H_3O_3$	−340.733 [14]
2	Acetate	ac	$C_2H_3O_2^-$	−237.775 [14]
3	Coenzyme A	coa	$C_{21}H_{32}N_7O_{16}P_3S$	−4.38624 [14]
4	Acetyl CoA	accoa	$C_{23}H_{34}N_7O_{17}P_3S$	−48.1257 [14]
5	Carbon Dioxide	co2	CO_2	−394.4 [13]
6	Water	h2o	H_2O	−228.6 [13]
7	Phosphate	pi	HO_4P	−1055.12 [14]
8	Acetyl Phosphate	actp	$C_2H_3O_5P$	−1094.02 [14]
9	Phosphonate	PO3	PO_3^{3-}	−866.93‡

†Using nomenclature in BiGG [26]

‡Difference between ATP and ADP, using ΔG_f^0's from [14]

References

1. Varma, A., Palsson, B.O.: Metabolic flux balancing: basic concepts, scientific and practical use. Nat. Biotechnol. **12**, 994–998 (1994)
2. Kauffman, K.J., Prakash, P., Edwards, J.S.: Advances in flux balance analysis. Curr. Opin. Biotechnol. **14**, 491–496 (2003)
3. Holzhutter, H.G.: The principles of flux minimization and its application to estimate stationary fluxes in metabolic networks. Eur. J. Biochem. **271**, 2905–2922 (2004)
4. Julius, A.A., Imielinski, M., Pappas, G.J.: Metabolic networks analysis using convex optimization. In: Proceedings of the 47th IEEE Conference on Decision and Control, p. 762 (2008)
5. Smallbone, K., Simeonidis, E.: Flux balance analysis: a geometric perspective. J. Theor. Biol. **258**, 311–315 (2009)

6. Murabito, E., Simeonidis, E., Smallbone, K., Swinton, J.: Capturing the essence of a metabolic network: a flux balance analysis approach. J. Theor. Biol. **260**(3), 445–452 (2009)
7. Lee, S., Phalakornkule, C., Domach, M.M., Grossmann, I.E.: Recursive MILP model for finding all the alternate optima in LP models for metabolic networks. Comput. Chem. Eng. **24**, 711–716 (2000)
8. Henry, C.S., Broadbelt, L.J., Hatzimanikatis, V.: Thermodynamic metabolic flux analysis. Biophys. J. **92**, 1792–1805 (2007)
9. Mahadevan, R., Edwards, J.S., Doyle, F.J.: Dynamic flux balance analysis in diauxic growth in Escherichia coli. Biophys. J. **83**, 1331–1340 (2002)
10. Patane, A., Santoro, A., Costanza, J., Nicosia, G.: Pareto optimal design for synthetic biology. IEEE Trans. Biomed. Circuits Syst. **9**(4), 555–571 (2015)
11. Angione, C., Costanza, J., Carapezza, G., Lio, P., Nicosia, G.: Multi-target analysis and design of mitochondrial metabolism. PLOS One **9**, 1–22 (2015)
12. Alberty, R.A.: Thermodynamics of Biochemical Reactions. Wiley, Hoboken (2003)
13. Elliott, J.R., Lira, C.T.: Introductory Chemical Engineering Thermodynamics, 2nd edn. Prentice Hall, Upper Saddle (2012)
14. Kummel, A., Panke, S., Heinemann, M.: Systematic assignment of thermodynamic constraints in metabolic network models. BMC Bioinform. **7**, 512–523 (2006)
15. Flamholz, A., Noor, E., Bar-Even, A., Milo, R.: eQuilibrator - the biochemical thermodynamics calculator. Nucleic Acids Res. **40** (2011). doi:10.1093/nar/gkr874
16. Yuan, Y.: A trust region algorithm for nash equilibrium problems. Pac. J. Optim. **7**, 125–138 (2011)
17. Facchinei, F., Kanzow, C.: Generalized nash equilibrium problems. Ann. Oper. Res. **175**, 177–211 (2010)
18. von Heusinger, A.: Numerical methods for the solution of generalized nash equilibrium problems. Ph.D. thesis, Universitat Wurzburg, Wurzburg, Germany (2009)
19. Lucia, A., Feng, Y.: Global terrain methods. Comput. Chem. Eng. **26**, 529–546 (2002)
20. Lucia, A., Feng, Y.: Multivariable terrain methods. AIChE J. **49**, 2553–2563 (2003)
21. Lucia, A., DiMaggio, P.A., Depa, P.: A geometric terrain methodology for global optimization. J. Global Optim. **29**, 297–314 (2004)
22. Orth, J.D., Conrad, T.M., Na, J., Lerman, J.A., Nam, H., Feist, A.M., Palsson, B.O.: A comprehensive genome-scale reconstruction of Escherichia coli metabolism-2011. Mol. Syst. Biol. **11**(7), 535 (2011). doi:10.1038/msb.2011.65
23. Zhang, Z., Shen, T., Rui, B., Zhou, W., Zhou, X., Shang, C., Xin, C., Liu, X., Li, G., Jiang, J., Li, C., Li, R., Han, M., You, S., Yu, G., Yi, Y., Wen, H., Liu, Z., Xie, X.: CeCaFDB: a curated database for the documentation, visualization and comparative analysis of central carbon metabolic flux distributions explored by 13C-fluxomics. Nucleic Acids Res. **43** (2015). doi:10.1093/nar/gku1137
24. Holms, H.: Flux analysis and control of the central metabolic pathways in Escherichia coli. FEMS Microbiol. Rev. **19**, 85–116 (1996)
25. Klein, A.H., Shulla, A., Reimann, S.A., Keating, D.H., Wolfe, A.J.: The intracellular concentration of acetyl phosphate in Escherichia coli is sufficient for direct phosphorylation of two-component response regulators. J. Bacteriol. **189**(15), 5574–5581 (2007)
26. King, Z.A., Lu, J.S., Drager, A., Miller, P.C., Federowicz, S., Lerman, J.A., Ebrahim, A., Palsson, B.O., Lewis, N.E.: BiGG models: a platform for integrating, standardizing, and sharing genome-scale models. Nucleic Acids Res. (2015). doi:10.1093/nar/gkv1049

A Blocking Strategy for Ranking Features According to Probabilistic Relevance

Gianluca Bontempi(✉)

Machine Learning Group, Computer Science Department,
Interuniversity Institute of Bioinformatics in Brussels (IB)2,
ULB, Université Libre de Bruxelles, Bruxelles, Belgium
gbonte@ulb.ac.be
http://mlg.ulb.ac.be

Abstract. The paper presents an algorithm to rank features in "small number of samples, large dimensionality" problems according to probabilistic feature relevance, a novel definition of feature relevance. Probabilistic feature relevance, intended as expected weak relevance, is introduced in order to address the problem of estimating conventional feature relevance in data settings where the number of samples is much smaller than the number of features. The resulting ranking algorithm relies on a blocking approach for estimation and consists in creating a large number of identical configurations to measure the conditional information of each feature in a paired manner. Its implementation can be made embarrassingly parallel in the case of very large n. A number of experiments on simulated and real data confirms the interest of the approach.

1 Introduction

Selecting relevant features in high dimensional supervised learning tasks is known to be a challenging problem because of the curse of dimensionality. The curse manifests itself as ill-conditioning in the parameters computation, high variance of the estimators, risk of false positives and, in particular, as context-dependent relevance of features. By context-dependent relevance we mean that it is very difficult to assess the relevance (or importance) of a feature since it strongly depends on the subset of variables to which it belongs. This implies that a feature can be irrelevant if taken alone but very informative once combined with others (e.g. because of complementarity or synergetic effects, like in the XOR problem) or that the relevance of a feature is strongly reduced by the presence of correlated ones (e.g. because of redundancy) [5].

The context-dependent nature of the relevance has been formalized by [6] who distinguished between strongly relevant, weakly relevant and irrelevant features. This definition has been then related to the notion of Markov Blanket in [13] where the authors showed that the set of variables forming the Markov Blanket is indeed the set of strongly relevant features. The estimation of the strong relevance of all input features and their consequent ranking should be then the ultimate goal of every feature selection technique. The definitions of

P.M. Pardalos et al. (Eds.): MOD 2016, LNCS 10122, pp. 59–69, 2016.
DOI: 10.1007/978-3-319-51469-7_5

Kohavi *et al.* [6] suggest that if we wish to assess the relevance (and in particular the strong relevance) of a feature we need to model the multivariate dependency between the entire set of features and the target. This is however difficult if not impossible in large feature-to-sample ratio datasets. We are then facing a bias/variance dilemma where keeping low the dimensionality of the input set reduces the estimation variance but prevents us from addressing the correct quantity (i.e. the strong relevance of a variable) while dealing with large dimensional subsets reduces the bias at the cost of imprecise and brittle estimations. If the first strategy (trading variance for bias) is typically implemented by filters selection techniques (e.g. univariate ranking, mRMR [10], mIMR [2]), the second is used by wrapper techniques, which, by using a learner to assess a number of multivariate subsets, return context based assessment of the strong relevances of the features. A third alternative is represented by embedded algorithms which return, as part of the learning process, information about the importance of variables. A well-known example is represented by Random Forests [3] which implement two measures of importances, notably the mean increase in accuracy and the mean increase in node purity.

In this paper we address the problems related to inaccurate estimates of feature relevance in large variate settings. In particular the paper has two contributions: first we introduce a new notion of relevance, called *probabilistic feature relevance*, and, second, we propose a blocking strategy for ranking features according to this quantity. Blocking is a well-known experimental design strategy [9] to perform hypothesis testing and selection which consists in producing similar experimental conditions to compare alternative configurations. Blocking increases the confidence that observed differences are statistically significant and not due to fluctuations and noise effects. Note that the blocking strategy has been already used in [1] to design a wrapper strategy which assesses with a number of paired learning algorithms the quality of a given feature subset. In this paper the blocking approach is justified by the fact that the probabilistic relevance of a feature is defined as the expectation of its weak relevance. This implies that the difference of probabilistic relevance of two variables can be quantified by defining a number of paired configurations for which the conditional information of the features is computed and compared. By using the same configurations for all features, the statistical power of the comparison is increased with beneficial impact on the accuracy of the final ranking.

Experimental results on synthetic and real-world data show that the proposed approach (denoted *RelRank*) is competitive with existing feature selection techniques and can sometimes yield superior ranking accuracy.

2 Notions of Information Theory

Let us consider three continuous random variables $\mathbf{y}$, $\mathbf{x}_j$ and $\mathbf{x}_k$ having a joint Lebesgue density[1]. Let us start by considering the relation between $\mathbf{x}_j$ and $\mathbf{y}$.

[1] Boldface denotes random variables.

The mutual information [4] between $\mathbf{x}_j$ and $\mathbf{y}$ is defined in terms of their probabilistic density functions $p(x_j)$, $p(y)$ and $p(x_j, y)$ as

$$I(\mathbf{x}_j; \mathbf{y}) = H[\mathbf{y}] - H[\mathbf{y}|\mathbf{x}_j] \tag{1}$$

where $H[\mathbf{y}]$ denotes the entropy of $\mathbf{y}$ and

$$H[\mathbf{y}|\mathbf{x}_j] = \int_{\mathcal{X}_j} H[\mathbf{y}|\mathbf{x}_j = x_j] p(x_j) dx_j \tag{2}$$

is the *conditional entropy.* This quantity measures the amount of stochastic dependence between $\mathbf{x}_j$ and $\mathbf{y}$ [4] or in other words the univariate relevance of $\mathbf{x}_j$. Note that, if $\mathbf{y}$ is Gaussian distributed the following relation holds

$$H[\mathbf{y}] = \frac{1}{2}(1 + \ln(2\pi V[\mathbf{y}])) \tag{3}$$

where $V[\mathbf{y}]$ is the variance of $\mathbf{y}$.

Let us now consider a third variable $\mathbf{x}_k$. The *conditional mutual information* [4] between $\mathbf{x}_j$ and $\mathbf{y}$ once $\mathbf{x}_k$ is given is defined by

$$I(\mathbf{x}_j; \mathbf{y}|\mathbf{x}_k) = H[\mathbf{y}|\mathbf{x}_j] - H[\mathbf{y}|\mathbf{x}_j, \mathbf{x}_k] \tag{4}$$

The conditional mutual information differs from the term (1) and is null if and only if $\mathbf{x}_j$ and $\mathbf{y}$ are conditionally independent given $\mathbf{x}_k$.

On the basis of the previous definition it is possible to define in information-theoretic terms what is a relevant variable in a supervised learning task where $\mathbf{X} = \{\mathbf{x}_1, \dots, \mathbf{x}_n\}$ is the set of the n input features and $\mathbf{y}$ is the target. These definitions are obtained by interpreting in information theoretic terms the definitions introduced in [6].

Definition 1 (Strong Relevance). *A feature* $\mathbf{x}_j \in \mathbf{X}$ *is* strongly relevant *to the target* $\mathbf{y}$ *if*

$$I(\mathbf{x}_j; \mathbf{y}|\mathbf{X}_{-j}) > 0$$

where $\mathbf{X}_{-j}$ *is the set of* $n-1$ *features obtained by removing the variable* $\mathbf{x}_j$ *from* $\mathbf{X}$.

In other words, the strong relevance of $\mathbf{x}_j$ indicates that this feature is always necessary for creating an optimal subset or, equivalently, that it carries some specific information about $\mathbf{y}$ that no other variable does. In the following we will denote by

$$R_j = I(\mathbf{x}_j; \mathbf{y}|\mathbf{X}_{-j}), \qquad j = 1 \dots, n \tag{5}$$

the amount of relevance of the jth input variable.

Definition 2 (Weak Relevance). *A feature is* weakly relevant *to the target* $\mathbf{y}$ *if it is not strongly relevant and*

$$\exists \mathbf{S} \subseteq \mathbf{X}_{-j} : I(\mathbf{x}_j; \mathbf{y}|\mathbf{S}) > 0$$

Given a context $\mathbf{S} \subset \mathbf{X}_{-j}$, the degree of weakly relevance of a feature $\mathbf{x}_j \notin \mathbf{S}$ is then

$$W_j(\mathbf{S}) = I(\mathbf{x}_j; \mathbf{y}|\mathbf{S}) \tag{6}$$

In other words, a feature is weakly relevant when it exists a certain context $\mathbf{S} \subseteq \mathbf{X}_{-j}$ in which it carries information about the target. Weak relevance suggests that the feature is not always necessary but may become necessary in certain conditions.

Definition 3 (Irrelevance). *A variable is* irrelevant *if it is neither strongly or weakly relevant.*

Irrelevance indicates that the feature is not necessary at all.

A related notion which can be described in terms of conditional mutual information is the notion of the Markov Blanket (MB). The Markov Blanket of the target $\mathbf{y}$ is the smallest subset of variables belonging to $\mathbf{X}$ which makes $\mathbf{y}$ conditionally independent of all the remaining ones. Let us consider the subset $\mathbf{M} \subset \mathbf{X}$. This subset is said to be a *Markov blanket* of $\mathbf{y}$ if it is the minimal subset satisfying

$$I(\mathbf{y}; (\mathbf{X} \setminus \mathbf{M})|\mathbf{M}) = 0$$

Effective algorithms have been proposed in literature to infer a Markov Blanket from observed data [14]. Feature selection algorithms are also useful to construct a Markov blanket of a given target variable once they rely on notions of conditional independence to select relevant variables [8].

3 Probabilistic Feature Relevance

Let us consider a supervised learning task where $\mathbf{y} \in \mathbb{R}$ is the continuous target and $\mathbf{X} \in \mathcal{X} \subset \mathbb{R}^n$ is the set of n input variables where $n >> N$ and N is the number of observed input-output pairs.

If we were able to estimate accurately the quantities (5) for $j = 1, \ldots, n$ we could rank the set of variables and determine the Markov blanket of the target $\mathbf{y}$. Unfortunately this is not feasible since the estimate of large variate conditional entropies is poor when the ratio n/N is large.

In this paper we assume that it is not recommended to rely on R_j for ranking since we are not able to estimate it accurately. It is better to consider an alternative measure which captures the degree of relevance of the variables and can be estimated also in high variate settings. This measure is based on the notion of weak relevance and extends it by giving more weight to the weak relevance terms (6) whose context has an higher probability of being informative.

Definition 4 (Probabilistic Feature Relevance). *Let us consider a supervised learning problem where*

$$\mathbf{S}_j^* = \arg \max_{\mathbf{S} \subseteq \mathbf{X}_{-j}} I(\mathbf{S}; \mathbf{y})$$

is the most informative subset of features among the ones not including $\mathbf{x}_j$. *The degree of probabilistic relevance of the* j*th feature is*

$$P_j = E_{\mathbf{S}:\mathbf{x}_j \notin \mathbf{S}}[W_j(\mathbf{S})] = E_{\mathbf{S}:\mathbf{x}_j \notin \mathbf{S}}[I(\mathbf{x}_j;\mathbf{y}|\mathbf{S})] = E_{\mathbf{S}:\mathbf{x}_j \notin \mathbf{S}}[H(\mathbf{y}|\mathbf{S}) - H(\mathbf{y}|\{\mathbf{x}_j,\mathbf{S}\})] = \int_{\mathbf{S}:\mathbf{x}_j \notin \mathbf{S}} [H(\mathbf{y}|\mathbf{S}) - H(\mathbf{y}|\{\mathbf{x}_j,\mathbf{S}\})] \mathit{Prob}\{\mathbf{S} = \mathbf{S}^*_j\}\, d\mathbf{S} \quad (7)$$

Note that if we have no uncertainty about the information content of each subset, i.e. we know with certainty what is the best subset $\mathbf{S}^*_j$, the probabilistic relevance P_j boils down to the strong relevance $R_j = P_j$. In other words, probabilistic relevance converges to strong relevance for $N \to \infty$.

In general, the term Prob $\{\mathbf{S} = \mathbf{S}^*_j\}$ quantifies the importance of $\mathbf{S}$ in defining the relevance of a feature j. In other words, the terms $I(\mathbf{x}_j;\mathbf{y}|\mathbf{S})$ for which we have little evidence that $\mathbf{S} = \mathbf{S}^*_j$ should be weighted less than the terms $I(\mathbf{x}_j;\mathbf{y}|\mathbf{S})$ for which $\mathbf{S}$ is a high informative subset.

In a realistic high-dimensional setting ($n >> N$) the distribution Prob $\{\mathbf{S} = \mathbf{S}^*_j\}$ is unknown and we are again in front of a difficult estimation problem. However, what matters in ranking is not estimating the single values P_j but rather the differences $P_j - P_k$, $1 \le j \ne k \le n$, where P_j and P_k are expectation terms. This is a well known problem in statistics, known as paired difference estimation [9], for which a blocking implementation is recommended.

4 Relevance Ranking Algorithm

This section describes an algorithm for paired estimation of probabilistic feature relevance which relies on the definition (7). In a conventional paired difference analysis (e.g. estimating the difference of means between control and cases) a pairing between the samples of the two populations is established. In our case, for each pair of variables $\mathbf{x}_j$ and $\mathbf{x}_k$ we have to create a pairing between the samples used to estimate P_j and the ones of P_k.

We propose then a ranking algorithm (denoted *RelRank*) whose pseudocode is detailed in Algorithm 1. This blocking estimation approach is composed of three parts.

The first part (lines 1–4) consists in exploring the space of feature sets $\mathbf{S}$ and returning candidates $\mathbf{S}^{(r)}, r = 1, \dots, R$, where R is a parameter of the algorithm.

The second part (lines 6–12), given a candidate $\mathbf{S}^{(r)}$, computes a paired estimation of the weak relevance term $W_j(\mathbf{S}^{(r)})$ for all features $\mathbf{x}_j, j = 1, \dots, n$. Now we have a problem since according to (6), $W_j(\mathbf{S}^{(r)})$ is not defined for all $\mathbf{x}_j \in \mathbf{S}^{(r)}$. So, in order to implement a blocking procedure we need to extend the definition of $W_j(\mathbf{S}^{(r)})$ to $\mathbf{x}_j \in \mathbf{S}^{(r)}$ as follows

$$W_j(\mathbf{S}^{(r)}) = \begin{cases} H(\mathbf{y}|\mathbf{S}^{(r)}) - H(\mathbf{y}|\{\mathbf{x}_j, \mathbf{S}^{(r)}\}) & \text{if } \mathbf{x}_j \notin \mathbf{S}^{(r)} \\ H(\mathbf{y}|\mathbf{S}^{(r)}_{-j}) - H(\mathbf{y}|\mathbf{S}^{(r)}) & \text{if } \mathbf{x}_j \in \mathbf{S}^{(r)} \end{cases}$$

where $\mathbf{S}^{(r)}_{-j}$ is the set of features obtained by removing $\mathbf{x}_j$ from $\mathbf{S}^{(r)}$. In other terms, given a context $\mathbf{S}^{(r)}$, the weak relevance of a feature belonging to $\mathbf{S}^{(r)}$

can be measured by the increase of the entropy that occurs when $\mathbf{x}_j$ is removed from $\mathbf{S}^{(r)}$.

In order to estimate the entropy and the information terms we make an hypothesis of Gaussian additive noise, so we can use Eqs. (1) and (3) where the V conditional variance terms are estimated by K fold cross-validation. It is worth to mention that, for the sake of blocking, the same cross-validation folds (line 2) are used to estimate all the terms W_j, $j = 1, \ldots, n$.

In the third part (lines 14–22) we estimate the probabilistic relevance (7) by *importance sampling*. The idea of importance sampling [12] is to adjust samples drawn from a *proposal* distribution in order to obtain samples from a *target* distribution, potentially known but impossible to be sampled directly. Consider the target distribution in (7) for which we would like to estimate the expectation by Monte Carlo sampling. Given the impossibility to sample it directly we generate samples from a different *proposal distribution*. In general the proposal distribution can be arbitrary with the only requirement that its support contains the support of the target. Since the samples drawn are incorrect we cannot simply average them but we have to weight them according to the importance they have in representing the target distribution. In our algorithm the proposal distribution is obtained by exploring the space of subsets $\mathbf{S}$. For each candidate subset, the related weak relevance is therefore weighted according to the probability that this subset is the most informative one.

It follows that probabilistic relevance is estimated by a weighted average (lines 22–24) of the terms $\hat{W}_j(\mathbf{S}^{(r)})$

$$\hat{P}_j = \sum_{r=1}^{R} \hat{p}_r \hat{W}(\mathbf{S}^{(r)}), \quad j = 1, \ldots, n \tag{8}$$

where the weights are given by a non parametric (bootstrap) estimation (lines 14–21) of the probability terms $p_r = \text{Prob}\left\{\mathbf{S}^{(r)} = \mathbf{S}^*\right\}$.

The outcome of the *RelRank* algorithm is then the estimation of n probabilistic relevances $\hat{P}_j$.

5 Discussion

The *RelRank* algorithm has a number of interesting properties which are worth to be discussed in detail:

- it can be made massively parallel at two different levels: (i) the loop over the n features (line 6 in the pseudo-code) and (ii) the loop over the R candidate sets (line 1). Note that parallelizing the loop over the R sets implies that no sequential information about the exploration of the feature space can be taken into account. This can be detrimental when the value R is small and n is large. This is the reason why in our experiments we parallelize only the loop over the features.

Algorithm 1. *RelRank*

Require: Training dataset X, Y, number R of context sets $\mathbf{S}^{(r)}$, number K of cross-validation folds, learner $\mathcal{L}$
{*N: number observations, n: number of features*}
1: **for** $r = 1$ to R **do**
2: define K folds for cross-validation
3: select a subset $\mathbf{S}^{(r)} \subset \{\mathbf{x}_1, \ldots, \mathbf{x}_n\}$
4: estimate $\hat{I}(\mathbf{S}^{(r)}; \mathbf{y})$ by K fold cross-validation of the learner $\mathcal{L}$ and store cross-validated errors in matrix $E[, r]$ of size $[N, R]$
5: $\hat{p}[r] = 0$
6: **for** $j = 1$ to n **do**
7: **if** $\mathbf{x}_j \in \mathbf{S}^{(r)}$ **then**
8: $\hat{W}_j(\mathbf{S}^{(r)}) = \hat{H}(\mathbf{y}|\mathbf{S}^{(r)}_{-j}) - \hat{H}(\mathbf{y}|\mathbf{S}^{(r)})$
9: **else**
10: $\hat{W}_j(\mathbf{S}^{(r)}) = \hat{H}(\mathbf{y}|\mathbf{S}^{(r)}) - \hat{H}(\mathbf{y}|\{\mathbf{x}_j, \mathbf{S}^{(r)}\})$
{$H(\cdot|\cdot)$ *terms are estimated by cross-validation of the learner* $\mathcal{L}$ *by using the same* K *folds defined in line 2*}
11: **end if**
12: **end for**
13: **end for**
14: **for** $b = 1$ to B **do**
15: set I_b to bootstrap subsample of $1, \ldots, N$
16: $r^* = \arg\min_r \sum_{i \in I_b} E[i, r]^2$
17: $\hat{p}[r^*] = \hat{p}[r^*] + 1$
18: **end for**
19: **for** $r = 1$ to R **do**
20: $\hat{p}[r] = \hat{p}[r]/B$
21: **end for**
22: **for** $j = 1$ to n **do**
23: $\hat{P}_j = \sum_{r=1}^{R} \hat{p}[r] \hat{W}_j(\mathbf{S}^{(r)})$;
24: **end for**
25: **return** $\hat{P}_j, j = 1, \ldots, n$;

- it addresses the variance issue related to the small number of samples by implementing an intensive blocking pairwise strategy (i.e. using the same cross-validation folds in lines 4, 8 and 10 of the algorithm) and by using the averaging step (8);
- an outcome of the algorithm is the sampling distribution of the probabilistic relevance estimator which can be used not only for point estimation (e.g. by taking the average for creating a ranking) but also in a meta analysis context (e.g. for merging the rankings deriving from several datasets or studies).
- another side effect of the algorithm is that it can be used to return detailed information about context dependent redundancies of the input features, e.g. estimates of the conditional information $I(\mathbf{x}_j; \mathbf{y}|\mathbf{x}_j)$.

6 Experiments

In order to test the *RelRank* algorithm and compare it with the state of the art we carried out two kinds of experiments. The first one uses simulated regression data and assesses the accuracy of *RelRank* by measuring the position of the Markov Blanket variables in the returned ranking. The second experiments uses a set of microarray data to assess the prediction accuracy of a classifier using the top ranked features returned by *RelRank*.

6.1 Simulated Data

We used the following simulation procedure in order to define a number of supervised learning tasks with different number of samples, variance and amount of noise. We first choose the number N of samples and the size m of the Markov Blanket. Then we set the number n of features to a multiple of N. Eventually we generate the input matrix $[N, n]$ according to a multivariate gaussian distribution with non diagonal covariance matrix in order to emulate tasks with collinearity. Then we consider a set of nonlinear dependencies (detailed in Table 1) with additive Gaussian noise. Note that each time only a subset of the n measured inputs $\mathbf{X}$ is randomly chosen to be part of the Markov Blanket $\mathbf{M} \subset \mathbf{X}$. The caption of the table details how bivariate dependencies are used to implement multivariate functions.

Table 1. Set of nonlinear bivariate dependencies. In our experiments a nonlinear multivariate input output dependency is obtained by setting $z_1 = \sum_{\mathbf{x}_k \in \mathbf{X}_1} |x_k|$, $z_2 = \sum_{\mathbf{x}_k \in \mathbf{X}_2} |x_k|$ where $\mathbf{X}_1 \cap \mathbf{X}_2 = \emptyset$ and $\mathbf{X}_1 \cup \mathbf{X}_2 = \mathbf{M} \subset \mathbf{X}$

$y = \log(z_1^2 + z_2^2)$
$y = \sqrt{\lvert \sin(z_1^2 + z_2^2) \rvert}$
$y = \log(z_1 z_2^2 + z_1^2 z_2)$
$y = \sqrt{\lvert \frac{z_1^2}{z_2+1} \rvert}$
$y = \frac{1}{z_1^2+z_2^2+1}$
$y = \frac{z_1 \sin(z_1 z_2)}{z_1^2+z_2^2+1)}$
$y = z_2 \exp(2z_1^2)$
$y = z_2 \sin(z_1) + z_1 \sin(z_2)$
$y = \frac{z_1^3 - 2z_1 z_2 + z_2^2}{z_1^2+z_2^2+1}$
$y = z_1 + z_2$
$y = \sin(z_1) + \log(z_2)$

100 training sets (generated by randomly chosing the dependencies in the Table 1) have been used to compare the accuracy of the ranking returned by *RelRank* to the accuracy of five state-of-the-art techniques: an univariate ranking based on correlation, a filter mRMR [10] where the mutual information

is computed on the basis of the correlation, an embedded lasso technique, an importance Random Forest (RF) ranking based on the increase of the prediction error and an importance RF ranking based on the increase of purity [7]. In each dataset N is chosen randomly in $48 \leq N \leq 120$, n is given by vN times where v is random in $5 \leq v \leq 10$, the number of features in the Markov Blanket is randomly chosen between 5 and 10 and the noise is Gaussian additive. We used a Random Forest as learner $\mathcal{L}$ and we set $R = 50$ in the Algorithm 1.

Figure 1 reports the distribution of the position of the Markov Blanket features in the rankings of the five techniques. Note that in all these comparisons the differences are statistically significant (paired pvalue smaller than 0.001).

The results show that the *RelRank* technique outperforms consistently the considered state-of-the art methods in the task of retrieving the members of the Markov Blanket. This holds for different sizes of the training set and for different ratios between the number of samples and number of variables.

	Rank	mRMR	Lars	RF IncMSE	RF IncPurity	RelRank
Min	83	83	83	30	24	18
Avg.	350	361	351	263	228	199
Max	626	662	626	569	503	482

Fig. 1. Distribution (min/mean/max) of the position (the lower the better) of the Markov Blanket features in the returned rankings averaged over 100 datasets. In each dataset N is chosen randomly in $48 \leq N \leq 120$, n is given by Nv where v is random in $5 \leq v \leq 10$, the number of features in the Markov Blanket is randomly chosen between 5 and 10 and the additive noise is Gaussian.

6.2 Real Data

The second part of the experiments focuses on assessing the predictive power of classifiers using top ranked features returned by *RelRank*. We consider 19 microarray classification tasks based on the datasets made available in the R package [11][2]. The classification tasks are either binary or converted to binary by considering the majority class vs. all the others. *RelRank* is compared to the same strategies considered in the previous section. The assessment is done by performing a 3-fold Monte Carlo cross-validation. For each fold, first the ranking techniques use two third of the dataset to return the top 15 features and then three classifiers (Naive Bayes, Linear Discriminant Analysis and SVM) use such features to predict the test classes. The AUC of the predictions for the different subsets of variables are reported in Fig. 2.

From a paired significance test it appears that *RelRank* is not significantly different from any other techniques.

[2] All details on the datasets (number of samples, number of variables, number of classes) are available in https://github.com/ramhiser/datamicroarray/blob/master/README.md.

File	Rank	mRMR	RF IncMSE	RF IncPurity	RelRank
alon	0.761	0.827	0.836	0.813	0.823
burczynski	0.868	0.840	0.839	0.853	0.872
chiaretti	0.830	0.861	0.860	0.867	0.860
chin	0.915	0.895	0.916	0.916	0.902
chowdary	0.971	0.976	0.921	0.921	0.957
christensen	0.984	0.998	0.993	0.992	0.995
golub	0.931	0.932	0.947	0.953	0.939
gordon	0.971	0.971	0.963	0.964	0.977
gravier	0.716	0.761	0.700	0.758	0.730
khan	1.000	1.000	1.000	1.000	0.999
pomeroy	0.711	0.713	0.695	0.733	0.703
shipp	0.606	0.641	0.615	0.662	0.609
singh	0.875	0.898	0.929	0.914	0.951
sorlie	0.943	0.956	0.953	0.952	0.938
su	0.856	0.874	0.857	0.859	0.8743
subramanian	0.958	0.966	0.958	0.958	0.960
tian	0.875	0.867	0.874	0.869	0.875
west	0.625	0.603	0.648	0.646	0.624
yeoh	0.999	0.999	0.999	0.999	0.999
avg	0.863	0.873	0.869	0.875	0.873

Fig. 2. Averaged cross-validated AUC obtained with 3 classifiers using the first 15 features ranked by the 5 techniques. The last line reports the average AUC value

As final consideration, we can state that the *RelRank* technique is definitely competitive with state-of-the-art techniques in terms of accuracy and deserves the attention of the community not only for its accuracy but also for the interesting characteristics described in Sect. 5.

7 Conclusion

Feature selection in large dimensional settings with small number of samples is a difficult task which is highly exposed to the risk of excessive variance and instability. For that reason, state-of-the-art methods tend to trade bias for variance by using regularization techniques (e.g. embedding methods like lasso) or by focusing on low variate statistics (e.g. low variate filters). This paper aims to bridge the gap between the multivariate nature of the task and the risk of large variance by introducing a new notion of relevance and by adopting blocking, a well known strategy in statistics which defines appropriate paired experimental conditions to reduce variance and then maximise precision. The paper shows that under weak assumptions, it is possible to define a blocking scheme to assess in parallel and in paired manner a large number of features and then derive a reliable ranking. Preliminary results show that this method is competitive with state of the art techniques. Future work will take advantage of the massively parallel nature of the algorithm to implement a big data version of the algorithm and will explore the added value of the approach in a meta analysis context.

Acknowledgements. The author acknowledges the support of the "BruFence: Scalable machine learning for automating defense system" project (RBC/14 PFS-ICT 5), funded by the Institute for the encouragement of Scientific Research and Innovation of Brussels (INNOVIRIS, Brussels Region, Belgium).

References

1. Bontempi, G.: A blocking strategy to improve gene selection for classification of gene expression data. IEEE/ACM Trans. Comput. Biol. Bioinf. **4**(2), 293–300 (2007)
2. Bontempi, G., Meyer, P.E.: Causal filter selection in microarray data. In: Proceeding of the ICML 2010 Conference (2010)
3. Breiman, L.: Random forests. Mach. Learn. **45**, 5–32 (2001)
4. Cover, T.M., Thomas, J.A.: Elements of Information Theory. Wiley, New York (1990)
5. Guyon, I., Elisseeff, A.: An introduction to variable and feature selection. J. Mach. Learn. Res. **3**, 1157–1182 (2003)
6. Kohavi, R., John, G.H.: Wrappers for feature subset selection. Artif. Intell. **97**(1–2), 273–324 (1997)
7. Liaw, A., Wiener, M.: Classification and regression by randomforest. R News **2**(3), 18–22 (2002)
8. Meyer, P.E., Bontempi, G.: Information-theoretic gene selection in expression data. In: Biological Knowledge Discovery Handbook. IEEE Computer Society (2014)
9. Montgomery, D.C.: Design and Analysis of Experiments. Wiley, Hoboken (2001)
10. Peng, H., Long, F., Ding, C.: Feature selection based on mutual information: criteria of max-dependency, max-relevance, and min-redundancy. IEEE Trans. Pattern Anal. Mach. Intell. **27**(8), 1226–1238 (2005)
11. Ramey, J.A.: Datamicroarray: Collection of Data Sets for Classification (2013). R package version 0.2.2
12. Robert, C.P., Casella, G.: Monte Carlo Statistical Methods. Springer, New York (1999)
13. Tsamardinos, I., Aliferis, C.: Towards principled feature selection: relevancy. In: Proceedings of the 9th International Workshop on Artificial Intelligence and Statistics (2003)
14. Tsamardinos, I., Aliferis, C.F., Statnikov, A.: Algorithms for large scale Markov blanket discovery. In: Proceedings of the 16th International FLAIRS Conference (FLAIRS 2003) (2003)

A Scalable Biclustering Method for Heterogeneous Medical Data

Maxence Vandromme[1,2,3(✉)], Julie Jacques[1], Julien Taillard[1], Laetitia Jourdan[2,3], and Clarisse Dhaenens[2,3]

[1] Alicante, Seclin, France
[2] CRIStAL, UMR 9189, University of Lille, CNRS, Centrale Lille, Villeneuve d'ascq, France
[3] INRIA Lille - Nord Europe, Villeneuve d'ascq, France
maxence.vandromme@inria.fr

Abstract. We define the problem of biclustering on heterogeneous data, that is, data of various types (binary, numeric, etc.). This problem has not yet been investigated in the biclustering literature. We propose a new method, HBC (Heterogeneous BiClustering), designed to extract biclusters from heterogeneous, large-scale, sparse data matrices. The goal of this method is to handle medical data gathered by hospitals (on patients, stays, acts, diagnoses, prescriptions, etc.) and to provide valuable insight on such data. HBC takes advantage of the data sparsity and uses a constructive greedy heuristic to build a large number of possibly overlapping biclusters. The proposed method is successfully compared with a standard biclustering algorithm on small-size numeric data. Experiments on real-life data sets further assert its scalability and efficiency.

1 Introduction

Biclustering can be seen as *two-way clustering*, where clustering is applied to both the rows and columns of a data matrix. Biclustering methods produce a set of *biclusters*, each corresponding to a subset of the rows and a subset of the columns of the data matrix. The goal of such methods is to group rows which are similar according to *some* of the columns, and vice-versa. Biclustering detects correlations among both rows and columns, hence giving more insight on the studied data. In this work, we consider, without loss of generality, rows as *instances* of the data, and columns as *attributes* describing each instance. Therefore, a bicluster composed of rows R and columns C can be interpreted as "instances in R are similar on all attributes in C". Biclustering has most of its applications in biological data analysis, and more specifically gene expression analysis. In the present study, we apply biclustering to medical data extracted from hospitals' information systems. Such data include various types of information to describe each patient or each hospital stay: personal information

C. Dhaenens—This work was partially supported by project ClinMine - ANR-13-TECS-0009.

P.M. Pardalos et al. (Eds.): MOD 2016, LNCS 10122, pp. 70–81, 2016.
DOI: 10.1007/978-3-319-51469-7_6

(age, gender, address, etc.), biological measures, acts, diagnoses, and others. Biclustering is usually applied to homogeneous data, i.e. data where all attributes have the same type. A large part of biclustering applications focus on binary data. Others focus on numeric data with normalized values. Thus, the first major contribution of this work is a biclustering method able to handle heterogeneous data. A second distinctive characteristic of medical data is their size: hundreds of thousands of records with thousands of possible acts or diagnoses. This high dimensionality is coupled with high sparsity in the data matrix, as only a small fraction of all possible medical events happen in a given hospital stay. This fact brings the twofold challenge of handling large data matrices with a lot of missing values, while allowing the desired method to exploit the high sparsity in its design. Biclustering on hospital data can provide valuable insight, by identifying groups of similar patients (i.e. row clustering), detecting correlations amongst medical events and other attributes (i.e. column clustering), and identifying groups of patients sharing common attributes (which is the main feature of biclustering). In Sect. 2, existing work on the subject is presented. Section 3 details the method proposed to answer the identified problems. Section 4 presents experimental results on synthetic and real-life data sets, in order to show the efficiency and scalability of the proposed method. Section 5 presents conclusions on this work and possible improvements or evolutions.

2 Related Work

Many biclustering methods have been proposed since the seminal work of Cheng and Church on the subject [4]. However, the overwhelming majority of those focus on biological data analysis: detection of gene expression, protein interaction, microarray data analysis, etc. Gene expression was the original field of study from which originated the first biclustering methods, and these methods have achieved considerable success in helping understanding the studied biological systems. Therefore, most biclustering algorithms are designed to handle only one type of data, either numeric or binary (see [8] for a recent review on the subject). Indeed, these are the usual types of data available from biological applications, representing for example the activation of gene (binary) or its expression level (numeric). One algorithm, SAMBA [9], deals with heterogeneous data by integrating data from different sources. It remains, however, focused on numeric data and does not handle other data types. As mentioned in [3], biclustering has also been applied to other data mining problems, such as text mining [5] or recommendation systems [11]. As far as we know, none of the existing applications to data mining tasks involve heterogeneous data. Focusing on homogeneous data means that biclustering methods can be designed to extract biclusters describing several types of correlations on rows and columns. As explained in Sect. 3.1, all of these correlations are not applicable on heterogeneous data, which may partly explain the lack of studies on this topic. The other main characteristic of the medical data we deal with is high dimensionality and data sparsity. Most biclustering methods work well on data matrices with up to hundreds or thousands of rows and columns. Scaling to larger data remains an open problem, with interesting approaches being developed, usually involving parallel computing to provide suitable speedup [12]. We focus here on large data matrices, with around 100,000 rows and 10,000 columns, which have also high sparsity. Sparsity is also a seldom explored characteristic in biclustering. A notable work on this topic presents a method for data with around 20% of non-missing values [10]. In this study, we consider very sparse data matrices, with around 1% non-missing values, a case that has not yet been explored in the literature. A recent study on this problem [6] focuses on problems close to those presented here, notably sparsity and large scale data. However, this study studies biological data (gene and protein interaction), and therefore does not handle heterogeneous data.

3 Proposed Model

3.1 Heterogeneous Biclusters

Data Types. We define here the various types of attributes handled by HBC:

- numeric: either integer or real values. Examples: *age* or *blood pressure*.
- ordered symbolic: symbolic values with an ordering relation. Example: *severity* $\in \{low, medium, high\}$.
- unordered symbolic: other symbolic values. Example: *blood type* $\in \{A, B, AB, O\}$.
- binary: a special case of unordered symbolic, but separated for more flexibility. Example: *high blood pressure* $\in$ {*true*, *false*}.

Type of Biclusters. The first question one should answer when choosing or designing a biclustering algorithm is "What type(s) of biclusters does it need to be able to detect ?". Several types of biclusters exist (see [8] for a recent review), and most algorithms only focus on a sub-set of all possible types, depending on the application and precise goals of the method. In this work, we focus on medical data and try to detect sets of patients or hospital stays sharing the same characteristics for some of the attributes. Therefore, we want a method able to detect biclusters with constant values on columns. Because the data is heterogeneous, detecting biclusters with constant values on rows does not make sense. Coherent values on rows and/or columns do not either. Note that detecting biclusters with constant values on columns implies the ability to detect biclusters with constant values on both rows and columns (i.e. same value in the entire bicluster). This type of bicluster is not relevant on heterogeneous data, but will be used for experimental validation and comparison in Sect. 4, since it is the simplest case and is supported by all biclustering methods.

Quality Measure. Most biclustering methods use quality measures to evaluate the internal coherence of biclusters. As we will see in Sect. 3.2, the algorithm proposed in this work does not require such an evaluation during the biclustering process. However, computing the quality of a bicluster remains useful, as it may be used in the post-processing phase (keeping only the biclusters of best quality, for example) and in the evaluation process. Since we focus on heterogeneous data, most of the standard bicluster quality measures, such as *mean-squared residue error* [4], are not applicable. Instead, we split the bicluster on columns and use one of the following homogeneity measures for each column, depending on the attribute type:

- variance: statistical measure, for numeric and ordered symbolic attributes.
- homogeneity: defined as the number of occurrences of the most present value, over the total number of values in this column of the bicluster. Used for unordered symbolic and binary attributes.

One measure is computed for each column; the global quality measure for the bicluster is the average of all these values. Since the variance and homogeneity measures always take values in $[0, 1]$, we do no need further normalization and can simply use the average of these measures as composite quality measure.

3.2 HBC - Method Details

Since we focus on discovering biclusters with identical values in the columns, we propose a heuristic method, which takes advantage of this and of the expected data sparsity. The idea is to build biclusters by adding, at each step, the column with the largest number of identical values for the rows of the current bicluster.

Evaluating the similarity of values is easy on binary or symbolic data, but may be problematic on numeric data, where we can not expect many values to be identical. To alleviate this problem, we choose to apply discretization to numeric data as a preprocessing step. By doing this, we ensure that every column can only contain a fixed, finite number of different possible values. An additional benefit and motivation for using discretization is that it has been shown to often improve performance in data mining tasks. We use equal-length binning (with 5 bins) as a simple, unsupervised discretization method. We are restricted to unsupervised methods since biclustering is itself an unsupervised task. Note that the number of bins has a direct effect on the biclusters found by the algorithm: using more bins leads to smaller but more homogeneous biclusters. The algorithm is best described by the pseudo-code presented below.

This heuristic builds "perfect" biclusters on the discretized data, i.e. biclusters where all the values in each column are identical. These biclusters are only "perfect" on the simplified data; values for numeric attributes need not be exactly identical. In a way, the threshold for considering numeric values identical is controlled by the discretization process. These biclusters are then translated on real data and adjusted in a post-processing phase. Note that the simplified data is only used in the first phase of process for building biclusters; quality evaluation and any further operation is done on the real biclusters (i.e. translated back to original data). Iterating over all the columns and all their possible values as starting points ensures that even relatively small biclusters can be found.

It also means that the same bicluster may be discovered several times, starting from several of its columns on different iterations. This problem is easily dealt with in the post-processing phase (see "filter($realBiclus$)" in the pseudo-code), where duplicate biclusters are removed. Other post-processing operations are implemented and may be used to further reduce the (usually large) number of biclusters produced:

- remove biclusters with too few lines: the threshold for "too few" obviously depends on the size of the data.
- remove biclusters with too few columns: the method is forced to add at least $nbColsMin$ columns in each bicluster, regardless of whether it decreases the size of the bicluster or not. This mechanism was added to ensure that the algorithm would not stop after only one column. Experimental observations show that biclusters that stop their expansion as soon as this constraint is withdrawn are less interesting. Therefore, biclusters with $nbColsMin$ columns may be removed by this filter.
- only keep the top k biclusters, ranked by quality.
- only keep the top k biclusters, ranked by size.

```
build discretized data matrix SD from real data matrix D;
foreach column C in data matrix SD do
    foreach possible value V occurring in C do
        R ← rows in C with value V;
        create new bicluster B ← (R, C);
        while (NOT(stop) OR nbCols(B) < nbColsMin) do
            foreach column C_j ≠ C do
                vals_j ← values in column C_j at rows R;
                H_j ← count(most represented value in vals_j);
            end
            C_best ← C_j with max(H_j);
            size_t ← size(B);
            add column C_best to B;
            remove rows so that B remains perfectly homogeneous;
            size_{t+1} ← size(B);
            if size_{t+1} < size_t then
                stop = true;
            end
        end
        add B to biclus;
    end
end
realBiclus ← translate biclus on real data D;
filter(realBiclus);
return realBiclus;
```

There is no guarantee that the method finds the "best" biclusters, since the column selection process at each step is greedy. However, it gives a fair chance to all columns and all possible values. The most delicate part of the process lies in the post-processing phase, with several possible filters (with added parameters). The biclustering heuristic in itself finds a large number of biclusters in an unsupervised way. Then, depending on what kind of insight is most wanted on the data, these various filters may be applied in order to prune the full set of results and make it easier and faster to read.

3.3 Data Storage and Complexity

Space Complexity. HBC aims at handling large data matrices with high sparsity (i.e. very few non-missing values). Such data matrices ought not be loaded into memory using a regular 2D array, since this would induce heavy memory requirements for real data. Instead, it is advisable to use compressed data formats. Since the proposed method only requires fast column access, we use the *Compressed Sparse Column* (CSC) format [2]. Using this format, data storage has a space complexity of $O(2n)$, where n is the number of non-missing values.

Time Complexity. The core operation of the heuristic described above is *add column*, in which a column is added to a bicluster and rows are removed to keep this bicluster perfectly homogeneous. The cost of this operation is that of searching which rows of the column to add are present in the current bicluster. This translates into $O(i.log(i))$ worst case complexity, since looking for an element in a list has $O(log(i))$ complexity, with i the number of rows in the data matrix. At each step, the best column is added, which means all the possible columns have to be added in order to determine which one is the optimal choice. Therefore, $O(j^2)$ *add column* operations are performed during the construction of each bicluster. Finally, this whole process is performed for each column and each possible value in this column as starting points, hence $O(j.k)$ iterations, with k the number of possible values for an attribute. This leads to a worst case time complexity of $O(i.log(i).j^3.k)$. However, in practice, way less than $O(j^2)$ *add column* operations are performed, since the method usually reaches the stopping criterion after adding a few dozens columns at most (even on real data). This complexity analysis, along with experimental observations, show that the proposed method scales mostly with the number of columns and the data sparsity; the total number of rows does not have too much of an impact on the runtime.

4 Experimental Results

4.1 Comparison on Synthetic Data

Experimental Protocol. First we compare the proposed method with another existing biclustering algorithm, in order to assess its validity and efficiency. We choose the well-known CC algorithm [4], which was one of the first biclustering methods to be developed, and still remains a standard algorithm in most biclustering tools or softwares. It also has the interesting property of being an iterative constructive heuristic, i.e. it has the same underlying algorithmic structure as HBC. We used the implementation available in the 'biclust' R package. Note that several biclustering methods are available in this package, but none other than CC seemed to give any result on the data used for this study. These other methods either find no bicluster at all, or one single bicluster containing the whole data matrix.

The efficiency of a biclustering method can be evaluated in several ways. A first way is to compare the results (i.e. a set of biclusters) produced by an algorithm, using measures such as quality and size. A second way is to inject into the data one (or more) bicluster(s), and see if the algorithm is able to find these biclusters among the generated solutions. These two ways complement each other nicely, since they focus on different properties of the biclustering methods.

We use synthetic numeric data with 500 rows and 100 columns, as such dimensions are standard in the biclustering literature. Values for each cell of the data matrix are generated randomly using a uniform distribution on $[0, 100]$. Injected biclusters have a fixed size and are composed of identical values on every cell. The most basic case is to add one such bicluster in the data matrix, replacing the former values by the bicluster values at the adequate row and column positions. However, we also want to evaluate the ability of biclustering methods to deal with more complex situations (e.g. more than one bicluster). Therefore, for each dataset, a number of variants are generated, using different parameters for biclusters creation and insertion, as described below:

- number of biclusters: discovering several high-quality biclusters is a valuable property, but not every biclustering method performs evenly in this regard. Possible values are $[1, 3, 5]$.
- bicluster size: small biclusters are usually more difficult to find than larger ones. One of the possible reason is that biclustering methods try to maximize the size of biclusters, since these are often deemed more valuable. Thus, it is not uncommon for small, homogeneous biclusters to be discarded and replaced by larger, slightly less homogeneous biclusters. Possible values are [20X5, 50X10, 100X15] (written as "nb.rows X nb.columns").
- noise level: the ideal case is when all the bicluster values are identical. In this case, internal quality measures such as variance or homogeneity are optimal and the biclustering algorithm is more easily led towards such biclusters. However, real data usually has noisy or missing values. In order to take this into account, bicluster values are noised, by adding a random value drawn from a gaussian distribution with zero mean. Standard deviation of this distribution determines the noise level. Possible values are $[0, 2, 5, 10]$.
- overlapping: detecting partly overlapping biclusters may be more difficult for some methods. It is, however, a desirable behavior, especially on medical data where a same act or diagnosis can be used to define several different patient groups. We define the overlapping rate as the percentage of data of a bicluster that also belongs to another bicluster. Possible values are $[0, 20]$.

For each data set, 72 variants are generated, with every possible parameter configuration. 20 data sets are generated, each with 72 variants. Because the two considered algorithms are deterministic, only one run is done on each data set, which amounts to 1,440 runs in total for each method. All experiments are performed on the same computer, running Ubuntu 14.0.1, with four cores at 3.30 GHz and 4 GB of RAM. HBC implementation is done in C++, version 4.8.2.

Phase 1 - Biclusters Quality. In this phase, we focus on the overall quality of the set of biclusters discovered by each method. A notable difference here is that CC produces a fixed number of biclusters (default value is 20 in the R implementation), while HBC generates an undefined number of biclusters. Because of this and to ensure fairness in the comparison, we only consider the 20 biclusters of best quality for the second case. Two measures are used in the comparison: quality (to minimize) and size (nb.rows X nb.columns, to maximize). Table 1 presents the average and standard deviation over the 20 data sets for each parameter configuration over *noise* and *bicluster size*. Due to space constraints, only the table for 5 inserted biclusters with overlapping, which is the most difficult case, is presented here. All the tables exhibit roughly the same behavior, which is discussed below. In the tables, the notation $nXsY$ refers to a dataset where the biclusters have noise level X and size Y (*s0* is 20X5, *s1* is 50X10, *s2* is 100X15).

Table 1. Bicluster quality and size, 5 biclusters with overlapping

	Quality		Size	
	HBC	CC	HBC	CC
n0s0	**2.392 ± 0.787**	5.011 ± 3.049	**221.456 ± 76.156**	6.58 ± 2.207
n0s1	**0.405 ± 0.082**	3.536 ± 3.715	**262.305 ± 97.237**	38.68 ± 65.697
n0s2	**0.229 ± 0.052**	3.373 ± 4.15	**677.477 ± 274.265**	268.59 ± 380.721
n2s0	**2.505 ± 0.709**	5.272 ± 3.536	**226.191 ± 67.198**	6.35 ± 1.801
n2s1	**0.673 ± 0.12**	2.984 ± 3.56	**409.39 ± 74.215**	26.71 ± 30.141
n2s2	**0.373 ± 0.075**	0.492 ± 1.111	**955.397 ± 323.922**	117.095 ± 69.435
n5s0	**2.712 ± 0.477**	5.626 ± 3.466	**243.009 ± 60.428**	6.267 ± 1.704
n5s1	**1.096 ± 0.195**	2.535 ± 3.021	**389.465 ± 88.809**	11.157 ± 5.125
n5s2	**0.658 ± 0.112**	0.778 ± 0.264	**850.082 ± 306.033**	21.89 ± 5.672
n10s0	**2.942 ± 0.1**	5.123 ± 3.198	**261.402 ± 19.605**	6.193 ± 1.691
n10s1	**2.001 ± 0.336**	3.791 ± 3.05	**397.339 ± 104.94**	7.382 ± 2.282
n10s2	**1.138 ± 0.117**	1.728 ± 0.856	**702.155 ± 185.492**	10.492 ± 2.673

From these results, one may observe some interesting behaviors that are shared by both methods. A first observation is that, for datasets with smaller biclusters, the algorithms produce noticeably lower-quality biclusters. Several reasons may explain this, so we postpone the interpretation for this point to the second phase described below. There is also a progressive degradation of the average quality as the noise increases. Average size of the detected biclusters tends to increase with the size of the inserted biclusters, which seems logical and indicates that both methods find (at least part of) these biclusters. The results also show a strong dominance of HBC over the CC algorithm, both on quality and size. CC performs best on the two variants *n2s2* and *n5s2* with regard to quality, but the biclusters remain much smaller than those produced by HBC. Variance on the quality for the heuristic is smaller that this of CC, which ensures consistently better average performance. Variance on the size is quite high, although it seems to be in line with the large average size. Overall, HBC produces better biclusters on every data set and for both measures.

Phase 2 - Finding Inserted Biclusters. In this second phase, we focus on retrieving the inserted biclusters among the solutions found by the algorithms. Like in the first phase, we only consider the 20 biclusters of best quality for the proposed method. We use two measures (see [1]) to evaluate how efficient each algorithm is at discovering the hidden biclusters:

- U: part of the hidden bicluster present in the bicluster found by the algorithm (to maximize).
- E: part of the bicluster that does not belong to the hidden bicluster (to minimize).

For each hidden bicluster, we only consider, among the biclusters found by the algorithm, the one with the highest U for this hidden bicluster. When several hidden biclusters are inserted, the "best-matching" bicluster for each hidden one is extracted, and the averages of U and E are computed. Table 2 presents the

Table 2. Biclusters recovery measures, 5 biclusters with overlapping

	U		E	
	HBC	CC	HBC	CC
n0s0	**0.924 ± 0.094**	0.014 ± 0.037	**0.023 ± 0.055**	0.401 ± 0.2
n0s1	**0.922 ± 0.068**	0.216 ± 0.189	**0.0 ± 0.0**	0.141 ± 0.244
n0s2	**0.914 ± 0.075**	0.571 ± 0.16	**0.0 ± 0.0**	0.012 ± 0.087
n2s0	**0.739 ± 0.364**	0.012 ± 0.029	**0.157 ± 0.261**	0.39 ± 0.204
n2s1	**0.858 ± 0.159**	0.107 ± 0.082	**0.001 ± 0.005**	0.122 ± 0.234
n2s2	**0.84 ± 0.192**	0.132 ± 0.047	**0.002 ± 0.02**	0.034 ± 0.126
n5s0	**0.606 ± 0.419**	0.006 ± 0.011	**0.29 ± 0.312**	0.419 ± 0.164
n5s1	**0.771 ± 0.306**	0.022 ± 0.017	**0.052 ± 0.152**	0.133 ± 0.212
n5s2	**0.691 ± 0.287**	0.013 ± 0.009	**0.006 ± 0.036**	0.166 ± 0.282
n10s0	**0.075 ± 0.137**	0.004 ± 0.007	0.71 ± 0.127	**0.427 ± 0.186**
n10s1	**0.567 ± 0.304**	0.011 ± 0.009	**0.112 ± 0.217**	0.186 ± 0.233
n10s2	**0.398 ± 0.172**	0.005 ± 0.004	**0.013 ± 0.053**	0.213 ± 0.331

average and standard deviation over the 20 data sets for each parameter configuration over *noise* and *bicluster size*, for 5 inserted biclusters with overlapping.

A first observation on the results for U is that HBC finds (at least part of) the hidden biclusters on all the variants, except the *n10s0* (highest noise, smallest biclusters), which is understandably the hardest case. On the other hand, CC only finds large parts of the hidden biclusters on the simplest cases (low or no noise, largest biclusters). Note that a relatively high variance indicates that the proposed heuristic does not always find all the hidden biclusters, especially as the noise increases. Results for the E measure show that few rows and columns external to the hidden biclusters are added. Overall, HBC is able to find, in most cases, a significant part of the inserted biclusters (high U values), and to clearly isolate these among the other data (low E values). Note that no definitive conclusion can be drawn about the influence of the size of the hidden biclusters. HBC seems to perform better on the medium-sized ones, but the high variance does not allow any assertion on the influence of this parameter.

Wilcoxon statistical test indicates that HBC performs significantly better than CC on the four considered measures, with $p - values < 0.05$.

4.2 Scalability on Real Data

So far, we have assessed the efficiency of the proposed biclustering method on relatively small, numeric, homogeneous data. This first validation phase was done to allow for comparison with other biclustering algorithms. However, as stated in the motivations of this work, such data is not the primary focus of this method. Thus, in a second phase, we run the biclustering process on real-life medical data sets as extracted from a hospital's information system. These data sets form a good sample of the data we wish to be able to deal with, exhibiting heterogeneity, high sparsity and large size [7]. Table 3 presents the main characteristics of the three studied data sets, Table 4 presents the average

Table 3. Real data sets characteristics

	Nb rows	Nb columns	Nb non-missing	% non-missing
MRB	194,715	11,450	2,584,252	0.12%
PMSI	102,896	11,413	1,413,057	0.12%
PMSI_2013	49,231	7,941	746,566	0.19%

Table 4. Biclustering results on real data sets

	Quality	Size	Runtime (s)
MRB	0.048 ± 0.179	1446.446 ± 4388.173	5863
PMSI	0.035 ± 0.065	624.433 ± 1354.666	1047
PMSI_2013	0.014 ± 0.04	736.741 ± 1625.09	519

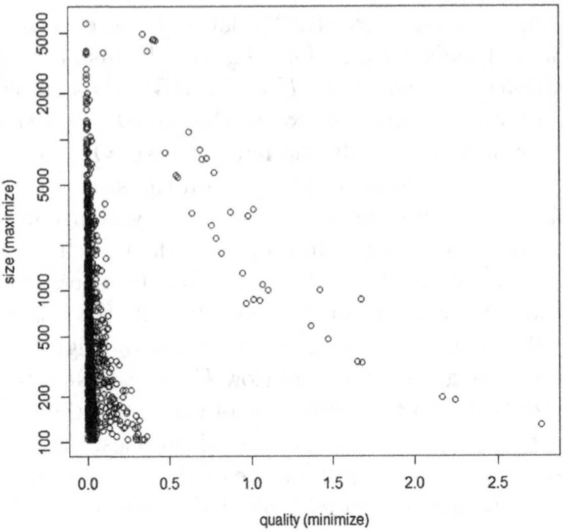

Fig. 1. Size and quality of biclusters found for data set MRB

quality and size of the biclusters discovered. Figure 1 shows a graph plotting the quality and size of all the biclusters found on data set MRB.

Note that datasets PMSI and PMSI_2013 both contain a large majority of binary attributes (medical events), and only *age* and *gender* as non-binary. In that regard, the MRB data set is more interesting, since it includes a few dozens numeric and unordered symbolic attributes, describing additional data on antibiotics. The first observation that may be done from the results is that the average quality of all the discovered biclusters is very good (i.e. near 0), with relatively small variance. On the other hand, average size is also good (i.e. large), but with high variance. On the whole, this means that the biclusters found all have high homogeneity (and therefore quality), and cover a very wide range of sizes. This observation is supported by Fig. 1, which shows most of the biclusters as having high quality and very diverse sizes (hence the logarithmic scale). This shows the method's ability to discover smaller biclusters that may still be of interest, in addition to larger ones. Runtimes remain realistic, taking less than two hours on the largest data set and less than ten minutes on the smallest (which still includes data from one year of activity of a medium size hospital).

5 Conclusions

In this study, we present a new biclustering method designed to handle heterogeneous, large-scale, sparse data matrices. The method uses a preprocessing step to simplify data through discretization, and a constructive greedy heuristic to build biclusters by iteratively adding columns. The goal is to detect biclusters with constant values on columns. Results on synthetic data show that the proposed method outperforms the standard CC biclustering algorithm, on both the general quality of solutions and the ability to find hidden biclusters. Experiments on heterogeneous real-life data show that the method scales well, exhibiting good performance and reasonable runtime. Since this is, to the best of our knowledge, the first method to support this type of data, no comparison could be done with other algorithms. However, biclusters found on these data sets are currently under evaluation by medical experts, to assess their correctness and relevance. The proposed method, HBC, could be improved to handle temporal data in addition to the already supported data types. Indeed, each medical event (act, diagnosis, etc.) has an associated timestamp, and the set of these timestamps for a patient describe its real course through the various hospital units. Using this additional information may help producing more insightful and precise biclusters, and is therefore the main axis of development under consideration at the present time.

References

1. Bozdağ, D., Kumar, A.S., Catalyurek, U.V.: Comparative analysis of biclustering algorithms. In: Proceedings of the First ACM International Conference on Bioinformatics and Computational Biology, pp. 265–274. ACM (2010)
2. Buluc, A., Fineman, J.T., Frigo, M., Gilbert, J.R., Leiserson, C.E.: Parallel sparse matrix-vector and matrix-transpose-vector multiplication using compressed sparse blocks. In: SPAA, pp. 233–244 (2009)
3. Busygin, S., Prokopyev, O., Pardalos, P.M.: Biclustering in data mining. Comput. Oper. Res. **35**(9), 2964–2987 (2008)
4. Cheng, Y., Church, G.M.: Biclustering of expression data. ISMB **8**, 93–103 (2000)
5. Dhillon, I.S.: Co-clustering documents and words using bipartite spectral graph partitioning. In: Proceedings of the Seventh ACM SIGKDD International Conference on Knowledge Discovery and Data Mining, pp. 269–274. ACM (2001)
6. Henriques, R., Madeira, S.C.: BicNET: flexible module discovery in large-scale biological networks using biclustering. Algorithms Mol. Biol. **11**(1), 1 (2016)
7. Jacques, J., Taillard, J., Delerue, D., Dhaenens, C., Jourdan, L.: Conception of a dominance-based multi-objective local search in the context of classification rule mining in large and imbalanced data sets. Appl. Soft Comput. **34**, 705–720 (2015)
8. Pontes, B., Giráldez, R., Aguilar-Ruiz, J.S.: Biclustering on expression data: a review. J. Biomed. Inform. **57**, 163–180 (2015)
9. Tanay, A., Sharan, R., Kupiec, M., Shamir, R.: Revealing modularity and organization in the yeast molecular network by integrated analysis of highly heterogeneous genomewide data. Proc. Natl. Acad. Sci. U.S.A. **101**(9), 2981–2986 (2004)
10. van Uitert, M., Meuleman, W., Wessels, L.: Biclustering sparse binary genomic data. J. Comput. Biol. **15**(10), 1329–1345 (2008)
11. Yang, J., Wang, W., Wang, H., Yu, P.: δ-clusters: capturing subspace correlation in a large data set. In: Proceedings of the 18th International Conference on Data Engineering, pp. 517–528. IEEE (2002)
12. Zhou, J., Khokhar, A.: ParRescue: scalable parallel algorithm and implementation for biclustering over large distributed datasets. In: 26th IEEE International Conference on Distributed Computing Systems, ICDCS 2006, pp. 21–21. IEEE (2006)

Neural Learning of Heuristic Functions for General Game Playing

Leo Ghignone and Rossella Cancelliere(✉)

Department of Computer Science, University of Turin, v. Pessinetto 12, 10149 Torino, Italy
leo.ghignone@edu.unito.it, rossella.cancelliere@unito.it

Abstract. The proposed model represents an original approach to general game playing, and aims at creating a player able to develop a strategy using as few requirements as possible, in order to achieve the maximum generality. The main idea is to modify the known minimax search algorithm removing its task-specific component, namely the heuristic function: this is replaced by a neural network trained to evaluate the game states using results from previous simulated matches. A method for simulating matches and extracting training examples from them is also proposed, completing the automatic procedure for the setup and improvement of the model. Part of the algorithm for extracting training examples is the Backward Iterative Deepening Search, a new original search algorithm which aims at finding, in a limited time, a high number of leaves along with their common ancestors.

Keywords: Game playing · Neural networks · Reinforcement learning · Online learning

1 Introduction

A general game player is an agent able to develop a strategy for different kinds of games, based on only the rules of the game to be learned; differently from a specialized player (a notable example of which can be found in [1]), properties and particularities of the played games cannot be hardcoded into the player. Most of the literature on general game playing focuses on the usage of the Game Description Language (GDL) which is part of the General Game Playing (GGP) project of Stanford University, and has become the standard for general game playing during the last years. It is a logical language which describes the rules of a game with a set of logical rules and facts, using specific constructs in order to define properties such as initial states, player rewards, turn order, etc. This approach however causes a double loss in generality: a general game player is unable to learn games not describable by GDL (like games with hidden information or randomic elements), and a direct access to all the rules of the game is required. As a consequence, most of the usual general game players focus on logically inferring features and strategies from the rules of the game (see, e.g., [2–5]) even when they do admit some sort of learning.

P.M. Pardalos et al. (Eds.): MOD 2016, LNCS 10122, pp. 82–93, 2016.
DOI: 10.1007/978-3-319-51469-7_7

The goal of this work is to propose a system, named Neural Heuristic Based player (NHB-player) that, in the framework of the classical two-players board games, gradually learns a strategy; for this purpose a neural network is used whose training set is collected through multiple simulated matches. Its main advantage is the possibility of learning from experience, which can give better results than rule analysis when the rules are complex or tricky, and allows the development of a strategy even if the rules are not given according to a logic-based description. The proposed general game player is immediately able to play any game showing the following properties:

- zero-sum, i.e. gain for a player always means loss for the other(s)
- turn-based
- deterministic, that is no random elements are permitted
- all players have perfect information, (no information is hidden to any player)
- termination is guaranteed.

State-space searches are a powerful tool in artificial intelligence, used in order to define a sequence of actions that can bring the system to a desired state. Searching through all possible evolutions and actions is almost always impractical, so *heuristic functions* are used to direct the search. A heuristic function is a function that, given a state, returns a value that is proportional to the distance from that state to a desired goal. These functions usually have to be implemented by human experts for each specific task, thus turning a general-purpose algorithm such as a state space search into a specific one. For this reason, state-of-the-art general game players [6] usually avoid heuristic functions, relying instead on Monte-Carlo searches to find the values of different states.

The first contribution of our work is to explore the possibility that a feedforward neural network can be automatically trained to compute heuristic functions for different game states rather than using Monte-Carlo methods for state evaluation.

The moves of our NHB-player in the space of all possible game states are chosen using a minimax algorithm with alpha-beta pruning and iterative deepening: this method is described in Sect. 2. The neural model is a single hidden layer feedforward neural network (SLFN), described in Sect. 3. Over a few decades, methods based on gradient descent have mainly been used for its training; among them there is the large family of techniques based on backpropagation, widely studied in its variations. The start-up of these techniques assigns random initial values to the weights connecting input, hidden and output nodes, that are then iteratively adjusted.

Some non iterative procedures have been proposed in literature as learning algorithms for SLFNs based on random assignation of input weights and output weights evaluation through pseudoinversion; some pioneering works on the subject are [7,8]. Recently Huang et al. [9] proposed the Extreme Learning Machine (ELM) which has the property of reducing training to an analytic one-step procedure while maintaining the universal approximation capabilities of the neural network; we use for training our model an online formulation called OS-ELM proposed by Liang et al. [10] and described in Sect. 3.1, which allows to divide the training examples into multiple sets incrementally presented to the network. This choice comes from the fact that the search algorithm and the neural heuristic assignment in our case are complementary to each other during all the phases of player's training, so that the training instances, each composed by a state and the heuristic function value for that state, are subsequently collected during different training matches and are not entirely available from the beginning.

A crucial step for neural training is the selection of a sufficient number of training examples. Since we aim at creating a system which autonomously learns a heuristic function, we don't want to turn to records of previously played games or to external opponents to train against. We therefore propose an original method for extracting the training examples directly from matches simulated by the system itself through a procedure described in Sect. 4. Part of this procedure is the Backward Iterative Deepening Search (Backward IDS), a new search algorithm presented in Sect. 4.1, which aims at finding, in a limited time, a high number of leaves along with their common ancestors.

Performance is evaluated on two classical board games of different complexity: *C*onnect Four and *R*eversi. Implementation details, experimentation description and results are reported in Sect. 5.

2 Search Algorithm

The state-space search algorithm classically used for this kind of games is the minimax algorithm: it was originally developed for two-players zero-sum games, but it can also be adapted to other kinds of games. The interested reader can find a description of the minimax search and of other algorithms cited in this Section (alpha-beta search, iterative deepening) in [11].

The idea behind this algorithm is that a player wants to minimize its opponent's maximum possible reward at the end of the game. In order to know what is the best outcome reachable for the opponent at each state of the game, the algorithm builds a tree of all the possible evolutions of the game starting from the current state and then backtracks from the terminal states to the root. Since it is usually impractical to expand the whole tree to the end of the game, the search is stopped at a certain level of the tree and the heuristic function is applied to the nodes of that level in order to estimate how favorable they are. For games without a final score, reward values are set to -1 for a loss, 0 for a tie and 1 for a win.

In our model we utilise an improvement over the minimax algorithm, i.e. the so-called *alpha-beta pruning*: if at some points during the exploration of the tree a move is proven to be worse than another, there is no need to continue expanding the states which would be generated from it.

When using this pruning, the complexity of the search depends on the order in which nodes are explored: in our implementation nodes are ordered by their value of heuristic function, given by the neural network, thus resulting in searches that are faster the better the network is trained.

One of the main factors influencing performance is the depth of the search. The playing strength can be greatly improved increasing the depth at which the search is stopped and the heuristic function is applied, but this objective is reached at the cost of an exponentially increasing reasoning time.

With the aim of finding a good tradeoff between these two requirements we applied an iterative deepening approach: this choice has also the advantage of automatically adapting the depth during a game, increasing it when states with few successors are encountered (for example when there are forced moves for some player), and decreasing it when there are many moves available. Setting a time limit instead of a depth limit is easier since the duration of the training can be directly determined, and a finer tuning is allowed.

3 Neural Model and Pseudo-Inversion Based Training

In this section we introduce notation and we recall basic ideas concerning the use of pseudo-inversion for neural training. The core of the NHB-player learning is a standard SLFN with P input neurons, M hidden neurons and Q output neurons, non-linear activation functions $\phi(x)$ in the hidden layer and linear activation functions in the output layer. In our model the input layer is composed by as many nodes as are the variables in a state of the considered game, while the hidden layer size is determined through a procedure of k-fold cross validation described in the following section. All hidden neurons use the standard logistic function $\phi(x) = \frac{1}{1+e^{-x}}$ as activation function. The output layer is composed by a single node because only a single real value must be computed for each state.

From a general point of view, given a dataset of N distinct training samples of (input, output) pairs $(\mathbf{x}_j, \mathbf{t}_j)$, where $\mathbf{x}_j \in \mathbb{R}^P$ and $\mathbf{t}_j \in \mathbb{R}^Q$, the learning process for a SLFN aims at producing the matrix of desired outputs $\mathbf{T} \in \mathbb{R}^{N \times Q}$ when the matrix of all input instances $\mathbf{X} \in \mathbb{R}^{N \times P}$ is presented as input.

As stated in the introduction, in the pseudoinverse approach input weights c_{ij} (and hidden neurons biases) are randomly sampled from a uniform distribution in a fixed interval and no longer modified. After having determined input weights $\mathbf{C}$, the use of linear output units allows to determine output weights w_{ij} as the solution of the linear system $\mathbf{H}\,\mathbf{W} = \mathbf{T}$, where $\mathbf{H} \in \mathbb{R}^{N \times M}$ is the hidden layer output matrix of the neural network, i.e. $\mathbf{H} = \Phi(\mathbf{X}\,\mathbf{C})$.

Since $\mathbf{H}$ is a rectangular matrix, the least square solution $\mathbf{W}^*$ that minimises the cost functional $E_D = ||\mathbf{HW} - \mathbf{T}||_2^2$ is $\mathbf{W}^* = \mathbf{H}^+\mathbf{T}$ (see e.g. [12,13]).

$\mathbf{H}^+$ is the Moore-Penrose generalised inverse (or pseudoinverse) of matrix $\mathbf{H}$.

Advantage of this training method is that the only parameter of the network that needs to be tuned is the dimension of the hidden layer. Section 4.2 will show the procedure utilized to do so.

3.1 Online Sequential Version of the Training Algorithm

The training algorithm we implemented to evaluate $\mathbf{W}^*$ is the variation proposed in [10], called Online Sequential Extreme Learning Machine (OS-ELM):

its main advantage is the ability to incrementally learn from new examples without keeping previously learned examples in memory. In addition to the matrices $\mathbf{C}$ and $\mathbf{W}$, containing the hidden and output layer weights, a network trained with this algorithm also utilises an additional matrix $\mathbf{P}$.

The starting values for the net's matrices $\mathbf{P}_0$ and $\mathbf{W}_0$ are computed in the OS-ELM algorithm with the following formulas:

$$\mathbf{P}_0 = (\mathbf{H}_0^T\mathbf{H}_0)^{-1}, \quad \mathbf{W}_0 = \mathbf{P}_0\mathbf{H}_0^T\mathbf{T}_0 \tag{1}$$

where $\mathbf{H}_0$ and $\mathbf{T}_0$ are respectively the hidden layer output matrix and the target matrix relative to the initial set of training examples.

The update is done in two steps, according to the following equations:

$$\mathbf{P}_{k+1} = \mathbf{P}_k - \mathbf{P}_k\mathbf{H}_{k+1}^T(\mathbf{I} + \mathbf{H}_{k+1}\mathbf{P}_k\mathbf{H}_{k+1}^T)^{-1}\mathbf{H}_{k+1}\mathbf{P}_k \tag{2}$$

$$\mathbf{W}_{k+1} = \mathbf{W}_k + \mathbf{P}_{k+1}\mathbf{H}_{k+1}^T(\mathbf{T}_{k+1} - \mathbf{H}_{k+1}\mathbf{W}_k) \tag{3}$$

The network weights $\mathbf{W}$ are thus updated basing on the hidden layer output matrix for the new set of training examples $\mathbf{H}_k + 1$, the target matrix of the new set of training examples $\mathbf{T}_k + 1$, and the previous evaluated matrices $\mathbf{P}_k$ and $\mathbf{W}_k$. $\mathbf{I}$ is the identity matrix.

The network can be trained with sets of examples of varying and arbitrary magnitude: it is however better to accumulate examples obtained from some different matches, instead of allowing learning after every game, in order to be able to discard duplicates, so reducing the risk of overfitting on the most frequent states. This technique can anyway cause a considerable decrease of the learning speed, since the matrix to be inverted in Eq. (2) has a dimension equal to the square of the number of training examples presented. In this case or if the dimension of this matrix exceeds the available memory space it is sufficient to split the training set in smaller chunks and iteratively train the network on each chunk.

4 Training Examples Selection

As already mentioned, multiple training examples are required, each composed by a state and the desired heuristic function value for that state. Since we aim at creating a NHB-player which autonomously learns the heuristic function, we do not want to rely on records of previously played games or on external opponents to train against. Therefore, examples are extracted directly from simulated matches.

The optimal candidates for building training examples are the terminal states of the game, whose value of heuristic function can be determined directly by the match outcome: we assign to these states a value 1 for a win, −1 for a loss and 0 for a tie. Using the search tree, the values of terminal states can be propagated unto previous states that lead to them; in this way we are able to build examples using non-terminal states, which are the ones for which a heuristic function is actually useful.

In addition, for each state of the played match a new example is generated: the heuristic value of these examples is computed as increasing quadratically during the match from 0 to the value of the final state. This way of assigning scores is arbitrary and it is based on the assumption that the initial condition is neutral for the two players and that the winning player's situation improves regularly during the course of the game. This assumption is useful because it allows to collect a certain number of states with heuristic function values close to 0, which can prevent the net from becoming biased towards answers with high absolute value (for games in which drawing is a rare outcome). Even if some games are not neutral in their initial state, they are nevertheless complex enough to be practically neutral for any human (and as such imperfect) player; moreover, the value of the initial state is never utilized in the search algorithm, so its potentially incorrect estimation doesn't affect the performance of the system. Finally, duplicate states within a training set are removed, keeping only one example for each state; if in this situation a value assigned for a played state and a value coming from propagation of a terminal state are conflicting, the second one is kept since it's the most accurate one.

After enough training matches, the system can propagate heuristic values from intermediate states, provided that these values are the result of previous learning and are not completely random. This can be estimated using the concept of *consistency* of a node, discussed in [14]: a node is considered consistent if its value of heuristic function is equal[1] to the value computed for that node by a minimax search with depth 1. All consistent nodes can be used for propagation in the same way as terminal nodes, maintaining the rule that, when conflicts arise, values propagated from terminal nodes have the priority.

4.1 Backward Iterative Deepening Search

When in the complete tree[2] of a game the majority of the leaves are positioned flatly in one or a small number of levels far from the root, they are not reachable with a low-depth search from most of the higher nodes. In games like *R*eversi, for example, no terminal states are encountered during the search until the very last moves, so the number of training examples extracted from every match is quite small.

In order to find more terminal states in a game of this kind, a deeper search is needed starting from a deep enough node. Good candidates for this search are the states occurred near the end of a match, but it is difficult to determine how near they must be in order to find as many examples as possible in a reasonable time. The solution is once again to set a time limit instead of a depth limit and utilize an iterative deepening approach: in this case we talk about a *Backward*

[1] In the case of the proposed system, since the output is real-valued, the equality constraint must be softened and all values within a margin ε are accepted.

[2] The *complete* tree of a game is the tree having as root the initial state and as children of a node all the states reachable with a single legal move from that node. The leaves of this tree are the terminal states of the game.

Iterative Deepening Search (or *Backward IDS*). The increase in depth at each step is accompanied by a raising of the root of the search, resulting in sub-searches all expanding to the same level but starting from nodes increasingly closer to the root (see Fig. 1). Differently from the standard iterative deepening search, in the Backward IDS each iteration searches a sub-tree of the following iteration, so the value of the previous sub-search can be passed to the following in order to avoid recomputing; this iterative search has then the same complexity as the last fixed-depth sub-search done, while automatically adapting itself to different tree configurations.

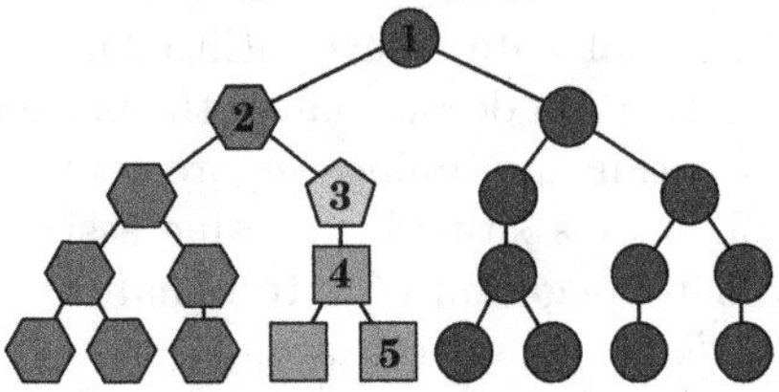

Fig. 1. An example of application of the Backward IDS. The played game is the sequence of states 1-2-3-4-5. The search starts from state 4 with depth one, then goes back to state 3 with depth two and so on until timeout. Nodes with the same shape are explored during the same iteration.

4.2 Training Procedure

Each training procedure starts with a certain number of games played between agents which both use random value heuristics. The states explored during these games by both players are collected, multiple copies of the same state are eliminated, and the resulting set is used as initial training set for the neural network.

Since the more trained a network is the better training examples will it generate, further training has to be performed by alternating phases of game simulation (where training examples are extracted and accumulated through simulating games played by the trained system) and phases of neural network training (where the system improves basing on the training examples extracted during the previous phase).

So doing the extraction of post-initial training examples is performed in the framework of games played against a similar opponent which applies the same search strategy but whose heuristic function comes from a randomly initialized neural network. Most games will (hopefully) be won by the trained player, but the system will still improve itself through the exploration of new states.

It can also be useful to have the system play against itself, but for no more than one match in each training cycle. The deterministic structure of the search algorithm implies in fact that, unless the heuristic function is modified, multiple matches against the same opponent will be played exactly in the same way.

Aiming at finding the optimal number of hidden neurons that allows to avoid overfitting, we determine hidden layer size and hidden layer weights through a 10-fold cross validation: hidden layers of increasing sizes are explored until for 10 iterations in a row the best average validation error doesn't decrease. During each iteration three different random initialization for the hidden weights are tried, and only the best one is kept.

5 Implementation Details, Game Descriptions and Results

The system structure described in the previous sections is completely independent from the game to be played. A modular structure is adopted to maintain this independence: four game-dependent functions, separated from the rest of the system, completely describe each game and contain all its rules so that different implementations of these functions permit to play different games.

init_state: This function returns the initial state of the game, taking as argument the starting player (1 or −1 for the two-player games implemented).

expand: This function returns a list of all the moves available from a given state received as input, along with the list of states they lead to.

test_terminal: This function takes as input a state and checks if it is a terminal one, returning in this case also the final outcome.

apply: This function gets as input a state and a move, then applies the chosen move to the state and returns the new state reached.

We implemented two different games to test the learning capabilities of our artificial player: *Connect Four* and *Reversi*. These games have different degrees of complexity and, while they both possess the properties described in the Introduction, they have some peculiarities that make them apt to test different aspects of the system. We will now briefly present their characteristics and discuss the obtained results.

5.1 Connect Four

This game is played on a grid with 6 rows and 7 columns, where players alternatively place their tokens. A token can only be placed on the lowest unoccupied slot of each column[3], and the first player which puts four of its tokens next to each other horizontally, vertically, or diagonally is the winner.

Allis [15] demonstrated that the first player always wins, under the condition of perfect playing on both sides; this level of playing, almost impossible for a human, is difficult even for a machine, since artificial perfect players usually require a database of positions built with exhaustive search in order to successfully choose the best move.

[3] The most common commercial versions of this game are played on a suspended grid, where tokens fall until the lowest free position.

The branching factor[4] for this game is constant and equal to 7, until a column is filled. This makes the search depth almost constant, thus reducing the advantage of an iterative deepening search.

We evaluate the performance of our model looking at the percentage of matches won against an untrained opponent. We also explored the influence of the time limit for each move, varying the move time available during training: three different networks were trained, with move times of 0.25 seconds (s), 0.5 s and 1 s respectively. Aiming at comparing networks trained for the same **total** amount of time, the number of training matches was proportionally halved.

Table 1. Varying move time in Connect Four training. T = Training matches. L, D, W = Losses, Draws, Wins over 200 testing matches

move time=0.25s 96 hidden neurons				move time=0.5s 112 hidden neurons				move time=1s 113 hidden neurons			
T	L	D	W	T	L	D	W	T	L	D	W
20	79	11	110	10	54	14	132	5	59	18	123
40	53	9	138	20	68	12	120	10	64	13	123
60	46	13	141	30	88	13	99	15	57	13	130
80	68	18	114	40	44	12	144	20	69	16	115
100	46	12	142	**50**	**43**	**9**	**148**	25	61	19	120
120	**42**	**10**	**148**	60	69	13	118	**30**	**47**	**18**	**135**
140	66	11	123	70	50	8	142	35	84	18	98
160	59	9	132	80	54	13	133	40	77	12	111
180	60	14	126	90	58	7	135	45	62	14	124
200	47	13	140	100	55	11	134	50	61	10	129

As shown in Table 1, the best performance is obtained with a move time equal to 0.25 s and 120 training matches. The number of won matches clearly demonstrates the effectiveness of our proposed approach in developing a successfull playing stategy.

Looking at the results obtained with move time = 0.5 s we see that a comparable performance is reached with about half the number of training matches, i.e. 50; it is interesting to note that in this way the total training time remains almost unchanged. Besides, also the network trained with move time equal to 1 s reaches the best performance in approximately half the number of training games with respect to the case of move time = 0.5 s, but now it appears to be slightly worse: this indicates the possible existence of an optimal move time that can produce the best trained network in a fixed total training time. Developing a way to find this optimal time will be an objective for future work.

[4] The branching factor is the average number of branches (successors) from a (typical) node in a tree. It indicates the bushiness and hence the complexity of a tree. If a tree branching factor is B, then at depth d there will be approximately B^d nodes.

5.2 Reversi

*R*eversi, also known as Othello because of its most popular commercial variant, is a more complex game played on a 8 by 8 board where two players alternate placing disks whose two faces have different colours, one for each player. When the board is full or no player has a legal move to play, disks are counted and whoever owns the highest number of them wins the match. A more detailed description of this game can be found in [16].

This game is characterised by high average branching factor and game length (for almost all matches 60 moves, except for the rare cases where no player has a legal move); finding available moves and checking for terminal states is very expensive from a computational point of view, resulting in a high time required to complete a search.

While since 1980 programs exist that can defeat even the best human players (see [17] for a review of the history of Othello computer players), the game has not yet been completely solved: performance of the top machine players indicates that perfect play by both sides probably leads to a draw, but this conjecture has not yet been proved. The branching factor varies during the game, usually reaching the highest value (about 15) during the middle phase.

99	-8	8	6	6	8	-8	99
-8	-24	-4	-3	-3	-4	-24	-8
8	-4	7	4	4	7	-4	8
6	-3	4	0	0	4	-3	6
6	-3	4	0	0	4	-3	6
8	-4	7	4	4	7	-4	8
-8	-24	-4	-3	-3	-4	-24	-8
99	-8	8	6	6	8	-8	99

Fig. 2. The positional evaluation table used for state evaluation in old Microsoft Windows Reversi programs.

For the game of *R*eversi many heuristic functions for artificial players have been defined in the past (e.g. in [18]). In order to compare our proposed model with some baseline performance, we chose two human-defined heuristics, the piece-advantage and the positional ones: the first one is extremely simple while the second one is actually used in artificial playing.

Piece-advantage: this heuristic simply assigns state values equal to the difference between the number of owned disks and the number of opponent's disks.

Positional: this heuristic assigns to each disk a value which depends on its position; the values of all disks owned by each player are then summed and the difference between the two players' scores is returned. For our implementation we chose the same evaluation table used by some Microsoft Windows *R*eversi programs (as noted in [19]), shown in Fig. 2.

Table 2 compares performance obtained by the NHB-player and by two players which choose their moves basing on the heuristics described above.

Table 2. Performance comparison among NHB, Piece-advantage and Positional heuristics over 100 matches against opponents who play using random heuristics.

Heuristic	Training Matches	Losses	Draws	Wins
NHB (131 hidden neurons)	50	36	5	59
	100	**24**	**4**	**72**
	150	39	2	59
	200	36	0	64
	250	30	6	64
	300	41	4	55
	350	38	5	57
	400	42	2	56
	550	39	2	59
	600	36	3	61
Piece-advantage	-	53	3	44
Positional	-	26	2	72

We can see that the Piece-advantage heuristic performes even worse than the random one, losing more than half of the matches. This result emphasizes that the development of a heuristic function is not a simple task: the Piece-advantage heuristic, that may seem good in theory because it tries to maximise the score at each stage, turns out to be harmful in practice.

Because of the higher number of matches lost, the Positional heuristic performes slightly worse with respect to the NHB-player one: we stress the fact that this happens although it is a game-dependent human-defined heuristic.

6 Conclusions

Results discussed in the previous section show that our main objective has been achieved: for each implemented game the NHB-player is able to develop a successful playing strategy, without the necessity of tailor-made human contributions but only thanks to the knowledge of game rules.

The strength of the system lies in its adaptability, which allows to deal with both other games with the discussed properties and games characterized by different features.

Future work will focus on one hand in enabling the exploitation of additional information that may be given to the player (for example a score given turn-by-turn, as is the case in [20]), on the other hand in expanding our model to make it able to learn other interesting set of games, some of which are:

Chance-based games, that can be managed by simply modifying the minimax search algorithm.
Games with repeated states, that are hard to control since they can transform into never-ending ones. In this case a modification of the search algorithm in the training phase is necessary to avoid infinite loops.

References

1. Silver, D., Huang, A., et al.: Mastering the game of Go with deep neural networks and tree search. Nature **529**(7587), 484–489 (2016)
2. Draper, S., Rose, A.: Sancho GGP player. http://sanchoggp.blogspot.com
3. Michulke, D.: Neural networks for high-resolution state evaluation in general game playing. In: IJCAI-11 Workshop on General Game Playing (GIGA11), pp. 31–37 (2011)
4. Schiffel, S., Thielscher, M.: Fluxplayer: a successful general game player. In: 22nd National Conference on Artificial intelligence, pp. 1191–1196. AAAI Press, Menlo Park (2007)
5. Świechowski, M., Mańdziuk, J.: Specialized vs. multi-game approaches to AI in games. In: Angelov, P., et al. (eds.) Intelligent Systems 2014. AISC, vol. 322, pp. 243–254. Springer, Heidelberg (2015). doi:10.1007/978-3-319-11313-5_23
6. Świechowski, M., Park, H., Mańdziuk, J., Kim, K.-J.: Recent advances in general game playing. Sci. World J. **2015**, Article ID 986262, 22 p. (2015). doi:10.1155/2015/986262
7. Schmidt, W.F., Kraaijveld, M., Duin, R.P.W., et al.: Feedforward neural networks with random weights. In: International Conference on Pattern Recognition, Conference B: Pattern Recognition Methodology and Systems, pp. 1–4 (1992)
8. Pao, Y.H., Park, G.H., Sobajic, D.J.: Learning and generalization characteristics of the random vector functional-link net. Neurocomputing **6**, 163–180 (1994)
9. Huang, G.B., Chen, L., Siew, C.K.: Universal approximation using incremental constructive feedforward networks with random hidden nodes. IEEE Trans. Neural Netw. **17**, 879–892 (2006)
10. Liang, N.Y., Huang, G.B., Saratchandran, P., Sundararajan, N.: A fast and accurate online sequential learning algorithm for feedforward networks. IEEE Trans. Neural Netw. **17**, 1411–1423 (2006)
11. Russell, S., Norvig, P.: Artificial Intelligence: A Modern Approach. Prentice-Hall, Egnlewood Cliffs (1995)
12. Penrose, R.: On best approximate solutions of linear matrix equations. Math. Proc. Camb. Philos. Soc. **52**, 17–19 (1956)
13. Bishop, C.: Pattern Recognition and Machine Learning. Springer, Heidelberg (2006)
14. Gherrity, M.: A game-learning machine. Ph.D. thesis, University of California, San Diego (1993)
15. Allis, L.W.: A knowledge-based approach of connect-four. Technical report, Vrije Universiteit, Subfaculteit Wiskunde en Informatica (1988)
16. British Othello Federation: Game Rules. http://www.britishothello.org.uk/rules.html
17. Cirasella, J., Kopec, D.: The History of Computer Games. CUNY Academic Works, New York (2006)
18. Mitchell, D.H.: Using features to evaluate positions in experts' and novices' Othello games. Masters thesis, Northwestern University, Evanston (1984)
19. MacGuire, S.: Strategy Guide for Reversi and Reversed Reversi. www.samsoft.org.uk/reversi/strategy.htm#position
20. Mnih, V., Kavukcuoglu, K., Silver, D., et al.: Human-level control through deep reinforcement learning. Nature **518**(7540), 529–533 (2015)

Comparing Hidden Markov Models and Long Short Term Memory Neural Networks for Learning Action Representations

Maximilian Panzner(✉) and Philipp Cimiano

Semantic Computing Group, CITEC, Bielefeld University, Bielefeld, Germany
mpanzner@cit-ec.uni-bielefeld.de
https://cit-ec.de/

Abstract. In this paper we are concerned with learning models of actions and compare a purely generative model based on Hidden Markov Models to a discriminatively trained recurrent LSTM network in terms of their properties and their suitability to learn and represent models of actions. Specifically we compare the performance of the two models regarding the overall classification accuracy, the amount of training sequences required and how early in the progression of a sequence they are able to correctly classify the corresponding sequence. We show that, despite the current trend towards (deep) neural networks, traditional graphical model approaches are still beneficial under conditions where only few data points or limited computing power is available.

Keywords: HMM · LSTM · Incremental learning · Recurrent network · Action classification

1 Introduction

Representing and incrementally learning actions is crucial for a number of scenarios and tasks in the field of intelligent systems where agents (both embodied and unembodied) need to reason about their own and observed actions. Thus, developing suitable representations of actions is an important research direction. We are concerned with how to develop action models that allow intelligent systems to reason about observed actions in which some trajector is moved relative to some reference object. Such actions correspond to manual actions of humans that manipulate objects. We are concerned with finding models that support tasks such as: classifying observed actions according to their types, forward prediction and completion of actions. A particular focus lies on understanding how to incrementally learn manipulation actions from only few examples and how to enable models to classify actions which are still in progress. We compare two broad classes of models: generative sequence models (HMMs in particular) and discriminatively trained recurrent neural networks (LSTM recurrent networks in particular). We analyze their properties on the task of classifying action sequences in terms of how much training examples are required to make

P.M. Pardalos et al. (Eds.): MOD 2016, LNCS 10122, pp. 94–105, 2016.
DOI: 10.1007/978-3-319-51469-7_8

useful predictions and how early in a sequence they are able to correctly classify the respective sequence. Our models are trained on a dataset consisting of 1200 examples of action performances of human test subjects for four basic actions: *jumps over*, *jumps upon*, *circles around* and *pushes*. Our results are encouraging and show that action models can be successfully learned with both types of models, whereby the main advantage of HMMs is that they can learn with fewer examples and they show very favorable training times when compared to LSTMs (10 s for HMMs vs. 4 h for LSTMs) while displaying comparable performance of about 86% F-measure. Future work will involve scaling these models to more complex action types and model actions described in terms of richer features including sensor data.

2 Related Work

There has been a lot of work in the field of intelligent systems on developing formalisms for learning and representing actions, ranging from task-space representations [1] to high-level symbolic description of actions and their effects on objects [2]. Many approaches work directly on raw video streams, which is a challenging problem due to high variance in the video emanating from e.g. different lighting conditions and also due to high intra-class variance in the actions. These problems are often tackled by first finding robustly recognizable interest points to extract and classify features from their spatio-temporal neighborhood. Panzner et al. [3] detect points of interest using a Harris corner detector extended to 3D spatio-temporal volumes. They extract HOG3D/HOF features from the spatio-temporal neighborhood of these points which are then clustered using a variant of growing neural gas (GNG) to yield an incremental vocabulary of visual words. They represent action sequences as frequency histograms over these visual words (bag of words) which are then classified by another GNG layer. Veeriah et al. [4] use densely sampled HOG3D features which are directly classified by an extended LSTM network which considers the spatio-temporal dynamics of salient motions patterns. They use the Derivative of States $\frac{\partial s_t}{\partial t}$, where s_t is the state of memory cell s a time t, to gate the information flow in and out of the memory cell. In this paper we focus on mid-level representations to bridge the gap between low-level representations and high-level symbolic descriptions in a way that facilitates transfer of learned concepts between the representational levels.

3 Models

We compare two approaches to modeling sequences of positional relation between a trajector and a landmark. The first approach models the relation between the two objects as class-specific generative Hidden Markov Models [5]. The second approach utilizes a discriminative two-layer neural network with a LSTM (long-short term memory) recurrent network as the first layer followed by a fully connected linear output layer with softmax normalization to interpret the network output as class probabilities. We are particularly interested in how fast the models converge when there is only little training data and how early in a sequence they are able to correctly classify the respective sequence. As representation learning generally requires rather large training sets, we abstract away from the raw positions and velocities by discretizing them into qualitative basic relations between trajector and landmark following the QTC (qualitative trajectory calculus) [6] framework. This abstraction is especially beneficial for the HMM approach because the markov assumption is better satisfied when we encode position and velocity in a joint representation. LSTM networks are usually considered to work best when the input space is continuous but in this case preliminary experiments showed that the QTC discretization is also favorable for LSTM in terms of classification performance and training time. When the dataset is large enough it should also be possible to train both models on raw features like coördinates relative to a trajectory-intrinsic reference point as shown by Sugiura et al. [1] for HMMs.

3.1 Qualitative Trajectory Representation

To describe the relative position and movement between landmark and trajector we build on the qualitative trajectory calculus - double cross (QTC_{C1}) [6] as a formal foundation. In general, QTC describes the interaction between two moving point objects k and l with respect to the reference line RL that connects them at a specific point t in time. The QTC framework defines 4 different subtypes as a combination over different basic relations between the two objects. As we only have one actively moved object in our experiments, we decided on QTC_{C1} to give the best trade off between generalization and specificity of the qualitative relations. QTC_{C1} consists of a 4-element state descriptor (C_1, C_2, C_3, C_4) where each $C_i \in \{-, 0, +\}$ represents a so called constraint with the following interpretation:

C_1 Distance constraint: Movement of k with respect to l at time t_1:
- $-$ k is moving towards l
- 0 k is not moving relative to l
- $+$ k is moving away from l

C_2 Distance constraint: Movement of l with respect to k at time t_1: analogously to C_1

C_3 Side constraint: Movement of k with respect to RL at time t_1:
- $-$ k is moving to the left-hand side of RL
- 0 k is moving along RL or not moving at all
- $+$ k is moving to the right-hand side of RL

C_4 Side constraint: Movement of l with respect to RL at time t_1: analogously to C_3

As the positions in our dataset were sampled at a fixed rate, we could have missed some situations where one or more state descriptor elements transition through 0. These discretization artifacts are compensated by inserting the missing intermediate relations one at a time from left to right. QTC_{C1} is a rather coarse discretization, leading to situations where the qualitative relation between the two objects can hold for a longer portion of the trajectory and is, due to the fixed rate sampling, repeated many times. Unlike many spatial reasoning systems, where repeating states are simply omitted, we use a logarithmic compression of repetitive subsequences:

$$|\hat{s}| = \min(|s|, 10 \ln(|s| + 1)) \tag{1}$$

where $|s|$ is the original number of repeated symbols in the sequence and $|\hat{s}|$ is the new number of repeated symbols. By applying this compression scheme, we preserve information about the acceleration along the trajectory, which increases the overall performance especially for very similar actions like *jumps over* and *jumps upon*, while still allowing to generalize over high variations in relative pace of the action performances. The logarithmic compression of repetitive symbols in a sequence is in line with findings from psychophysics known as the Weber-Fechner law [7].

3.2 HMM

For the HMM parameter estimation, we apply an incremental learning scheme utilizing the best first model merging framework [8,9]. Model merging is inspired by the observation that, when faced with new situations, humans and animals alike drive their learning process by first storing individual examples (memory based learning) when few data points are available and gradually switching to a parametric learning scheme to allow for better generalization as more and more data becomes available [10]. Our approach mimics this behavior by starting with simple models with just one underlying sequence, which evolve into more complex models generalizing over a variety of different sequences as more data is integrated. Learning a new sequence in this framework is realized by first constructing a maximum likelihood (ML) Markov Chain which exactly reproduces the respective sequence, which is then integrated between the start and the end state of the existing (possibly empty) model. When incorporating more and more sequences, the resulting models would constantly grow and consist only of ML chains connected to the start and end states of the model. To yield more compact models, which are able to generalize to similar but unseen sequences, we consecutively merge similar states and thus intertwine their corresponding paths through the model. This way learning as generalization over the concrete observed examples is driven by structure merging in the model in a way that we trade model likelihood against a bias towards simpler models. This is known as the Occam's Razor principle, which among equally well predicting hypothesis prefers the simplest explanation requiring the fewest assumptions. As graphical models, HMMs are particularly well suited for a model merging approach because integrating a new sequence, merging similar states and evaluating the model's likelihood given the constituting dataset are straightforward to apply in this framework and implemented as graph manipulation operations:

Data Integration: When a new sequence is to be integrated into a given model we construct a unique path between the initial and the final state of the model where each symbol in the sequence corresponds to a fresh state in the new path. Each of these states emits its respective symbol in the underlying sequence and simply transitions to the next state with probability 1, yielding a sub path in the model which exactly reproduces the corresponding sequence.

State Merging: The conversion of the memory based learning scheme with unique sub paths for each sequence in the underlying dataset into a model which is able to generalize to a variety of similar trajectories is achieved by merging states which are similar according to their emission and transition densities. Merging two states q_1 and q_2 means replacing these states with a new state $\hat{q}$ whose transition and emission densities are a weighted mixture of the densities of the two underlying states.

Model Evaluation: We evaluate the models resulting in the merging process using a mixture composed of a structural model prior $P(M)$ and the data dependent model likelihood $P(X|M)$:

$$P(M|X) = \lambda P(M) + (1 - \lambda)P(X|M) \tag{2}$$

The model prior $P(M)$ acts as a data independent bias. Giving precedence to simpler models with fewer states makes this prior the primary driving force in the generalization process:

$$P(M) = e^{-|M|}, \tag{3}$$

where the model size $|M|$ is the number of states in the model. It is also possible to include the complexity of the transitions and emissions per state. For our dataset we found that using only the number of states generates the best performing models. While the structural prior favors simpler models, its antagonist, the model likelihood, has its maximum at the initial model with the maximum likelihood sub-paths. The exact likelihood of the dataset X given the model M is computed as:

$$P(X|M) = \prod_{x \in X} P(x|M) \tag{4}$$

with

$$P(x|M) = \sum_{q_1 \dots q_l \in Q^l} p(q_I \to q_1)p(q_1 \uparrow x_1) \dots p(q_l \uparrow x_l)p(q_l \to q_F) \tag{5}$$

where l is the length of the sample and q_I, q_F denote the initial and final states of the model. The probability to transition from a state q_1 to q_2 is given as $p(q_1 \to q_2)$ and $p(q_1 \uparrow x_1)$ denotes the probability to emit the symbol x_1 while being in state q_1. As we do not want to store the underlying samples explicitly, we use an approximation, which considers only the terms with the highest contribution, the Viterbi path:

$$P(X|M) \approx \prod_{q \in Q} \left(\prod_{q\prime \in Q} p(q \to q\prime)^{c(q \to q\prime)} \prod_{\sigma \in \Sigma} p(q \uparrow \sigma)^{c(q \uparrow \sigma)} \right) \tag{6}$$

where $c(q \to q\prime)$ and $c(q \uparrow \sigma)$ are the total counts of transitions and emissions occurring along the Viterbi path associated with the samples in the underlying dataset (see [8] for more details).

The simplest model in our approach is a model which simply produces a single sequence. These models are called maximum likelihood models because they produce their respective sequences with the highest possible probability. Starting from maximum likelihood models over individual sequences we build more general HMMs by merging simpler models and iteratively joining similar states to intertwine sub-paths constructed from different sequences, allowing them to generalize across different instances of the same action class. The first model M_0 of the example in Fig. 1 can be seen as a joint model of two maximum likelihood sequences $\{ab, abab\}$. When generating from such a model, the actual sequence which will be generated is determined early by taking one of the possible paths emanating from the start state. Only the transitions from the start state display stochastic behavior, the individual sub-paths are completely deterministic and generate either ab or $abab$. Intertwining these paths is done trough state merging, where we first build a list of possible merge candidates using a measure of similarity between state emissions and transition probability densities. In this approach we use the symmetrized Kullback-Leibler (KL) divergence

$$D_{\mathrm{SKL}}(P, Q) = D_{\mathrm{KL}}(P, Q) + D_{\mathrm{KL}}(Q, P). \tag{7}$$

with

$$D_{\mathrm{KL}}(P, Q) = \sum_i P(i) \log \frac{P(i)}{Q(i)} \tag{8}$$

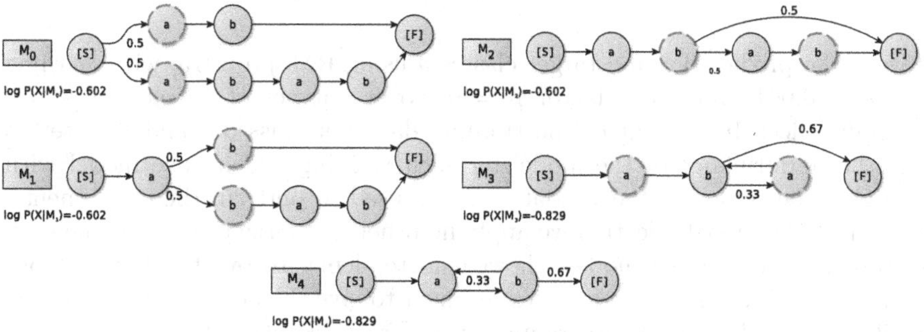

Fig. 1. Sequence of models obtained by merging samples from an exemplary language $(ab)^+$ and subsequently merging the highlighted states. Transitions without special annotations and all emissions have a probability of 1. The highlighted nodes are selected by the similarity of their transition and emission characteristics and consecutively merged to yield the subsequent model. (Color figure online) Reproduced from [8].

Then we greedily merge the best pair of states and re-evaluate the model likelihood. In the example above, the first two merges lead to model M_3 where we experienced a drop in log likelihood from -0.602 to -0.829. We continue the merging process until we reach a point where merging more states would deteriorate the model likelihood to a point where it is no longer compensated by the prior favoring simpler models (Eq. 3). The final model M_4 is now able to generate the whole set of sequences from the exemplary language $(ab)^+$ from which the two initial samples where generated from.

As neighboring QTC state IDs have no semantic similarity, we model emission densities as frequency distributions. We train a separate HMM for each of the four action classes. Sequences are classified according to the class specific HMM having the highest probability to have produced the respective sequence. When none of the HMMs assigns a probability higher than zero for a sequence, the model is unable to classify this sequence and assigns no class label (reject option). Training with 1100 sequences takes about 10 s on a 2 CPU – 20 core machine. See [11] for a more detailed analysis of the properties of this model.

3.3 LSTM

The discriminative recurrent network consists of a long-short term memory (LSTM) layer with 128 memory blocks followed by a fully connected linear layer with softmax normalization and one output unit per target class. The softmax normalization allows the network output to be interpreted as class probabilities. Unlike the HMM approach where we train a separate HMM for each class, this network learns a joint representation between the sequences of all 4 classes. Several variants of the block architecture of LSTM memory cells have been proposed [12, 13]. Our implementation is based on the architecture proposed by Hochreiter and Schmidhuber [14]. We optimize the cross entropy

$$H(p,q) = -\sum_{x} p(x)\log(q(x)) \tag{9}$$

between the predicted p and target classes q using RMSprop [15] with learning rate $lr = 0.001$ and decay factor $\rho = 0.9$ over batches of 50 sequences. We randomly select 10 percent of the training data per class as validation set to asses the performance of the model for each training epoch. The model with the highest accuracy on the validation set is used to perform the experiments. Other than the model selection we apply no other regularization technique such as dropout, because the models showed no tendency to overfit. The network layout was optimized in a separate experiment to give the best trade-off between classification performance and training time. Training with 1100 sequences over 150 epochs takes about 4 h on a single GTX-770 GPU.

4 Dataset

To acquire a dataset we implemented a simple game (Fig. 2) where the test subjects were asked to perform an action with two objects according to a given instruction. The game screen was divided into two parts. The upper part was the actual gamefield with the two freely movable objects and below the gamefield was a textfield, where the test subjects could see the instruction describing the desired action performance. We had a total of 12 test subjects (9 male, 3 female, mean age = 29,4 years) yielding a dataset with 1200 trajectory sequences balanced over the four action classes *jumps over*, *jumps upon*, *circles around* and *pushes*. The recorded trajectories are given as tuples with the current timestamp and the positions of trajector and landmark. The raw positions are then converted to QTC state sequences with an average length of 182 symbols. The compression scheme (Eq. 1) reduces the average length to 150 symbols which corresponds to a compression rate of $21,\overline{3}\%$. See [16] for a complete description and download [17] of the dataset.

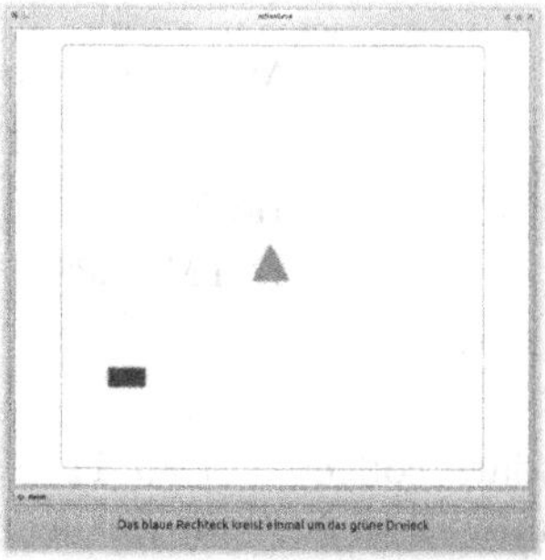

Fig. 2. Left: Simple game with two geometric objects which can be freely moved on the gamefield. In this screen test subjects are tasked to circle the blue rectangle around the green triangle (instruction in the lower part of the screen). (Color figure online)

5 Experiments

In this section we evaluate the overall performance of both models against each other and specifically explore their performance regarding their ability to make useful predictions when there is only little training data per class. We also explore how early in the progression of a sequence the models are able to correctly classify the respective sequence. Both models receive sequences of QTC state IDs. The IDs are simply derived by interpreting the QTC descriptors as a number in a ternary number system. For the LSTM network state IDs are normalized to 1 to better fit the range of the activation functions. We also tried a 4 element vector representation of the QTC state descriptors as input for the LSTM Network, which slightly reduced the number of epochs the training needed to converge but left the overall accuracy unchanged. Thus, for the sake of comparability we decided to use the same input data representation for both approaches.

5.1 HMM Vs LSTM Network

In this experiment we compare the performance of the HMM approach to the performance of the LSTM network. The dataset was divided such, that we trained

Table 1. Results of the 12-fold cross-validation for the HMM and the LSTM approach with and without sequence compression (Eq. 1). Results are given as F_1 score, precision, recall and standard deviation σ.

Model	Compressed				Not compressed			
	F_1	P	R	σ	F_1	P	R	σ
HMM	0.86	0.82	0.90	0.12	0.82	0.75	0.91	0.10
LSTM	0.88	0.88	0.88	0.18	0.88	0.88	0.88	0.18

on data collected from 11 test subjects and let the models classify sequences from a 12-th test subject as a novel performer. As can be seen in Table 1 both approaches perform comparably well. The LSTM network had a considerably higher standard deviation between the results of the 12 folds. We assume that the LSTM network gives higher precedence to differences in relative pace of the action performance, which is the most varying factor in our dataset. Only the HMM approach benefits from compressing long repetitive subsequences, while the LSTM approach is unaffected. Note that only the HMM approach had a reject option, when none of the class specific HMMs assigned a probability higher than zero to the respective sequence. As the LSTM approach learns a joint representation across all 4 classes there is no implicit reject option, because the model will always assign softmax normalized confidences to each class. Figure 3 shows the class confusion matrices for both approaches. The HMM approach frequently misclassifies *jumps over* sequences as sequences corresponding to *circles around* actions, for which the first half of the movement is identical to *jumps over*. This could be caused by the limitation of temporal context HMMs are able to take into consideration due to the markov assumption.

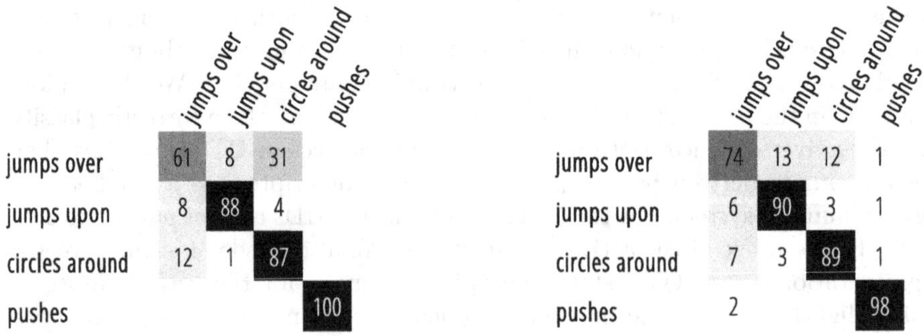

Fig. 3. Averaged class confusion matrix for the HMM (left) and the LSTM network (right).

5.2 Early Learning Behavior

In this experiment we evaluate the performance of the two models when they are presented with only few training sequences per class. Both networks are not trained from scratch when more training sequences are added to the training set, instead both are incrementally re-trained on the extended training set. Figure 4 shows that with an F_1 score of 0.44 the HMMs perform notably better than the LSTM network which starts at chance level after being trained on a single sequence per class. When trained with more than 23 sequences, the LSTM outperforms the HMMs clearly.

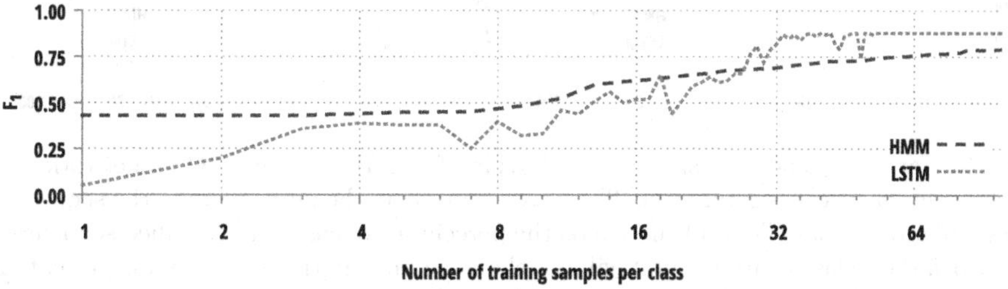

Fig. 4. Early learning behavior, averaged over 10 random folds. HMMs outperform the LSTM network where very little training data is available, while the latter performs better with more than 28 training sequences. Note that the x-axis is scaled logarithmically.

5.3 Early Sequence Classification

Applications in the field of human robot interaction often require the robot to produce action hypothesis when the respective action sequence is not yet completed. This experiment simulates this condition by letting the models classify incomplete sequences. We evaluate the classification performance with sequences truncated to 1 to 100 percent of their original length. The LSTM network has been trained for only 100 epochs to keep it from striving too much towards relying on late parts of the sequences. Apart from that both approaches were not specifically trained for the classification of truncated sequences. Because we trained one HMM per class they were quicker in capturing the essence of the *pushes* action, which was the only action where both objects moved. LSTM performed better on *jumps upon*, supporting the suspicion that LSTM assigns more weight on the pace of action performances, which is also the case for the *circles around* action. Because *jumps over* is the first part of an *circles around* action, both models need about half of the trajectory to separate these actions (Fig. 5).

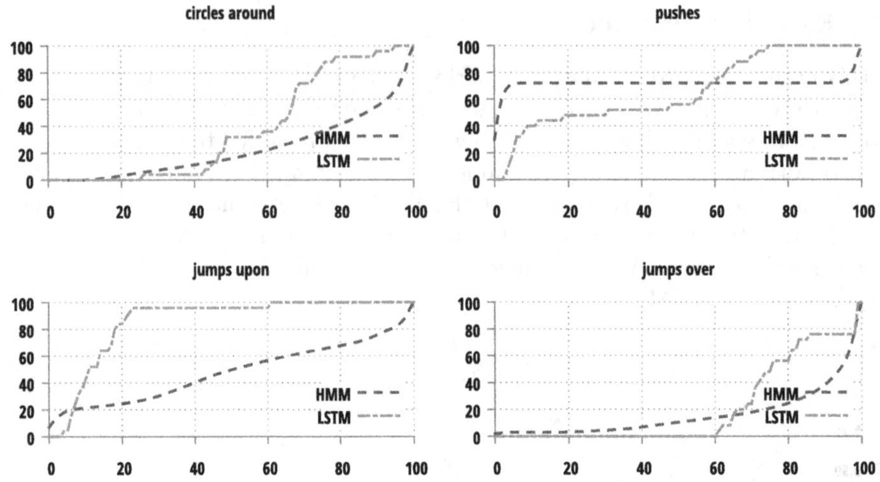

Fig. 5. Early sequence classification behavior of the HMM and LSTM network for each of the four action categories. The x-axis represents the percentage of the sequence presented to the classifier and the y-axis the percentage of correctly classified sequences for which their classification result will not change if more of the sequence was presented (stable classification results).

6 Conclusion

Our results show that HMMs can be used competitively on action modeling tasks compared to discriminatively trained neural networks in scenarios where training data is limited. In particular, our experiments show that the classification performance of HMMs is already useful when training with only a single example per class, consuming only a few milliseconds of CPU time. This is especially useful for example when a robot has to quickly adapt to a human interaction partner. The LSTM network required about 23 examples to reach the classification performance of a class dependent HMM and took a minimum of 50 s training time to converge which is too long for fluent interaction. When classifying incomplete actions, both models had their strength and weaknesses. The HMMs were better in capturing strong discriminative features, as in the *pushes* action where two objects are moving, while LSTM displayed a slightly better overall performance and was able to make decisions earlier throughout all four action categories. For our targeted field of application it could be beneficial to use both models in parallel. HMM for early learning (short-term action memory) or when quick reaction towards new stimuli is required and LSTM when enough data is available (long-term). In the future we would also like to explore whether deep neural networks can learn interpretable qualitative relations that are comparable to the QTC relations we used for this evaluation.

Acknowledgement. This research/work was supported by the Cluster of Excellence Cognitive Interaction Technology 'CITEC' (EXC 277) at Bielefeld University, which is funded by the German Research Foundation (DFG).

References

1. Sugiura, K., Iwahashi, N., Kashioka, H., Nakamura, S.: Learning, generation and recognition of motions by reference-point-dependent probabilistic models. Adv. Robot. **25**(6–7), 825–848 (2011)
2. Tenorth, M., Beetz, M.: KnowRob - knowledge processing for autonomous personal robots. In: IEEE/RSJ International Conference on Intelligent Robots and Systems, pp. 4261–4266 (2009)
3. Panzner, M., Beyer, O., Cimiano, P.: Human activity classification with online growing neural gas. In: Workshop on New Challenges in Neural Computation (NC2) (2013)
4. Veeriah, V., Zhuang, N., Qi, G.-J.: Differential recurrent neural networks for action recognition. In: Proceedings of the IEEE International Conference on Computer Vision, pp. 4041–4049 (2015)
5. Rabiner, L., Juang, B.: An introduction to hidden Markov models. IEEE ASSP Mag. **3**(1), 4–16 (1986)
6. Weghe, N., Kuijpers, B., Bogaert, P., Maeyer, P.: A qualitative trajectory calculus and the composition of its relations. In: Rodríguez, M.A., Cruz, I., Levashkin, S., Egenhofer, M.J. (eds.) GeoS 2005. LNCS, vol. 3799, pp. 60–76. Springer, Heidelberg (2005). doi:10.1007/11586180_5
7. Bruss, T., Rüschendorf, L.: On the perception of time. Gerontology **56**(4), 361–370 (2010)
8. Omohundro, S.: Best-first model merging for dynamic learning and recognition. In: Advances in Neural Information Processing Systems 4, pp. 958–965. Morgan Kaufmann (1992)
9. Stolcke, A., Omohundro, S.: Inducing probabilistic grammars by Bayesian model merging. In: Carrasco, R.C., Oncina, J. (eds.) ICGI 1994. LNCS, vol. 862, pp. 106–118. Springer, Heidelberg (1994). doi:10.1007/3-540-58473-0_141
10. Shepard, R.N.: Toward a universal law of generalization for psychological science. Science **237**(4820), 1317–1323 (1987)
11. Panzner, M., Cimiano, P.: Incremental learning of action models as HMMs over qualitative trajectory representations. In: Workshop on New Challenges in Neural Computation (NC2) (2015). http://pub.uni-bielefeld.de/publication/2775414
12. Greff, K., Srivastava, R.K., Koutník, J., Steunebrink, B.R., Schmidhuber, J.: LSTM: a search space odyssey. arXiv preprint arXiv:1503.04069 (2015)
13. Gers, F.A., Schmidhuber, J., Cummins, F.: Learning to forget: continual prediction with LSTM. Neural Comput. **12**(10), 2451–2471 (2000)
14. Hochreiter, S., Schmidhuber, J.: Long short-term memory. Neural Comput. **9**(8), 1735–1780 (1997)
15. Tieleman, T., Hinton, G.: Lecture 6.5-rmsprop: divide the gradient by a running average of its recent magnitude. COURSERA: Neural Netw. Mach. Learn. **4**, 2 (2012)
16. Panzner, M., Gaspers, J., Cimiano, P.: Learning linguistic constructions grounded in qualitative action models. In: IEEE International Symposium on Robot and Human Interactive Communication (2015). http://pub.uni-bielefeld.de/publication/2733058
17. Panzner, M.: TLS Dataset (2016). doi:10.4119/unibi/2904362

Dynamic Multi-Objective Optimization with jMetal and Spark: A Case Study

José A. Cordero[1], Antonio J. Nebro[2(✉)], Cristóbal Barba-González[2], Juan J. Durillo[3], José García-Nieto[2], Ismael Navas-Delgado[2], and José F. Aldana-Montes[2]

[1] European Organization for Nuclear Research (CERN), Geneva, Switzerland
[2] Khaos Research Group, Ada Byron Research Building, Departamento de Lenguajes y Ciencias de la Computación, University of Málaga, Málaga, Spain
antonio@lcc.uma.es
[3] Distributed and Parallel Systems Group, University of Innsbruck, Innsbruck, Austria

Abstract. Technologies for Big Data and Data Science are receiving increasing research interest nowadays. This paper introduces the prototyping architecture of a tool aimed to solve Big Data Optimization problems. Our tool combines the jMetal framework for multi-objective optimization with Apache Spark, a technology that is gaining momentum. In particular, we make use of the streaming facilities of Spark to feed an optimization problem with data from different sources. We demonstrate the use of our tool by solving a dynamic bi-objective instance of the Traveling Salesman Problem (TSP) based on near real-time traffic data from New York City, which is updated several times per minute. Our experiment shows that both jMetal and Spark can be integrated providing a software platform to deal with dynamic multi-optimization problems.

Keywords: Multi-objective optimization · Dynamic optimization problem · Big data technologies · Spark · Streaming processing · jMetal

1 Introduction

Big Data is defined in a generic way as dealing with data which are too large and complex to be processed with traditional database technologies [1]. The standard *de facto* plaftorm for Big Data processing is the Hadoop system [2], where its HDFS file system plays a fundamental role. However, another basic component of Hadoop, the MapReduce framework, is loosing popularity in favor of modern Big data technologies that are emerging, such as Apache Spark [3], a general-purpose cluster computing system that can run on a wide variety of distributed systems, including Hadoop.

Big Data applications can be characterized by no less than four V's: Volume, Velocity, Variety, and Veracity [4]. In this context, there are many scenarios that

P.M. Pardalos et al. (Eds.): MOD 2016, LNCS 10122, pp. 106–117, 2016.
DOI: 10.1007/978-3-319-51469-7_9

can benefit from Big Data technologies although the tackled problems do not fulfill all the V's requirements. In particular, many applications do not require to process amounts of data in the order of petabytes, but they are characterized by the rest of features. In this paper, we are going to focus on one of such applications: dynamic multi-objective optimization with data received in streaming. Our purpose is to explore how Big Data and optimization technologies can be used together to provide a satisfactory solution for this kind of problems.

The motivation of our work is threefold. First, the growing availability of Open Data by a number of cities is fostering the appearance of new applications making use of them, as for example Smart city applications [5,6] related to traffic. In this context, the open data provided by the New York City Department of Transportation [7], which updates traffic data several times per minute, has led us to consider the optimization of a dynamic version of the Traveling Salesman Problem (TSP) [8] by using real data to define it.

Second, from the technological point of view, Spark is becoming a dominant technology in the Big Data context. This can be stated in the Gartner's Hype Cycle for Advanced Analytics and Data Science 2015 [9], where Spark is almost at the top of the Peak of Inflated Expectations.

Third, metaheuristics are popular algorithms for solving complex real-world optimization problems, so they are promising methods to be applied to deal with the new challenging applications that are appearing in the field known as Big Data Optimization.

With these ideas in mind, our goal here is to provide a software solution to a dynamic multi-objective optimization problem that is updated with a relative high frequency with real information. Our proposal is based on the combination of the jMetal optimization framework [10] with the Spark features to process incoming data in streaming. In concrete, the contributions of this work can be summarized as follows:

- We define a software solution to optimize dynamic problems with streaming data coming from Open Data sources.
- We validate our proposal by defining a dynamic bi-objective TSP problem instance and testing it with both, synthetic and real-world data.
- The resulting software package is freely available[1].

The rest of the paper is organized as follows. Section 2 includes a background on dynamic multi-objective optimization. The architecture of the software solution is described in Sect. 3. The case study is presented in Sect. 4. Finally, we present the conclusions and lines of future work in Sect. 5.

2 Background on Dynamic Multi-Objective Optimization

A multi-objective optimization problem (MOP) is composed of two or more conflicting objective or functions that must be minimized/maximized at the same

[1] https://github.com/jMetal/jMetalSP.

time. When the features defining the problem do not change with time, many techniques can be used to solve them. In particular, multi-objective evolutionary algorithms (EMO) have been widely applied in the last 15 years to solve these kinds of problems [11,12]. EMO techniques are attractive because they can find a widely distributed set of solutions close to the Pareto front (PF) of a MOP in a single run.

Many real world applications are not static, hence the objective functions or the decision space can vary with time [13]. This results in a Dynamic MOP (DMOP) that requires to apply some kind of dynamic EMO (DEMO) algorithm to solve it.

Four kinds of DMOPs can be characterized [13]:

- Type I: The Pareto Set (PS) changes, i.e. the set of all the optimal decision variables changes, but the PF remains the same.
- Type II: Both PS and PF change.
- Type III: PS does not change whereas PF changes.
- Type IV: Both PS and PF do not change, but the problem can change.

Independently of the DMOP variant, traditional EMO algorithms must be adapted to transform them into some kind of DEMO to solve them. Our chosen DMOP is a bi-objective formulation of the TSP where two goals, distance and traveling time, have to be minimized. We assume that the nodes remain fixed and that there are variations in the arcs due to changes in the traffic, such as bottlenecks, traffic jams or cuts in some streets that may affect the traveling time and the distance between some places. This problem would fit in the Type II category.

3 Architecture Components

In this section, the proposed architecture is described, giving details of the two main components; jMetal Framework and Spark.

3.1 Proposed Architecture

We focus our research on a context in which: (1) the data defining the DMOP are produced in a continuous, but not necessarily constant rate (i.e. they are produced in streaming), (2) they can be generated by different sources, and (3) they must be processed to clear any wrong or inconsistent piece of information. These three features cover the velocity, variety and veracity of Big Data applications. The last V, the volume, will depend on the amount of information that can be obtained to be processed.

With these ideas in mind we propose the software solution depicted in Fig. 1. The dynamic problem (multi-objective TSP or MSTP) is being continuously dealt with a multi-objective algorithm (dynamic NSGA-II), which stores in an external file system the found Pareto front approximations. In parallel, a component reads information from data sources in streaming and updates the problem.

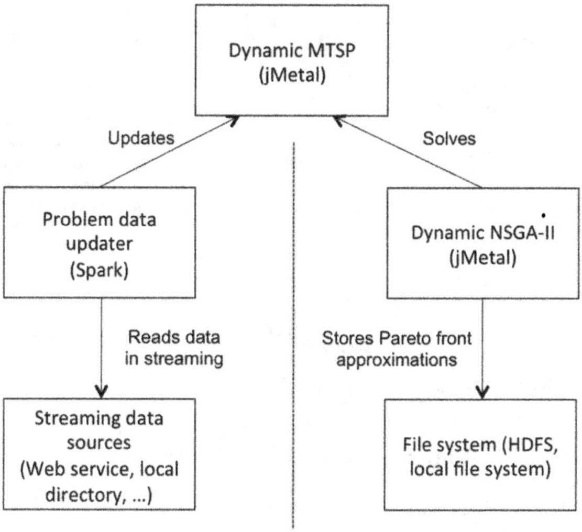

Fig. 1. Architecture of the proposed software solution.

To implement this architecture, the problem and the algorithm are jMetal objects, and the updater component is implemented in Spark. We provide details of all these elements next.

3.2 The jMetal Framework

jMetal is Java object-oriented framework aimed at multi-objective optimization with metaheuristics [10]. It includes a number of algorithms representative of the state-of-the-art, benchmark problems, quality indicators, and support for carrying out experimental studies. These features have made jMetal a popular tool in the field of multi-objective optimization. In this work, we use jMetal 5 [14], which has the architecture depicted in Fig. 2. The underlying idea in this framework is that an algorithm (metaheuristic) manipulates a number of solutions with some operators to solve an optimization problem.

Focusing on the `Problem` interface, it provides methods to know about the basic problem features: number of variables, number of objectives, and number of constraints. The main method is `evaluate()`, which contains the code implementing the objective functions that are computed to evaluate a solution.

Among all the MOPs provided by jMetal there exist a `MultiobjectiveTSP` class, having two objectives (distance and cost) and assuming that the input data are files with TSPLIB [15] format. We have adapted this class to be used in a dynamic context, which has required three changes:

1. Methods for updating part or the whole data matrices have to be incorporated. Whenever one of them is invoked, a flag indicating a data change must be set.

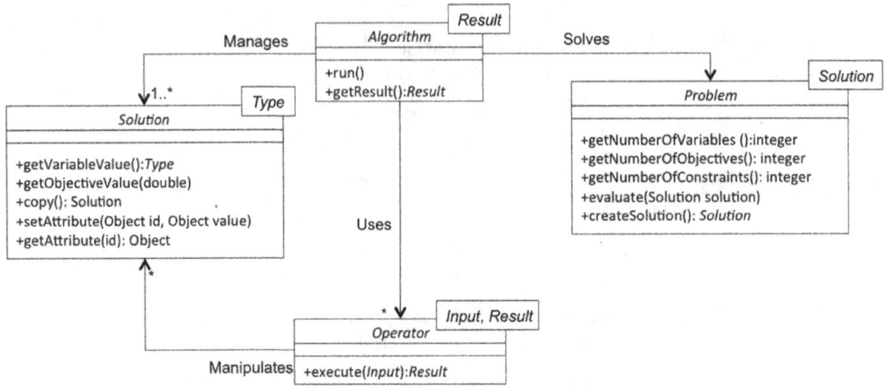

Fig. 2. jMetal 5.0 architecture.

2. The aforementioned methods and the `evaluate()` method have to be tagged as synchronized to ensure the mutual exclusion when accessing to the problem data.
3. A method is needed to get the status of the data changed flag and another one to reset it.

In order to explain how to adapt an existing EMO algorithm to solve a DMOP, we describe next the steps that are required to develop a dynamic version of the well-known NSGA-II algorithm [16]. A feature of jMetal 5 is the inclusion of algorithm templates [14] that mimic the pseudo-code of a number of multi-objective metaheuristics. As an example, the `AbstractEvolutionaryAlgorithm` template (an abstract class) contains a `run()` method that includes the steps of a generic EMO algorithm as can be observed in the following code:

```
@Override public void run() {
  List<S> offspringPopulation;
  List<S> matingPopulation;

  population = createInitialPopulation();
  population = evaluatePopulation(population);
  initProgress();
  while (!isStoppingConditionReached()) {
    matingPopulation = selection(population);
    offspringPopulation = reproduction(matingPopulation);
    offspringPopulation =
                   evaluatePopulation(offspringPopulation);
    population = replacement(population, offspringPopulation);
    updateProgress();
  }
}
```

The implementation of NSGA-II follows this template, so it defines the methods for creating the initial population, evaluating the population, etc. To implement a dynamic variant of NSGA-II (DNSGA-II), the class defining it can inherit from the NSGA-II class and only two methods have to be redefined:

- `isStoppingConditionReached()`: when the number of function evaluations reaches its limit (stopping condition), a file with the Pareto front approximation found is written, but instead of terminating the algorithm, a re-start operation is carried out and the algorithm begins again.
- `updateProgress()`: after an algorithm iteration, a counter of function evaluations is updated. In the DNSGA-II code, the data changed flag of the problem is checked. If the result is positive, the population is re-started and evaluated and the flag is reset.

3.3 Apache Spark

Apache Spark is a general-purpose distributed computing system [3] based on the concept of Resilient Distributed Datasets (RDDs). RDDs are collections of elements that can be operated in parallel on the nodes of a cluster by using two types of operations: transformations (e.g. map, filter, union, etc.) and actions (e.g. reduce, collect, count, etc.). Among the features of Spark (high level parallel processing programming model, machine learning algorithms, graph processing, multi-programming language API), we use here its support for streaming processing data. In this context, Spark manages the so called `JavaDStream` structures, which are later discretized into a number of RDDs to be processed.

There are a number of streaming data sources that Spark can handle. In our proposal, we choose as a source a directory where the new incoming data will be stored. We assume that a daemon process is iteratively fetching the data from a Web service and writing it to that source directory. The data will have the form of text files where each line has the following structure: a symbol 'd' (distance) or 't' (travel time), two integers representing the coordinates of the point, and the new distance/time value.

The pseudo-code of the Spark+jMetal solution is described next:

```
 1. DynamicMultiobjectiveTSP problem <- intializeProblem();
 2. File outputDirectory <- createOutputDirectory();
 3. DynamicNSGAII algorithm = initilizeAlgorithm(problem, outputDirectory);
 4. startAlgorith(algorithm);
 5. SparkConf sparkConf = new SparkConf().setAppName("SparkClient");
 6. JavaStreamingContext streamingContext =
      new JavaStreamingContext(sparkConf, Durations.seconds(5));
 7. JavaDStream<String> lines =
      streamingContext.textFileStream(inputDataDirectory);
 8. JavaDStream<Map<>> routeUpdates = lines.map(s -> {return parsed lines});
 9. routeUpdates.foreachRDD(
      s -> {list = s.collect();
      list.foreach(items -> {updateProblem(problem, items.next()})});

10. streamingContext.start();
11. streamingContext.awaitTermination();
```

The first steps initialize the problem with the matrices containing data of distance and travel time, and to create the output directory where the Pareto front approximations will be stored (lines 1 and 2). Then the algorithm is created and its execution is started in a thread (lines 3 and 4). Once the algorithm is running, it is the turn to start Spark, what requires two steps: creating a `SparkConf` object (line 5) and a `JavaStreamingContext` (line 6), which indicates the polling frequency (5 s in the code).

The processing of the incoming streaming files requires three instructions: first, a text file stream is created (line 7), which stores in a `JavaDStream<String>` list all the lines of the files arrived to the input data directory since the last polling. Second, a map transformation is used (line 8) to parse all the lines read in the previous step. The last step (line 9) consists in executing a `foreachRDD` instruction to update the problem data with the information of the parsed lines.

The two last instructions (lines 10 and 11) start the Spark streaming context and await for termination. As a result, the code between lines 7 and 9 will be iteratively executed whenever new data files have been written in the input data directory since the last polling.

We would like to remark that this pseudo-code mimics closely the current Java implementation, so no more than a few lines of code are needed.

4 Case Study: Dynamic Bi-objective TSP

To test our software architecture in practice we apply it to two scenarios: an artificial dynamic TSP (DTSP) with benchmark data and another version of the problem with real data. We analyze both scenarios next.

4.1 Problem with Synthetic Data

Our artificial DTSP problem is built from the data to two 100 TSP instances taken from TSPLIB [15]. One instance represents the distances and the other one the travel time. To simulate the modification of the problem, we write every 5 s a data file containing updated information.

The parameter settings of the dynamic NSGA-II algorithm are the following: the population size is 100, the crossover operator is PMX (applied with a 0.9 probability), the mutation operator is swap (applied with a probability of 0.2), and the algorithm computes 250,000 function evaluations before writing out the found front and re-starting. As development and target computer we have used a MacBook Pro laptop (2,2 GHz Intel Core i7 processor, 8 GB RAM, 256 GB SSD) with MacOS 10.11.3, Java SE 1.8.0_40, and Spark 1.4.1.

Figure 3 depicts some fronts produced by the dynamic NSGA-II throughout a given execution, starting from the first one (`FUN0.tsv`) up to the 20th one (`FUN20.tsv`). We can observe that the shape of the Pareto front approximations change in time due to the updating operation of the problem data. In fact, each new front contains optimized solutions with regards to the previous

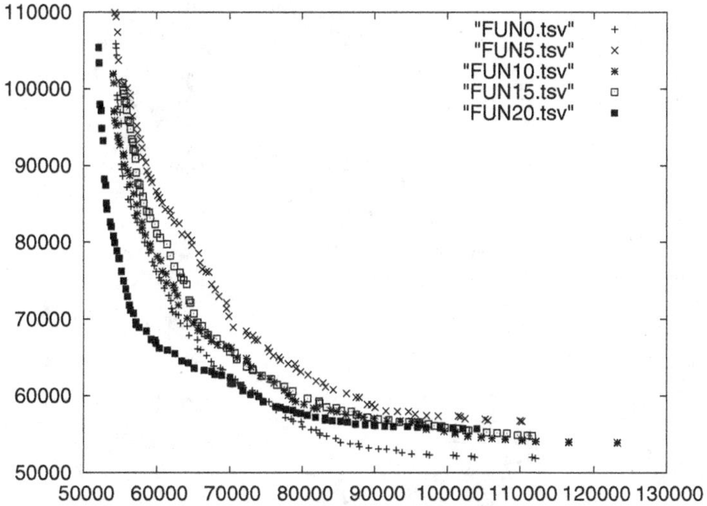

Fig. 3. Evolution of the Pareto front approximations yielded by the dynamic NSGA-II algorithm (FUNx.tsv refers to the xth front that have been computed).

ones, which lead us to suggest that the learning model of the optimization procedure is kept through different re-starts, although detecting the changes in the problem structure.

Once we have tested our approach on a synthetic instance, we now aim at managing real-world data, in order to show whether the proposed model is actually applicable or not.

4.2 Problem with Real Data

The DTSP with real-world data we are going to define is based on the Open Data provided by the New York City Department of Transportation, which updates the traffic information several times per minute[2]. The information is provided as a text file where each line includes, among other data:

- `Id`: link identifier (an integer number)
- `Speed`: average speed a vehicle traveled between end points on the link in the most recent interval (a real number).
- `TravelTime`: average time a vehicle took to traverse the link (a real number).
- `Status`: if the link is closed by accidents, works or any cause (a boolean).
- `EncodedPolyLine`: an encoded string that represents the GPS Coordinates of the link. It is encoded using the Google's Encoded Polyline Algorithm [17].
- `DataAsOf`: last time data was received from link (a date).
- `LinkName`: description of the link location and end points (a string).

[2] At the time of writing this paper, the data can be obtained from this URL: http://207.251.86.229/nyc-links-cams/LinkSpeedQuery.txt.

Fig. 4. Real dynamic TSP instance: 93 nodes of the city of New York.

As the information is given in the form of links instead of nodes, we have made a pre-processing of the data to obtain a feasible TSP. To join the routes, we have compared the GPS coordinates. A communication between two nodes is created when there is a final point of a route and a starting point of another in the same place (with a small margin error). After that, we iterate through all the nodes removing those with degree <2. The list of GPS coordinates is decoded from the `EncodedPolyLine` field of each link.

As as result, the DTSP problem is composed of 93 locations and 315 communications between them, which are depicted in Fig. 4. We have to note that the links are bi-directional, so the resulting DTSP is asymmetric. In this regard, it is worth mentioning that we have approached the TSP by following a vector permutation optimization model, as commonly done in population based metaheuristics.

To determine the distance between points, we have used a Google service [18] that, given two locations, it returns the distance between them; this step only is carried out to initialize the graph. The initial travel time and speed are obtained from the first data file read.

As in the synthetic DTSP, a daemon process is polling the data source every 30 s, parsing the received information, and writing the updates directly into the cost and distance matrices that Spark is using to calculate the results. If a route has a status 1, meaning that it has been closed to circulation, we assume an infinite cost and distance for that route.

After running our software for 40 min, the number of data requests to the traffic service was 77, from which 37 returned no information. In those cases

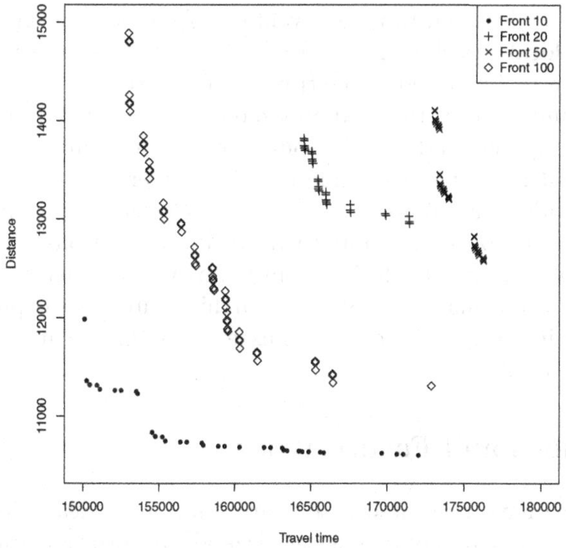

Fig. 5. Pareto front approximations obtained when solving the real data DTSP (Front x refers to the xth front that have been obtained).

where there was new information, the number of updates ranged between 141 and 238.

Figure 5 shows the Pareto front approximations obtained after 10, 20, 50, and 100 runs of the dynamic NSGA-II algorithm. Similarly to the obtained results in the synthetic instance, in this case the fronts of solutions are refined through the optimization process. Therefore, the learning model also is kept when dealing with real-world data.

4.3 Discussion

An analysis at a glance of our architecture could indicate that the use of Spark does not justify selecting it to be included in our system, because we are not taking advantage of many of its features. However, we would like to note what we are presenting here is a first approximation to the problem and many extensions are possible. For example, thinking in Smart City applications and our considered optimization problem, more data sources could be used for a more precise problem formulation, such as weather forecasting, social networks, GPS data from mobile phones in cars, etc. This would require more demanding storage (e.g. in HDFS) and more computing power for processing and integrating all the data (e.g. a Hadoop cluster), and in this scenario not only the streaming feature of Spark will be useful, but its high performance computing capabilities.

From the optimization point of view, we have presented a simple adaptation of a well-known algorithm, NSGA-II, to solve the dynamic TSP problem. In case of defining a more realistic problem, with more data sources, a more sophisticated algorithm could be designed to take advantage of problem-specific data, e.g. by incorporating a local search or new variation operators. Furthermore, we must consider that in a real scenario, if the traveler starts to visit the nodes while the application is running, the traversed arcs should be removed from the problem, so it would be simplified and then it would not be a TSP anymore.

It is clear that a complete solution of the considered optimization problem would require additional components, such as a visualization module able of displaying the evolution of the Pareto front approximations and enabling to choose a particular trade-off solution from them. We consider that this kind of components are orthogonal to our software architecture. As the produced fronts are being stored in secondary storage, an external program can be developed to display them as needed.

5 Conclusions and Future Work

In this paper, we have presented a software solution to deal with a Big Data Optimization problem by combining the jMetal optimization framework with the Spark cluster computing system and we have demonstrated how to apply it to solve a concrete example: a dynamic multi-objective TSP.

Our motivation has been driven by the availability of Open Data sources, the raise of Spark as distributed computing platform in clusters and Hadoop systems, and the utilization of the jMetal framework to provide the infrastructure to deal with multi-objective optimization problems.

We have presented two case studies that consider a bi-objective formulation of the problem, where the total distance and time travel are goals to be minimized. First, a synthetic version of the problem, based on benchmark data, to test the working of the system; second, a real instance created from Open Data of the city of New York. In this latter case, the TSP nodes correspond to real locations, and the problem data is updated in streaming.

Defining more realistic problems, including additional data sources, as well as considering other Smart city related problems are matters of future works.

Acknowledgments. This work is partially funded by Grants TIN2011-25840 (Ministerio de Ciencia e Innovación) and P11-TIC-7529 and P12-TIC-1519 (Plan Andaluz de Investigación, Desarrollo e Innovación). Cristóbal Barba-González is supported by Grant BES-2015-072209 (Ministerio de Economía y Competitividad).

References

1. Editorial: Community cleverness required. Nature **455**, 1 (2008)
2. White, T.: Hadoop: The Definitive Guide, 1st edn. O'Reilly Media Inc., Sebastopol (2009)
3. Zaharia, M., Chowdhury, M., Franklin, M., Shenker, S., Stoica, I.: Spark: cluster computing with working sets. In: Proceedings of the 2nd USENIX Conference on Hot Topics in Cloud Computing, Berkeley, CA, USA, HotCloud 2010, pp. 10. USENIX Association (2010)

4. Marr, M.: Big Data: Using SMART Big Data Analytics and Metrics to Make Better Decisions and Improve Performance. Wiley, Hoboken (2015)
5. Nam, T., Pardo, T.: Smart city as urban innovation: focusing on management, policy, and context. In: Proceedings of the 5th International Conference on Theory and Practice of Electronic Governance, ICEGOV 2011, pp. 185–194. ACM (2011)
6. Garcia-Nieto, J., Olivera, A., Alba, E.: Optimal cycle program of traffic lights with particle swarm optimization. IEEE Trans. Evol. Comput. **17**, 823–839 (2013)
7. NYCDOT: New York City traffic speed detectors data set (2016). http://nyctmc.org
8. Papadimitriou, C.H.: The Euclidean travelling salesman problem is NP-complete. Theor. Comput. Sci. **4**, 237–244 (1977)
9. Gartner Inc.: Gartner's hype cycle for advanced analytics and data science (2015). https://www.gartner.com/doc/3087721/hype-cycle-advanced-analytics-data
10. Durillo, J., Nebro, A.: jMetal: a java framework for multi-objective optimization. Adv. Eng. Softw. **42**, 760–771 (2011)
11. Deb, K.: Multi-objective Optimization Using Evolutionary Algorithms. Wiley, New York (2001)
12. Coello, C., Lamont, G., van Veldhuizen, D.: Multi-objective Optimization Using Evolutionary Algorithms, 2nd edn. Wiley, New York (2007)
13. Farina, M., Deb, K., Amato, P.: Dynamic multiobjective optimization problems: test cases, approximations, and applications. IEEE Trans. Evol. Comput. **8**, 425–442 (2004)
14. Nebro, A., Durillo, J.J., Vergne, M.: Redesigning the jMetal multi-objective optimization framework. In: Proceedings of the Companion Publication of the 2015 Annual Conference on Genetic and Evolutionary Computation, GECCO Companion 2015, pp. 1093–1100. ACM, New York (2015)
15. Reinelt, G.: TSPLIB - a traveling salesman problem library. INFORMS J. Comput. **3**, 376–384 (1991)
16. Deb, K., Pratap, A., Agarwal, S., Meyarivan, T.: A fast and elitist multiobjective genetic algorithm: NSGA-II. IEEE Trans. Evol. Comput. **6**, 182–197 (2002)
17. Google Inc.: Encoded polyline algorithm format (2016). https://developers.google.com/maps/documentation/utilities/polylinealgorithm
18. Google Inc.: Google maps distance matrix API (2016). https://developers.google.com/maps/documentation/distance-matrix

Feature Selection via Co-regularized Sparse-Group Lasso

Paula L. Amaral Santos(✉), Sultan Imangaliyev, Klamer Schutte, and Evgeni Levin

TNO Research, The Hague, The Netherlands
paula.amrl@gmail.com, {sultan.imangaliyev,klamer.schutte, evgeni.levin}@tno.nl
http://www.tno.nl

Abstract. We propose the co-regularized sparse-group lasso algorithm: a technique that allows the incorporation of auxiliary information into the learning task in terms of "groups" and "distances" among the predictors. The proposed algorithm is particularly suitable for a wide range of biological applications where good predictive performance is required and, in addition to that, it is also important to retrieve all relevant predictors so as to deepen the understanding of the underlying biological process. Our cost function requires related groups of predictors to provide similar contributions to the final response, and thus, guides the feature selection process using auxiliary information. We evaluate the proposed method on a synthetic dataset and examine various settings where its application is beneficial in comparison to the standard lasso, elastic net, group lasso and sparse-group lasso techniques. Last but not least, we make a python implementation of our algorithm available for download and free to use (Available at www.learning-machines.com).

Keywords: Sparse models · Co-regularized learning · Systems biology

1 Introduction

The advent of high-throughput technologies has been redefining the approach used to describe and understand biological mechanisms. As the amount of available data increases, it encourages a transition from knowledge- to data-driven modeling of biological phenomena. In this context, machine learning techniques and especially supervised learning methods have been widely used to elucidate biological processes and to generate models that are able to predict a certain health outcome based on a set of predictors. There are, however, still many challenges in building meaningful predictive models based on biological data. A well-known example is the great unbalance between the number of predictors p and the number of examples n ($p >> n$), which makes feature selection both a desire and a necessity in order to reduce the amount of required data to be collected, facilitate visualization and understanding of results, and improve

P.M. Pardalos et al. (Eds.): MOD 2016, LNCS 10122, pp. 118–131, 2016.
DOI: 10.1007/978-3-319-51469-7_10

predictive performance [1,2]. One interesting fact related to feature selection in the biological domain is that the available predictors, despite of being many, are often not able to fully describe the targeted response (y). This lack of information about the complete system leads to an "irreducible error" [11], which raises the challenge of making sure that all relevant predictors are selected, even when their contribution to y is of the order of the irreducible error (or any other error source). One way of tackling this problem is to make use of auxiliary information when learning the feature selection model. This can be done for example by introducing into the model information about the relationship among sets of predictors, such as arranging them in groups, as it was proposed initially by Yuan et al. [3] and later refined by Simon et al. [4], leading to the implementation of the group-lasso and the sparse-group lasso, respectively. When it comes to relationships among predictors, there is often more information available that could be used to guide the learning process such as expected "related behavior" of a set of groups. In the data-rich "-omics" fields (genomics, proteomics and metabolomics), for example, predictors can be organized in groups that are expected (or not) to be related to a given phenomenon in the same way. In metagenomics, specifically, microorganisms can be grouped based on a phylogenetic tree that depicts their similarities regarding genetic or physical characteristics [8].

Motivated by the possibility of including more detailed relationships among a set of predictors, we propose a modification to the sparse-group lasso by imposing co-regularization of related groups. We assume that there is a known relationship among the predictors and use it to (a) divide them into groups and (b) determine the relative "distance" or similarity between each pair of groups. We then impose that each group contributes individually to the prediction of the response variable and that groups that are closely related should provide more similar contributions. This constraint is introduced as a bias in the cost function by enforcing the predictions of pairs of groups to be similar under L_2 norm.

The contributions of this work include the proposal of a novel regression method that allows the incorporation of auxiliary information in the form of relationships between groups of predictors. We make the implementation of our algorithm in python available for download and free to use and evaluate the proposed co-regularization method on a synthetic dataset.

This paper is organized as follows: We present the notation that will be used throughout the paper in Sect. 2, discuss the related work in more details in Sect. 3, present the co-regularization method in Sect. 4 and describe the evaluation of the proposed method in Sect. 5. Finally, we summarize our contributions and discuss future work in Sect. 6.

2 Notation

2.1 General Notation

Let $S = (\mathbf{X}, \mathbf{y})$ denote a data set where $\mathbf{y}$ is the $n \times 1$ response vector and $\mathbf{X}$ is the $n \times p$ predictor matrix, where n is the number of observations and p the number of predictors. After a location and scale transformation, we can assume that the response vector is centered and the predictors are standardized. We will seek to generate a model fitting procedure that produces the vector of coefficients $\boldsymbol{\beta}$ ($p \times 1$) that relates the predictor matrix $\mathbf{X}$ to the response vector $\mathbf{y}$.

Depending on the methodology being discussed, different types of vector norms will be introduced in the cost function. In this case, the L_a norm of a vector $\mathbf{v}$ is defined as:

$$\|\mathbf{v}\|_a = \left(\sum_i |v_i|^a\right)^{1/a}, \; for\ (a = 1, 2, ...) \tag{1}$$

For all discussed methods, α and λ are non-negative regularization parameters that control the amount of induced sparsity in the vector of coefficients and the complete regularization term, respectively.

2.2 Group Notation

In the case in which the predictors are divided into M groups, we re-write:

- the predictor matrix $\mathbf{X} = (\mathbf{X}^1|...|\mathbf{X}^M)$, where each $\mathbf{X}^{(v)}$ is a $n \times p^{(v)}$ sub-matrix, where $p^{(v)}$ is the number of predictors in group v
- the vector of coefficients: $\boldsymbol{\beta} = (\beta_1, ..., \beta_p) = (\boldsymbol{\beta}^{(1)}|...|\boldsymbol{\beta}^{(M)})$, where $\boldsymbol{\beta}^{(v)}$ ($p^{(v)} \times (1)$) is the vector of coefficients of the predictors of group v.

3 Previous Work

In this section, we present a brief review of four methods currently available for performing feature selection on data sets: the lasso [12], the elastic net [13], the group lasso [3] and the sparse group lasso [4,6].

The lasso, proposed by Tibshirani [12] in 1996, performs both shrinkage and variable selection simultaneously due to the presence of the L_1-norm in its cost function:

$$L_{lasso}(\alpha, \boldsymbol{\beta}) = \frac{1}{2n}\|\mathbf{y} - \mathbf{X}\boldsymbol{\beta}\|_2^2 + \alpha\|\boldsymbol{\beta}\|_1 . \tag{2}$$

The lasso has two major limitations: (a) in the case where $p > n$, the lasso is said to "saturate", leading to the selection of maximum n variables; (b) when the dataset is composed by groups of predictors that have strong pair-wise correlation, the lasso has the tendency to keep one of the predictors (chosen at random) in the final model and eliminate the others, which is often not desirable in biological applications.

The elastic net is a compromise between the lasso and the ridge penalties [12–14] and has the form:

$$L_{EN}(\alpha, \lambda, \boldsymbol{\beta}) = \frac{1}{2n}\|\mathbf{y} - \mathbf{X}\boldsymbol{\beta}\|_2^2 + \lambda\alpha\|\boldsymbol{\beta}\|_1 + \lambda(1-\alpha)\|\boldsymbol{\beta}\|_2 , \tag{3}$$

where the term $||\boldsymbol{\beta}||^2$ encourages highly correlated features to have similar coefficients, whereas $||\boldsymbol{\beta}||_1$ encourages a sparse solution. Another advantage of the elastic net penalty over the lasso is that it does not saturate when the number of predictors is larger than the number of samples. Therefore, it does not limit the selected predictors in the final model to $min(n, p)$. The elastic net penalty has been widely used for performing feature selection in biological data [2,9,10]. However, it does not allow for the incorporation of any auxiliary information about the relationship among predictors.

Yuan and Lin proposed in 2007 the group lasso penalty [3], which assumes that predictors belong to predefined groups (e.g. genes that belong to the same pathway) and supposes that it might be desirable to shrink and select the predictors in the same group together. The cost function of the group lasso penalty is given by:

$$L_{GL}(\lambda, \boldsymbol{\beta}) = \frac{1}{2n}\left\|\mathbf{y} - \sum_{l=1}^{M}\mathbf{X}^{(l)}\boldsymbol{\beta}^{(l)}\right\|_2^2 + \lambda\sum_{l=1}^{M}\sqrt{p^{(l)}}\left\|\boldsymbol{\beta}^{(l)}\right\|_2. \tag{4}$$

One of the drawbacks of the group lasso lies on the impossibility of "within group sparsity": if one group of predictors is selected, all the coefficients within the group are non-zero.

The sparse-group lasso combines the lasso [12] and the group lasso [3] in order to provide sparse effects both on a group and within group level:

$$\begin{aligned} L_{SGL}(\alpha, \lambda, \boldsymbol{\beta}) = \frac{1}{2n}\left\|\mathbf{y} - \sum_{l=1}^{M}\mathbf{X}^{(l)}\boldsymbol{\beta}^{(l)}\right\|_2^2 \\ + \alpha\lambda||\boldsymbol{\beta}||_1 + (1-\alpha)\lambda\sum_{l=1}^{M}\sqrt{p^{(l)}}\left\|\boldsymbol{\beta}^{(l)}\right\|_2, \end{aligned} \tag{5}$$

where $\alpha \in [0, 1]$ and $\alpha = 0$ gives the group lasso fit and $\alpha = 1$ gives the lasso fit. The sparse group lasso and elastic net penalties look very similar, differing mostly due to the presence of the L_2-norm in the third term of the sparse group lasso penalty, which is not differentiable at $\mathbf{0}$, allowing complete groups to be zeroed out. It was shown, however, that within each non-zero group, the sparse group lasso gives an elastic net fit [4].

Other methods have also been proposed to allow more flexibility in terms of the definition of the relationship between predictors. One of them is the overlapping group lasso, which, as in the group lasso and the sparse-group lasso, divides the predictions into groups that can potentially overlap [7].

In short, the methods mentioned above either do not make use of any auxiliary information or do so by imposing a penalty that depends on the contribution of individual groups to the response. None of them, however, allows to incorporate auxiliary information about the relationship between different groups. Our contribution is to propose a method that exploits the relationship among groups of predictors in an attempt to improve the ability of algorithms to retrieve relevant features based on the assumption that predictors belonging to groups that are closely related should provide similar contributions to the final response.

4 Co-regularized Sparse Group Lasso

To allow for the co-regularization of closely related groups, we propose the following cost function:

$$L_{CSGL}(\boldsymbol{\beta}, \alpha, \lambda, \boldsymbol{\Gamma}) = \frac{1}{2n}\left\| \mathbf{y} - \sum_{l=1}^{M} \mathbf{X}^{(l)}\boldsymbol{\beta}^{(l)} \right\|_2^2 + \frac{1}{2n}\sum_{l,v=1}^{M} \gamma_{l,v}\left\| \mathbf{X}^{(l)}\boldsymbol{\beta}^{(l)} - \mathbf{X}^{(v)}\boldsymbol{\beta}^{(v)} \right\|_2^2 + \alpha\lambda\|\boldsymbol{\beta}\|_1 + (1-\alpha)\lambda\sum_{l=1}^{M}\sqrt{p^{(l)}}\left\|\boldsymbol{\beta}^{(l)}\right\|_2, \tag{6}$$

where $\gamma_{l,v}$ is the co-regularization coefficient between groups l and v, which can assume any value equal or greater than zero.

Given M groups of predictors, the co-regularization coefficients can be arranged in the matrix $\boldsymbol{\Gamma}$ $(M \times M)$ defined as:

$$\boldsymbol{\Gamma} = \begin{pmatrix} \gamma_{1,1} & \gamma_{1,2} & \cdots & \gamma_{1,M} \\ \gamma_{2,1} & \gamma_{2,2} & \cdots & \gamma_{2,M} \\ \vdots & \vdots & \ddots & \vdots \\ \gamma_{M,1} & \gamma_{M,2} & \cdots & \gamma_{M,M} \end{pmatrix} \tag{7}$$

where $\gamma_{1,1}, \gamma_{2,2}, ..., \gamma_{M,M}$ have no effect, since the co-regularization term will always go to zero when $l = v$.

The proposed cost function is an extension of the sparse-group lasso by including the co-regularization term: $\frac{1}{2n}\sum_{l,v=1}^{M}\gamma_{l,v}\left\| \mathbf{X}^{(l)}\boldsymbol{\beta}^{(l)} - \mathbf{X}^{(v)}\boldsymbol{\beta}^{(v)} \right\|_2^2$. This term is analogous to a multi-view problem [16,17], in which the prediction given by different views representing the same phenomenon should be similar to each other. In the case of groups defined by domain knowledge, this term works by encouraging the intrinsic relationship among predictors to be transferred to their contribution in the final prediction. The value chosen for $\gamma_{l,v}$ will therefore reflect the extent to which two groups l and v are expected to give similar contributions to $\mathbf{y}$, where $\gamma_{l,v} = 0$ represents the case where no similarity at all is expected. Ideally, domain knowledge should be the main driver to determine the value of each element of the matrix $\boldsymbol{\Gamma}$. Fine-tuning of values can be done by methods such as grid search, but this procedure can become computationally extensive depending of the number of groups of predictors. Therefore, the use of domain knowledge can help decrease the number of degrees of freedom by setting to zero the co-regularization term between two groups that are not expected to be related.

As for the regularization parameters α and λ, their roles remain the same as in the sparse-group lasso method: α controls the balance between group sparsity and within group sparsity, whereas λ controls the contribution of the regularization terms to the cost function.

In order to fit the co-regularized sparse-group lasso, we chose to follow the same procedure suggested by Simon et al. [4] to fit the sparse-group lasso. This can be done since the change introduced in the cost function do not alter its properties. Firstly, it is not differentiable due to the presence of the L_1-norm. Therefore, proximal gradient methods [15] can be used in the weight-update procedure. Secondly, the penalty remains convex and separable between groups and, thus, a global minimum can be found by block-wise gradient descent [4–6]. Due to the similarity to the sparse-group lasso, we chose to omit detailed derivations that can be found elsewhere [4,6] and focus on what has to be adapted to take the co-regularization term into account.

4.1 The Proximity Operator for the Co-regularized SGL

The proximal gradient method consists of splitting the cost function in two terms f and g, the former being differentiable. The proximal gradient method for our cost function is given by:

$$\boldsymbol{\beta}_{k+1} := \mathbf{prox}_g(\boldsymbol{\beta}_k - t\nabla f(\boldsymbol{\beta}_k)) \,. \tag{8}$$

$$f = \frac{1}{2n}\left\|\mathbf{y} - \sum_{l=1}^{M} \mathbf{X}^{(l)}\boldsymbol{\beta}^{(l)}\right\|_2^2 + \frac{1}{2n}\sum_{l,v=1}^{M} \gamma_{l,v}(\mathbf{X}^{(l)}\boldsymbol{\beta}^{(l)} - \mathbf{X}^{(v)}\boldsymbol{\beta}^{(v)})^2 \,. \tag{9}$$

$$g = \lambda(\alpha\|\boldsymbol{\beta}\|_1 + (1-\alpha)\|\boldsymbol{\beta}\|^2) \,. \tag{10}$$

where $k, t > 0$ represent respectively each iteration of the minimization problem and an arbitrary step size. Because the co-regularization term is differentiable, $\mathbf{prox}_g(b)$ is the same as for the standard sparse group lasso, which is given by the soft-thresholding operator multiplied by a shrinkage factor [15]:

$$\mathbf{prox}_g(b) = \frac{1}{1+\lambda(1-\alpha)} \times S(b, \alpha\lambda) \,. \tag{11}$$

$$S(b, \alpha\lambda) = \begin{cases} b - \lambda\alpha sign(b) & \text{if } |b| > \lambda\alpha \,. \\ 0 & \text{if } |b| \leqslant \lambda\alpha \,. \end{cases} \tag{12}$$

As for the differentiable part, its gradient is composed by the gradient of the unpenalized cost function (the same for the GL and the SGL) and the gradient of the co-regularization term, both being group-dependent. Therefore, it is useful to re-write Eq. 8 for the predictors of each group l:

$$\boldsymbol{\beta}_{k+1}^{(l)} := \mathbf{prox}_g(\boldsymbol{\beta}_k^{(l)} - t\nabla_{(l)} f(\boldsymbol{\beta}_k)) := \mathbf{prox}_g(\mathbf{b}_{k+1}^{(l)}) \tag{13}$$

$$\nabla_{(l)} f(\boldsymbol{\beta}_k) = \nabla_{(l)} f_{ls}(\boldsymbol{\beta}_k) + \nabla_{(l)} f_{cR}(\boldsymbol{\beta}_k)\,. \tag{14}$$

$$\nabla_{(l)} f_{ls}(\boldsymbol{\beta}_k) = -\frac{1}{n}(\mathbf{X}^{(l)T}\mathbf{r}_{(-l)} - \mathbf{X}^{(l)T}\mathbf{X}^{(l)}\boldsymbol{\beta}^{(l)})\,. \tag{15}$$

$$\nabla_{(l)} f_{cR}(\boldsymbol{\beta}_k) = \frac{2}{n}\mathbf{X}^{(l)T}\sum_{v=1}^{M}\gamma_{l,v}(\mathbf{X}^{(l)}\boldsymbol{\beta}^{(l)} - \mathbf{X}^{(v)}\boldsymbol{\beta}^{(v)})\,. \tag{16}$$

where $\mathbf{r}_{(-l)}$ is the residual of $\mathbf{y}$ due to the fit given by all groups, except group l:

$$\mathbf{r}_{(-l)} = \mathbf{y} - \sum_{v \neq l}^{M}\mathbf{X}^{(v)}\boldsymbol{\beta}^{(v)}. \tag{17}$$

In Eqs. 15 and 16, $\nabla_{(l)} f_{ls}$ and $\nabla_{(l)} f_{cR}$ represent the gradients of the standard least-squares penalty and of the co-regularization term, respectively. Combining Eqs. 13–16, the final update rule is obtained:

$$\boldsymbol{\beta}^{(l)}_{k+1} = \left(1 - \frac{t(1-\alpha)\lambda}{\left\|S(\mathbf{b}^{(l)}_{k+1}, t\alpha\lambda)\right\|_2}\right)_+ (S(\mathbf{b}^{(l)}_{k+1}, t\alpha\lambda))_+. \tag{18}$$

4.2 Criterion to Eliminate a Given Group

In Sect. 4.1, we presented the weight-update procedure. The formulation of the block-wise update, however, offers a shortcut to eliminate complete groups directly by applying the soft-thresholding operator to the residual $\mathbf{r}_{(-l)}$ [4,6]. At a given iteration, if the residual without a group l is smaller than a threshold, the whole group is eliminated. This condition can be formulated as:

$$\left\|S(\mathbf{X}^{(l)T}\mathbf{r}_{(-l)}, \alpha\lambda)\right\|_2 \leqslant (1-\alpha)\lambda \Rightarrow \boldsymbol{\beta}^{(l)} \leftarrow \mathbf{0}. \tag{19}$$

Below we summarize the procedure to minimize the proposed cost function.

Algorithm 1. Co-regularized Sparse Group Lasso

Require: $\mathbf{y}$, $\mathbf{X} = (\mathbf{X}^{(1)}, ..., \mathbf{X}^{(M)})$, regularization parameters λ, α, co-regularization matrix $\boldsymbol{\Gamma}$ and maximum number of iterations k_{max}.
Ensure: $\boldsymbol{\beta}_0 = (\boldsymbol{\beta}^{(1)}_0|...|\boldsymbol{\beta}^{(M)}_0) \leftarrow \mathbf{0}$.
1: **while** number of iterations $k < k_{max}$ **do**
2:     **for** each group $l = 1, ..., M$ **do**
3:         Check if group l can be eliminated by checking condition (19)
4:         **if** yes **then**
5:             $\boldsymbol{\beta}^{(l)} \leftarrow \mathbf{0}$ and proceed to the next group
6:         **else**
7:             until convergence, update $\boldsymbol{\beta}^{(l)}$ using the update rule (18)

---

5 Experiments

We applied the proposed method to a synthetic dataset and compared it to the standard group lasso, to the standard lasso and to the elastic net. Since the problem we aim to address is the ability to retrieve all relevant predictors, different methods were compared primarily based on the selected predictors, where the aim was to select the groups of predictors known to be relevant and eliminate the predictors known to be irrelevant. Secondarily, the methods were also compared in terms of predictive performance via the computation of the mean squared error. Even though achieving higher predictive performance was never the main goal, a good feature selection procedure should not cause it to significantly decrease.

5.1 Determination of Regularization Parameters

For every method analyzed in this section, the optimal value of the regularization parameters was determined via 5-fold cross-validation. For the co-regularized method, the matrix $\boldsymbol{\Gamma}$ also needs to be specified. We chose for the simple case where $\gamma_{l,v} \in \{0, \gamma\}$, for any $l, v \in 1, 2, ..., M$, where $\gamma_{l,v} = 0$ means that groups v, l are not related and therefore should not be co-regularized and $\gamma_{l,v} = \gamma$ means that groups v, l are closely related and their co-regularization will be enforced to a degree dependent on the value of γ. The value of γ was determined as follows: a relatively small grid of values was chosen and a model was fitted for each of them. Finally, the quality of all models was analyzed regarding their ability to retrieve all relevant predictors in order to evaluate the effect of the co-regularization term.

It is important to mention that the algorithm for the co-regularized sparse-group lasso can be used to perform the standard sparse-group lasso and group lasso by simply setting $\gamma = 0$ and $\alpha = 0$, respectively, as it can be seen in Eq. 6.

5.2 Synthetic Data

We generated a predictor matrix of $n = 5000$ observations (4500 for training and 500 for testing) and $p = 50$ predictors divided into $M = 5$ groups. Each column of $\mathbf{X}$ was drawn from a normal distribution and later standardized to have zero mean and unit variance. The 5 groups were formed so that groups 1 and 2 are related to each other, groups 4 and 5 are related to each other and group 3 is not related to any other group. In this case, the matrix $\boldsymbol{\Gamma}$ is given by:

$$\boldsymbol{\Gamma} = \begin{pmatrix} \gamma\, \gamma\, 0\; 0\; 0 \\ \gamma\, \gamma\, 0\; 0\; 0 \\ 0\; 0\; \gamma\, 0\; 0 \\ 0\; 0\; 0\; \gamma\, \gamma \\ 0\; 0\; 0\; \gamma\, \gamma \end{pmatrix}. \tag{20}$$

Predictors were set to have 0.4 correlation with other predictors within the same group, 0.2 correlation with predictors of related groups and zero correlation with predictors of unrelated groups. Therefore, the correlation between two predictors $p_i^{(v)}, p_j^{(l)}$ from groups v and l is given by:

$$c_{v,l} = \begin{cases} 0.4 & \text{if } v = l\ . \\ 0.2 & \text{if } v \neq l\ . \end{cases} \tag{21}$$

The response vector was constructed as follows:

$$\mathbf{y} = (1-\sigma)\sum_{l=1}^{5} \mathbf{X}^{(l)}\boldsymbol{\beta}^{(l)} + \sigma\epsilon, \tag{22}$$

where ϵ is a noise term also drawn from a normal distribution and $\sigma \in [0,1]$ controls the balance between noise and signal. In this case, we chose to work with a quite high value of $\sigma = 0.5$, since we want to test the ability of the proposed algorithm to retrieve all relevant groups in the presence of high levels of (irreducible) noise. To test the ability of our algorithm to select the right groups of predictors, the vector of coefficients were such that: $\boldsymbol{\beta}^{(1)} = [0.9, 0.9, ..., 0.9]$, $\boldsymbol{\beta}^{(2)} = [0.1, 0.1, ..., 0.1]$, $\boldsymbol{\beta}^{(3)} = [1, 1, ..., 1]$ and $\boldsymbol{\beta}^{(4)} = \boldsymbol{\beta}^{(5)} = [0, 0, ..., 0]$. Those vectors of coefficients make clear that predictors of groups 4 and 5, despite of being related to one another, do not contribute to the response, groups 1 and 3 are the most important groups for predicting the response and group 2 is relevant, but in a lower degree, which might lead to it not being selected due to the high noise level. It is also important to notice that in this construction there is no within-group sparsity. This choice was made to keep the experiments simple and focus on the selection of relevant groups, since the added value of the sparse-group lasso has already been shown elsewhere [4,6].

In the experiments that were conducted we aimed at showing that high levels of noise can mask the relevance of groups of predictors that contribute to the response. By means of co-regularization, we expect related groups to share their "prediction strength" so that groups that would have small, but relevant contribution to the response would gain more weight in the final model.

5.3 Results

By construction of the dataset, we would expect the perfect feature selection method to select the predictors of groups 1, 2 and 3 and eliminate the predictors from groups 4 and 5. In addition to that, we expect the selection of the predictors of group 2 to be challenging, since their contribution to the final outcome is of the order of the irreducible error.

The overall effect of the co-regularization parameter in this feature selection process can be seen in Fig. 1(a), which shows the weight of all 50 coefficients for the group lasso (GL) and for 2 values of γ (0.05 and 0.175). The weights obtained for the elastic net and the lasso were omitted, since they were overlapping with the ones given by the GL. For increasing values of γ, the computed weights for the predictors of group 2 increase, which shows that the imposing co-regularization leads to the intended effect of strengthening the contribution of groups of predictors that could be masked by noise.

Regarding performance, the increase of the co-regularization parameter leads to an increase of the mean-squared error (MSE) in the cross-validation, as it is shown in Fig. 1(b). This is not surprising, since γ introduces a balance in the

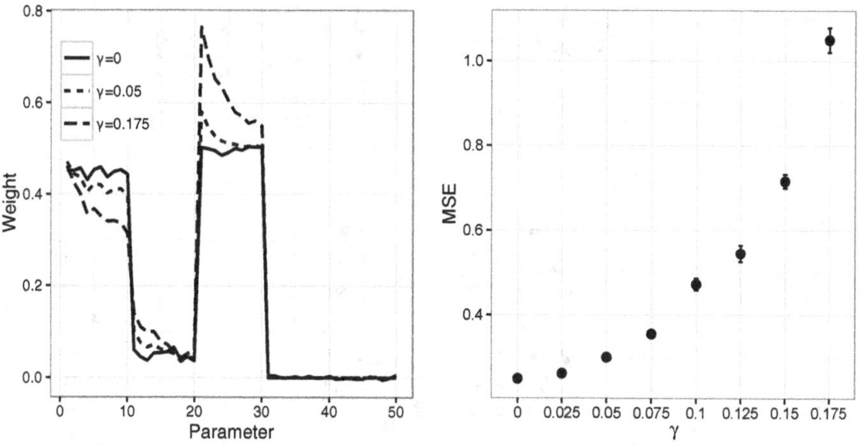

(a) (b)

Fig. 1. (a) Computed weights of all 50 predictors for the group lasso ($\gamma = 0$) and for 2 values of γ (0.05 and 0.175). (b) Mean-squared error (MSE) in the cross-validation as a function of the co-regularization parameter γ.

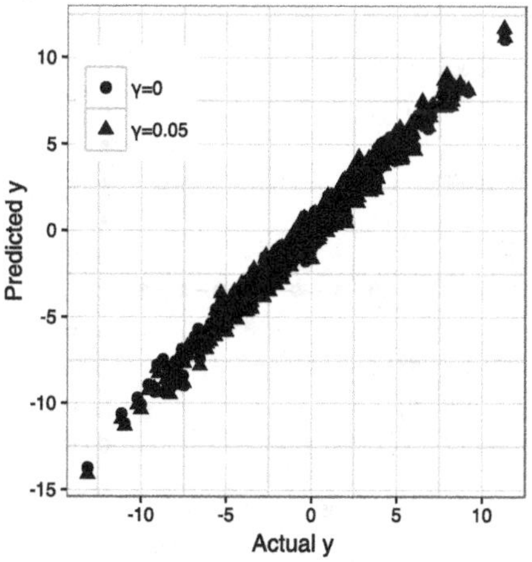

Fig. 2. Scatter plots of Actual vs Predicted values of y for the samples of the test set for the fits given by the GL and the co-regularized method with $\gamma = 0.05$ (true y' refers to the true values of y for those samples).

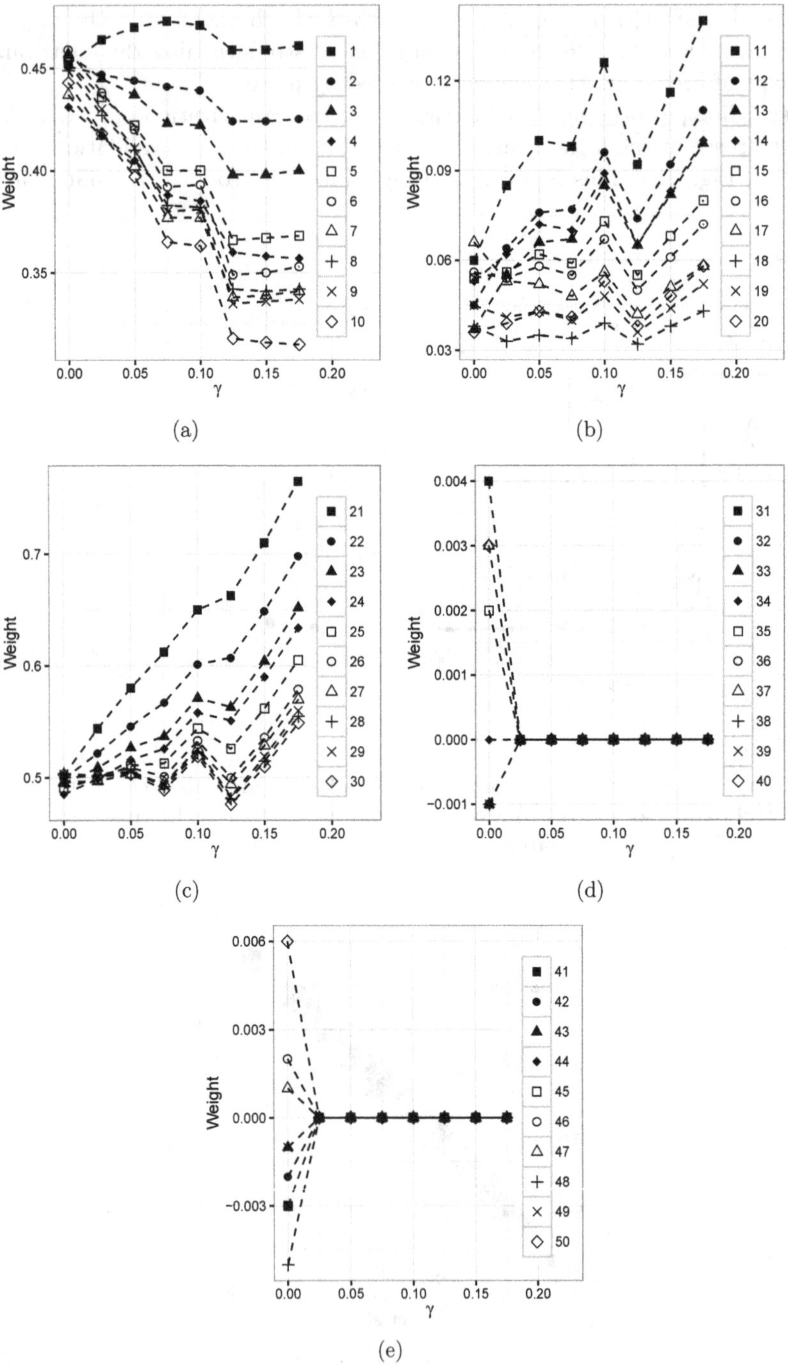

Fig. 3. Computed weights for all 50 predictors of the synthetic dataset. (a–e): coefficient paths for each group of predictors as a function of the co-regularization parameter γ.

priority of fitting $\mathbf{y}$ and keeping the contribution of related groups similar. This shows that, in the situation where we wish to use co-regularization to recover a group of predictors that is relevant but do not contribute strongly to the response, the value of γ should be non-zero, but small, so that the performance is not strongly affected by this extra penalty term. The optimal value of γ will therefore depend on each specific data set. In our experiments, we found $\gamma = 0.05$ to be a good compromise between a clear increase in the relevance of group 2 with still good predictive performance. Figure 2 shows a scatter plot of actual *vs* predicted values of the outcome y in the test set for the fits given by the GL and the co-regularized method for $\gamma = 0.05$.

In order to illustrate further how co-regularization affects the weights of all groups of predictors, we plot the coefficient paths for the predictors of all 5 groups as a function of γ (Figs. 3(a)–(e)). By construction, the predictors of group 2 contribute to the response with much less strength than groups 1 and 3. Imposing co-regularization between the predictions of groups 1 and 2 leads to an overall increase on the weights of the predictors of group 2 and a consequent decrease of the weights of the predictors of group 1: the prediction given by each individual group is re-adjusted and brought closer to the other. As a consequence, the weights of the predictors of group 3 are also altered, even though this group is not explicitly co-regularized with any other group. This occurs due to the fact that the re-adjustment of the weights of group 1 changes the fit given to $\mathbf{y}$, which is then compensated by an increase in the weights of group 3. Regarding groups 4 and 5, the values of their coefficients were already quite small for the standard group lasso, but they were still being selected. It can be seen in Figs. 3(d) and (e) that a small value of γ already led to the elimination of those two groups. This shows that, due to co-regularization, groups that are related to each other but not to the response are eliminated even more easily, since they will be jointly encouraged to have zero weight.

6 Conclusions

Biological applications often pose a set of challenges to machine learning approaches: high predictive performance is required, but as important is the retrieval of all relevant predictors so that the results given by the final model can also be used to deepen the understanding of complex biological processes. In this context, it is important to focus not only on selecting a small set of predictors, but also on guiding the feature selection process so that all relevant predictors are retrieved. We believe that such a balance can only be achieved by the incorporation of as much domain knowledge/auxiliary information as possible when learning a model.

In this paper, we worked under the assumption that domain knowledge concerning the relationships among predictors is available such that they can be divided into groups and that groups can be related to each other. We proposed a modification to the cost function of the sparse-group lasso that allows the incorporation of the relationship between groups into the feature selection process.

Our term works by imposing that groups that are closely related in the "predictor" domain should also provide similar contributions to the final prediction of the response.

In our experiments, we focused on the case where one group of predictors (group 2), despite of being relevant to the response, could have its importance masked due to the presence of (irreducible) noise. By imposing that this group should be co-regularized to another more strongly relevant group (group 1), we were able to raise its weights, making groups 1 and 2 jointly relevant. This, of course leads to two predictable side-effects. Firstly, the predictive performance decreases as a function of the co-regularization parameter. Secondly, the weights of other not co-regularized, but relevant groups (group 3) are also affected to account for the decrease in predictive performance. Therefore, the optimal value of the co-regularization parameter is found through a balance between fitting $\mathbf{y}$ and strengthening "weakly-relevant" groups of predictors.

6.1 Future Work

For future work we foresee two main activities. Firstly, the co-regularized sparse group lasso will be applied to a real-world dataset. Secondly, we will also explore the application of the proposed method in the semi-supervised setting. This extension is possible due to the fact that the co-regularization term does not depend on the response $\mathbf{y}$ and can be very valuable in cases where the amount of annotated data is small, but unannotated data is abundant.

Acknowledgments. This work was funded by TNO Early Research Program (ERP) "Making sense of big data".

References

1. Guyon, I., Elisseeff, A.: An introduction to variable and feature selection. J. Mach. Learn. Res. **3**, 1157–1182 (2003)
2. Hea, Z., Weichuan Yub, W.: Stable feature selection for biomarker discovery. Comput. Biol. Chem. **34**, 215–225 (2010)
3. Yuan, M., Lin, Y.: Model selection and estimation in regression with grouped variables. J. R. Stat. Soc. B **68**(part 1), 49–67 (2006)
4. Simon, N., Friedman, J., Hastie, T., Tibshirani, R.: A sparse group lasso. J.Comput. Graph. Stat. **22**(2), 231–245 (2013)
5. Simon, N., Friedman, J., Hastie, T.: A Blockwise descent algorithm for group-penalized multiresponse and multinomial regression (2013)
6. Friedman, J., Hastie, T., Tibshirani, R.: A note on the group lasso and a sparse group lasso (2010)
7. Jacob, L., Obozinski, G., Vert, J.P.: Group lasso with overlap and graph lasso. In: Proceedings of the 26th International Conference on Machine Learning, Montreal, Canada (2009)
8. Rosselló-Móra, R.: Towards a taxonomy of Bacteria and Archaea based on interactive and cumulative data repositories. Taxon. Biodivers. **14**(2), 318–334 (2012)

9. Das, J., Gayvert, K.M., Bunea, F., Wegkamp, M.H., Yu, H.: ENCAPP: elastic-net-based prognosis prediction and biomarker discovery for human cancers. BMC Genomics **16**(1), 263 (2015)
10. Zhang, F., Hong, D.: Elastic net-based framework for imaging mass spectrometry data biomarker selection and classification. Stat. Med. **30**, 753–768 (2011)
11. James, G., Witten, D., Hastie, T., Tibshirani, R.: An Introduction to Statistical Learning. Morgan Kaufmann, San Francisco (1999)
12. Tibshirani, R.: Regression shrinkage and selection via the lasso. J. R. Stat. Soc. Ser. B **58**, 267–288 (1994)
13. Hastie, T., Zou, H.: Regularization and variable selection via the elastic net. J. R. Stat. Soc. B **67**(part 2), 301–320 (2005)
14. Hoerl, A., Kennard, R.: Ridge regression. In: Encyclopedia of Statistical Sciences, vol. 8, pp. 129–136. Wiley, New York (1988)
15. Parikh, N., Boyd, S.: Proximal Algorithms. Now Publishers Inc., Breda (2013). ISBN 978-1601987167
16. Ruijter, T., Tsivtsivadze, E., Heskes, T.: Online co-regularized algorithms. In: Ganascia, J.-G., Lenca, P., Petit, J.-M. (eds.) DS 2012. LNCS (LNAI), vol. 7569, pp. 184–193. Springer, Heidelberg (2012). doi:10.1007/978-3-642-33492-4_16
17. Sindhwani, V., Niyogi, P., Belkin, M.: A co-regularization approach to semisupervised learning with multiple views. In: Proceedings of ICML Workshop on Learning with Multiple Views (2005)

Economic Lot-Sizing Problem with Remanufacturing Option: Complexity and Algorithms

Kerem Akartunalı and Ashwin Arulselvan(✉)

Department of Management Science, University of Strathclyde, 130 Rottenrow, Glasgow G4 0GE, UK
{kerem.akartunali,ashwin.arulselvan}@strath.ac.uk

Abstract. In a single item dynamic lot-sizing problem, we are given a time horizon and demand for a single item in every time period. The problem seeks a solution that determines how much to produce and carry at each time period, so that we will incur the least amount of production and inventory cost. When the remanufacturing option is included, the input comprises of number of returned products at each time period that can be potentially remanufactured to satisfy the demands, where remanufacturing and inventory costs are applicable. For this problem, we first show that it cannot have a fully polynomial time approximation scheme (FPTAS). We then provide a pseudo-polynomial algorithm to solve the problem and show how this algorithm can be adapted to solve it in polynomial time, when we make certain realistic assumptions on the cost structure. We finally give a computational study for the capacitated version of the problem and provide some valid inequalities and computational results that indicate that they significantly improve the lower bound for a certain class of instances.

1 Introduction

Remanufacturing is the process of recovering used products by repairing and replacing worn out components so that a product is created at the same quality level as a newly manufactured product saving tonnes of landfill every year by providing an environmentally friendlier alternative to classical manufacturing. It also offers industries the potential to significantly save money by exploiting used product inventories and reusing many precious raw materials that are becoming increasingly scarcer. With this motivation, we study the single item production planning problem over a finite horizon with the option of remanufacturing. Like in classical lot-sizing problem, the problem is defined over a finite planning horizon, where demand in each time period is provided as input. With the remanufacturing option, the demand can be satisfied either by manufacturing new items or remanufacturing returned items, which are also provided as input to the problem. The problem consists of separate inventory costs for carrying remanufactured items and manufactured items (sometimes referred to as serviceable inventory), and there is also a cost incurred for manufacturing or remanufacturing.

P.M. Pardalos et al. (Eds.): MOD 2016, LNCS 10122, pp. 132–143, 2016.
DOI: 10.1007/978-3-319-51469-7_11

The classical lot-sizing problem was introduced in [17] by Wagner and Whitin, with a dynamic program that can solve the uncapacitated problem in polynomial time. Various variants of it have been thoroughly studied over the last 6 decades, see [1] for a recent review. Later, the capacitated version was introduced and the problem was shown to be NP-hard, see [6]. A dynamic program was provided in [5] which runs in polynomial time for unary encoding. A fully polynomial time approximation scheme (FPTAS) was provided in [9]. The problem with remanufacturing option was first studied in [8] and proved as NP-hard in [15]. A dynamic program with polynomial running time was provided for a special case of when the cost involved are time invariant and there is a joint set-up cost involved for both manufacturing and remanufacturing [14]. A polynomial time algorithm was provided when all costs are linear by solving it as a flow problem [8]. Since then, very little progress has been made for polynomial special cases.

We first show in Sect. 2 that the general case of this problem cannot have an FPTAS unless P=NP. We refer the reader to [7] for concepts about NP-hardness and [16] for concepts about FPTAS. We then provide a straightforward dynamic program for the general case that runs in pseudopolynomial time in Sect. 3. We use this dynamic program as an ingredient to design an algorithm that runs in polynomial time to solve a special case, where the inventory cost of the returned items is at least as much as the inventory cost of the manufactured items. In addition, we assume that the concave costs involved in manufacturing has a fixed cost and variable cost component. We also assume that the costs are time invariant. The algorithm and the proof would work for time varying costs but they need to increase over time. No other assumptions are required for the polynomial solvability of this special case. Finally, in Sect. 4 we study the effects of two families of valid inequalities for the capacitated version of the problem. These are based on the well known flow cover cuts and they are NP-hard to separate in general. We show that a class of polynomially separable cases of these cuts are quite effective to close the gaps down for many instances. We test them on several instances that were generated to have a gap. Creating these instances are of interest, which have substantial gaps and hence are of concern to improve performance on such instances, whereas instances with small gaps will be handled effectively by standard optimization packages.

In the single item economic lot-sizing problem with remanufacturing option (ELSR), we are given a time horizon T. For each time period $t = 1 \dots T$, we are given a demand D_t, and the amount of returned products R_t that is available for remanufacturing. W.l.o.g., we assume that manufacturing and remanufacturing can be both completed for an item in a single period. We also define the following cost functions for each time period $t = 1 \dots T$: (1) manufacturing cost $f_t^m : T \to \mathbb{R}_{\geq 0}$, (2) remanufacturing cost $f_t^r : T \to \mathbb{R}_{\geq 0}$, (3) holding cost of manufactured items $h_t^m : T \to \mathbb{R}_{\geq 0}$, and (4) holding cost of returned items $h_t^r : T \to \mathbb{R}_{\geq 0}$. We will assume that cost could be time variant but we have a linear inventory cost and the concave cost structure associated with remanufacturing and manufacturing involves in a fixed cost and linear variable cost component, i.e., $f_t^i(x) = f_t^i + l_t^i(x)$, when $x > 0$ and 0 otherwise, for

$i = r, m$. $f_t^r, f_t^m(l_t^r, l_t^m)$ are the fixed costs (variable costs) incurred in period t for remanufacturing and manufacturing respectively. We slightly abuse the notation here to denote both the fixed cost component and the function by the same notation, but this is easy to distinguish from the context. In each time period, we have the option to remanufacture the returned item, manufacture the item new, or use serviceable inventory from previous period to satisfy the demand. The problem aims to generate a production plan that details the amount of products to be manufactured p_t^m, remanufactured p_t^r, the returned items carried in inventory q_t^r, and manufactured items carried in the inventory q_t^m, for each time period $t = 1, \ldots, T$ such that the demand is met in each time period and we minimize the total cost incurred.

2 Complexity

The problem is known to be NP-hard in general [15]. We extend the reduction provided in [15] to show the following theorem.

Theorem 1. *ELSR does not have FPTAS unless P=NP.*

Proof. We will show this through a reduction from a variation of the partition problem, wherein we are given $2n$ integer $a_1, a_2, \ldots, a_{2n}$ such that $\sum_{i=1}^{2n} a_i = 2A$ and for all $i = 1, \ldots, 2n$, we have $\frac{A}{n+1} < a_i < \frac{A}{n-1}$. We can also assume $a_i \geq 2, i = 1, \ldots, 2n$. We are asked the question, whether there is a subset of these integers that add to A. Even with this restriction, the partition problem is NP-hard (see [10]). Also note that, if there exists a subset that adds up to exactly A, then such a subset must have exactly n items due to the assumption we made on the integers.

In our reduction, we first take the time horizon $T = 2n$ and the demand for each time period $i = 1, \ldots, T = 2n$ as a_i. We incur a fixed cost of 1 for both manufacturing and remanufacturing. While the remanufacturing does not incur a variable cost, manufacturing incurs a variable cost of $\frac{k}{n}$ for some constant k. The inventory cost of the manufactured items is some big M and for returned items is 0. The amount of returns in period 1 is $R_1 = A$ and there are no returns for all other time periods, $R_i = 0, i = 2, \ldots, 2n$. If there is a solution to the partition problem, then it is easy to see that the optimal solution to ELSR is less than $2n + k$. Otherwise, the optimal solution to ELSR is at least $2n + k + \frac{k}{n}$. In order to see this case, note that all subsets of n items either add up to at least $A + 1$ or at most $A - 1$. If we choose to use the A returned products to satisfy a subset of n demands that add up to $A + 1$ or more, then we need to manufacture for the remaining n periods and at least in one of the n periods in which we used remanufactured goods to satisfy demands (as they could only satisfy a total demand of A). In this case, we incur a cost of at least $2n + k + \frac{k}{n} + 1$. In the event we use the remanufactured goods to satisfy some n items that add up to $A - 1$ or less, then the remaining n items that add up to $A + 1$ or more needs to come from manufacturing in all n periods (in addition to the possibly remanufacturing of the residual return goods). In this case, we incur a cost of at least $2n + k + \frac{k}{n}$. This also rules out an FPTAS for the problem, since we can choose $\epsilon = \frac{k}{n(2n+k)}$ and an algorithm that runs in $O(f(n, \frac{1}{\epsilon}))$, $f(n, \frac{1}{\epsilon})$ being a polynomial function in n and $\frac{1}{\epsilon}$, to obtain an $(1 + \epsilon)$-approximation for the ELSR can then distinguish YES and NO instances of the partition problem in polynomial time. □

3 Dynamic Program for the General Case

We now provide a dynamic program that runs in pseudopolynomial time to solve the general case exactly. This is an extension of Wagner and Whitin's solution that incorporates the remanfacturing option, and we present it here as we will be needing it as an ingredient of our special case. We define the following function $W_t(p, q)$ as the minimum cost of obtaining an inventory level of p for the returned products and q for the manufactured products at the end of period t, such that all demands are met for the periods $i = 0, \ldots, t$ either through manufacturing new items or remanufacturing returns. We define for $t = 1, \ldots, T$, $\mathcal{D}_t = \sum_{i=1}^{t} D_i$ and $\mathcal{R}_t = \sum_{i=1}^{t} R_i$. We will now do a forward recursion. It is easy to compute the value for $W_1(p, q)$, $p = 0, \ldots R_1$, $q = 0, \ldots, \mathcal{D}_T - \mathcal{D}_1$. For a specific value of p and q, there is exactly one way of obtaining the solution, so it is easy to compute $W_1(p, q)$ for all possible values of p and q. For infeasible solutions, we set $W_1(p, q) = \infty$. Then, the recursive function is:

$$W_t(p,q) = \min_{\substack{0 \leq \tilde{p} \leq \mathcal{R}_{t-1} \\ 0 \leq \tilde{q} \leq \mathcal{D}_T - \mathcal{D}_{t-1}}} [W_{t-1}(\tilde{p}, \tilde{q}) + f_t^r(p - R_t - \tilde{p}) + f_t^m(D_t + q - \tilde{q} - (p - R_t - \tilde{p})) + h_t^r(\tilde{p}) + h_t^m(\tilde{q}))]$$

The size of the state space of the problem is $T \cdot \mathcal{D}_T \cdot \mathcal{R}_T$, making the above algorithm pseudopolynomial in running time.

Theorem 2. *$W_t(p, q)$ is the optimal value of the ELSR problem for periods $1, \ldots, t$, when we need p (q) as the return (manufacturing) inventory level at the end of time period t*

The proof is omitted as it is a straightforward extension from Wagner and Whitin [17] for the dynamic lot-sizing problem. We will provide it in the full version of the paper. As a consequence of Theorem 2, we get the following result.

Corollary 3. $\max_{p,q} W_T(p, q)$ *is the optimal solution to ELSR.*

3.1 Dynamic Program for the Special Case: Return Inventory Cost Is Higher Than Serviceable Inventory Cost

We now investigate the special case where $h^r(p) \geq h^m(p)$, for all $p \in \mathbb{R}_{\geq 0}$. This is a reasonable assumption in practice, where the value of the returned products depreciate faster than a newly manufactured product and in terms of storage space, there is no difference between a returned or a manufactured product. Note that we also omitted the time index as we are assuming the costs are time invariant. If the cost are time variant, the algorithm requires that cost (both set up and variable costs) increases over time, and all arguments in the proof remain valid. In practice, this assumption is also realistic. For this special case, we now provide a dynamic program that runs in polynomial time. The following three lemmas help us reduce the state space for this special case. Remember the possible values of p and q in our function $W_t(p,q)$ are $0,\ldots,\mathcal{R}_t$ and $0,\ldots,\mathcal{D}_T-\mathcal{D}_t$ respectively. Let t^* be the last time period in an optimal solution where we use the returned products to satisfy demand and for all $t > t^*$ we only manufacture new items or use inventory of serviceable goods to satisfy the demands and let $(\mathbf{p}^*,\mathbf{q}^*)$ be the corresponding optimal solution. We define the following notation for compactness: $(a-b)^+ := \max\{a-b,0\}$.

Lemma 1. *For the optimal solution $\mathbf{p}^*,\mathbf{q}^*$, let us say we have $\tilde{R}_t$ of returned goods available for some time period $t < t^*$, then there is an optimal solution (by possibly re-writing the solution $\mathbf{p}^*,\mathbf{q}^*$ from time t and onwards) in which the amount of remanufactured items in time t will only be from the set $\{0,\tilde{R}_t\}$. If such a choice of return inventory is not possible, then we can create a new optimal solution with t being the last remanufactured period.*

Proof is provided in Appendix A

Lemma 2. *For each time period $t < t^*$, there exists an optimal solution for ELSR where the possible inventory level of the returned products right after period t takes a value only from the set $\{0, R_t, \bigcup_{i=1}^{t}\{\mathcal{R}_t - \mathcal{R}_i\}\}$, where t^* is the last time period of remanufacturing in that optimal solution.*

Proof Sketch: The proof can be easily obtained through induction by invoking Lemma 1 and the induction hypothesis. □

An alternative way of interpreting the above lemma is that whenever we choose to remanufacture at time t, we remanufacture all return inventory available or nothing. This fact also helps us bound the serviceable goods inventory level up until a certain point. Just like we guess the last time period of remanufacturing, we will also make a guess on the last time period of manufacturing before time t^* in an optimal solution and we shall denote it by ℓ^*. We can now bound the state space of the serviceable goods inventory level for all time period $t \leq \ell^*$, where $\ell^* \leq t^*$.

Lemma 3. *There exists an optimal solution, where the inventory level of the manufactured goods in time period $t < \ell^*$ is in the set $\{0, \bigcup_{i=t+1}^{T}\{\sum_{k=t+1}^{i} D_k\}, \bigcup_{i=t+1}^{\ell^*}\bigcup_{j=t+1}^{i}\{(\sum_{k=t+1}^{i} D_k - \sum_{k=t+1}^{j} R_k)^+\}, \bigcup_{i=1}^{t}\{(\sum_{k=1}^{i} R_k - \mathcal{D}_t)^+\}\}$.*

Proof provided in Appendix B.

Now that we have bounded the size of the state space for all our time periods before t^*, we are ready to provide our algorithm. Since we can make all possible guesses for our ℓ^* and t^*, we will be providing the algorithm assuming that we made the correct guess. We just need to repeat the algorithm for all possible guesses and pick the guess with the cheapest solution.

Definition 4 (Uncapacitated minimum concave cost flow problem (UMCCF)). *We are given a directed network* $G(V, A)$ *and a demand function* $d : V \rightarrow \mathbb{Q}$, *such that* $\sum_{i \in V} d_i = 0$ *and a concave cost function* $c : A \rightarrow \mathbb{Q}_+$. *Find a set of assignment of flows* $f : A \rightarrow \mathbb{Q}_+$, *such that for each node* $i \in V$, *we have* $\sum_{(i,j) \in A} f_{ij} + d_i = \sum_{(j,i) \in A} f_{ji}$ *and* $\sum_{(i,j) \in A} c_{ij}(f_{ij})$ *is minimised.*

For the UMCCF problem defined on a planar graph, a send-and-split (SAS) dynamic program was provided in [4] which runs in a time which is exponential in the number of distinct faces in which the terminal (source and supply) nodes lie. If the number of distinct faces on which the terminal nodes lie is bounded, then the algorithm has a polynomial time complexity.

Algorithm DP-SAS

- Run the dynamic program for the general case provided in Sect. 3 until ℓ^* using the reduced state space.
- Set up a minimum concave cost flow for planar problem from time period ℓ^* and onwards, where all the demand nodes lie on exactly two different faces and solve the problem for all possible input inventory levels at ℓ^* (to solve this problem in polytime use the SAS algorithm in [4], where the algorithm runs in a time that is an exponential in the number of distinct faces on which all the terminal nodes are lying).

The Algorithm DP-SAS has to guess the values of ℓ^* and t^* and for every possible guess, all possible inventory levels entering period ℓ^* are enumerated. For each such level, we set up a minumum concave cost flow problem by first creating two source nodes, one for manufactured products and one for return products, with the corresponding inventory levels as their surplus flow values and the demand and return values occuring after time ℓ^* are taken as the rest of terminal nodes (acting as sink and source nodes respectively). Since we know from assumption that there is no manufacturing between periods ℓ^* and t^* and there is no remanufacturing after t^*, all terminal nodes lie on at most three distinct faces. This along with Lemmas 2 and 3 guarantees polynomial running time of DP-SAS. We would like to point out that this is a highly impractical algorithm as we are running the SAS algorithm (which runs in a time that is a cubic function of T) for every possible guess of ℓ^* and t^* and for all possible inventory levels for the corresponding guesses. We are mainly interested in the answering the question of tractability. Designing a faster algorithm or improving upon the above algorithm to solve this special case will be a good research question for the future. We will now focus on a more practical exact algorithm for solving the general version of the problem in the following section.

4 Valid Inequalities and Numerical Computation

In this section, we consider a more general variation of the problem, wherein we are given a capacity restriction of $\mathcal{C}_j$ on how much we can manufacture in each time period $j = 1, \ldots, T$. For this general version, we give an integer programming formulation and provide two families of valid inequalities and demonstrate the strength of these inequalities on some test instances. The inequalities are based on the popular flow-cover cuts introduced in [12], which in turn can be viewed as a generalisation of Dicuts introduced in [13] for uncapacitated fixed charge flow problems. We follow the same line of Carnes and Shmoys [3], where they introduce these inequalities for capacitated single item dynamic lot sizing problem and show how these inequalities help in bounding the gap. In this work, we also show how to tighten these inequalities. We will show the inequalities' validity and the special case where the separation time is polynomial. In the general case, the separation problem is NP-hard, see [12]. We perform the numerical computation only for the family of inequalities where the separation time is polynomial and we will show through numerical computations that these are enough to close most of the gap.

We will cast the whole problem as an uncapacitated fixed charge flow problem and in order to balance the supply and demand, we will increase the planning horizon by 1 and this last time period will have a demand of $\mathcal{R}_T = \sum_{i=1}^{T} r_i$. We will first give the following MIP formulation that will solve the ELSR. The decision variables are:

- x^i_{jk}: The percentage of d_k satisfied by returns from i that was remanufactured in j. Note by this definition, for $x^i_{jk} > 0$, we need $i \leq j \leq k$.
- y_{jk}: The percentage of d_k satisfied by the order that was manufactured in period k. Again, we need $j \leq k$ for $y_{jk} > 0$
- u_j: Binary variable indicating whether we remanufactured in period j
- v_j: Binary variable indicating whether we manufactured in period j

We will re-write the cost functions in a more convenient format for our model: $g^r_{i,jk}(x) := \sum_{t=i}^{j} h^r_j(x) + \sum_{t=j}^{k} h^m_t(x) + l^r_j(x), g^m_{jk}(x) := \sum_{t=j}^{k} h^m_t(x) + l^m_j(x)$

$$\mathcal{P}_1 : \min \sum_{j=1}^{T} f^r_j u_j + \sum_{j=1}^{T} f^m_j v_j + \sum_{k=1}^{T}\sum_{j=1}^{k}\sum_{i=1}^{j} g^r_{i,jk} d_k x^i_{jk} + \sum_{k=1}^{T}\sum_{j=1}^{k} g^m_{jk} d_k y_{jk}$$

$$\sum_{j=1}^{k}\sum_{i=1}^{j} x^i_{jk} + \sum_{j=1}^{k} y_{jk} \geq 1, \qquad \forall k \in [T]$$

$$\sum_{k=1}^{T}\sum_{j=i}^{k} d_k x^i_{jk} \leq r_i, \qquad \forall i \in [T]$$

$$\sum_{k=j}^{T} d_k y_{jk} \leq \mathcal{C}_j, \qquad \forall j \in [T]$$

$$\sum_{i=1}^{j} x_{jk}^{i} \leq u_j, \qquad \forall k \in [T], j = 1, \ldots, k$$
$$y_{jk} \leq v_j, \qquad \forall k \in [T], j = 1, \ldots, k$$
$$\mathbf{x}, \mathbf{y} \geq 0$$
$$\mathbf{u}, \mathbf{v} \in \{0,1\}^T$$

For the subsets of return, demand and manufacturing periods $A, B, C \subseteq [T]$ respectively, we define $d(A, B, C)$ as the demand in B that cannot be satisfied by just the returns in A and the manufacturing capacity provided in periods in C. This demand, $d(A, B, C)$, needs to be satisfied by either manufacturing in the periods $\bar{C} = [T] \backslash C$ or by using the returns in $\bar{A} = [T] \backslash A$. The value of $d(A, B, C)$ can be determined by solving the following LP.

$$\text{LP-U:} \qquad \sum_{k \in B} d_k - \left(\max \sum_{k \in B} \sum_{i \in A} \sum_{j: i \leq j \leq k} d_k x_{jk}^{i} + \sum_{j \in C} \sum_{k \in B} d_k y_{jk} \right)$$
$$\sum_{j=1}^{k} \sum_{i=1}^{j} x_{jk}^{i} + \sum_{j=1}^{k} y_{jk} \leq 1, \qquad \forall k \in [T]$$
$$\sum_{j=i}^{k} \sum_{k=j}^{T} d_k x_{jk}^{i} \leq r_i, \qquad \forall i \in [T]$$
$$\sum_{k=j}^{T} d_k y_{jk} \leq C_j, \qquad \forall j \in [T]$$
$$\mathbf{x}, \mathbf{y} \geq 0$$

For $j \in [T]$, we define $w_j^r(A, B, C)$ as the maximum amount of $d(A, B, C)$ that could be satisfied by returns in $\bar{A}$, if they are allowed to be remanufactured only during j. We calculate this value by setting up a suitable LP similar to LP-U. For each $j \notin C$, we define $w_j^m(A, B, C) := d(A, B, C) - d(A, B, C \cup \{j\})$ which denotes the decrease in unsatisfiable demand in B by C if j were to be added to C. Let $\mathcal{T}_C$ be the set of all 2-tuple sets that partitions $[T] \backslash C$ for some set C and and let $\mathcal{T}$, be the set of all 2-tuple sets that partitions $[T]$. Let $\{T_1^m, T_2^m\} \in \mathcal{T}_C$ and $\{T_1^r, T_2^r\} \in \mathcal{T}$. Now consider the following inequality.

$$\sum_{k \in B} \sum_{j \in T_1^r} \sum_{i \notin A}^{j} d_k x_{jk}^{i} + \sum_{j \in T_2^r} w_j^r(A, B, C) u_j + \sum_{k \in B} \sum_{j \in T_1^m} d_k y_{jk}$$
$$+ \sum_{j \in T_2^m} w_j^m(A, B, C) v_j \geq d(A, B, C), \forall (T_1^r, T_2^r) \in \mathcal{T}, (T_1^m, T_2^m) \in \mathcal{T}_C, B \subseteq [T] \tag{1}$$

For a fixed A, B and C it is pretty straightforward to realise that the separation problem is polynomially solvable. Note that, even for a fixed A, B and C we have exponentially many constraints, as we need to enumerate all the partitions of $[T]$. For a fixed A, B and C, and a given LP solution $\hat{\mathbf{x}}, \hat{\mathbf{u}}, \hat{\mathbf{y}}, \hat{\mathbf{v}}$, the partition could be determined in the following way

- If $\sum_{k \in B} \sum_{i \notin A}^{j} d_k \hat{x}_{jk}^i \geq w_j^r(A, B, C)\hat{u}_j$ then $j \in T_2^r$ otherwise $j \in T_1^r$.
- Similarly, if $\sum_{k \in B} d_k \hat{y}_{jk} \geq w_j^m(A, B, C)\hat{v}_j$ then $j \in T_2^m$ otherwise $j \in T_1^m$.

Valid inequalities of type 1 are not very strong. We illustrate this through the following example.

Example 5. *We have uniform capacities $\mathcal{C}$ and our fixed sets are as follows: $A = \{1, \ldots, t\}, B = \{1, \ldots, t\}, C = \emptyset$, for some t.*

In Example 5, it is clear that $d(A, B, C) \geq \max(\mathcal{D}_t - \mathcal{R}_t, 0)$. Our inequality 1 for this example will be

$$\mathcal{C} \sum_{i=1}^{t} u_i \geq d(A, B, C)$$

This gives us a straight forward mechanism to tighten it. We can divide by $\mathcal{C}$ and round down the right hand side to obtain the valid inquality:

$$\sum_{i=1}^{t} u_i \geq \left\lfloor \frac{d(A, B, C)}{\mathcal{C}} \right\rfloor$$

Drawing on this, from inequalities 1 we derive MIR inequalities. We briefly describe MIR inqualities here. For the set $\{(x, y) \in \mathbb{R} \times \mathbb{Z} : x + cy \geq b, x \geq 0\}$, it is easy to see that $x + fy \geq f\lceil \frac{b}{c} \rceil$ is a valid inequality, where $f = b - c\lfloor \frac{b}{c} \rfloor$ (see [11] for more details).

The next set of inequalities are quite similar to the flow cut inequalities but are given with respect to the returned products. Since the flows are balanced in our network, we can reverse all arc orientations and take the nodes with returns as the demand nodes with demands as the returns and write our flow cuts based on these as sinks. In other words, it is a generalisation of the following simple idea: For every $i \in [T]$ with $r_i > 0$, we need to have either $u_i \geq 1$ or $\sum_{j \geq i} \sum_{k \geq j} x_{jk}^i d_k \geq r_i$. We then get the disjunctive inequality:

$$u_i r_i + \sum_{j \geq i} \sum_{k \geq j} x_{jk}^i d_k \geq r_i \tag{2}$$

Once again for every partition (T_1, T_2) of $\{i, i+1, \ldots T\}$, we have the following inequality:

$$\sum_{i' \in T_1} u_{i'} r_i + \sum_{\substack{j \geq i \\ j \in T_2}} \sum_{k \geq j} x_{jk}^i d_k \geq r_i \tag{3}$$

The validity of the above inequality is straightforward and it is easy to see that 2 is a special case of 3 with $T_1 = \{i\}$.

5 Experiments

The experiments were carried out on an Intel core i5-3320 @2.6Ghz CPU with a 8GB RAM and the implementation was done in Gurobi 6.0.4 using Python API. It was difficult to directly adapt the instances available for capaciated lot sizing problem [2] to the remanufacturing setting in a way that the new instances could have large root gaps. So we randomly generated instances with special return and demand structures in order to create instances with gap. We took the manufacturing capacities to be uniform. For varying cost and demand structures we generated nine instances with 100 time periods and three instances with 200 time periods. We implemented the cover and MIR cuts from 1 and as we could see from Table 1, for the gap instances we generated, we were able to close the gap at the root when all other Gurobi cuts were turned off. The default settings of Gurobi could not solve the problem when we increased the time horizon to 200 within 1000 s that we set as a time limit for all our test runs. As one could see the new cuts proposed drastically reduced the running time for solving these gap instances. In addition to that the lower bounds were considerably improved at the root. As we moved on to larger instance (with time horizon = 200), the problem was getting increasingly harder to solve with just the default GUROBI settings.

Table 1. Gaps from test runs with Gurobi default setting and cover and MIR derived from 3

	Size	Gurobi			Cuts from 1		
		Root gap (%)	Final gap (%)	Time(s)	Root gap (%)	Final gap (%)	Time (s)
1	100	14.28	0.5	341	0.5	0.5	26
2	100	14.20	0.5	350	0.5	0.5	29
3	100	9.81	0.5	330	0.5	0.5	24
4	100	12.89	0.5	294	0.5	0.5	27
5	100	14.26	0.5	330	0.5	0.5	23
6	100	15.11	0.5	416	0.5	0.5	697
7	100	21.24	0.5	612	0.74	0.74	1000
8	100	24.00	0.5	670	0.5	0.5	23
9	100	22.13	0.5	666	0.56	0.56	1000
10	200	17.20	17	1000	0.5	0.5	471
11	200	24.31	24.31	1000	0.5	0.5	613
12	200	23.32	23.32	1000	0.5	0.5	460

6 Conclusion and Open Problems

In this work, we studied the ELSR problem. We first provided a hardness proof that rules out FPTAS for this problem. We then provided a dynamic program with a pseudopolynomial running time to solve the general version of the problem. We later showed how this can be used to design a polynomial running time algorithm, when we make some assumptions on the cost structure.

We finally performed a computational study on the capacitated version of the problem, where we introduced two families of valid inequalities and our preliminary results were promising. Our future work would include obtaining an approximation algorithm for the general version of the problem. In the negative side, although we have ruled out a possibility of FPTAS, we have no proofs for APX-hardness (see [16]) and this is still open. We are also interested in performing a more comprehensive experimental study of the capacitated version of the problem. It will be interesting to verify whether the proofs presented above could be extended to multiple items (taking the number of items to be a constant will be a good starting point) and general concave functions.

A Proof of Lemma 1

Proof. Suppose in $(\mathbf{p}^*, \mathbf{q}^*)$, we produce something not from this set $\{0, \tilde{R}_t\}$. Hence, some intermediate return stock of $0 < a < \tilde{R}_t$ is carried, which also means that we are remanufacturing at time t in the optimal solution. Since $t < t^*$, there exist a time period after t in the optimal solution where we remanufacture. Let the $\tilde{t}$ be the first time period after t, when we remanufacture in the optimal solution. We are also carrying a non-zero return inventory until this time period. If we remanufacture at least a in time $\tilde{t}$, then we could have remanufactured this a in time t and carried a units of manufactured inventory until time $\tilde{t}$ with no additional cost, since return inventory cost is higher than manufactured inventory cost. If we produced less than a in time $\tilde{t}$, say $\tilde{a}$, then we could have produced $\tilde{a}$ in time t and produced nothing in time $\tilde{t}$ and continue with our argument. If $\tilde{t} = t^*$ and $\tilde{a} < a$, then we would have new optimal solution with t being the last time period of remanufacturing. □

B Proof of Lemma 3

Proof. We prove this lemma again through induction. For $t = 1$, the lemma's claim is that the inventory level of serviceable goods after time $t = 1$ will belong to the set $\{0, \bigcup_{i=2}^{T}\{\sum_{k=2}^{i} D_k\}, \bigcup_{i=2}^{\ell^*}\bigcup_{j=2}^{i}\{(\sum_{k=2}^{i} D_k - \sum_{k=2}^{j} R_k)^+\}, (R_1 - D_1)^+\}$. In order to show this, we do a case analysis:

Case 1 **Manufacturing takes place at $t = 1$:** In this case, either remanufacturing does not take place or it does take place and we have $R_1 < D_1$, otherwise we could manufacture everything we manufactured in time period 1 in time period 2 and get a cheaper solution by saving on the inventory cost. Let us take k, where $1 < k \leq \ell^*$, as the first time period after time $t = 1$ when we manufacture again. Now the demand for all the intermediate periods $\sum_{i=2}^{k-1} D_i$ needs to come from either the manufactured goods in period 1 or the return products from periods 1 to $k - 1$. From Lemma 2, we know that the only possible return inventory levels between the time periods 2 to $k - 1$ are $\{0, \bigcup_{i=2}^{k-1}\{\mathcal{R}_i - R_1\}\}$ and we know from lemma 1 that at any of these time periods, we either remanufacture all available inventory or none of them. The remaining demand then needs to be manufactured at time period 1. This has to be true for all values of $k = 2 \ldots \ell^*$. So we get the claim.

Case 2 **Manufacturing does not take place at** $t = 1$**:** This would mean that $R_1 \geq D_1$, otherwise, we would not have a feasible solution and the possible inventory levels in this case is $(R_1 - D_1)^+$.

In order to see that the lemma is true, we invoke the induction hypothesis and do a similar case analysis as above. □

References

1. Akartunalı, K., Miller, A.: A computational analysis of lower bounds for big bucket production planning problems. Comput. Optim. Appl. **53**(3), 729–753 (2012)
2. Atamtürk, A., Muñoz, J.C.: A study of the lot-sizing polytope. Math. Program. **99**, 443–465 (2004)
3. Carnes, T., Shmoys, D.: Primal-dual schema for capacitated covering problems. Math. Program. **153**(2), 289–308 (2015)
4. Erickson, R., Monma, C., Veinott, J.A.F.: Send-and-split method for minimum-concave-cost network flows. Math. Oper. Res. **12**(4), 634–664 (1987)
5. Florian, M., Klein, M.: Deterministic production planning with concave costs and capacity constraints. Manage. Sci. **18**, 12–20 (1971)
6. Florian, M., Lenstra, J., Rinnooy Kan, H.: Deterministic production planning: algorithms and complexity. Manag. Sci. **26**(7), 669–679 (1980)
7. Garey, M., Johnson, D.: Computers and intractability: a guide to the theory of NP-completeness. W. H. Freeman & Co., New York (1979)
8. Golany, B., Yang, J., Yu, G.: Economic lot-sizing with remanufacturing options. IIE Trans. **33**(11), 995–1003 (2001)
9. Hoesel, C.V., Wagelmans, A.: Fully polynomial approximation schemes for single-item capacitated economic lot-sizing problems. Math. Oper. Res. **26**, 339–357 (2001)
10. Korte, B., Schrader, R.: On the existence of fast approximation schemes. In: Magasarian, S.R.O., Meyer, R. (eds.) Nonlinear Programming, vol. 4, pp. 415–437. Academic Press, New York (1981)
11. Nemhauser, G., Wolsey, L.: Integer and Combinatorial Optimization. Wiley-Interscience, New York (1988)
12. Padberg, M., van Roy, T., Wolsey, L.: Valid linear inequalities for fixed charge problems. Oper. Res. **33**(4), 842–861 (1985)
13. Rardin, R., Wolsey, L.: Valid inequalities and projecting the multicommodity extended formulation for uncapacitated fixed charge network flow problems. Eur. J. Oper. Res. **71**(1), 95–109 (1993)
14. Teunter, R., Bayındır, Z., van den Heuvel, W.: Dynamic lot sizing with product returns and remanufacturing. Int. J. Prod. Res. **44**(20), 4377–4400 (2006)
15. van den Heuvel, W.: On the complexity of the economic lot-sizing problem with remanufacturing options. Econometric Institute Research Papers EI 2004-46, Erasmus University Rotterdam, Erasmus School of Economics (ESE), Econometric Institute (2004)
16. Vazirani, V.: Approximation Algorithms. Springer-Verlag New York, Inc., New York (2001)
17. Wagner, H., Whitin, T.: Dynamic version of the economic lot size model. Manage. Sci. **5**, 89–96 (1958)

A Branch-and-Cut Algorithm for a Multi-item Inventory Distribution Problem

Agostinho Agra[1], Adelaide Cerveira[2,3](✉), and Cristina Requejo[1]

[1] University of Aveiro and CIDMA, 3810-193 Aveiro, Portugal
{aagra,crequejo}@ua.pt
[2] University of Trás-os-Montes e Alto-Douro, 5001-801 Vila Real, Portugal
cerveira@utad.pt
[3] INESC TEC, Porto, Portugal

Abstract. This paper considers a multi-item inventory distribution problem motivated by a practical case occurring in the logistic operations of an hospital. There, a single warehouse supplies several nursing wards. The goal is to define a weekly distribution plan of medical products that minimizes the visits to wards, while respecting inventory capacities and safety stock levels. A mathematical formulation is introduced and several improvements such as tightening constraints, valid inequalities and an extended reformulation are discussed. In order to deal with real size instances, an hybrid heuristic based on mathematical models is introduced and the improvements are discussed. A branch-and-cut algorithm using all the discussed improvements is proposed.

Finally, a computational experimentation is reported to show the relevance of the model improvements and the quality of the heuristic scheme.

Keywords: Multi-item inventory · Hospital logistics · Branch-and-cut · Supply chain management

1 Introduction

We consider a Multi-Item Inventory Distribution (MIID) problem. The motivation for this study is to investigate logistics activities in respect to supply chain management in an hospital. There is a central warehouse that receives goods from suppliers and delivers those goods to nursing wards regularly. The ward's demand comprise a wide variety of products which include, amongst others, medicines, hospital supplies and medical devices. The main focus of this work is the planning of the delivery scheme of goods to meet the demand while keeping the stock levels between the desired bounds. The planning horizon is one week and should be repeated every week. The MIID problem is to find the delivery plan that satisfies the inventory and delivery capacity constraints that minimizes the number of visits to wards.

From the inventory management point of view, we consider a multi-item problem where one warehouse supplies several locations (wards), which are typically retailers in literature. Each location is visited at most once per day (except

P.M. Pardalos et al. (Eds.): MOD 2016, LNCS 10122, pp. 144–158, 2016.
DOI: 10.1007/978-3-319-51469-7_12

for extraordinary deliveries not considered here). As the number of time periods is small, inventory aspects are considered only at the supplied locations, since the supply decisions for the warehouse are taken with months in advance. Only a part of items (goods) are consumed at each location. For those items, safety stocks are mandatory at each location. Furthermore, the stock level in the first and last period must coincide, in order to ensure that the delivery plan can be replicated in subsequent weeks. A global stock upper bound is imposed at each location. Given the proximity of locations, the routing aspects can be ignored. Only the capacity of the vehicle is considered to bound the amount to be delivered in each time period.

Operations Research has been widely used in health care issues. Many operations research papers on health care subjects have been published in both Operations Research and Health Care journals. In [19] is presented a survey which points out a variety of studied problems in health care using operational research techniques. They include, among others, health care planning aspects, management and logistics health problems and the health care practices with treatment planning and preventive care. In [23] is presented the state-of-the-art of the research on material logistics management in an hospital. In [14,22] are considered pharmaceutical supply chain and inventory management issues in health care industry. The same topic of research is covered in [9] where the tasks and management approaches of hospital materials are considered. In [16] a case study is reported that looks into logistics activities of hospitals in Singapore. From the point of view of inventory management, complex problems combining inventory management and distribution decisions are receiving an increased attention in the last years. In [10] an overall introduction into inventory management is presented. In [7] a review of replenishment policies for hospital inventory systems and two models to deal with capacity limitations and service requirements are presented. In [18] inventory models are discussed for hospital problems. Practical complex inventory problems occur in other industries, from maritime transportation [2,3] to land transportation [8]. A recent review on complex inventory problems combined with production and distribution aspects is given in [1]. In relation to other practical problems, the problem considered in this paper has the particularity of including a huge number of variables and the distribution costs are negligible.

We provide a mixed integer formulation for MIID problem and discuss several approaches to improve that formulation, such as, tightening of constraints; derivation of valid inequalities; use of an extended reformulation. Several such techniques have been used for related inventory management problems and are of main importance to allow the improved formulation to solve practical problems through branch-and-cut and branch-and-bound algorithms. For exact branch-and-cut approaches see [2,3,5,20]. The MIID problem is NP-hard, since it generalizes well-known NP-hard problems, such as the capacitated lot-sizing problem [17]. Therefore we also propose an heuristic scheme to find good feasible solutions. This scheme works in two steps. In the first step we solve a relaxation obtained by aggregating all products and ignoring some constraints.

In the second step a local search is conducted in the neighborhood of the first step solution. Finally, an exact branch-and-cut algorithm is proposed to the MIID problem. This algorithm uses the improved formulation and it is feeded up with the upper bound given by the heuristic scheme.

The outline of the paper is as follows. In Sect. 2 we introduce a mixed integer formulation to the MIID problem. In Sect. 3 we discuss the model improvements. In Sect. 4 we introduce the heuristic scheme and in Sect. 5 a branch-and-cut algorithm based on the improved model is proposed. In Sect. 6 the details of benchmark instances generation are given, and computational results to test the model, the model improvements, the heuristic scheme and the branch-and-cut algorithm are presented. Finally, the conclusions are stated in Sect. 7.

2 Mathematical Model

In this section we introduce a mixed integer formulation to the MIID problem. Consider the sets $N = \{1,\ldots,n\}, L = \{1,\ldots,m\}$, and $T = \{1,\ldots,r\}$, representing the set of items, the set of locations (wards) to visit, and the set of time periods, respectively. Additionally, N_j represents the set of items consumed in location j and L_i represents the set of locations that consume item i. We define the following variables: x_{ijt} is the amount of item i to be delivered at location j in time period t; s_{ijt} is the stock level of item i in location j at time period t; and the binary variables y_{jt} take value 1 if there is a delivery in location j at time period t, and 0 otherwise. Additionally, we consider the following parameters. The demand for item i in location j at time period t is represented by d_{ijt}. The minimum stock level (safety stock) of item i in location j at time period t is $\underline{S}_{ijt}$. There is a single vehicle with capacity C, and we assume item i uses c_i units of that capacity. We also assume that only K locations can be visited per time period, and each location has a limited stock capacity of V_j items. The initial stock level of product i at location j is given by S_{ij}. M is an upper-bound for the number of items delivered at each location and each period. The MIID model is as follows:

$$\min \quad \sum_{j\in L}\sum_{t\in T} y_{jt} \tag{1}$$

$$s.t. \quad x_{ijt} + s_{ij,t-1} = d_{ijt} + s_{ijt}, \qquad i\in N, j\in L, t\in T, \tag{2}$$

$$\sum_{i\in N_j} x_{ijt} \le M y_{jt}, \qquad j\in L, t\in T, \tag{3}$$

$$\sum_{j\in M} y_{jt} \le K, \qquad t\in T, \tag{4}$$

$$\sum_{i\in N}\sum_{j\in L_i} c_i x_{ijt} \le C, \qquad t\in T, \tag{5}$$

$$\sum_{i\in N_j} s_{ijt} \le V_j, \qquad j\in L, t\in T, \tag{6}$$

$$s_{ijt} \geq \underline{S}_{ijt}, \qquad i \in N, j \in L_i, t \in T, \tag{7}$$
$$s_{ij0} = S_{ij}, \qquad i \in N, j \in L_i, \tag{8}$$
$$x_{ijt} \in \mathbb{Z}_+, \qquad i \in N, j \in L_i, t \in T, \tag{9}$$
$$y_{jt} \in \{0, 1\}, \qquad j \in L, t \in T. \tag{10}$$

The objective function (1) is to minimize the total number of visits to locations. Constraints (2) are the inventory flow balance at the locations. Constraints (3) ensure that if any item is delivered to location j at period t, then variable y_{jt} must be one. Constraints (4) and (5) impose limits on the deliveries in each time period. Constraints (4) state that only a limited number of locations can be visited while the vehicle capacity is guaranteed by constraints (5). Constraints (6) impose a storage capacity limit at each location. Safety stocks are imposed by the inventory lower bound constraints (7). Constraints (8), (9) and (10) are the variable domain constraints.

In this practical case it is desirable that the distribution plan for the planning week could be repeated for other weeks with similar demand rates. For this purpose, we need to ensure that the stock at the end of the time horizon is equal to the initial stock, that is, $\underline{S}_{ijr} = S_{ij}$.

Table 1. Example of a supply scheme. For each item and each ward is displayed the initial stock, and for each time period is displayed the safety stock, the demand and the optimal delivery quantities.

Ward (j)	Item (i)	S_{ij}	$\underline{S}_{ijt}$					Demand (d_{ijt})					Deliver (x_{ijt})				
			$t=1$	$t=2$	$t=3$	$t=4$	$t=5$	$t=1$	$t=2$	$t=3$	$t=4$	$t=5$	$t=1$	$t=2$	$t=3$	$t=4$	$t=5$
1	1	7	2	2	2	2	7	3	3	5	2	9	-	22	-	-	-
	2	44	11	11	11	11	44	7	8	9	10	21	-	19	36	-	-
	3	27	7	7	7	7	27	4	7	6	4	12	-	33	-	-	-
	4	25	6	6	6	6	25	4	2	4	3	18	-	31	-	-	-
2	2	37	9	9	9	9	37	5	6	8	10	18	-	-	-	2	45
	4	27	7	7	7	7	27	5	3	5	2	18	-	-	-	-	33
	5	23	6	6	6	6	23	3	2	4	2	18	-	-	-	28	1

Example 1. In Table 1 we present an example having 5 items, 5 periods and 2 locations (wards) with $N_1 = \{1, 2, 3, 4\}$ and $N_2 = \{2, 4, 5\}$. The optimal delivery quantities are presented in the last five columns with $y_{12} = y_{13} = y_{24} = y_{25} = 1$ and the optimal value is 4.

3 Model Improvements

It is well-known that the performance of exact algorithms based on mathematical models, such as the branch-and-bound and the branch-and-cut depend greatly on the quality of the model, see [15] for general integer programming problems, and [17] for lot-sizing problems. In this section we discuss model improvements that aim at tightening the formulation, that is, deriving a formulation whose linear relaxation is closer to the convex hull of the set of feasible solutions.

In order to ease the presentation we start with the introduction of the concept of net demand (see [2] for a related use of this concept). The net demand of item i in location j at time period t is denoted by nd_{ijt} and represents the minimum amount of item i that must be delivered in location j at time period t taking into account the initial stock level and the safety stocks (lower bound on the inventory levels). In this case it can be computed as follows:

$$nd_{ijt} = \max\{0, \sum_{\ell=1}^{t} d_{ij\ell} + \underline{S}_{ijt} - S_{ij} - \sum_{\ell=1}^{t-1} nd_{ij\ell}\}.$$

The net demand is computed iteratively for each time period. For time period $t = 1$, the net demand is just the demand in period 1, d_{ij1}, plus the safety stock at the end of time period 1, $\underline{S}_{ij1}$, minus the initial inventory level. If the initial inventory level is enough to cover the demand and the safety stock, then the net demand is zero. For the remaining periods, the net demand is computed as the net demand until period t ($\sum_{\ell=1}^{t} d_{ij\ell} + \underline{S}_{ijt} - S_{ij}$), minus the accumulated net demand until the previous time period ($\sum_{\ell=1}^{t-1} nd_{ij\ell}$).

Example 2. Consider the case of item 1 and location 1 given in Example 1, where the initial inventory level is $S_{11} = 7$, the safety stock vector for 5 periods is given by $\underline{S}_{11\cdot} = [2, 2, 2, 2, 7]$, and the demand vector is $d_{11\cdot} = [3, 3, 5, 2, 9]$. The net demands are given by: $nd_{111} = \max\{0, 3+2-7\} = 0, nd_{112} = \max\{0, 6+2-7-0\} = 1, nd_{113} = \max\{0, 11+2-7-1\} = 5, nd_{114} = \max\{0, 13+2-7-6\} = 2$, $nd_{115} = \max\{0, 22 + 7 - 7 - 8\} = 14$.

Net demands can be used to strengthen inequalities (3). Constant M can be replaced by

$$M_{jt} = \sum_{i \in N_j} \sum_{\ell=t}^{r} nd_{ij\ell}$$

which is the minimum upper bound on the delivery quantity at each location j and on the time period t. The strengthened inequalities are

$$\sum_{i \in N_j} x_{ijt} \leq M_{jt} y_{jt}, \qquad j \in L, t \in T. \tag{11}$$

Next, we discuss two types of approaches to further strengthen the model: the inclusion of valid inequalities and the use of extended formulations.

3.1 Valid Inequalities

A first family of valid inequalities can be derived by disaggregation of inequalities (11) (or the weaker version (3)), as follows:

$$x_{ijt} \leq \sum_{\ell=t}^{r} nd_{ij\ell} y_{jt}, \qquad i \in N, j \in L_i, t \in T. \tag{12}$$

As a large number of items may be considered, introducing all such inequalities can make the model too large. Several approaches are possible. An approach is to add these inequalities dynamically, that is, add each inequality when it is violated by the linear fractional solution. Another approach is to consider only a representative item for each location and add these inequalities for this item. For each location $j \in L$, we choose the representative item, denoted by $i(j)$, as follows:

$$i(j) = argmax_{i \in N_j} \{ \sum_{t \in T} nd_{ijt} \}. \tag{13}$$

Another family of valid inequalities is a type of cut set inequalities which have been introduced for related, although simpler, lot-sizing problems with upper bounds on the inventory levels, see [6]. For each $j \in L$ define the set $T_j = \{(t_1, t_2) \in (T, T) \mid \sum_{i \in N_j} \sum_{t=t_1}^{t_2} nd_{ijt} > V_j\}$. Thus (t_1, t_2) belongs to T_j if the accumulated net demand during the periods t_1 to t_2 is greater than the upper stock capacity, which implies that this demand cannot be fully met using only inventory. Thus, at least a delivery is necessary to the location j during the periods t_1 to t_2:

$$\sum_{t=t_1}^{t_2} y_{jt} \geq 1, \qquad j \in L, (t_1, t_2) \in T_j. \tag{14}$$

When $t_1 = 1$ we obtain a particular case since the initial inventory level can be used instead of the upper bound capacity V_j. As we are already using this initial stock to compute the net demands, we just need to check in which time period at least one net demand is positive in order to enforce the first visit to that location. For each location j let $t(j)$ denote the first time period where the net demand is positive, that is, $t(j) = \min\{t \in T \mid \sum_{i \in N_j} \sum_{\ell=1}^{t-1} nd_{ij\ell} = 0 \wedge \sum_{i \in N_j} \sum_{\ell=1}^{t} nd_{ij\ell} > 0\}$. Then, the following inequalities are valid for the set of feasible solutions of the MIID problem:

$$\sum_{\ell=1}^{t(j)} y_{j\ell} \geq 1, \qquad j \in L. \tag{15}$$

3.2 Extended Formulation

A common approach to derive tighter models in lot-sizing problems is to use extended formulations that use additional variables, see for instance [11,20,21]. Here we consider additional variables $v_{ijt\ell}$ to indicate the fraction of the net demand for item i in location j at time period ℓ that is delivered in time period t, for $\ell \geq t$.

The following constraints are added to the model MIID:

$$x_{ijt} = \sum_{\ell=t}^{r} nd_{ij\ell} v_{ijt\ell}, \qquad i \in N, j \in L_i, t \in T, \tag{16}$$

$$v_{ijt\ell} \leq y_{jt}, \qquad i \in N, j \in L_i, t, \ell \in T | t \leq \ell, \tag{17}$$

$$\sum_{t=1}^{\ell} v_{ijt\ell} = 1, \qquad i \in N, j \in L_i, \ell \in T, \tag{18}$$

$$v_{ijt\ell} \geq 0, \qquad i \in N, j \in L_i, t, \ell \in T | t \leq \ell. \tag{19}$$

Constraints (16) relate the new variables with the x_{ijt} variables. Constraints (17) ensure that if $v_{ijt\ell} > 0$ for at least one $\ell \geq t$, then y_{jt} must be one. Constraints (18) ensure that demand must be completely satisfied, and constraints (19) define the new variables as nonnegative. Constraints (18) together with constraints (19) ensure $v_{ijt\ell} \leq 1$.

The resulting model (with constraints (2)–(10), (16)–(19)), denoted by MIID-EF, can become too large for practical size instances. In order to use such type of reformulation in practical instances we tested two weaker approaches. One is to introduce constraints (16)–(19) only for the representative item $i(j)$ of each location j. The resulting model will be denoted by MIID-EF-R. The other approach is to introduce the extended formulation for the aggregated model. For the aggregated case we define $V_{jt\ell}$ to indicate the fraction of the total net demand for period ℓ in location j delivered in time period t, for $\ell \geq t$. The extended formulation for the aggregated model, denoted by MIID-A-EF, is given by (2)–(10), (20)–(23), with:

$$\sum_{i \in N_j} x_{ijt} = \sum_{i \in N_j} \sum_{\ell=t}^{r} nd_{ij\ell} V_{jt\ell}, \qquad j \in L, t \in T, \tag{20}$$

$$V_{jt\ell} \leq y_{jt}, \qquad j \in L, t, \ell \in T | t \leq \ell, \tag{21}$$

$$\sum_{t=1}^{\ell} V_{jt\ell} = 1, \qquad j \in L, \ell \in T, \tag{22}$$

$$V_{jt\ell} \geq 0, \qquad j \in L, t, l \in T | t \leq \ell. \tag{23}$$

4 Heuristic Scheme

As the MIID problem is NP-hard it is important to derive heuristic schemes that provide good quality solutions within reasonable amount of time. Here we describe an hybrid heuristic scheme that uses mathematical models. Hybrid schemes have been successfully employed in the past [4].

First we solve a relaxation of MIID-A-EF where the integrality of the x variables is relaxed (constraints (9)) and the vehicle capacity constraints (5) are ignored. This relaxation is tightened with inequalities (15). This relaxation of the MIID problem is denoted RMIID.

A feasible solution $(\overline{x}, \overline{y})$ to the RMIID problem may lead to an infeasible solution to the MIID problem. That is, fixing $y_{jt} = \overline{y}_{jt}$ in model (2)–(10) may originate an infeasible solution. Nevertheless such solution is expected to be "close" to a good solution of MIID. The next step is a local search heuristic that searches for solutions in a neighborhood of $\overline{y}$. In order to do that search we add the following inequality,

$$\sum_{j \in L, t \in T | \overline{y}_{jt}=0} y_{jt} + \sum_{j \in L, t \in T | \overline{y}_{jt}=1} (1 - y_{jt}) \leq \Delta. \tag{24}$$

Inequality (24) counts the number of variables that are allowed to flip their value from the value taken in the solution.

We denote the Local Search model which is the MIID model with the additional constraint (24) by $LSP(\Delta)$. Notice that by increasing the value of Δ the model becomes less restrictive, that is, the neighborhood becomes larger. For larger values of Δ the constraint (24) becomes ineffective and we obtain the original model MIID. This local search can be seen as a particular case of the Local Branching approach introduced in [13].

The heuristic scheme is given in Algorithm 1.

Algorithm 1. An hybrid heuristic scheme for the MIID problem.

1: Solve the RMIID model through branch-and-cut, using a solver, for α seconds.
2: Let $\overline{y}_{jt}$ denote the best solution found.
3: Set $\Delta \leftarrow \beta$.
4: **repeat**
5:    do a local search by solving the $LSP(\Delta)$ for α seconds;
6:    $\Delta \leftarrow \Delta + 1$;
7: **until** a feasible solution is found.

In the computational tests we used $\alpha = 100$ and $\beta = 5$.

5 Branch and Cut Algorithm

The discussed model improvements to tighten the formulation and the proposed heuristic were used together within a branch and cut algorithm. Several combinations were tested. For completeness, the exact approach that provided best results is given below in Algorithm 2.

Algorithm 2. The Branch and Cut algorithm.

1: Solve the heuristic described in Algorithm 1.
2: Set BFS to the value obtained in Step 1.
3: Solve the improved model (1)–(2), (4)–(10), (11), (12) for the representative item only, (14)–(15) with a solver and with the inclusion of the cutoff point BFS.

6 Computational Results

In order to test the performance of the heuristic and the exact approaches, in this section we report some computational experiments. All the computational tests were performed using a processor Intel(R) Core(TM) i7-4750HQ CPU @

2.00 GHz with 8 GB of RAM and using the software Xpress 7.6 (Xpress Release July 2015 with Xpress-Optimizer 28.01.04 and Xpress-Mosel 3.10.0) [12].

First we provide details on the generation of benchmark instances. We generated 12 instances with $n = 4200$ items and $m = 20$ wards that resemble the dimensions of the real problem. The demands were randomly generated following patterns of some representative items. Except in special cases, there are no deliveries on weekends and holidays. Thus we considered $r = 5$ time periods. The first four periods are a single week day, corresponding to Monday, Tuesday, Wednesday and Thursday. The last one, Friday, is deemed to correspond to three days, Friday, Saturday and Sunday. There are items required in all nursing wards and there are specific items that are only required in some nursing wards. The data was generated such that approximately 0.6% of the items are required in all the nursing wards and each ward requires about 10% of the items. The demand of each item, at each ward on the first four periods are randomly generated between 0 and 60 and in the fifth period can be up to 180. The capacity c_i of each item i is a random number between 1 and 5 and the capacity C of the vehicle is approximately 30% of the capacity of the total demand during all the five periods. The safety stock $\underline{S}_{ijt}$ is obtained by rounding up the average demand by period. For the initial stock level S_{ij} of each product i at location j two cases were considered. In case 1 this value is approximately four times the corresponding average demand per period while in case 2 this value is approximately three times of the average demand. The value of stock capacity V_j is a randomly generated number between 30% and 80% of the total demand at this nursing ward plus the initial stock level. These instances can be found in http://www.cerveira.utad.pt/datamod2016.

In the computational experiments were considered two values for parameter K, $K = 15$ and $K = 12$, which bounds the number of locations that can be visited per time period, constraints (4). The instances $A11$, $A21$, $A31$, $A12$, $A22$, $A32$ consider parameter $K = 15$ while instances $B11$, $B21$, $B31$, $B12$, $B22$, $B32$ consider parameter $K = 12$. Instances ending with 1 correspond to case 1 of the initial stock level while instances ending with 2 correspond to case 2 of the initial stock level. For each pair of these parameters we have three instances. The middle number of the name of the instance differentiates these three instances.

In Table 2 are shown the days when there are deliveries for an optimal solution considering the two cases for initial stock level. This illustrates how the initial stock levels influence the weekly number of deliveries. In this instance, the number of items required at each nursing ward ranges between 409 and 477, being 436 on average. Amongst the 4200 items, 28 are required in all wards and approximately 45% are specific items of a nursing ward. In case 1, the initial stock S_{ij} on average is equal to 124.46 and the optimal value is 34 while, in case 2, the average initial stock is equal to 93.36 and the optimal value decreased to 38. With higher values of initial stock, case 1, there are more wards visited only once. By a detailed analysis of the solutions it can be seen that even in the cases with two delivers per week, there are some products delivered once.

Table 2. Optimal supply scheme

	Case 1						Case 2					
Ward	$t=1$	$t=2$	$t=3$	$t=4$	$t=5$	# visits	$t=1$	$t=2$	$t=3$	$t=4$	$t=5$	# visits
w1			✓		✓	2	✓				✓	2
w2	✓				✓	2	✓			✓		2
w3		✓				1	✓	✓				2
w4		✓				1		✓				1
w5	✓				✓	2	✓				✓	2
w6			✓			1		✓			✓	2
w7		✓				1		✓				1
w8	✓				✓	2	✓				✓	2
w9	✓				✓	2		✓			✓	2
w10			✓			1	✓				✓	2
w11	✓			✓		2	✓			✓		2
w12		✓			✓	2	✓				✓	2
w13		✓		✓		2		✓		✓		2
w14		✓			✓	2		✓			✓	2
w15	✓			✓		2	✓				✓	2
w16			✓			1	✓	✓				2
w17	✓				✓	2		✓			✓	2
w18		✓	✓			2		✓			✓	2
w19	✓			✓		2		✓		✓		2
w20	✓				✓	2		✓			✓	2
	Total delivers = 34						Total delivers = 38					
	Average of S_{ij} = 124.46						Average of S_{ij} = 93.36					

The computational results obtained for the 12 instances of the MIID problem are displayed in Tables 3, 5, 6, and 7. All the computational times are given in seconds.

Table 3 displays the computational results for the model with no improvements (model (1)–(10)) solved using the branch-and-cut procedure from the Xpress solver, with a time limit of 7200 s. Column "*bb*" indicates the value of the best bound found when the algorithm terminates. The corresponding gap, $gap = \frac{mip-bb}{mip} \times 100$, is displayed in column "*gap*", and the value of the best integer solution found is indicated in column "*mip*". We can see that for all the instances the algorithm has stopped without proving optimality.

Tables 5 and 6 display the computational results for the linear relaxation of model (1)–(10), denoted by lp, and its improvements from Sect. 3. For readability, in Table 4 the tested relaxations are summarized and characterized in terms of their constraints.

Table 5 displays the computational results obtained for lp, lp-A, lp-D, lp-R and lp-I. Each pair of columns (except the first one) give the value of the linear relaxation and running time. The best linear programming relaxation values

Table 3. Computational results for the model (1)–(10) solved using the branch-and-cut procedure.

Instance	*bb*	*gap*	*mip*
A11	29.0	12.0	33.0
A21	30.2	18.3	37.0
A31	29.1	17.0	35.0
A12	26.5	30.1	38.0
A22	30.4	21.9	39.0
A32	27.6	25.4	37.0
B11	31.2	8.2	34.0
B21	31.2	15.7	37.0
B31	30.0	16.6	36.0
B12	28.3	25.6	38.0
B22	32.2	17.3	39.0
B32	38.0	4.9	40.0

Table 4. List of relaxed models with indication of the constraints added to the linear relaxation lp.

Relaxed model	Added constraints
lp-A	(11)
lp-D	(12)
lp-R	(12) for the representative item given by (13)
lp-I	(14)–(15)
lp-MIID-EF	(16)–(19)
lp-MIID-EF-R	(16)–(19) for the representative item
lp-MIID-A-EF	(20)–(23)

were obtained for lp-D and lp-I relaxations. However lp-D has the disadvantage of using more execution time since it adds many more inequalities.

Table 6 displays the computational results for the linear relaxation of the extended formulations lp-MIID-EF, lp-MIID-EF-R, and lp-MIID-A-EF, see Table 4. As for Table 5, each pair of columns give the value of the linear relaxation of the improved model and the corresponding running time. The best value was obtained when considering model MIID-EF.

Overall, one can verify that the best approaches to improve the linear relaxation bound are the inclusion of the extended formulation, the disaggregated inequalities (12), lp-D relaxation, and the addition of inequalities (14) and (15), lp-I relaxation. While the two first cases imply a large increase on the size of the model, with inequalities (14) and (15) the size is kept under control.

Table 5. Computational results for the linear relaxation of the model (1)–(10) and the improvements from Sect. 3.1.

Instance	lp		lp-A		lp-D		lp-R		lp-I	
	Value	Time	Value	Time	Value	Time	Value	Time	Value	Time
A11	1.4	0.9	20.0	1.4	20.9	91.1	20.0	0.9	20.5	0.9
A21	1.4	0.8	20.0	1.3	21.0	97.7	20.0	0.9	20.5	1.0
A31	1.4	0.9	20.0	1.4	20.9	109.3	20.0	0.9	20.5	0.9
A12	1.4	0.7	20.3	3.7	22.7	79.8	20.1	0.8	25.7	0.7
A22	1.4	0.7	20.2	3.4	22.5	101.1	20.0	0.6	25.8	0.7
A32	1.4	0.7	20.2	3.7	22.7	101.2	20.0	0.6	23.9	0.7
B11	1.4	0.9	20.0	1.4	20.9	93.6	20.0	0.9	20.5	0.9
B21	1.4	0.8	20.0	1.3	21.0	106.7	20.0	0.8	20.5	1.0
B31	1.4	0.9	20.0	1.4	20.9	120.1	20.0	0.9	20.5	0.9
B12	1.4	0.7	20.3	3.7	22.7	79.1	20.1	0.8	25.7	0.7
B22	1.4	0.7	20.2	3.3	22.5	91.6	20.0	0.6	25.9	0.7
B32	1.4	0.7	20.3	3.3	22.7	105.3	20.0	0.6	29.6	0.7

Table 6. Computational results for the linear relaxations of the extended formulations from Sect. 3.2.

Instance	lp-MIID-EF		lp-MIID-EF-R		lp-MIID-A-EF	
	Value	Time	Value	Time	Value	Time
A11	23.3	3.9	20.0	0.9	20.0	0.9
A21	23.7	3.8	20.0	0.9	20.0	0.9
A31	23.4	5.6	20.0	0.9	20.0	0.9
A12	25.7	8.3	20.3	1.0	20.3	1.0
A22	26.2	7.6	20.4	1.1	20.4	1.1
A32	26.4	7.4	20.4	1.0	20.4	1.0
B11	23.3	3.9	20.0	0.8	20.0	0.8
B21	23.7	3.8	20.0	0.8	20.0	0.9
B31	23.4	5.5	20.0	0.8	20.0	0.9
B12	25.6	8.1	20.4	1.0	20.4	1.0
B22	26.2	6.7	20.4	1.2	20.4	1.2
B32	25.7	9.5	20.4	1.1	20.4	1.0

Table 7 displays the computational results for Algorithm 1 and for the improved branch-and-cut algorithm introduced in Sect. 5. The second and third columns give the objective value and running time of the first part of the heuristic scheme, that is, solving the relaxation model *RMIID* as described in Step 1 of Algorithm 1. The fourth column gives the value of the best solution

Table 7. Computational results for the heuristic procedure, and for the improved branch-and-cut.

Instance	Heuristic				Improved branch-and-cut		
	Relaxation RMIID		Local branching				
	Value	Time	Value	Time	Value	Time	Gap
A11	34	100.4	34	57.7	34	5.9	0.0
A21	37	99.6	37	100.0	37	6.5	0.0
A31	35	19.6	35	80.2	35	59.9	0.0
A12	38	99.9	38	100.2	38	218.9	0.0
A22	39	71.9	39	100.5	39	1200	8.3
A32	37	38.0	37	103.2	37	204.8	0.0
B11	34	99.5	34	26.2	34	5.7	0.0
B21	37	100.3	37	101.4	37	6.1	0.0
B31	35	20.4	35	72.3	35	47.9	0.0
B12	39	99.4	38	100.0	38	366.7	0.0
B22	39	40.7	39	100.4	39	1200	8.5
B32	41	99.5	40	100.1	40	136.2	0.0

found with the second part of the heuristic scheme as described in Steps 2 to 7 of Algorithm 1 and the corresponding running time is given in column five. The last three columns give the best solution value, the running time, and the final gap of the improved branch-and-cut procedure, Algorithm 2, solved with the Xpress solver and imposing a time limit of 1200 s.

With the improved branch-and-cut, the optimal value is obtained for 10 out of 12 instances. Only for the instances A22 and B22, the obtained solutions have a lower bound gap of 8.3 and 8.5, respectively. These results confirm both the quality of the heuristic and the importance of the improvements to derive optimal or near optimal solutions within reasonable running times.

7 Conclusions

The paper presents exact and heuristic approaches for a multi-item problem that combines lot-sizing and distribution decisions, occurring in the logistics of an hospital. All the approaches are based on a mathematical formulation which is improved with valid inequalities and an extended formulation. The heuristic scheme combines the resolution of a relaxation with a local search. The exact approach is a branch-and-cut algorithm based on the improved model and using as cut-off the upper bound provided with the heuristic scheme. Computational results show the heuristic and the exact approach are efficient in solving the tested instances.

For future research we intend to integrate the current weekly distribution planning models into a larger horizon planning problem where the orders from suppliers and the inventory management at the warehouse are considered.

Acknowledgements. The research of the first and third authors was partially supported through CIDMA and FCT, the Portuguese Foundation for Science and Technology, within project UID/MAT/ 04106/2013. The research of the second author is financed by the ERDF - European Regional Development Fund through the Operational Programme for Competitiveness and Internationalisation - COMPETE 2020 Programme within project "POCI-01-0145-FEDER-006961", and by National Funds through the FCT as part of project UID/EEA/50014/2013.

References

1. Adulyasak, Y., Cordeau, J., Jans, R.: The production routing problem: a review of formulations and solution algorithms. Comput. Oper. Res. **55**, 141–152 (2015)
2. Agra, A., Andersson, H., Christiansen, M., Wolsey, L.: A maritime inventory routing problem: discrete time formulations and valid inequalities. Networks **62**, 297–314 (2013)
3. Agra, A., Christiansen, M., Delgado, A.: Mixed integer formulations for a short sea fuel oil distribution problem. Transp. Sci. **47**, 108–124 (2013)
4. Agra, A., Christiansen, M., Delgado, A., Simonetti, L.: Hybrid heuristics for a short sea inventory routing problem. Eur. J. Oper. Res. **236**, 924–935 (2014)
5. Archetti, C., Bertazzi, L., Laporte, G., Speranza, M.G.: A branch-and-cut algorithm for a vendor-managed inventory-routing problem. Transp. Sci. **41**(3), 382–391 (2007)
6. Atamtürk, A., Küçükyavuz, S.: Lot sizing with inventory bounds and fixed costs: polyhedral study and computation. Oper. Res. **53**(4), 711–730 (2005)
7. Bijvank, M., Vis, I.F.A.: Inventory control for point-of-use locations in hospitals. J. Oper. Res. Soc. **63**, 497–510 (2012)
8. Brown, G., Keegan, J., Vigus, B., Wood, K.: The Kellogg company optimizes production, inventory, and distribution. Interfaces **31**, 1–15 (2001)
9. Dacosta-Claro, I.: The performance of material management in health care organizations. Int. J. Health Plann. Manag. **17**(1), 69–85 (2002)
10. de Vries, J.: The shaping of inventory systems in health services: a stakeholder analysis. Int. J. Prod. Econ. **133**, 60–69 (2011)
11. Eppen, G.D., Martin, R.K.: Solving multi-item capacitated lot-sizing problems using variable redefinition. Oper. Res. **35**, 832–848 (1997)
12. FICO Xpress Optimization Suite. http://www.fico.com/en/Products/DMTools/Pages/FICO-Xpress-Optimization-Suite.aspx
13. Fischetti, M., Lodi, A.: Local branching. Math. Program. Ser. B **98**, 23–47 (2003)
14. Kelle, P., Woosley, J., Schneider, H.: Pharmaceutical supply chain specifics and inventory solutions for a hospital case. Oper. Res. Health Care **1**(2–3), 54–63 (2012)
15. Nemhauser, G., Wolsey, L.A.: Integer and Combinatorial Optimization. Wiley, Hoboken (1988)
16. Pan, Z.X., Pokhareal, S.: Logistics in hospitals: a case study of some Singapore hospitals. Leadersh. Health Serv. **20**(3), 195–207 (2007)
17. Pochet, Y., Wolsey, L.A.: Production Planning by Mixed Integer Programming. Springer, Berlin (2006)

18. Rosales, C.R., Magazine, M., Rao, U.: Point-of-use hybrid inventory policy for hospitals. Decis. Sci. **45**(5), 913–937 (2014)
19. Rais, A., Viana, A.: Operations research in healthcare: a survey. Int. Trans. Oper. Res. **18**, 1–31 (2011)
20. Solyal, O., Sural, H.: A branch-and-cut algorithm using a strong formulation and an a priori tour-based heuristic for an inventory-routing problem. Transp. Sci. **45**(3), 335–345 (2011)
21. Solyal, O., Sural, H.: The one-warehouse multi-retailer problem: reformulation, classification and computational results. Ann. Oper. Res. **196**(1), 517–541 (2012)
22. Uthayakumar, R., Priyan, S.: Pharmaceutical supply chain and inventory management strategies: optimization for a pharmaceutical company and a hospital. Oper. Res. Health Care **2**, 52–64 (2013)
23. Volland, J., Fügener, A., Schoenfelder, J., Brunner, J.O.: Material logistics in hospitals: a literature review, Omega (2016, in press)

Adaptive Targeting in Online Advertisement: Models Based on Relative Influence of Factors

Andrey Pepelyshev[1(✉)], Yuri Staroselskiy[2], Anatoly Zhigljavsky[1,3], and Roman Guchenko[2,4]

[1] Cardiff University, Cardiff, UK
{pepelyshevan,ZhigljavskyAA}@cardiff.ac.uk
[2] Crimtan, London, UK
{yuri,rguchenko}@crimtan.com
[3] Lobachevskii State University of Nizhnii Novgorod, Nizhnii Novgorod, Russia
[4] St. Petersburg State University, Saint Petersburg, Russia

Abstract. We consider the problem of adaptive targeting for real-time bidding for internet advertisement. This problem involves making fast decisions on whether to show a given ad to a particular user. For demand partners, these decisions are based on information extracted from big data sets containing records of previous impressions, clicks and subsequent purchases. We discuss several criteria which allow us to assess the significance of different factors on probabilities of clicks and conversions. We then devise simple strategies that are based on the use of the most influential factors and compare their performance with strategies that are much more computationally demanding. To make the numerical comparison, we use real data collected by Crimtan in the process of running several recent ad campaigns.

Keywords: Online advertisement · Real-time bidding · Adaptive targeting · Big data · Conversion rate

1 Introduction

During the last decade online advertisement became a significant part of the total advertisement market. Many companies including Google, Facebook and online news portals provide possibilities for online advertisement of their webpages to generate revenue. With high penetration of internet, online advertisement has gained attraction from marketers due to its specific features like scalability, measurability, ability to target individual users and relatively low cost per ad shown.

There are three main forms of online advertisement: search advertising (occurrs when a user conducts an online search), classified advertising (ads appear on websites of particular types, e.g. jobs and dating websites), and display advertising (banner ads on websites which are not search engines). During the last five years search and display advertising have moved from direct relationship between seller and buyer of ads to an advanced and flexible auction-based

P.M. Pardalos et al. (Eds.): MOD 2016, LNCS 10122, pp. 159–169, 2016.
DOI: 10.1007/978-3-319-51469-7_13

model [12]. In this model, there is a seller of an ad space and several buyers - technology companies, who specialize in efficient ad delivery. Typically, demand partners pay per view, and prices are defined as cost per thousand ad exposures.

Display advertisement via actions has empowered the growth of the so-called programmatic buying, that is buying when decisions are made by machines based on algorithms and big data sets, rather than people. Demand partners typically collect databases with logs of all previous requests from auctions, impressions, clicks, conversions and users who visited a website which is currently advertised. These logs usually contain an anonimized user id, a browser name, an OS name, a geographical information derived from the IP address and a webpage link where an auction is run. Merging these datasets with third party data sources provides possibilities for contextual, geographical and behavioural targeting.

We consider the problem of online advertisement via auctions holding by independent ad exchanges from the position of a demand partner which wants to optimise the conversion rate. The demand partner has to decide how reasonable is it showing an ad in regard to a request from an auction and then possibly suggest a bid.

Demand partners put a special code on an advertised site to record users who made conversions. After few weeks of monitoring and running an ad campaign, demand partners collect a database with several thousands of conversions with just few records from this database occurring due to impressions. Also demand partners collect another database with requests on possibility to show an ad. By comparing these databases, demand partners have to develop procedures for estimating the conversion rate for new requests and subsequent bidding. Since demand partners should suggest a bid in few milliseconds, these procedures must be fast.

The demand partner has to solve the problem of maximizing either the click through rate (CTR) or the conversion rate by targeting a set of requests under several constraints: (a) budget (total amount of money available for advertising), (b) number of impressions N_{total} (the total amount of ad exposures), and (c) time (any ad campaign is restricted to a certain time period).

The problem of adaptive targeting for ad campaigns was recently addressed in quite a few papers, see e.g. [4,5,7,13]. In 2014 two contests were organized in Kaggle portal, see [14] and [15] on algorithms for predicting the CTR using a dataset with subsampled non-click records so that the CTR for the dataset is about 20% while for a typical advertising campaign the CTR is about 0.4% or less. The algorithms, which were proposed are publicly available and give approximately the same performance with respect to the logarithmic loss criterion

$$\text{logloss} = -1/N \sum_{i=1}^{N} (y_i \log(p_i) + (1 - y_i) \log(1 - p_i)), \quad (1)$$

where N is the size of the data set, p_i is the predicted probability of click for the i-th request, and $y_i = 1$ if the i-th leads to click and $y_i = 0$ otherwise. This criterion, however, does not look very sensible when the probabilities p_i

are very small as it pays equal weights to type I and type II error probabilities. Moreover, the criterion (1) and other loss functions are not much often used in the industry as the advertisers are not interested in approximating click (conversion) probabilities at the entire range of admissible values of these probabilities; they are interested in making a decision (whether to show an ad) and hence they are only interested in making a correct decision whether some p_i is smaller or larger than some threshold value p^* (so that if $p_i \geq p^*$ then the demand partner will propose a bid for the i-the user). The threshold value p^* should be small enough to ensure that we will get the total number of impressions in time. On the other hand, we cannot let p^* to be too small as otherwise the final CTR (or conversion rate) will be too small. Rather than reporting values of the logloss or other criteria for different strategies, we present graphs which display the conversion rate as a function of the size of the sample with largest predicted values of the conversion probabilities. These types of figures are very common in the industry for assessing performances of different strategies.

In previous two papers [8,9] we have made a critical analysis of several procedures for on-line advertisement, provided a unified point of view on these procedures and have had a close look at the so-called 'look-alike' strategies. In the present paper we study relative influence of different factors on the conversion rate and hence develop simple procedures which are very computationally light but achieve the same accuracy as computationally demanding algorithms like Gradient Boosting Machines (see Sect. 2.5) or Field-Aware Factorization Machines (FFM), see e.g. [10]. Note that the number of parameters in the simplest FFM models is the sum of all factor levels plus perhaps interactions between factor levels. It counts to millions and if at least some interactions are taken into account then the count takes to much larger numbers.

Our models proposed in Sect. 2.4 are entirely different, they have a relatively small number of parameters. In particular, we propose a sparse model where only a few most significant factors are used for predicting the conversion rate.

2 Relative Influence of Factors

2.1 Notation and Statement of the Problem

Databases of logs contain records with many factors. Therefore, it is important to find the relative influence of all available factors and then build a prediction model using only the most important factors (and perhaps their interactions) in order to keep the computational time of evaluating the model for a new request small. Let us start with a formal statement of the problem.

Suppose that we have a database with records $x_1, \ldots, x_N$ and a vector $y_1, \ldots, y_N$ of binary outputs such that $y_j = 1$ if the j-th record has led to a conversion and $y_j = 0$ otherwise.

Each record x_j is described by m factors, $x_j = (x_{j,1}, \ldots, x_{j,m})$. The list of factors typically includes a browser name, an OS name, a device type, a country, a region, a visited webpage and behaviour categories. Let $p(x)$ be an idealistic conversation rate for a request x; that is, $p(x) = \Pr\{y(x) = 1\}$. The knowledge of function $p(\cdot)$ would help us to construct an effective strategy of adaptive targeting for online advertisement. In practice, the function $p(\cdot)$ is unknown and even its existence is a mathematical model.

Suppose that the i-th factor has L_i levels $l_{i,1},\ldots,l_{i,L_i}$. A relationship between the i-th factor and the binary output can be described by the contingency table. Specifically, we define

$$n_{i,k,s} = \#\{j : y_j = s, x_{j,i} = l_{i,k}\}$$

as the number of records for which the output y_j equals s and the value $x_{j,i}$ takes the value at the k-th level for the i-th factor. Here $s \in \{0,1\}$, $i = 1,\ldots,m$ and $k = 1,\ldots,L_i$; note that $k = k_i$ depends on i.

For fixed i, the frequency table $(p_{i,k,s})_{k=1,\ldots,L_i}^{s=0,1}$ with

$$p_{i,k,s} = n_{i,k,s}/N$$

provides the joint empirical distribution for the pair of the i-th factor and the binary output, where N is the total number of records.

The row-sums for these frequency tables are $p_{i,k,*} = p_{i,k,0} + p_{i,k,1}$, so that the vector with frequencies $p_{i,k,*}$, $k = 1,\ldots,L_i$ gives the empirical distribution of levels for the i-th factor.

The column-sums for the frequency tables are

$$p_{i,*,s} = p_{i,1,s} + \ldots + p_{i,L_i,s}.$$

These values clearly do not depend on i so that

$$P = \frac{\sum_{j=1}^{N} y_j}{N} = p_{i,*,1} \text{ and } p_{i,*,0} = 1 - P$$

for all i where P is the overall frequency of 1 for the binary output; that is, the overall conversion rate for the database. Note, however, that P is not the conversion rate of an ad campaign because the database contains records of non-converted requests and converters which are not related to the active ad campaign (the converters recorded directly by the demand partners who put a special code on an advertised site to record users who made conversions).

To identify how the i-th factor affects the conversion rate $p(x)$, we consider several statistics which measure the dispersion or mutual information.

2.2 Relative Influence via Dispersion

To find the relative influence of the i-th factor on the conversion rate in the sense of the dispersion of the conversion rate for different levels of the i-th factor, we propose the statistic defined by

$$I_i^{(D)} = \sum_{k=1}^{L_i} p_{i,k,*}(q_{i,k} - P)^2$$

where

$$q_{i,k} = \frac{n_{i,k,1}}{n_{i,k,0} + n_{i,k,1}} = \frac{p_{i,k,1}}{p_{i,k,*}}$$

is the conversion rate for the records with k-th level for the i-th factor.

2.3 Relative Influence via Mutual Information

Mutual information is an information-theoretic measure of divergence between the joint distribution and the product of two marginal distributions, see the classical book [2]. If two random variables are independent, then the mutual information is zero. We apply mutual information to measure a degree of dependence between the i-th factor and the binary output.

To find the relative influence of the i-th factor in the sense the mutual information based on the Shannon entropy, we consider the statistic defined by

$$I_i^{(Sh)} = \sum_{k=1}^{L_i} \sum_{s=0}^{1} p_{i,k,s} \log_2 \frac{p_{i,k,s}}{p_{i,k,*}\, p_{i,*,s}}.$$

To find the relative influence of the i-th factor in the sense the mutual information based on the Renyi entropy of order α, we consider the statistic defined by

$$I_i^{(Re,\alpha)} = \log_2 \sum_{k=1}^{L_i} \sum_{s=0}^{1} \frac{p_{i,k,s}^{\alpha}}{p_{i,k,*}^{\alpha-1}\, p_{i,*,s}^{\alpha-1}}.$$

It is known that mutual information is not robust in the case in which there are levels with rare occurrence, see [1]. To regularize the above statistics $I_i^{(D)}, I_i^{(Sh)}$ and $I_i^{(Re,\alpha)}$, we perform a pre-processing of the database by replacing rare levels with $n_{i,k,*} \leq 9$ by a dummy level.

Note that the Renyi mutual information $I_i^{(Re,\alpha)}$ was used for factor selection in the literature, see e.g. [6]; however, the range of applications was entirely different. The Shannon mutual information $I_i^{(Sh)}$ is a standard in many areas.

2.4 MI-based Model for Estimating the Conversion Rate

Suppose that we are given a new request $X = (X_1, \ldots, X_m)$ and we want to estimate the conversion rate $p(X)$. As an estimator of $p(X)$, we propose

$$\hat{p}(X) = \frac{\sum_{i=1}^{m} I_i q_{i,k_i}}{\sum_{i=1}^{m} I_i} \tag{2}$$

where k_i is such that $X_i = l_{i,k_i}$ and I_i is a relative influence of the i-th factor.

Furthermore, if we want to use a sparse predictive model then we can set the values of I_i such that $I_i \leq \epsilon$ to zero, for some small $\epsilon > 0$.

The expression (2) resembles the form of the multi-factor multi-level ANOVA model. However, the model (2) uses totally different methods of estimating parameters than standard ANOVA regression.

The main advantage of the proposed model (2) is its simplicity and time efficiency. As we demonstrate below, precision of this model is basically identical to the precision of much more complicated models based on the use of the Gradient Boosting Machines (GBM).

2.5 Gradient Boosting Machines

GBM is a method of sequential approximation of the desired function $p(x)$ by a function of the form

$$p^{(k)}(x) = \sum_{i=1}^{k} \alpha_i h(x, \theta_i),$$

where at iteration k the coefficient α_k and the vector θ_k are estimated through minimizing some loss criterion $L(\cdot,\cdot)$; see e.g. [3,11]; the values of α_i and θ_i for $i < k$ being kept from previous iterations. Since many factors are categorical, we consider the special case of the so-called tree-based GBM, where the function $h(x,\theta)$ is called a regression tree and has the form

$$h(x, \theta) = \sum_{j=1}^{J} b_j \mathbf{1}_{R_j}(x)$$

where $R_1, \ldots, R_J$ are disjoint sets whose union is the whole space and these sets correspond to J terminal nodes of the tree. The indicator function $\mathbf{1}_R(x)$ equals 0 if x belongs to a set R and 0 otherwise. The vector θ for the regression tree $h(x,\theta)$ is a collection of $b_1, \ldots, b_J$ and $R_1, \ldots, R_J$, which parameterize the tree. Note that levels of categorical variables are encoded by integer numbers.

To build the GBM model for a real data, we take the `gbm` package in `R`. We use the function `gbm` which constructs the generalized boosted regression model has the following parameters, see [11]:

(i) `n.trees`, the total number of trees in the model,
(ii) `interaction.depth`, the maximal depth of factor interactions,
(iii) `n.minobsinnode`, the minimal number of records in the terminal nodes of trees,
(iv) `bag.fraction`, the fraction of records from the training set randomly selected to construct the next tree,
(v) `shrinkage`, the learning rate ν which is used to define $\alpha_i = \nu\gamma_i$, where

$$\gamma_i = \arg\min_{\gamma} \sum_{j=1}^{N} L(y_j, p^{(i-1)}(x_j) + \gamma h(x_j, \theta_i)).$$

The values used in industry are typically as follows `n.minobsinnode` = 100, `n.trees` = 500, `shrinkage` = 0.1, `interaction.depth` = 5, `bag.fraction` = 0.5.

3 Numerical Results

In the present section we analyze several ad campaigns which were executed by Crimtan.

To investigate the performance of different strategies for the database of requests for an ad campaign, we split the database of records into 2 sets: the training set of past records with dates until a certain time T and the test set of future records with dates from the time T. The training set contains 50,000 records but the test sets are much larger (their sizes are in the range of 1 million). We now compare GBM and the model based on the use of (2) by comparing the conversion rate for the samples of most favorable requests with the highest chances of conversion.

To form the sample of most favorable requests for the GBM approach, we construct the GBM model using the training set and then apply this model to predict the probability of conversion for each request from the test set. Now we can sort the predicted probabilities and create samples of requests with highest predicted probabilities of conversion.

In Fig. 1 we can see that all four considered versions of the relative influence of factors give somewhat similar orderings. We note that the factor 36 provides significant influence in some ad campaigns and small influence in others. However, factors 33 and 40 have large influence in all four ad campaigns.

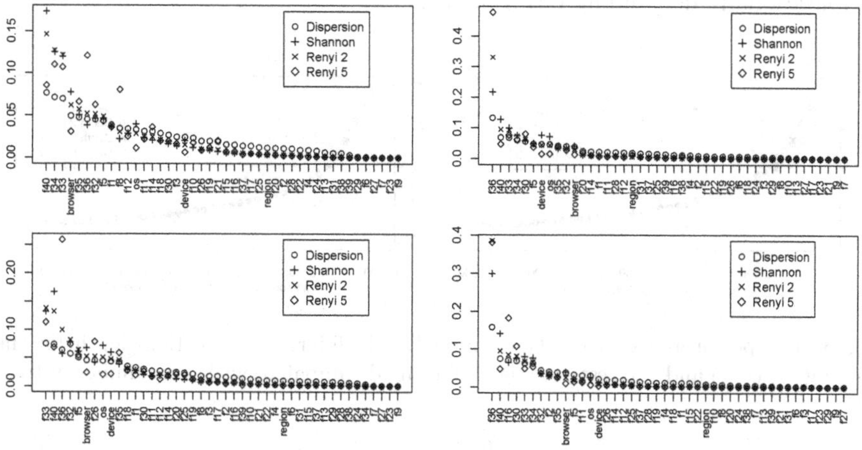

Fig. 1. Relative influence of factors for 4 ad campaigns.

In Fig. 2 we can see that the performance of the MI-based model is the same for the four considered versions of the relative influence of factors both for the training set and the test set for a chosen ad campaign. Since the performance for the test set is similar to the performance for the training set, we can conclude that there is no over-fitting in the MI-based model. This is not the case for the GBM, see Fig. 3. In particular, if the depth level is high then the GBM performance for the training set is visibly better than its performance for the test set.

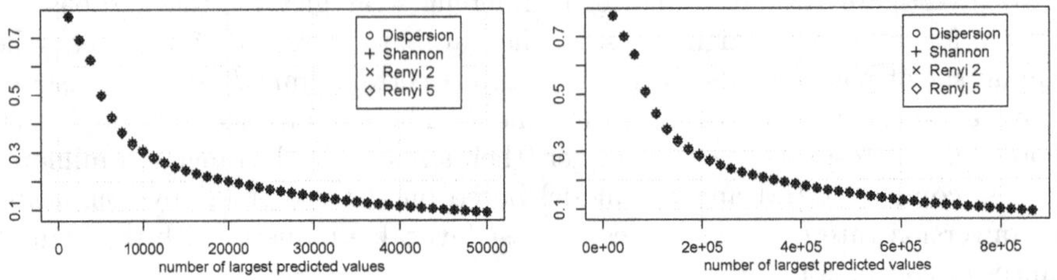

Fig. 2. The performance of the MI-based model with 4 forms of the relative influences of factors for the training set (left) and the test set (right) for an ad campaign. The y-scale is the conversion rate of samples of largest predicted values for various sample sizes.

In Fig. 3 we can also see that the performance of the GBM model does not depend on the interaction depth, when the number of trees is 500 and the bag fraction is 0.5. Comparing Figs. 2 and 3 we can see that the performance of the simple MI-based model is very close to the performance of the complex, computationally demanding GBM model.

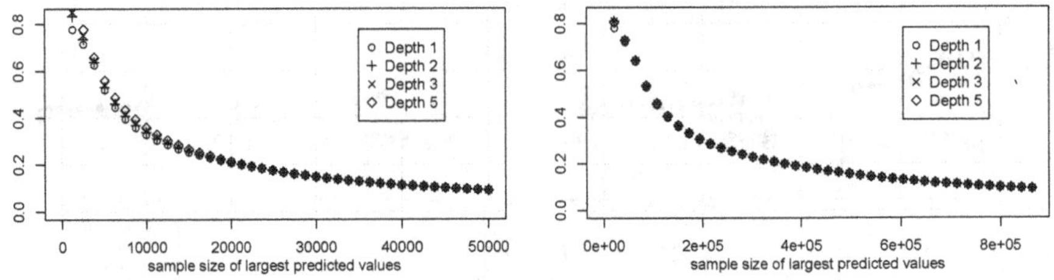

Fig. 3. The performance of the GBM model with different interaction depth for the training set (left) and the test set (right) for an ad campaign, with the number of trees 500.

In Fig. 4 we can see that the performance of the GBM model with larger number of trees on the training set is marginally better than with smaller number of trees. However, the performance of the GBM model with different number of trees for the test set is virtually the same.

In Figs. 5 and 6 we compare the performance of the proposed MI-based model with $I_i^{(D)}$ to the performance of the GBM model with 500 trees and the interaction depth 5 (which is a very time-consuming algorithm). For both ad campaigns, GBM performance on the training sets is slightly better than the performance

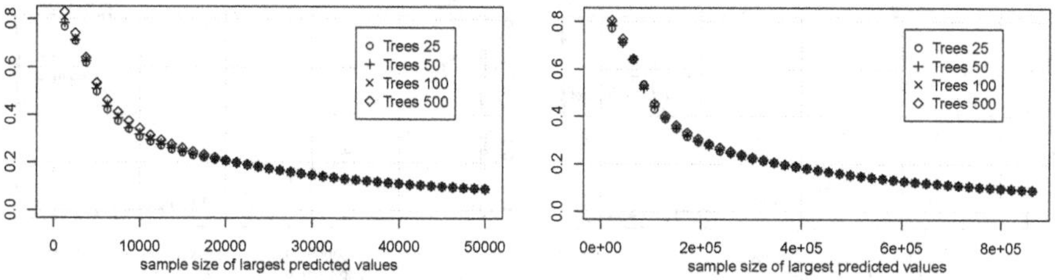

Fig. 4. The performance of the GBM model with different number of trees for the training set (left) and the test set (right) for an ad campaign, with the interaction depth 2.

of the proposed algorithm. This can be explained by the fact that GBM has thousands times more degrees of freedom than our model and, by the definition of the method, GBM tries to fit the data as best as it can.

GBM performances on the test sets are slightly worse than that on the training sets and they are very similar to the performance of the proposed algorithm. A slight advantage of GBM over the MI-based method for the records X that have high values of probabilities $p(X)$ is not important for the following two reasons: (a) high probabilities of $p(X)$ can only be observed for the supplementary part of the database containing the records which are not a part of the ad campaign, and (b) as mentioned above, we are interested in a good estimation of $p(X)$ such that $p(X) \simeq p^*$, where p^* is the threshold value, which is quite small (see http://cs.roanoke.edu/~thlux/).

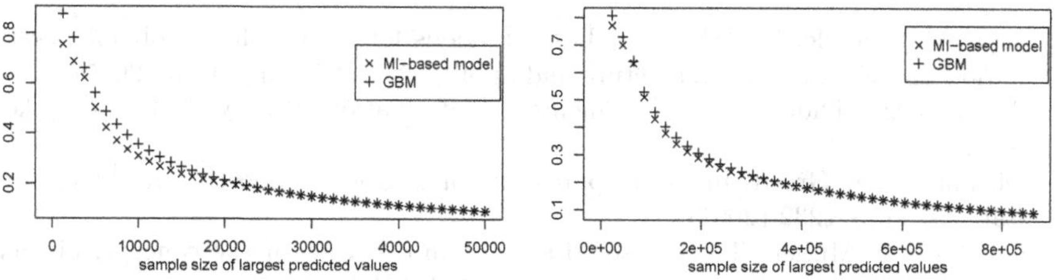

Fig. 5. The performance of the MI-based model with $I_i^{(D)}$ and the GBM model with 500 trees and the interaction depth 5 for the training set (left) and the test set (right) for an ad campaign.

We should notice that the MI-based model $\hat{p}(X)$ is not good for estimating the conversion rate $p(X)$ in view of some bias. We can only use the MI-based models for ranking the requests using predictive values and choosing the most promising ones. If one wishes to enhance the MI-based model and obtain a good estimator of $p(X)$, then we recommend to remove non-influential factors by computing the mutual information and build a logistic model using the most influential factors and possibly their interactions.

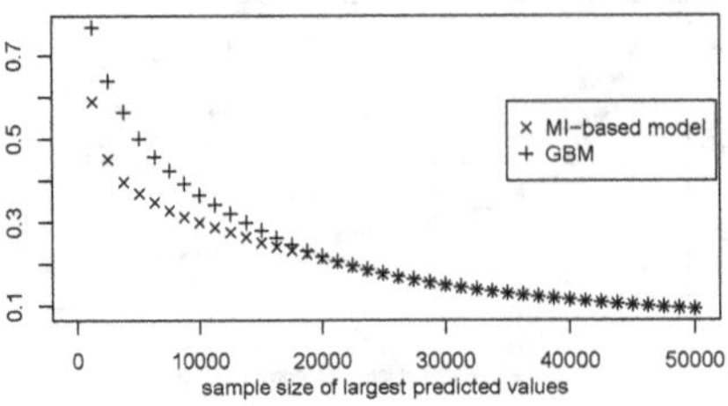

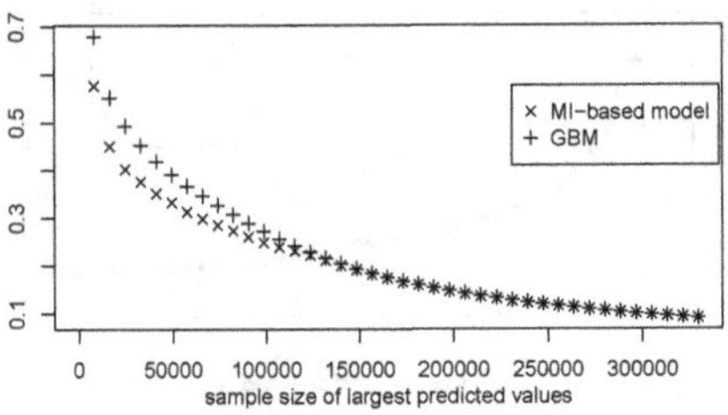

Fig. 6. The performance of the MI-based model with $I_i^{(D)}$ and the GBM model with 500 trees and the interaction depth 5 for the training set (left) and the test set (right) for another ad campaign.

Finally we would like to highlight the time efficiency of computations for the proposed model. Construction of the MI-based model and evaluating of the MI-based model for new requests is at least 10 times faster comparing with the GBM model. In fact, the MI-based model can be used in the regime of real-time learning; that is, the contingency tables and the predictive model can be easily updated as bunches of new requests arrived.

Acknowledgement. The paper is a result of collaboration of Crimtan, a provider of proprietary ad technology platform and University of Cardiff. Research of the third author was supported by the Russian Science Foundation, project No. 15-11-30022 "Global optimization, supercomputing computations, and application".

References

1. Buja, A., Stuetzle, W., Shen, Y.: Loss functions for binary class probability estimation and classification: structure and applications. Working draft (2005)
2. Cover, T.M., Thomas, J.A.: Elements of Information Theory. Wiley, Hoboken (2012)
3. Friedman, J.H.: Greedy function approximation: a gradient boosting machine. Ann. Stat. **29**, 1189–1232 (2001)
4. Jansen, B.J., Mullen, T.: Sponsored search: an overview of the concept, history, and technology. Int. J. Electron. Bus. **6**(2), 114–131 (2008)
5. He, X., Pan, J., Jin, O., Xu, T., Liu, B., Xu, T., Candela, J.Q.: Practical lessons from predicting clicks on ads at Facebook. In: Proceedings of the Eighth International Workshop on Data Mining for Online Advertising, pp. 1–9. ACM (2014)
6. Lima, C.F.L., de Assis, F.M., de Souza, C.P.: An empirical investigation of attribute selection techniques based on Shannon, Renyi and Tsallis entropies for network intrusion detection. Am. J. Intell. Syst. **2**(5), 111–117 (2012)
7. McMahan, H.B., Holt, G., Sculley, D., et al.: Ad click prediction: a view from the trenches. In: Proceedings of the 19th ACM SIGKDD International Conference on Knowledge Discovery and Data Mining, pp. 1222–1230. ACM (2013)
8. Pepelyshev, A., Staroselskiy, Y., Zhigljavsky, A.: Adaptive targeting for online advertisement. In: Pardalos, P., Pavone, M., Farinella, G.M., Cutello, V. (eds.) MOD 2015. LNCS, vol. 9432, pp. 240–251. Springer, Heidelberg (2015). doi:10.1007/978-3-319-27926-8_21

9. Pepelyshev, A., Staroselskiy, Y., Zhigljavsky, A.: Adaptive designs for optimizing online advertisement campaigns. Stat. Pap. **57**, 199–208 (2016)
10. Rendle, S.: Factorization machines. In: 2010 IEEE 10th International Conference on Data Mining (ICDM), pp. 995–1000. IEEE (2010)
11. Ridgeway, G.: Generalized boosted models: a guide to the gbm package. Update 1.1 (2007)
12. Wang, J., Yuan, S., Shen, X., Seljan, S.: Real-time bidding: a new frontier of computational advertising research. In: CIKM Tutorial (2013)
13. Yang, S., Ghose, A.: Analyzing the relationship between organic and sponsored search advertising: positive, negative or zero interdependence? Mark. Sci. **29**(4), 602–623 (2010)
14. https://www.kaggle.com/c/avazu-ctr-prediction
15. https://www.kaggle.com/c/criteo-display-ad-challenge

Design of Acoustic Metamaterials Through Nonlinear Programming

Andrea Bacigalupo[1], Giorgio Gnecco[1(✉)], Marco Lepidi[2], and Luigi Gambarotta[2]

[1] IMT School for Advanced Studies, Lucca, Italy
{andrea.bacigalupo,giorgio.gnecco}@imtlucca.it
[2] DICCA, University of Genoa, Genoa, Italy
{marco.lepidi,luigi.gambarotta}@unige.it

Abstract. The dispersive wave propagation in a periodic metamaterial with tetrachiral topology and inertial local resonators is investigated. The Floquet-Bloch spectrum of the metamaterial is compared with that of the tetrachiral beam lattice material without resonators. The resonators can be designed to open and shift frequency band gaps, that is, spectrum intervals in which harmonic waves do not propagate. Therefore, an optimal passive control of the frequency band structure can be pursued in the metamaterial. To this aim, suitable constrained nonlinear optimization problems on compact sets of admissible geometrical and mechanical parameters are stated. According to functional requirements, sets of parameters which determine the largest low-frequency band gap between selected pairs of consecutive branches of the Floquet-Bloch spectrum are soughted for numerically. The various optimization problems are successfully solved by means of a version of the method of moving asymptotes, combined with a quasi-Monte Carlo multi-start technique.

Keywords: Metamaterials · Wave propagation · Passive control · Relative band gap optimization · Nonlinear programming

1 Introduction

An increasing interest has been recently attracted by the analysis of the transmission and dispersion properties of the elastic waves propagating across periodic materials [1,2,4,5,11,14,15,19]. In particular, several studies have been developed to parametrically assess the dispersion curves characterizing the wave frequency spectrum and, therefrom, the amplitudes and boundaries of frequency band gaps lying between pairs of consecutive non-intersecting dispersion curves.

In this background, a promising improvement with respect to conventional beam lattice materials, realized by a regular pattern of stiff disks/rings connected by flexible ligaments, consists in converting them into *inertial metamaterials*. To this aim, intra-ring inertial resonators, elastically coupled to the microstructure of the beam lattice material, are introduced [3,7,12,18]. If properly optimized, the geometrical and mechanical parameters of the metamaterial may allow the

P.M. Pardalos et al. (Eds.): MOD 2016, LNCS 10122, pp. 170–181, 2016.
DOI: 10.1007/978-3-319-51469-7_14

adjustment and enhancement of acoustic properties. For instance, challenging perspectives arise in the tailor-made design of the frequency spectrum for specific purposes, such as opening, enlarging, closing or shifting band gaps in target frequency ranges. Once completed, this achievement potentially allows the realization of a novel class of fully customizable mechanical filters.

Among the others, an efficient approach to the metamaterial design can be based on the formulation of suited constrained nonlinear optimization problems. In the paper, focus is made on the filtering properties of the tetrachiral periodic material and the associated metamaterial, by seeking for optimal combinations of purely mechanical and geometrical parameters. The relative maximum amplitude of band gaps between different pairs of dispersion curves is sought. This approach strengthens the results achieved in [6], where a similar optimization strategy was applied to the passive control of hexachiral beam lattice metamaterials, while the optimization was restricted to the lowest band gap of the Floquet-Bloch spectrum [8] (namely, that lying between the second acoustic branch, and the first optical branch). For nonlinearity reasons, the resulting optimization problems are solved numerically by combining a version of the method of moving asymptotes [16] with a quasi-Monte Carlo initialization technique.

The paper is organized as follows. Section 2 describes the physical-mathematical model of the metamaterial. Section 3 states the relative band gap optimization problem, describes the solution approach adopted, and reports the related numerical results. Finally, Sect. 4 presents some conclusions. Mechanical details about the physical-mathematical model are reported in the Appendix.

2 Physical-Mathematical Model

A planar cellular metamaterial, composed of a periodic tesselation of square cells along two orthogonal periodicity vectors $\mathbf{v}_1$ and $\mathbf{v}_2$, is considered. In the absence of an embedding soft matrix, the internal microstructure of each cell, as well as the elastic coupling between adjacent cells, are determined by a periodic pattern of central rings connected to each other by four elastic ligaments, spatially organised according to a tetrachiral geometric topology (see Fig. 1a).

Focusing on the planar microstructure with unit thickness of the generic cell (Fig. 1b), the central massive and highly-stiff ring is modelled as a rigid body (in red), characterized by mean radius R and width w_{an}. The light and highly-flexible ligaments (in black) are modelled as massless, linear, extensible, unshearable beams, characterized by natural length L (between the ring-beam joints), transversal width w, and inclination β (with respect to the ε-long line connecting the centres of adjacent rings). By virtue of the periodic symmetry, the cell boundary crosses all the ligaments at midspan, halving their natural length. A heavy internal circular inclusion with external radius r (blue circle in Fig. 1b), is located inside the ring through a soft elastic annulus (in grey). This inclusion, modelled as a rigid disk with density ρ_r, plays the role of a low-frequency resonator. The beam material is supposed linearly elastic, with Young's modulus E_s, and uniform mass density, assumed as negligible with respect to the density

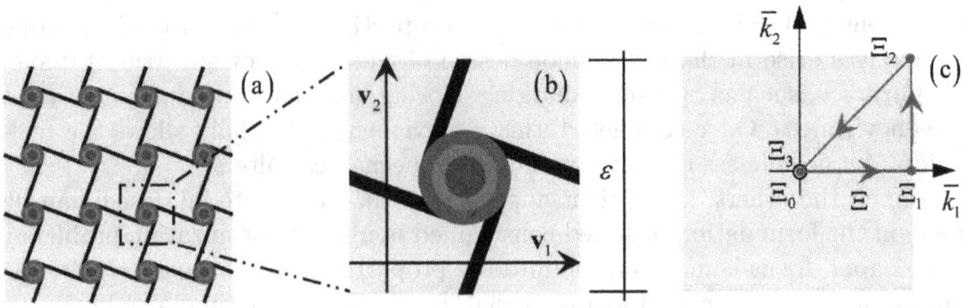

Fig. 1. Tetrachiral cellular material equipped with resonators: (a) pattern; (b) details of the single cell for the case of tangent ligaments, corresponding to $\beta = \arcsin\left(2\frac{R}{\varepsilon}\right)$. (Color figure online)

ρ_{s_an} of the highly-stiff ring. Hence, the whole mass of the lattice is assignable to the heavy rings. The deformable annulus between the ring and the resonator is considered a homogeneous, linearly elastic and isotropic solid, with Young's modulus E_r and Poisson's ratio ν_r. It is worth noting that the ligament natural length L is a (ε, β, R)-dependent parameter, obeying to the geometric relation

$$L = \varepsilon \left(\cos(\beta) - \sqrt{(2R/\varepsilon)^2 - (\sin(\beta))^2} \right). \tag{1}$$

Specializing the approach proposed in [6] to deal with the case-study under investigation, in which the tetrachiral cell topology is featured by two periodicity vectors, the generalized eigenproblem governing the free propagation of harmonic waves (with normalized angular frequency $\bar{\omega}$ and dimensionless wave vector $\bar{\mathbf{k}}$) in the metamaterial reads

$$\left(\bar{\mathbf{K}}(\bar{\boldsymbol{\mu}}, \bar{\mathbf{k}}) - \bar{\omega}^2 \bar{\mathbf{M}}(\bar{\boldsymbol{\mu}})\right) \bar{\boldsymbol{\psi}} = \mathbf{0} \tag{2}$$

where the dimensionless six-by-six matrices $\bar{\mathbf{K}}(\bar{\boldsymbol{\mu}}, \bar{\mathbf{k}})$ and $\bar{\mathbf{M}}(\bar{\boldsymbol{\mu}})$ are Hermitian and diagonal, respectively, and explicitly depend on the minimal dimensionless vector $\bar{\boldsymbol{\mu}}$ of independent geometrical and mechanical parameters

$$\bar{\boldsymbol{\mu}} = \left(\frac{w}{\varepsilon}, \frac{w_{an}}{w}, \frac{R}{\varepsilon}, \beta, \frac{r}{\varepsilon}, \frac{E_r}{E_s}, \nu_r, \frac{\rho_r}{\rho_{s\_an}} \right) \in \mathbb{R}^8 \tag{3}$$

with components $\bar{\mu}_l$, $l = 1, \ldots, 8$. Fixed $\bar{\boldsymbol{\mu}} \in \mathbb{R}^8$ and $\bar{\mathbf{k}} \in \mathbb{R}^2$, the eigenproblem solution is composed by six real-valued eigenvalues $\bar{\omega}_h^2(\bar{\boldsymbol{\mu}}, \bar{\mathbf{k}})$ $(h = 1, \ldots, 6)$, and the corresponding complex-valued eigenvectors $\bar{\boldsymbol{\psi}}_h(\bar{\boldsymbol{\mu}}, \bar{\mathbf{k}}) \in \mathbb{C}^6$. Here, $\bar{\omega}_h(\bar{\boldsymbol{\mu}}, \bar{\mathbf{k}})$ is the h-th normalized angular frequency, which is related to the unnormalized angular frequency ω_h through $\bar{\omega}_h = (\omega_h \varepsilon) / (E_s/\rho_{s\_an})^{1/2}$. The parametric dispersion relations $\bar{\omega}_h(\bar{\boldsymbol{\mu}}, \bar{\mathbf{k}})$ are the six roots of the nonlinear equation imposing the singularity condition on the matrix governing the linear eigenproblem (2).

Distinguishing between the ring and the resonator degrees-of-freedom through the partition $\bar{\psi} = (\bar{\psi}^s, \bar{\psi}^r)$, the matrices $\bar{\mathbf{K}}(\bar{\boldsymbol{\mu}}, \bar{\mathbf{k}})$ and $\bar{\mathbf{M}}(\bar{\boldsymbol{\mu}})$ have the form

$$\bar{\mathbf{K}}(\bar{\boldsymbol{\mu}}, \bar{\mathbf{k}}) = \begin{bmatrix} \bar{\mathbf{K}}^s(\bar{\boldsymbol{\mu}}, \bar{\mathbf{k}}) & \bar{\mathbf{K}}^{sr}(\bar{\boldsymbol{\mu}}, \bar{\mathbf{k}}) \\ \bar{\mathbf{K}}^{rs}(\bar{\boldsymbol{\mu}}, \bar{\mathbf{k}}) & \bar{\mathbf{K}}^r(\bar{\boldsymbol{\mu}}, \bar{\mathbf{k}}) \end{bmatrix}, \quad \bar{\mathbf{M}}(\bar{\boldsymbol{\mu}}) = \begin{bmatrix} \bar{\mathbf{M}}^s(\bar{\boldsymbol{\mu}}) & \mathbf{O} \\ \mathbf{O} & \bar{\mathbf{M}}^r(\bar{\boldsymbol{\mu}}) \end{bmatrix} \tag{4}$$

where the entries of the three-by-three submatrices are reported in the Appendix. The submatrices $\bar{\mathbf{K}}^{sr}(\bar{\boldsymbol{\mu}}, \bar{\mathbf{k}})$ and $\bar{\mathbf{K}}^{rs}(\bar{\boldsymbol{\mu}}, \bar{\mathbf{k}})$ describe the interaction between the resonator and the rest of the microstructure. In the absence of the resonator (i.e., when $r/\varepsilon = 0$), the parameter vector reduces to

$$\bar{\boldsymbol{\mu}}^s = \left(\frac{w}{\varepsilon}, \frac{w_{an}}{w}, \frac{R}{\varepsilon}, \beta\right) \in \mathbb{R}^4 \tag{5}$$

and the generalized eigenvalue problem reduces to

$$\left(\bar{\mathbf{K}}^s(\bar{\boldsymbol{\mu}}^s, \bar{\mathbf{k}}) - \bar{\omega}^2 \bar{\mathbf{M}}^s(\bar{\boldsymbol{\mu}}^s)\right) \bar{\psi}^s = \mathbf{0} \tag{6}$$

whose solution gives the eigenpairs $\bar{\omega}_h(\bar{\boldsymbol{\mu}}^s, \bar{\mathbf{k}})$ and $\bar{\psi}^s_h(\bar{\boldsymbol{\mu}}^s, \bar{\mathbf{k}}) \in \mathbb{C}^3$ $(h = 1, \ldots, 3)$.

For any fixed choice of the parameter vector $\bar{\boldsymbol{\mu}}$, the h-th dimensionless angular frequency locus along the closed boundary ∂B of the Brillouin irreducible zone B [8], spanned anticlockwise by the dimensionless curvilinear coordinate Ξ (Fig. 1c), is the h-th dispersion curve of the Floquet-Bloch spectrum. In particular, the B-vertices are $\bar{\mathbf{k}}_0 = (0, 0)$, $\bar{\mathbf{k}}_1 = (0, \pi)$, $\bar{\mathbf{k}}_2 = (\pi, \pi)$, and $\bar{\mathbf{k}}_3 = \bar{\mathbf{k}}_0$. The segments ∂B_1 and ∂B_3 of the boundary ∂B join, respectively, $\bar{\mathbf{k}}_0$ and $\bar{\mathbf{k}}_1$ (i.e., $\Xi \in [\Xi_0, \Xi_1]$, where $\Xi_0 = 0$ and $\Xi_1 = \pi$), and $\bar{\mathbf{k}}_2$ and $\bar{\mathbf{k}}_3$ (i.e., $\Xi \in [\Xi_2, \Xi_3]$, where $\Xi_2 = 2\pi$ and $\Xi_3 = 2\pi + \sqrt{2}\pi$). For $k = h + 1$, the *relative amplitude of the full band gap* (or, for short, the *full band gap*) between the h-th and k-th dispersion curves can be defined

$$\Delta\bar{\omega}_{hk,\partial B,\mathrm{rel}}(\bar{\boldsymbol{\mu}}) = \frac{\min\limits_{\bar{\mathbf{k}} \in \partial B} \bar{\omega}_k(\bar{\boldsymbol{\mu}}, \bar{\mathbf{k}}) - \max\limits_{\bar{\mathbf{k}} \in \partial B} \bar{\omega}_h(\bar{\boldsymbol{\mu}}, \bar{\mathbf{k}})}{\frac{1}{2}\left[\min\limits_{\bar{\mathbf{k}} \in \partial B} \bar{\omega}_k(\bar{\boldsymbol{\mu}}, \bar{\mathbf{k}}) + \max\limits_{\bar{\mathbf{k}} \in \partial B} \bar{\omega}_h(\bar{\boldsymbol{\mu}}, \bar{\mathbf{k}})\right]}. \tag{7}$$

It is worth noting that, when the right-hand side of (7) is non-positive, there is actually no band gap. Similarly, *partial (relative) band gaps* $\Delta\bar{\omega}_{h,\partial B_1,\mathrm{rel}}$ are obtained by replacing ∂B with ∂B_1 in (7), and are associated to waves characterized by $\bar{k}_2 = 0$ and variable $\bar{k}_1$. The *relative band amplitude* (or, for short, the *band amplitude*) of the h-th dispersion curve is defined as

$$\Delta_A\bar{\omega}_{h,\partial B,\mathrm{rel}}(\bar{\boldsymbol{\mu}}) = \frac{\max\limits_{\bar{\mathbf{k}} \in \partial B} \bar{\omega}_h(\bar{\boldsymbol{\mu}}, \bar{\mathbf{k}}) - \min\limits_{\bar{\mathbf{k}} \in \partial B} \bar{\omega}_h(\bar{\boldsymbol{\mu}}, \bar{\mathbf{k}})}{\frac{1}{2}\left[\max\limits_{\bar{\mathbf{k}} \in \partial B} \bar{\omega}_h(\bar{\boldsymbol{\mu}}, \bar{\mathbf{k}}) + \min\limits_{\bar{\mathbf{k}} \in \partial B} \bar{\omega}_h(\bar{\boldsymbol{\mu}}, \bar{\mathbf{k}})\right]}. \tag{8}$$

To preserve the structural meaning of the solution with proper bounds fixed a priori, the following geometrical constraints on the parameters are introduced

$$\frac{1}{10}\frac{R}{\varepsilon} \leq \frac{w_{an}}{w}\frac{w}{\varepsilon} \leq \frac{R}{\varepsilon}, \tag{9}$$

$$\beta \leq \arcsin\left(2\frac{R}{\varepsilon}\right), \tag{10}$$

$$\frac{w}{\varepsilon} \leq \frac{2}{3}\left(\frac{R}{\varepsilon} + \frac{1}{2}\frac{w_{an}}{w}\frac{w}{\varepsilon}\right), \tag{11}$$

$$\frac{1}{5}\left(\frac{R}{\varepsilon} - \frac{1}{2}\frac{w_{an}}{w}\frac{w}{\varepsilon}\right) \leq \frac{r}{\varepsilon} \leq \frac{9}{10}\left(\frac{R}{\varepsilon} - \frac{1}{2}\frac{w_{an}}{w}\frac{w}{\varepsilon}\right). \tag{12}$$

The related admissible ranges of the parameters are summarized in Table 1. In the absence of the resonator, the definitions (7), (8) hold with $\bar{\boldsymbol{\mu}}$ replaced by $\bar{\boldsymbol{\mu}}^s$ and the constraint (12) is absent.

Table 1. Lower and upper bounds on the geometrical and mechanical parameters.

$\bar{\mu}_l$	$\frac{w}{\varepsilon}$	$\frac{w_{an}}{w}$	$\frac{R}{\varepsilon}$	β	$\frac{r}{\varepsilon}$	$\frac{E_r}{E_s}$	ν_r	$\frac{\rho_r}{\rho_{s\_an}}$
$\bar{\mu}_{l,\min}$	$\frac{3}{50}$	$\frac{1}{20}$	$\frac{1}{10}$	0	$\frac{1}{50}$	$\frac{1}{10}$	$\frac{1}{5}$	$\frac{1}{2}$
$\bar{\mu}_{l,\max}$	$\frac{1}{5}$	$\frac{10}{3}$	$\frac{1}{5}$	$\arcsin\left(\frac{2}{5}\right)$	$\frac{9}{50}$	1	$\frac{2}{5}$	2

3 Optimization Problems

Some optimization problems, imposed on the Floquet-Bloch spectrum $\bar{\omega}_h(\bar{\boldsymbol{\mu}}, \bar{\mathbf{k}})$ of the material/metamaterial, are considered. They are formulated as constrained nonlinear optimization problems, and solved by using the globally convergent method of moving asymptotes (GCMMA) [17], combined with a quasi-Monte Carlo multi-start technique. The solution method consists in tackling a sequence of concave-maximization subproblems, locally approximating the original nonlinear optimization problem[1] (a different approximation at each sequence element), whereas the quasi-Monte Carlo multi-start technique increases the probability of finding a global maximum point through a set of quasi-random initializations of the sequence. More details about the combined method are reported in [6].

3.1 Band Gap Between the Second and Third Dispersion Curves

Considering that the first two dispersion curves always meet at the origin for the selected choice of the parameter range (so that the case $h = 1$ and $k = 2$ has no interest), in the absence of the resonator the optimization problem reads

$$\begin{aligned} &\underset{\bar{\boldsymbol{\mu}}^s}{\text{maximize}}\ \Delta\bar{\omega}_{23,\partial B_1,\text{rel}}\left(\bar{\boldsymbol{\mu}}^s\right) \\ &\text{s.t.} \quad \bar{\mu}^s_{l,\min} \leq \bar{\mu}^s_l \leq \bar{\mu}^s_{l,\max}, \ l = 1, \ldots, 4, \\ &\qquad \text{and the constraints (9), (10), and (11).} \end{aligned} \tag{13}$$

[1] The *moving asymptotes* [16] are asymptotes of functions (changing when moving from one optimization subproblem to the successive one), which are used to approximate (typically nonlinearly) the original objective and constraint functions.

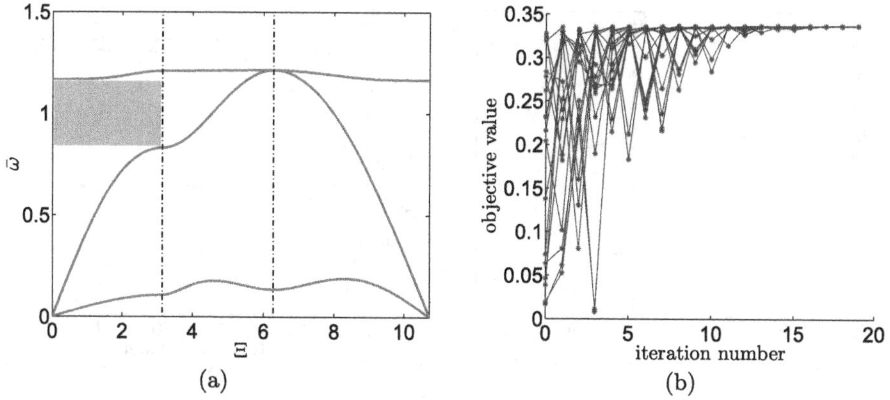

Fig. 2. Optimization of the objective $\Delta\bar{\omega}_{23,\partial B_1,\mathrm{rel}}$ in the absence of the resonator: (a) best Floquet-Bloch spectrum; (b) convergence of the sequence of objective values obtained in the GCMMA iterations, for each quasi-Monte Carlo initialization.

To obtain the numerical results, a quasi-random 100-points Sobol' sequence [13] in the parameter unit hypercube was generated, then all the points of the subsequence satisfying all the constraints were used as initial points for the GCMMA. Moreover, the partial band gap $\Delta\bar{\omega}_{23,\partial B_1,\mathrm{rel}}$ was approximated by replacing ∂B_1 with its uniform discretization, using 30 points. After each valid (constraints-compatible) quasi-Monte Carlo initialization, a number of iterations of the GCMMA sufficiently large to obtain convergence was performed. The numerical results of the optimization are reported in Fig. 2, and demonstrate the presence of a positive partial band gap (with amplitude approximately equal to 0.337) at the *best* obtained (higher-valued) objective value $\Delta\bar{\omega}^{s,*}_{23,\partial B_1,\mathrm{rel}}$ among all the suboptimal solutions generated during the optimization (for all the optimization problems examined in the paper, suboptimal solutions are considered, since an optimal solution cannot be computed exactly). The associated best parameter vector $\bar{\boldsymbol{\mu}}_1^{s,*}$ is listed in the first row of Table 2.

Then, the problem (13) has been extended to the optimization of the full band gap between the second and third dispersion curves. A uniform 90-point discretization of the boundary ∂B has been employed, and zero has been obtained as best value of the objective, corresponding to the absence of a full band gap. This result has been also confirmed by evaluating the objective function on a sufficiently fine grid in the parameter space (10 points for each component), considering only the admissible range of the constrained parameters.

The presence of a partial band gap between the second and third dispersion curves has been also obtained as result of the optimization problem

$$\begin{aligned} &\underset{\bar{\boldsymbol{\mu}}}{\text{maximize}}\ \Delta\bar{\omega}_{23,\partial B_1,\mathrm{rel}}(\bar{\boldsymbol{\mu}}) \\ &\text{s.t.} \quad \bar{\mu}_{l,\min} \le \bar{\mu}_l \le \bar{\mu}_{l,\max},\ l = 1,\dots,8, \\ &\qquad\ \text{and the constraints (9), (10), (11), and (12),} \end{aligned} \tag{14}$$

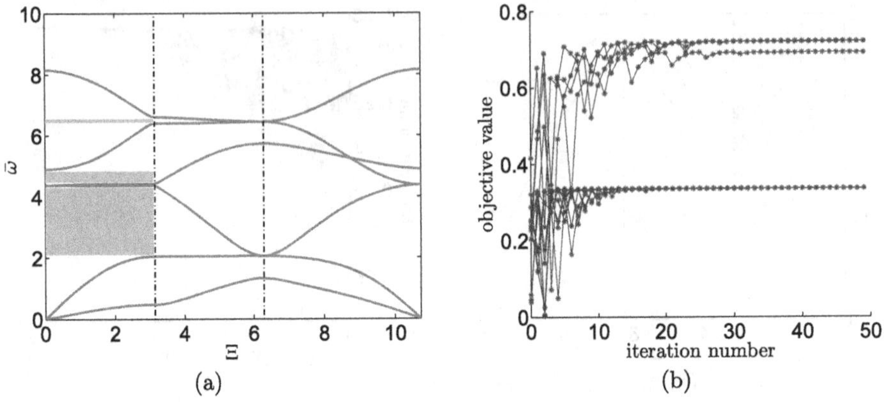

Fig. 3. Optimization of the objective $\Delta\bar{\omega}_{23,\partial B_1,\mathrm{rel}}$ in the presence of the resonator: (a) best Floquet-Bloch spectrum; (b) convergence of the sequence of objective values obtained in the GCMMA iterations, for each quasi-Monte Carlo initialization.

which rises up with the introduction of the resonator (see Fig. 3). In this case, by varying the quasi-Monte Carlo initialization, the GCMMA has converged to various solutions characterized by different objective values, and the best partial band gap has been found nearly equal to $\Delta\bar{\omega}^*_{23,\partial B_1,\mathrm{rel}} = 0.722$. This result demonstrates that the presence of the resonator can increase the optimal band gap amplitude. Two more partial band gaps, between the fourth-fifth and fifth-sixth pairs of dispersion curves have been obtained (Fig. 3a). The associated best parameter vector $\bar{\boldsymbol{\mu}}^*_2$ is listed in the second row of Table 2.

Finally, the problem (14) has been extended to the optimization of the *full* (with ∂B instead of ∂B_1) band gap between the second and third dispersion curves. Again, the best value of the objective obtained by the combined method has resulted to be zero, meaning that the presence of the resonator is unable to open a full band gap between the second and third dispersion curves.

3.2 Weighted Band Gap Between the Third and Fourth Dispersion Curves

The full band gap between the third and fourth dispersion curves has been considered in the presence of the resonator. To this aim, the optimization problem is re-formulated to maximize a trade-off (i.e., the product[2]) between the full band

[2] In a preliminary phase, we also considered as objective function a weighted sum, with positive weights, of the full band gap between the third and fourth dispersion curves and the band amplitude of the fourth dispersion curve. However, for various choices of the weights, the obtained solution was characterized by a negligible value either of the full band gap, or of the band amplitude, making an optimal choice of the weights difficult. For our specific goal, instead, the product of the two terms was more effective.

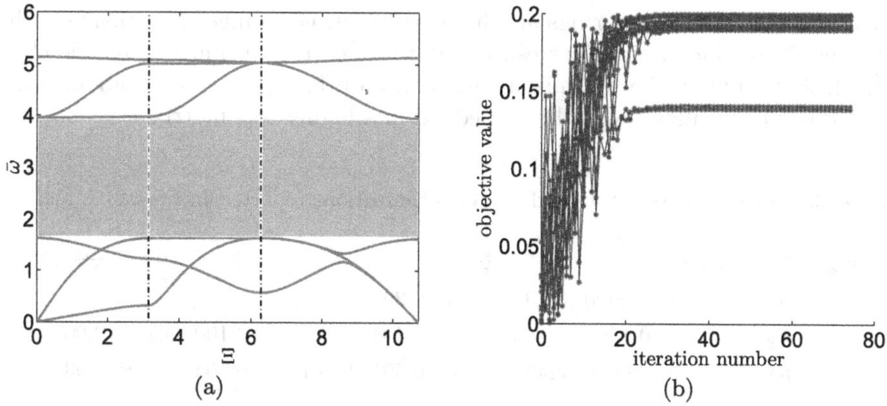

Fig. 4. Optimization of the objective $\Delta\bar{\omega}_{34,\partial B,\mathrm{rel}}\ \Delta_A\bar{\omega}_{4,\partial B,\mathrm{rel}}$ in the presence of the resonator: (a) best Floquet-Bloch spectrum; (b) convergence of the sequence of objective values obtained in the GCMMA iterations, for each quasi-Monte Carlo initialization.

gap $\Delta\bar{\omega}_{34,\partial B,\mathrm{rel}}$, and the band amplitude $\Delta_A\bar{\omega}_{4,\partial B,\mathrm{rel}}$ of the fourth dispersion curve (see Eq. (8)):

$$\begin{aligned} &\underset{\bar{\boldsymbol{\mu}}}{\text{maximize}}\ (\Delta\bar{\omega}_{34,\partial B,\mathrm{rel}}(\bar{\boldsymbol{\mu}})\ \Delta_A\bar{\omega}_{4,\partial B,\mathrm{rel}}(\bar{\boldsymbol{\mu}})) \\ &\text{s.t.} \quad \bar{\mu}_{l,\min} \leq \bar{\mu}_l \leq \bar{\mu}_{l,\max},\ l = 1,\ldots,8, \\ &\qquad\ \text{and the constraints (9), (10), (11), and (12).} \end{aligned} \tag{15}$$

The reason is that, in the absence of an elastic coupling between the resonator and the cell structure, three of the dispersion curves are Ξ-independent (null band amplitude), as they express the fixed frequencies of the free-standing resonator. Therefore, a large but definitely not significant full band gap would be obtained by separating such curves from the rest of the Floquet-Bloch spectrum as much as possible. The band amplitude of the fourth dispersion curve, acting as a weighting multiplier in the objective function, is expected to avoid this shortcoming. Indeed, by taking the product, preference is given to parametric designs $\bar{\boldsymbol{\mu}}$ of the beam-lattice metamaterial for which both factors $\Delta\bar{\omega}_{34,\partial B,\mathrm{rel}}(\bar{\boldsymbol{\mu}})$ and $\Delta_A\bar{\omega}_{4,\partial B,\mathrm{rel}}(\bar{\boldsymbol{\mu}})$ are simultaneously large. The optimization results are reported in Fig. 4. The best solution (whose parameter vector $\bar{\boldsymbol{\mu}}_3^*$ is listed in the third row of Table 2) is associated with a full band gap with amplitude approximately equal to $\Delta\bar{\omega}^*_{34,\partial B,\mathrm{rel}} = 0.830$, whereas the band amplitude of the fourth dispersion curve is about $\Delta_A\bar{\omega}^*_{4,\partial B,\mathrm{rel}} = 0.241$.

Summing up, the main results show that, for a periodic cell with fixed characteristic size, the presence of resonators is a mandatory condition for the existence of full band gaps in the low-frequency range. However, only a partial band gap can be opened between the second and third branches. Full band gaps can be obtained between the third and fourth branches, and – by virtue of the numerical optimization – the best band gap amplitude overcomes the best band amplitude of the fourth branch of the spectrum. The best-obtained results correspond to large-radius rings and highly-slender, non-tangent ligaments with quasi-negligible inclination (corresponding to a nearly-vanishing geometric chirality). Accordingly, the optimized resonators are found to possess about half the radius of the rings and to be embedded in a highly-soft matrix.

Table 2. Best parameters obtained in the optimizations reported in Figs. 2, 3, and 4.

Figs.	Param. vect	$\frac{w}{\varepsilon}$	$\frac{w_{an}}{w}$	$\frac{R}{\varepsilon}$	β	$\frac{r}{\varepsilon}$	$\frac{E_r}{E_s}$	ν_r	$\frac{\rho_r}{\rho_{s\_an}}$
2	$\bar{\mu}_1^{s,*}$	0.0600	3.19	0.200	0.295	–	–	–	–
3	$\bar{\mu}_2^*$	0.139	0.143	0.198	0.404	0.0864	0.100	0.200	2.00
4	$\bar{\mu}_3^*$	0.0950	0.439	0.200	0.00163	0.127	0.104	0.218	1.93

4 Conclusions

A parametric model of periodic metamaterial has been formulated, and its wave-propagation properties have been investigated. Then, some optimization problems related to such properties have been stated and solved numerically. From the physical viewpoint, the desirable target is a highly-performing material with marked filtering capacities for low-frequency signals. The best results of the optimizations demonstrate that a partial band gap can be obtained between the second and third dispersion curves, both in the absence and in the presence of a resonator inside the periodic cell. However, in both cases, no full band gaps are obtainable. On the contrary, the resonator allows the opening of a positive full band gap between the third and fourth dispersion curves, associated with a non-negligible elastic coupling between the cell structure and the resonator. This achievement implied the optimization of a multiplicative trade-off between the full band gap and the band amplitude of the fourth dispersion curve. All the optimization problems have been solved numerically by combining the GCMMA with a quasi-Monte Carlo multi-start technique.

As possible developments, the optimization framework could be extended to its regularized version, to reduce the sensitivity of the best solution with respect to changes in the nominal parameter values. To this aim, suitable regularization techniques, typical of machine learning problems, could be used [9,10]. Other nonlinear optimization methods could be also used, such as sequential linear or quadratic programming. Finally, an electromechanical extension of the physical-mathematical model would allow the design of smart metamaterials.

Appendix

Adapting to the present tetrachiral context the analysis made in [6] for the hexachiral case, one can show that the non vanishing components of the three-by-three positive definite diagonal submatrices $\bar{\mathbf{M}}^s$ and $\bar{\mathbf{M}}^r$, that make up the six-by-six dimensionless block diagonal matrix $\bar{\mathbf{M}}$, read

$$\begin{aligned} \bar{M}^s_{11} &= 2\,\pi\,\frac{R}{\varepsilon}\,\frac{w}{\varepsilon}\,\frac{w_{an}}{w}, \qquad \bar{M}^s_{22} = 2\,\pi\,\frac{R}{\varepsilon}\,\frac{w}{\varepsilon}\,\frac{w_{an}}{w}, \\ \bar{M}^s_{33} &= \frac{1}{2}\,\pi\,\frac{R}{\varepsilon}\,\frac{w}{\varepsilon}\,\frac{w_{an}}{w}\left(\left(\frac{w}{\varepsilon}\right)^2\left(\frac{w_{an}}{w}\right)^2+4\left(\frac{R}{\varepsilon}\right)^2\right), \\ \bar{M}^r_{11} &= \pi\left(\frac{r}{\varepsilon}\right)^2\frac{\rho_r}{\rho_{s\_an}}, \quad \bar{M}^r_{22} = \pi\left(\frac{r}{\varepsilon}\right)^2\frac{\rho_r}{\rho_{s\_an}}, \quad \bar{M}^r_{33} = \frac{1}{2}\,\pi\,\frac{\rho_r}{\rho_{s\_an}}\left(\frac{r}{\varepsilon}\right)^4. \end{aligned} \tag{16}$$

In order to express the components of the three-by-three Hermitian submatrix $\bar{\mathbf{K}}^s$, we need to introduce the dependent parameters k_d/E_s and $k_\theta/(\varepsilon^2 E_s)$, which are functions of the other parameters, respectively, of the forms $\frac{k_d}{E_s} = f_d\left(\frac{r}{\varepsilon}\frac{\varepsilon}{R}, \frac{E_r}{E_s}, \nu_r\right)$ and $\frac{k_\theta}{\varepsilon^2 E_s} = f_\theta\left(\frac{r}{\varepsilon}\frac{\varepsilon}{R}, \frac{E_r}{E_s}, \nu_r\right)$. The definitions of these functions are reported in [6]. Then, the components of $\bar{\mathbf{K}}^s$ are expressed as follows:

$$\bar{K}^s_{ij} = \frac{1}{\Lambda}\left(\bar{K}^{s\_3}_{ij}\left(\frac{w}{\varepsilon}\right)^3 + \bar{K}^{s\_1}_{ij}\frac{w}{\varepsilon} + \bar{K}^{s\_0}_{ij}\right), \tag{17}$$

where $i, j = 1, 2, 3$, $\Lambda = (\cos(\beta) - \Psi)^3$ and $\Psi = \sqrt{(\cos(\beta))^2 + 4\left(\frac{R}{\varepsilon}\right)^2 - 1}$. More precisely, one obtains

$$\begin{aligned} \bar{K}^{s\_3}_{11} &= \left(2\cos(\bar{k}_1) - 2\cos(\bar{k}_2)\right)(\cos(\beta))^2 - 2\cos(\bar{k}_1) + 2, \\ \bar{K}^{s\_1}_{11} &= \left(-2\cos(\bar{k}_1) + 2\cos(\bar{k}_2)\right)(\cos(\beta))^4 \\ &\quad + \left(4\cos(\bar{k}_1)\Psi - 4\cos(\bar{k}_2)\Psi\right)(\cos(\beta))^3 \\ &\quad + \left(-2\cos(\bar{k}_1)\Psi^2 + 2\cos(\bar{k}_2)\Psi^2 - 2\cos(\bar{k}_2) + 2\right)(\cos(\beta))^2 \\ &\quad + \left(4\cos(\bar{k}_2)\Psi - 4\Psi\right)\cos(\beta) - 2\cos(\bar{k}_2)\Psi^2 + 2\Psi^2, \\ \bar{K}^{s\_0}_{11} &= (\cos(\beta))^3\frac{k_d}{E_s} - 3(\cos(\beta))^2\Psi\frac{k_d}{E_s} + 3\cos(\beta)\Psi^2\frac{k_d}{E_s} - \Psi^3\frac{k_d}{E_s}, \\ \bar{K}^{s\_3}_{22} &= \left(-2\cos(\bar{k}_1) + 2\cos(\bar{k}_2)\right)(\cos(\beta))^2 - 2\cos(\bar{k}_2) + 2, \\ \bar{K}^{s\_1}_{22} &= \left(2\cos(\bar{k}_1) - 2\cos(\bar{k}_2)\right)(\cos(\beta))^4 \\ &\quad + \left(-4\cos(\bar{k}_1)\Psi + 4\cos(\bar{k}_2)\Psi\right)(\cos(\beta))^3 \\ &\quad + \left(2\cos(\bar{k}_1)\Psi^2 - 2\cos(\bar{k}_2)\Psi^2 - 2\cos(\bar{k}_1) + 2\right)(\cos(\beta))^2 \\ &\quad + \left(4\cos(\bar{k}_1)\Psi - 4\Psi\right)\cos(\beta) - 2\cos(\bar{k}_1)\Psi^2 + 2\Psi^2, \\ \bar{K}^{s\_0}_{22} &= (\cos(\beta))^3\frac{k_d}{E_s} - 3(\cos(\beta))^2\Psi\frac{k_d}{E_s} + 3\cos(\beta)\Psi^2\frac{k_d}{E_s} - \Psi^3\frac{k_d}{E_s}, \\ \bar{K}^{s\_3}_{33} &= \left(1/3\cos(\bar{k}_1) + 1/3\cos(\bar{k}_2) + 4/3\right)(\cos(\beta))^2 \\ &\quad + \left(-2/3\Psi + 1/3\cos(\bar{k}_1)\Psi + 1/3\cos(\bar{k}_2)\Psi\right)\cos(\beta) \\ &\quad - 1/6\cos(\bar{k}_1)\Psi^2 - 1/6\cos(\bar{k}_2)\Psi^2 + 1/3\Psi^2, \end{aligned} \tag{18}$$

$$
\begin{aligned}
\bar{K}_{33}^{s\_1} &= \left(-1/2\,\cos\left(\bar{k}_1\right) - 1/2\,\cos\left(\bar{k}_2\right) - 1\right)\left(\cos\left(\beta\right)\right)^4 \\
&+ \left(\cos\left(\bar{k}_1\right)\Psi + \cos\left(\bar{k}_2\right)\Psi + 2\,\Psi\right)\left(\cos\left(\beta\right)\right)^3 \\
&+ \left(1/2\,\cos\left(\bar{k}_2\right) + 1/2\,\cos\left(\bar{k}_1\right) + 1 - \Psi^2 - 1/2\,\cos\left(\bar{k}_1\right)\Psi^2\right. \\
&\qquad \left. - 1/2\,\cos\left(\bar{k}_2\right)\Psi^2\right)\left(\cos\left(\beta\right)\right)^2 \\
&+ \left(-\cos\left(\bar{k}_1\right)\Psi - \cos\left(\bar{k}_2\right)\Psi - 2\,\Psi\right)\cos\left(\beta\right) \\
&+ 1/2\,\cos\left(\bar{k}_1\right)\Psi^2 + 1/2\,\cos\left(\bar{k}_2\right)\Psi^2 + \Psi^2, \\
\bar{K}_{33}^{s\_0} &= -3\,\left(\cos\left(\beta\right)\right)^2\Psi\,\frac{k_\theta}{\varepsilon^2 E_s} + 3\,\cos\left(\beta\right)\Psi^2\frac{k_\theta}{\varepsilon^2 E_s} + \left(\cos\left(\beta\right)\right)^3\frac{k_\theta}{\varepsilon^2 E_s} - \Psi^3\frac{k_\theta}{\varepsilon^2 E_s}, \\
\bar{K}_{12}^{s\_3} &= 2\,\cos\left(\beta\right)\sin\left(\beta\right)\left(-\cos\left(\bar{k}_1\right) + \cos\left(\bar{k}_2\right)\right), \\
\bar{K}_{12}^{s\_1} &= 2\,\sin\left(\beta\right)\left(\cos\left(\bar{k}_1\right) - \cos\left(\bar{k}_2\right)\right)\left(\cos\left(\beta\right)\right)^3 \\
&+ 2\,\sin\left(\beta\right)\left(-2\,\cos\left(\bar{k}_1\right)\Psi + 2\,\cos\left(\bar{k}_2\right)\Psi\right)\left(\cos\left(\beta\right)\right)^2 \\
&+ 2\,\sin\left(\beta\right)\left(\cos\left(\bar{k}_1\right)\Psi^2 - \cos\left(\bar{k}_2\right)\Psi^2\right)\cos\left(\beta\right), \\
\bar{K}_{12}^{s\_1} &= 0, \\
\bar{K}_{13}^{s\_3} &= -\imath\sin\left(\bar{k}_2\right)\left(\cos\left(\beta\right)\right)^2\imath\sin\left(\beta\right)\sin\left(\bar{k}_1\right)\cos\left(\beta\right), \\
\bar{K}_{13}^{s\_1} &= -\imath\sin\left(\bar{k}_2\right)\left(\cos\left(\beta\right)\right)^4 - \imath\left(-\sin\left(\beta\right)\sin\left(\bar{k}_1\right) - 2\,\sin\left(\bar{k}_2\right)\Psi\right)\left(\cos\left(\beta\right)\right)^3 \\
&- \imath\left(2\,\sin\left(\bar{k}_1\right)\sin\left(\beta\right)\Psi + \sin\left(\bar{k}_2\right)\Psi^2 - \sin\left(\bar{k}_2\right)\right)\left(\cos\left(\beta\right)\right)^2 \\
&- \imath\left(-\sin\left(\bar{k}_1\right)\sin\left(\beta\right)\Psi^2 + 2\,\sin\left(\bar{k}_2\right)\Psi\right)\cos\left(\beta\right) + \imath\sin\left(\bar{k}_2\right)\Psi^2, \\
\bar{K}_{13}^{s\_1} &= 0, \\
\bar{K}_{23}^{s\_3} &= \imath\sin\left(\bar{k}_1\right)\left(\cos\left(\beta\right)\right)^2 + \imath\sin\left(\beta\right)\sin\left(\bar{k}_2\right)\cos\left(\beta\right), \\
\bar{K}_{23}^{s\_1} &= \imath\sin\left(\bar{k}_1\right)\left(\cos\left(\beta\right)\right)^4 + \imath\left(\sin\left(\beta\right)\sin\left(\bar{k}_2\right) - 2\,\sin\left(\bar{k}_1\right)\Psi\right)\left(\cos\left(\beta\right)\right)^3 \\
&+ \imath\left(-2\,\sin\left(\beta\right)\sin\left(\bar{k}_2\right)\Psi + \sin\left(\bar{k}_1\right)\Psi^2 - \sin\left(\bar{k}_1\right)\right)\left(\cos\left(\beta\right)\right)^2 \\
&+ \imath\left(\sin\left(\beta\right)\sin\left(\bar{k}_2\right)\Psi^2 + 2\,\sin\left(\bar{k}_1\right)\Psi\right)\cos\left(\beta\right) - \imath\sin\left(\bar{k}_1\right)\Psi^2, \\
\bar{K}_{23}^{s\_1} &= 0, \\
\bar{K}_{21}^{s} &= \bar{K}_{12}^{s}, \\
\bar{K}_{31}^{s} &= -\imath\,\mathrm{Im}\left(\bar{K}_{13}^{s}\right), \\
\bar{K}_{32}^{s} &= -\imath\,\mathrm{Im}\left(\bar{K}_{23}^{s}\right),
\end{aligned}
$$

where $\imath$ denotes the imaginary unit and $\mathrm{Im}(z)$ denotes the imaginary part of the complex number z. Finally, the non vanishing components of the diagonal submatrix $\bar{\mathbf{K}}^r$ are

$$
\bar{K}_{11}^{r} = \frac{k_d}{E_s}, \qquad \bar{K}_{22}^{r} = \frac{k_d}{E_s}, \qquad \bar{K}_{33}^{r} = \frac{k_\theta}{\varepsilon^2 E_s}, \tag{19}
$$

whereas the diagonal submatrices $\bar{\mathbf{K}}^{sr}$ and $\bar{\mathbf{K}}^{rs}$ satisfy the constraint $\bar{\mathbf{K}}^{sr} = \left(\bar{\mathbf{K}}^{sr}\right)^T = -\bar{\mathbf{K}}^{r}$.

References

1. Bacigalupo, A., Gambarotta, L.: Homogenization of periodic hexa- and tetra-chiral cellular solids. Compos. Struct. **116**, 461–476 (2014)
2. Bacigalupo, A., De Bellis, M.L.: Auxetic anti-tetrachiral materials: equivalent elastic properties and frequency band-gaps. Compos. Struct. **131**, 530–544 (2015)
3. Bacigalupo, A., Gambarotta, L.: Simplified modelling of chiral lattice materials with local resonators. Int. J. Solids Struct. **83**, 126–141 (2016)
4. Bacigalupo A., Lepidi M.: A lumped mass beam model for the wave propagation in anti-tetrachiral periodic lattices. In: XXII AIMETA Congress, Genoa, Italy (2015)
5. Bacigalupo, A., Lepidi, M.: High-frequency parametric approximation of the Floquet-Bloch spectrum for anti-tetrachiral materials. Int. J. Solids Struct. **97**, 575–592 (2016)
6. Bacigalupo, A., Lepidi, M., Gnecco, G., Gambarotta, L.: Optimal design of auxetic hexachiral metamaterials with local resonators. Smart Mater. Struct. **25**(5), 054009 (2016)
7. Bigoni, D., Guenneau, S., Movchan, A.B., Brun, M.: Elastic metamaterials with inertial locally resonant structures: application to lensing and localization. Phys. Rev. B **87**, 174303 (2013)
8. Brillouin, L.: Wave Propagation in Periodic Structures, 2nd edn. Dover, New York (1953)
9. Gnecco, G., Sanguineti, M.: Regularization techniques and suboptimal solutions to optimization problems in learning from data. Neural Comput. **22**, 793–829 (2010)
10. Gnecco, G., Gori, M., Sanguineti, M.: Learning with boundary conditions. Neural Comput. **25**, 1029–1106 (2013)
11. Lepidi M., Bacigalupo A.: Passive control of wave propagation in periodic anti-tetrachiral metamaterials. In: VII European Congress on Computational Methods in Applied Sciences and Engineering (ECCOMAS), Hersonissos, Crete Island (2016)
12. Liu, X.N., Hu, G.K., Sun, C.T., Huang, G.L.: Wave propagation characterization and design of two-dimensional elastic chiral metacomposite. J. Sound Vib. **330**, 2536–2553 (2011)
13. Niederreiter H.: Random number generation and Quasi-Monte Carlo methods. SIAM (1992)
14. Phani, A.S., Woodhouse, J., Fleck, N.A.: Wave propagation in two-dimensional periodic lattices. J. Acoust. Soc. Am. **119**, 1995–2005 (2006)
15. Spadoni, A., Ruzzene, M., Gonnella, S., Scarpa, F.: Phononic properties of hexagonal chiral lattices. Wave Motion **46**, 435–450 (2009)
16. Svanberg, K.: The method of moving asymptotes - a new method for structural optimization. Int. J. Numer. Meth. Eng. **24**, 359–373 (1987)
17. Svanberg, K.: A class of globally convergent optimization methods based on conservative convex separable approximations. SIAM J. Optim. **12**, 555–573 (2002)
18. Tan, K.T., Huang, H.H., Sun, C.T.: Optimizing the band gap of effective mass negativity in acoustic metamaterials. Appl. Phys. Lett. **101**, 241902 (2012)
19. Tee, K.F., Spadoni, A., Scarpa, F., Ruzzene, M.: Wave propagation in auxetic tetrachiral honeycombs. J. Vib. Acoust. ASME **132**, 031007–1/8 (2010)

Driver Maneuvers Inference Through Machine Learning

Mauro Maria Baldi[1(✉)], Guido Perboli[1,2], and Roberto Tadei[1]

[1] Politecnico di Torino, Turin, Italy
mauro.baldi@polito.it

[2] Centre interuniversitaire de recherche sur les reseaux d'entreprise, la logistique et le transport (CIRRELT), Montréal, Canada

Abstract. Inferring driver maneuvers is a fundamental issue in Advanced Driver Assistance Systems (ADAS), which can significantly increase security and reduce the risk of road accidents. This is not an easy task due to a number of factors such as driver distraction, unpredictable events on the road, and irregularity of the maneuvers. In this complex setting, Machine Learning techniques can play a fundamental and leading role to improve driving security. In this paper, we present preliminary results obtained within the Development Platform for Safe and Efficient Drive (DESERVE) European project. We trained a number of classifiers over a preliminary dataset to infer driver maneuvers of Lane Keeping and Lane Change. These preliminary results are very satisfactory and motivate us to proceed with the application of Machine Learning techniques over the whole dataset.

Keywords: Machine learning · Driving security · Advanced Driver Assistance Systems

1 Introduction

Machine Learning is a fundamental topic, which aims to classify data into different categories exploiting knowledge on previous measurements over known data. Nowadays, Machine Learning is becoming more and more important and widespread due to the number of applications which is able to address: computer science, traffic management, psychology, psychiatry, ethology, and cognitive science. A relevant application of Machine Learning is driving security, by inferring driver maneuvers.

In fact, inferring driver maneuvers is a fundamental issue in Advanced Driver Assistance Systems (ADAS), which can significantly increase security and reduce the risk of road accidents. This is not an easy task due to a number of factors such as driver distraction, unpredictable events on the road, and irregularity of the maneuvers [1–3]. In this complex setting, Machine Learning techniques can play a fundamental and leading role to improve driving security.

The Development Platform for Safe and Efficient Drive (DESERVE) European project [4] arises in the context of driving security. The aim of the

P.M. Pardalos et al. (Eds.): MOD 2016, LNCS 10122, pp. 182–192, 2016.
DOI: 10.1007/978-3-319-51469-7_15

DESERVE project is to design and implement a Tool Platform for embedded ADAS [5]. One of the objectives of the DESERVE project is the development of a Driver Intention Detection Module (DIDM). The DIDM is implemented according to three steps. First, a number of volunteers are recruited to drive on an Italian highway for about one hour on the demonstrator vehicle supplied by Centro Ricerche Fiat (CRF) [6]. Data measured from drivers make up the whole dataset, on which Machine Learning techniques are implemented to test a number of classifiers in Matlab through the Pattern Recognition Toolbox [7]. Finally, the best classifier is implemented within the platform through the RTMAPS software provided by INTEMPORA [8] to predict driver intentions, which can mainly be classified as Lane Keeping (LK) and Lane Change (LC).

In this paper, we present preliminary results of a number of classifiers trained over an initial dataset provided by CRF. The satisfactory results confirm the importance of Machine Learning techniques in the field of ADAS and driver security. This paper is organized as follows: in Sect. 2, we present an essential literature review of previous works in the field of improving driver security. In Sect. 3, we show how data were collected and the architecture of the DIDM. In Sect. 4, we introduce the classifiers used during the preliminary training and testing, which results are reported in Sect. 5. Finally, the conclusions are in Sect. 6.

2 Literature Review

Mandalia and Salvucci [9] described a technique for inferring the intention to change lanes using Support Vector Machines (SVMs). This technique was applied to experimental data from an instrumented vehicle that included both behavioral and environmental data. Butz and von Stryk [10] presented a two-level driver model for the use in real-time vehicle dynamics applications. Hayashi et al. [11] used Hidden Markov Model techniques to show how the difference of driving patterns affects prediction of driver stopping maneuvers. The proposed method consisted of two different driving models based on a driver state: one in normal state and one in hasty state. Hidden Markov Models were also used by Mandalia [1]. McCall et al. [12] presented robust computer vision methods and Sparse Bayesian Learning for identifying and tracking freeway lanes and driver head motion. Salvucci et al. [13] proposed a model tracing methodology for detecting driver lane changes. Dogan et al. [2] employed Neural Networks (NNs) and SVMs to recognize and predict drivers intentions as a first step towards a realistic driver model. Huang and Gao [14] employed SVMs to the classification of driving intentions using vehicle performance and driver-eye-gaze data from measurements of driving tasks (i.e., lane following and lane changing) into a well-designed simulation environment. Deng [15] performed recognition of lane-change behaviors using SVMs based on both simulation and in-vehicle data. Recent results can be found in [16,17,19,20]. Finally, there has been a recent evolution of SVMs algorithms in multiple channels utilization such as multiview Hessian regularization (mHR) [18].

3 The Driver Intention Detection Module

The DIDM is one of the two modules of the Driver Model in the DESERVE European project. The aim of the DIDM is to predict driver maneuvers of Lane Keeping and Lane Change.

The development of the DIDM was possible thanks to the demonstrator vehicle provided by CRF. A noteworthy technology was installed in the demonstrator vehicle. In particular: sensors to perform dynamic measurements, interior camera to detect driver head, external camera, lasers, and radar. The aim of the demonstrator vehicle is twofold. First, it allowed us to get data from a number of volunteers recruited by CRF and Politecnico di Torino (Polytechnic University of Turin) who drove for about one hour and a half on an Italian highway. Second, it will host the DIDM module in the final phase of the DESERVE project.

The overall structure of the DIDM is shown in Fig. 1. Inputs of the DIDM are the vehicle dynamics data (through the CAN bus), the user-head position (i.e., where the driver is looking at) from internal camera, additional information from external camera (lane position detection), lasers, and radar. The outputs of the DIDM are represented by a probability of the next maneuver the driver intends to perform or by a score representing a level of confidentiality of the next maneuver.

At this stage of the project, the DIDM has not been implemented yet on the demonstrator vehicle supplied by CRF. The DIDM will be implemented on the demonstrator vehicle through the RT-MAPS framework provided by INTEMPORA [8].

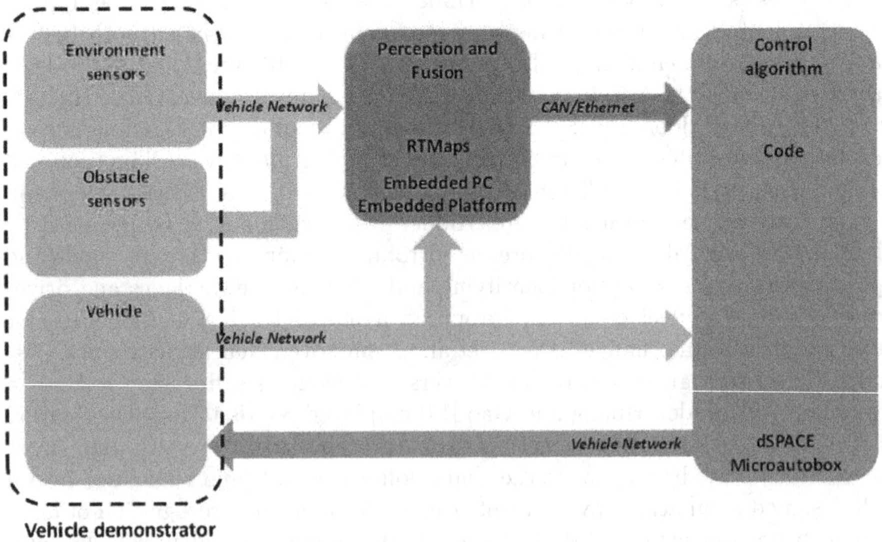

Fig. 1. The integration of the DIDM in the vehicle demonstrator

4 The Classifiers

In this section, we briefly recall the main principles of the classifiers used. These are:

1. Relevance Vector Machines (RVMs)
2. k-Nearest Neighbor (kNN)
3. Anfis
4. Hidden Markow Models (HMMs)
5. Support Vector Machines (SVMs).

All these classifiers work on a training and test set. Relevance Vector Machines are based on a linear Bayesian approach. The goal of a number of Machine-Learning problems is to interpolate an unknown function linking the N row-vectors of features of the training set (and gathered into matrix $\mathbf{X}$) with the corresponding target variables gathered into column vector $\mathbf{t}$. To interpolate this unknown function, we use M parameters making up vector $\mathbf{w}$ and M basis functions $\phi_j(\mathbf{x})$, with $j \in \{1, \ldots, M\}$. The *ideal* function interpolating data to be used for predictions is given by

$$y(\mathbf{x}, \mathbf{w}) = \sum_{j=0}^{M-1} w_j \phi_j(\mathbf{x}) = \mathbf{w}^T \mathbf{x}. \tag{1}$$

This function is ideal because it does not take into account the noise coming from measurements of data. The Bayesian approach consists in finding the optimal values for parameters $\mathbf{w}$ applying Bayes's Theorem as follows:

$$P(\mathbf{w}|\mathbf{t}) = \frac{P(\mathbf{t}|\mathbf{w}) \cdot P(\mathbf{w})}{P(\mathbf{t})} \tag{2}$$

where $P(\mathbf{w})$ is the so-called *prior* distribution of the parameters $\mathbf{w}$. In some applications, the prior distribution has a single parameter (actually called hyper-parameter) α to optimize. Vice versa, in RVMs, instead of a single hyper-parameter α, we have a vector of hyper-parameters $\boldsymbol{\alpha}$. In particular, we associate a single hyper-parameter α_j to each parameter w_j. As proved in [21], RVMs for classification require the resolution of a recursive algorithm where, at each iteration, the hyper-parameters $\boldsymbol{\alpha}$ are re-estimated.

The k-Nearest Neighbor (kNN) classifier exploits the following principle: given a vector of features $\mathbf{x}$ to classify, the decision is taken looking at the classes of the k closest vectors (i.e., the neighbors) in the data set. In classification problems, k is usually set to an odd value, to avoid parity when deciding the corresponding symbol $\hat{t}$. The best value for k is detected through calibration, varying k within a given range. For each value of k, a 3-cross validation is performed and the corresponding score (proportional to the accuracy of the cross validation) is detected. We select the value of k providing the highest score.

Adaptive Neural Fuzzy Inference System (ANFIS) is another powerful learning algorithm where the input data is performed through Sugeno-type fuzzy inference system [22,26]. In this inference systems, the input data is transformed in input characteristics. Input characteristics are then transformed to input membership functions, and input membership functions to rules. At this point, rules are transformed to a set of output characteristics. Output characteristics are transformed to output membership functions, and output membership functions to output. In particular, the algorithm works with a combination of least-squares and back-propagation gradient descent methods [23].

Hidden Markov Models are an effective technique for predicting new values given previous observations. Intuitively, recent observations provide more information than historical ones. Moreover, it would be infeasible to consider in a model the impressive number of dependences of future observations with all the past observations. Indeed, the resulting complexity would be clearly impractical. Markov models assume that future predictions are independent of all but the most recent observations. This also leads to a simplification of the product rule of probability. Given two random variables A and B, the conditional probability $P(A|B)$ and the joint probability $P(A, B)$ are linked through the following relationship:

$$P(A, B) = P(A|B) \cdot P(B). \tag{3}$$

If we add a third random variable and we consider the conditional probability of A having observed B and C, then (3) becomes

$$P(A, B, C) = P(A|B, C) \cdot P(B, C). \tag{4}$$

But, from (3), $P(B, C) = P(B|C)P(C)$, and (4) becomes

$$P(A, B, C) = P(A|B, C) \cdot P(B|C) \cdot P(C). \tag{5}$$

Therefore, given N observations $\mathbf{x}_1, \ldots, \mathbf{x}_N$ in a sequence of data, the joint probability $P(\mathbf{x}_1, \ldots, \mathbf{x}_N)$ is as follows

$$P(\mathbf{x}_1, \ldots, \mathbf{x}_N) = P(\mathbf{x}_1) \prod_{n=2}^{N} P(\mathbf{x}_n|\mathbf{x}_{n-1}, \ldots, \mathbf{x}_1). \tag{6}$$

Hidden Markov Models work with a simplification of (6) because, as mentioned before, taking into account all the dependencies lead to an exaggerated complexity of the model. For instance in first-order Markov chains it is assumed that each conditional distribution is independent of all previous observations but the most recent one, and (6) becomes

$$P(\mathbf{x}_1, \ldots, \mathbf{x}_N) = P(\mathbf{x}_1) \prod_{n=2}^{N} P(\mathbf{x}_n|\mathbf{x}_{n-1}). \tag{7}$$

Finally, HMMs assume the introduction of latent variables, which, as the name reveals, are not observable [24,25].

Finally, Support Vector Machines (SVMs) require the resolution of an optimization problem [27] involving kernel parameters and a further parameter

$C > 0$ representing the penalty paid if the hyperplane does not exactly separate the vectors belonging to the two categories. An issue in the resolution of this optimization problem is the choice of the kernel. A kernel is a scalar product involving the aforementioned basis functions $\phi_j(\mathbf{x})$. The aim of these functions is to map any vector $\mathbf{x}$ into a higher (in principle also infinite) dimensional space to foster hyperplane separability. In principle, there is an infinite number of choices for the kernel functions. One of the most popular kernels is the so-called Radial Basis Function (RBF) kernel, which expression is:

$$K(\mathbf{x}, \mathbf{x}') = \exp\left(-\gamma \|\mathbf{x} - \mathbf{x}'\|^2\right), \tag{8}$$

where γ is a parameter to calibrate. Therefore, we have two parameters to calibrate in total using a RBF kernel: the parameter γ associated to the exponential function and the penalty cost C. Similarly to the kNN classifier, the technique consists in varying parameters within a given set. For each combination of parameters, a 3-cross validation is performed and the corresponding score is detected. Finally, we select the combination providing the highest score.

5 Preliminary Results

In this section, we report preliminary results for the development of the DIDM in the DESERVE project. In particular, we show how we preprocessed the dataset and then how we calibrated and trained the proposed classifiers. All the classifiers were trained and tested on a preliminary dataset provided by CRF. This dataset consists of 720 rows with three features and corresponding target variables. 495 rows make up the training set, while the remaining 225 rows make up the test set. The features are: Mean Speed (MS), Mean Steering Angle (MSA) and Mean Time To Collision (MTTC). The reason of these features is that they are easily related to the dynamic of the vehicle, including the lateral one, and to the driver intention. Moreover, the driver was free to perform a Lane Change at any time he/she felt to.

Preprocessing of data is a crucial issue in learning algorithms. In fact, raw-data from direct measurements is full of noise and this can lead to bad performances of the training and the consequent testing of the classifier. As suggested by Mandalia [1], a way to reduce the consequences of noisy measurements is to use overlapping time windows containing a reasonable number of samples. Data inside a time window is averaged and the last feature in the time window is considered. The original sampling time interval for this preliminary dataset was 50 ms. We used time windows of 0 s, 0.5 s and 1 s, corresponding to time windows of 1, 10 and 20 elements.

Calibration consisted in the following procedure. For each classifier, we performed a threefold-cross-validation procedure, each time with a different combination of parameters associated to a particular classifier [27]. This means to divide the training set into three balanced parts and train each classifier in any of these parts. The mean of the three accuracies gives the score of that cross validation for that particular combination of parameters. At the end of the process, we retained the best combination of parameters. Figure 2 shows an example of calibration of a SVM with the Pattern Recognition Toolbox [7] in Matlab performed during our computational tests. The meaning of Fig. 2 is the following: as stated above, a threefold cross-validation is performed for each combination of values for C and γ and the result is a score. Each square in Fig. 2 represents a pair of C and γ values, where black correspond to the lowest score and white to the highest score. For RVMs, the best results were found with a maximum number of iterations of the internal algorithm equal to 1000 and with a tolerance of 1e-5. For the kNN classifier, it is common rule to set k to low and odd values. To calibrate the kNN classifier, we tried different odd values of k from 1 to 11 and we performed a threefold cross validation for each value of k, as done for the SVM. At the end of the procedure, the best value for k turned out to be 5. Anfis was used with 2 training epochs. SVM parameters C and γ were both varied in the set $\{1, \ldots, 10\}$. The best choice for these parameters turned out to be $C = \gamma = 8$. Finally, HMMs were designed with 2 states, one for LK and one for LC. Transition and estimation matrices were computed based on the training set features and target variables.

Table 1 reports the accuracies of our tests. According to statistical definitions, given a binary experiment with two possible outcomes (i.e., LC or LK, where LC is associated to a TRUE value and LK to FALSE), we can define the overall accuracy, the true positive rate (TPR) and the true negative rate (TNR) of each classifier for different kinds of preprocessing, i.e., for times windows of 0, 0.5 and 1 second. The accuracy, the true positive rate and the true negative rate were computed as follows: let i denote the index associated to a target variable (or symbol) t_i of the test set, assuming that this symbol corresponding to one row of the test set is equal to 1 if that row of features corresponds to a LK maneuver (i.e., a FALSE) and 2 to a LC maneuver (i.e., a TRUE), the accuracy η is the percentage of the corrected symbols $\hat{t}_i$ estimated by the classifier, over the overall number of symbols:

$$\eta = 100 \frac{\left|\{\hat{t}_i = t_i | t_i \in \mathbf{t}_{test}\}\right|}{\left|\{\mathbf{t}_{test}\}\right|}, \tag{9}$$

where $\mathbf{t}_{test}$ denotes the vector of target variables associated to the test set. Similarly, the true positive rate (accuracy η_{LC} associated to lane changes) and the true negative rate (accuracy η_{LK} associated to lane keepings) are defined as follows:

$$TPR = \eta_{LC} = 100 \frac{\left|\{\hat{t}_i = 2 \wedge t_i = 2 | t_i \in \mathbf{t}_{test}\}\right|}{\left|\{t_i = 2 | t_i \in \mathbf{t}_{test}\}\right|}. \tag{10}$$

and

$$TNR = \eta_{LK} = 100 \frac{\left|\{\hat{t}_i = 1 \wedge t_i = 1 | t_i \in \mathbf{t}_{test}\}\right|}{\left|\{t_i = 1 | t_i \in \mathbf{t}_{test}\}\right|}, \tag{11}$$

False positive rate (FPR) and False negative rate (FNR) can easily be defined and deducted from the above definitions.

Table 1. Results from the tests of the proposed classifiers.

Time window	Classifier	% TNR	% TPR	% Accuracy
0 s (1 sample)	RVM	80.67	62.67	74.67
	kNN	>99	>99	>99
	ANFIS	82.67	61.33	75.56
	HMM	98	92	80
	SVM	89.33	73.33	84.00
0.5 s (10 samples)	RVM	97.96	94.20	96.76
	kNN	>99	0.00	68.06
	ANFIS	>99	98.56	>99
	HMM	>99	86.95	95.37
	SVM	>99	>99	>99
1 s (20 samples)	RVM	98.59	98.44	98.54
	kNN	>99	0.00	68.93
	ANFIS	>99	>99	>99
	HMM	94.37	>99	96.12
	SVM	>99	>99	>99

Table 1 shows is the importance of preprocessing: if the length of the overlapping time windows increases, then the undesired noise effects are reduced and the overall performance increases. Table 1 also tells us that we can avoid the *k*NN classifier. Its performance is excellent for the dataset without preprocessing, but the *k*NN classifier becomes biased when the size of the time windows are increased. In fact, the accuracy for LK maneuvers (True Negative Rate) is very high while it is 0% for LC maneuvers (True Positive Rate). This means that, independently on the input vector of features, this classifier always provides the same prediction: lane keeping. This is clearly an undesired behavior. The reason of this behavior also relies on the fact that *k*NN classifiers are very simple to implement, but at the same time very sensitive to data. HMMs provide good-quality results, which, however, are outperformed by RVMs, Anfis and SVMs. A concrete issue encountered with the RVMs when using just partial data of the real dataset is the problem of available RAM memory. Although RVMs are a very promising tool, which also offer a number of advantages, in more than one circumstance the algorithm crashed because of lack of available memory. It is likely that this behavior is due to the matrix inversion required at every step of the RVM algorithm. The simplicity of calibration, the robust performance in terms of memory use and the very promising results shown in Table 1 motivated us to use the SVM classifier in the final part of the project, concerning the real dataset and the integration of the classifier into the DIDM. These results are also in line with other works in the literature dealing with SVM for detecting driver maneuvers. For instance, Mandalia and Salvucci [9] found an accuracy of 97.9% with a time window of 1.2. Dogan et al. [2] reported results of different classifiers and SVMs turned out to be the best classifier among the proposed ones. Clearly, these considerations do not mean that our SVM classifier is better than those presented by these authors, because the datasets and the approaches are different. Nevertheless, it is interesting to notice similar behaviors and conclusions with previous works in the literature.

Finally, we wish to point out again the importance of overlapping time windows. A time window of 0.5 s already reduces the problem of noisy data acquisition in a noteworthy manner. SVMs show high accuracy percentages, in addition to more stable implementations compared to Anfis, both in terms of software and hardware integration on board units. Thus, SVM is the methodology we suggest for these kinds of applications and for their scalability and portability. One limit of the present research is the use of a dataset with a limited size, a common issue for other datasets in the literature. For this reason, within the DESERVE project, a large number of tests is performed with a number of volunteers and increased number of features, such as more dynamics data, additional data from internal and external camera, turn indicators, and additional data from lasers and radar. At this stage of the project, computational tests are at an advanced development phase.

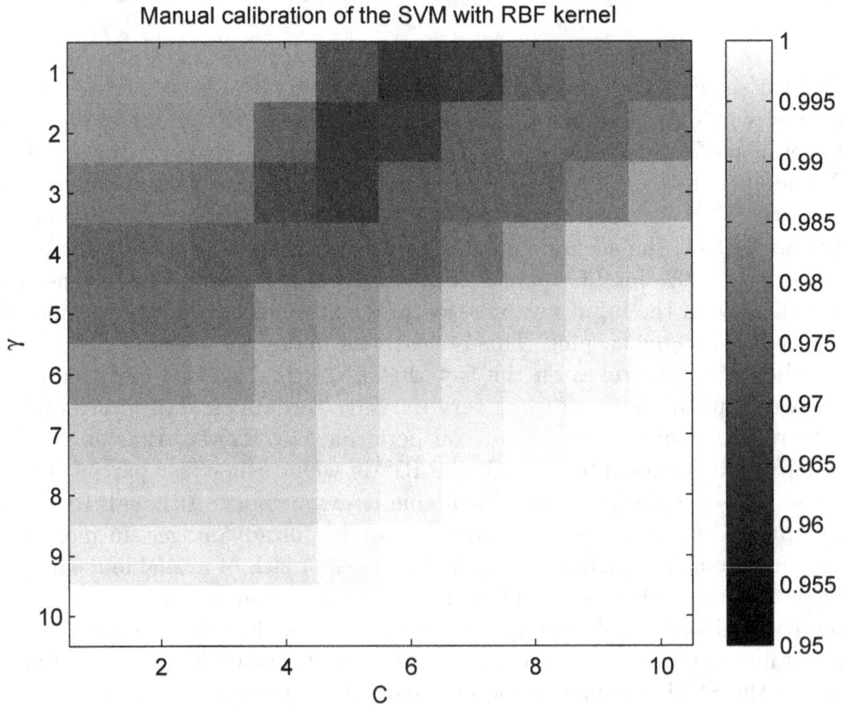

Fig. 2. An example of SVM calibration in matlab using the pattern recognition toolbox

6 Conclusions

In this paper we presented a set of classifiers used on a preliminary dataset for predicting driver maneuvers in the Development Platform for Safe and Efficient Drive (DESERVE) European project. Inferring driver maneuvers is a fundamental issue in Advanced Driver Assistance Systems. The aim of the DESERVE project is to provide an important contribution to this topic by also creating an in-vehicle Driver Intention Detection Module. We trained a number of classifiers over a preliminary dataset to infer driver maneuvers of Lane Keeping and Lane Change. The overall results are not available yet. Nevertheless, this preliminary study helped us to select a number of classifiers in spite of others. Considering also other issues encountered during computational tests over a part of the real dataset, we decided to use Support Vector Machines for the remaining part of the project. The complete results over the real dataset will be presented at the conference.

Acknowledgements. This research was developed under the European Research Project DESERVE, Development Platform for Safe and Efficient Drive, Project reference: 295364, Funded under: FP7-JTI. The authors are grateful to Fabio Tango, Sandro Cumani and Kenneth Morton for the support provided during the project.

References

1. Mandalia, H.M.: Pattern recognition techniques to infer driver intentions. Technical report DU-CS-04-08, Drexel University (2004). https://www.cs.drexel.edu/tech-reports/DU-CS-04-08.pdf
2. Dogan, U., Edelbrunner, H., Iossifidis, I.: Towards a driver model: preliminary study of lane change behavior. In: Proceedings of the XI International IEEE Conference on Intelligent Transportation Systems, pp. 931–937 (2008)
3. Burzio, G., Guidotti, L., Montanari, R., Perboli, G., Tadei, R.: A subjective field test on lane departure warning function - euroFOT. In: Proceedings of TRA-Transport Research Arena - Europe 2010 (2010)
4. DESERVE project. http://www.deserve-project.eu/
5. Calefato, C., Kutila, M., Ferrarini, C., Landini, E., Baldi, M.M., Tadei, R.: Development of cost efficient ADAS tool platform for automotive industry. In: The 22nd ITS World Congress in Bordeaux (France), 5–9 October 2015 (2015)
6. Centro Ricerche Fiat, Orbassano (TO), Italy. https://www.crf.it/IT
7. Torrione, P., Morton, K.: Pattern recognition toolbox. https://it.mathworks.com/matlabcentral/linkexchange/links/2947-pattern-recognition-toolbox
8. INTEMPORA, Issy-Les-Moulineaux, France. https://intempora.com
9. Mandalia, H.M., Salvucci, D.D.: Using support vector machines for lane change detection. In: Proceedings of the Human Factors and Ergonomics Society Annual Meeting, pp. 1965–1969. SAGE Publications (2005)
10. Butz, T., von Stryk, O.: Optimal control based modeling of vehicle driver properties. SAE Technical Paper 2005-01-0420 (2005). doi:10.4271/2005-01-0420
11. Hayashi, K., Kojima, Y., Abe, K., Oguri, K.: Prediction of stopping maneuver considering driver's state. In: Proceedings of the IEEE Intelligent Transportation Systems Conference, pp. 1191–1196 (2006)

12. McCall, J., Wipf, D., Trivedi, M., Rao, B.: Lane change intent analysis using robust operators and sparse bayesian learning. IEEE Trans. Intell. Transp. Syst. **8**(3), 431–440 (2007)
13. Salvucci, D.D., Mandalia, H.M., Kuge, N., Yamamura, T.: Lane-change detection using a computational driver model. Hum. Factors **49**(3), 532–542 (2007)
14. Huang, H., Gao, S.: Optimal paths in dynamic networks with dependent random link travel times. Transp. Res. B **46**, 579–598 (2012)
15. Deng, W.: A study on lane-change recognition using support vector machine. Ph.D. thesis, University of South Florida (2013)
16. Ly, M.V., Martin, S., Trivedi, M.M.: Driver classification and driving style recognition using inertial sensors. In: 2013 IEEE Intelligent Vehicles Symposium (IV), pp. 1040–1045 (2013)
17. Lin, N., Zong, C., Tomizuka, M., Song, P., Zhang, Z., Li, G.: An overview on study of identification of driver behavior characteristics for automotive control. Math. Probl. Eng. **2014**, 15. Article ID 569109 (2014). doi:10.1155/2014/569109
18. Liu, W., Tao, D.: Multiview hessian regularization for image annotation. IEEE Trans. Image Process. **22**(7), 2676–2687 (2013)
19. Ohn-Bar, E., Tawari, A., Martin, S., Trivedi, M.M.: Predicting driver maneuvers by learning holistic features. In: 2014 IEEE Intelligent Vehicles Symposium Proceedings, pp. 719–724 (2014)
20. Jain, A., Koppula, H.S., Raghavan, B., Soh, S., Saxena, A.: Car that knows before you do: anticipating maneuvers via learning temporal driving models (2015). http://arxiv.org/abs/1504.02789
21. Tipping, M.E.: Sparse Bayesian learning and the relevance vector machine. J. Mach. Learn. Res. **1**, 211–244 (2001)
22. Jang, J.S.R.: Anfis: adaptive-network-based fuzzy inference systems. IEEE Trans. Syst. Man Cybern. **23**(3), 665–685 (1993)
23. ANFIS. http://it.mathworks.com/help/fuzzy/anfis.html
24. Rabiner, L.R., Juang, B.H.: An introduction to hidden Markov models. IEEE ASSP Mag. **3**, 4–16 (1986)
25. Rabiner, L.R.: A tutorial on hidden Markov models and selected applications in speech recognition. Proc. IEEE **77**(2), 257–286 (1989)
26. Tagaki, T., Sugeno, M.: Fuzzy identification of systems and its applications to modeling and control. IEEE Trans. Syst. Man Cybern. **15**(1), 116–132 (1985)
27. Hsu, C.-W., Chang, C.-C., Lin, C.J.: A practical guide to support vector classification (2010). https://www.csie.ntu.edu.tw/~cjlin/papers/guide/guide.pdf

A Systems Biology Approach for Unsupervised Clustering of High-Dimensional Data

Diana Diaz[1(✉)], Tin Nguyen[1], and Sorin Draghici[1,2]

[1] Wayne State University, Computer Science, Detroit 48202, USA
dmd@wayne.edu
[2] Wayne State University, Obstetrics and Gynecology, Detroit 48202, USA

Abstract. One main challenge in modern medicine is the discovery of molecular disease subtypes characterized by relevant clinical differences, such as survival. However, clustering high-dimensional expression data is challenging due to noise and the curse of high-dimensionality. This article describes a disease subtyping pipeline that is able to exploit the important information available in pathway databases and clinical variables. The pipeline consists of a new feature selection procedure and existing clustering methods. Our procedure partitions a set of patients using the set of genes in each pathway as clustering features. To select the best features, this procedure estimates the relevance of each pathway and fuses relevant pathways. We show that our pipeline finds subtypes of patients with more distinctive survival profiles than traditional subtyping methods by analyzing a TCGA colon cancer gene expression dataset. Here we demonstrate that our pipeline improves three different clustering methods: k-means, SNF, and hierarchical clustering.

1 Introduction

Identifying homogeneous subtypes in complex diseases is crucial for improving prognosis, treatment, and precision medicine [1]. Disease subtyping approaches have been developed to identify clinically relevant subtypes. High-throughput technologies can measure the expression of more than ten thousand genes at a time. Subtyping patients using the whole-genome scale measurement is challenging due to the curse of high-dimensionality. Several clustering methods have been developed [2–5] to handle this type of high-dimensional data. Other approaches, such as iCluster [6], rely on feature selection to reduce the complexity of the problem.

There are many widely used feature selection methods [7–11]. The simplest way to perform unsupervised feature selection for subtyping is by ranking the list of genes and filtering out those with low rankings. For example, genes can be ranked using Fisher score-based methods [8,9] or t-test based methods [10]. Other methods, such us [11], use general purpose filtering metrics like Information Gain [12], Consistency [13], Chi-Squared [14] and Correlation-Based Feature Selection [15]. These filter-based methods are computationally efficient, but they

P.M. Pardalos et al. (Eds.): MOD 2016, LNCS 10122, pp. 193–203, 2016.
DOI: 10.1007/978-3-319-51469-7_16

do not account for dependency between genes or features. To address this, wrapper methods [16,17] use learning algorithms to find subsets of related features or genes. Even though these methods consider feature dependency, they have a high degree of computational complexity due to repeated training and testing of predictors. This makes them impractical for analyzing high-dimensional data.

Meanwhile, some approaches incorporate to gene-expression-based subtyping other types of data such us clinical variables [18–20] and multi 'omics' data [4,6,21]. These types of data are more and more available nowadays. Large public repositories, including the Cancer Genome Atlas (TCGA) (cancergenome.nih.gov), accumulate clinical and multi 'omics' data from thousands of patients. Clinical variables used for subtyping include survival data [18], epidemiological data [19], clinical chemistry evaluations and histopathologic observations [20]. These variables have shown to provide useful information for a better subtyping.

Subtyping patients using gene expression data has additional challenges because genes do not function independently. They function in synchrony to carry on complex biological processes. Knowledge of these processes is usually accumulated in biological pathway databases, such as KEGG [22] and Reactome [23]. Biological pathways are graphical representations of common knowledge about genes and their interactions on biological processes. This valuable information has been used to cluster related genes using gene expression [24–27] and should be used to identify disease subtypes as well. Clinical data and biological knowledge are complementary to gene expression and can leverage disease subtyping.

Here we present a disease subtyping pipeline that includes a new feature selection approach and any existing unsupervised clustering method. To the best of our knowledge, this is the first approach that integrates pathway knowledge and clinical data with gene expression for disease subtyping. Our framework is validated using gene expression and clinical data downloaded from the Cancer Genome Atlas (TCGA) and pathways from the Kyoto Encyclopedia of Genes and Genomes (KEGG). Using the features selected with our approach and three different clustering methods (k-means, SNF, and hierarchical clustering), our pipeline is able to identify subtypes that have significantly different survival profiles. This pipeline was developed in R programming language. The source code is available on github (http://datad.github.io/disSuptyper) to ease the reproducibility of the methods presented here [28,29].

2 Method

In this section, we introduce a new feature selection framework for disease subtyping. Figure 1 presents the overall pipeline of our framework. The input includes (i) gene expression data, (ii) survival data, and (iii) biological pathways (see Fig. 1a). The output is a set of selected genes (Fig. 1f) for finding subtypes with significantly distinct survival patterns (Fig. 1g).

Gene expression data can be represented as a matrix $D \in R^{M \times N}$, where the rows are different patients having the same disease and columns are different

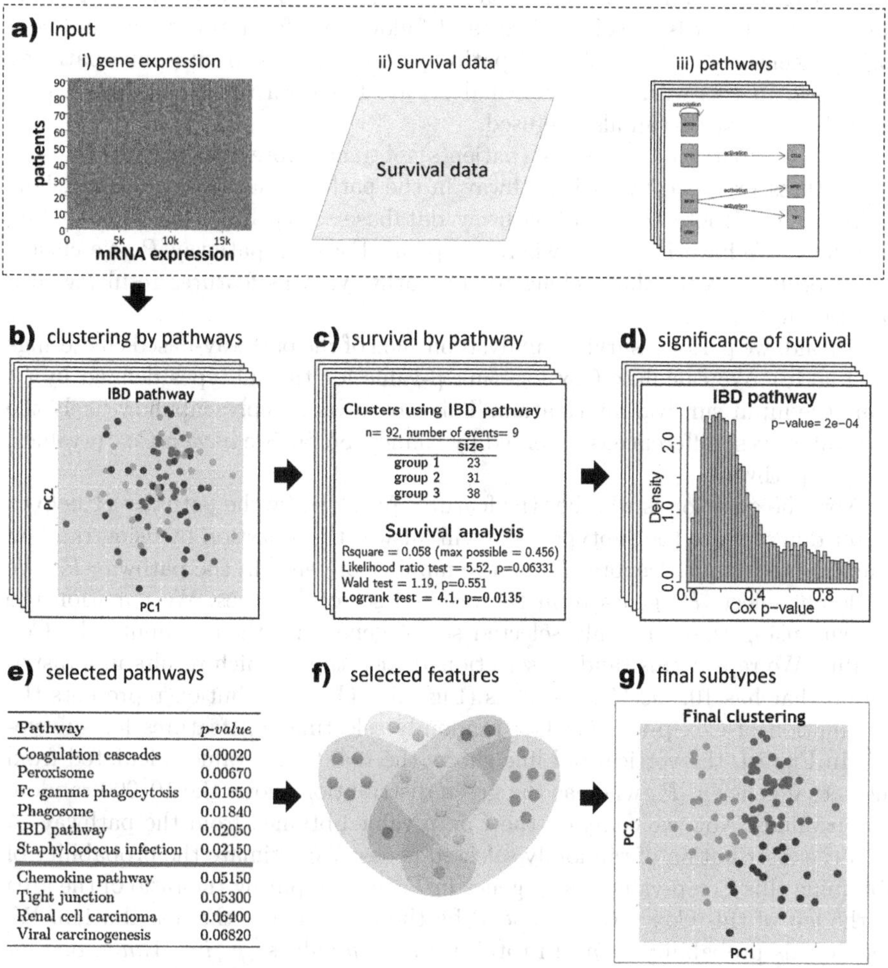

Fig. 1. New feature selection pipeline for disease subtyping using biological knowledge. (a) The input includes (i) gene expression data, (ii) survival data, and (iii) pathways downloaded from a database. (b) First, we partition the gene expression data using the set of genes in each pathway as features. (c) Second, we perform survival analysis on each resulting partition. (d) Third, we compute the p-value that represents how likely the pathway improves the subtyping. (e) Fourth, we rank the list of pathways by corrected p-value and select pathways that have a nominal p-value less than or equal to the significance threshold 5%. (f) Fifth, we merge the relevant pathways to construct the final set of features. (g) Finally, we subtype the patients using the selected features. The clustering is demonstrated in the first two principal components, but we use all dimensions/genes for clustering. Note: IBD pathway stands for Inflammatory Bowel Disease pathway. (Color figure online)

features (i.e. genes). M is the number of patients and N is the number of genes. For gene expression data, N can be as large as 20,000. The survival data include patient's vital status (dead or alive) and follow-up information (time and censored/uncensored). The biological pathways are collected from public pathway databases. In this work, our data analysis are based on KEGG pathways [22], but other databases can also be used.

First, we partition the rows (patients) of gene expression matrix D using the features provided by each pathway in the pathway database (Fig. 1b). Formally, let us denote $\mathbf{P}$ as the pathway database which has $n = |\mathbf{P}|$ signaling pathways. We have $\mathbf{P} = \{P_i\}$ where $i \in [1..n]$. For each pathway P_i, we cluster the rows using genes that belong to the pathway P_i as features resulting in a partitioning C_i.

Second, we perform survival analysis on each of the pathway-based clusterings C_i (Fig. 1c). We calculate Cox log-rank p-value for the subtypes defined by C_i using the input survival information. This Cox p-value represents how likely the survival curves' difference is observed by chance. So far, we have n Cox p-values, one per pathway.

Now the question is whether the features provided by the pathway P_i help to better differentiate the subtypes. We will answer this question by using random sampling technique. Denote $|P_i|$ as the number of genes in the pathway P_i. We randomly select $|P_i|$ genes from the original set of N genes. We partition the patients using this randomly selected set of genes and then compute the Cox p-value. We repeat this random selection 10,000 times which results in a distribution that has 10,000 Cox p-values (Fig. 1d). This distribution represents the distribution of Cox p-values when randomly selecting $|P_i|$ features for subtyping. In Fig. 1d, the vertical red line shows the real Cox p-value calculated from the actual genes in P_i, whereas the green distribution shows the 10,000 random Cox p-values. Now we compare the Cox p-value obtained from the pathway P_i with the distribution of randomly selected genes. We estimate the probability of obtaining this Cox p-value (using genes in P_i) by computing the ratio of the area to the left of this Cox p-value divided by the total area of the distribution. We denote this probability as p_i. In total, we have n values $\{p_i, i \in [1..n]\}$, one for each pathway. Each of these p-values p_i quantifies how likely it is to observe by chance a Cox log-rank statistic as extreme or more than the one observed. Therefore, this p-value of a pathway P_i represents how likely the features provided by the pathway help to improve the subtyping.

The third step is to choose a set of pathways that certainly help to improve the subtyping. To do this, we adjusted the p-values for multiple comparisons using False Discovery Rate (FDR), we rank the set of pathways and select those that have the corresponding nominal *p-values* less than or equal to the significance threshold of 5%. Let us name the pathways yielding significantly distinct survival curves as *relevant pathways*. For example, In Fig. 1e, the horizontal red line shows the significance threshold of 5%. In this example, the relevant pathways are *Coagulation cascades*, *Peroxisome*, *Fc gamma phagocytosis*, *Phagosome*, *Inflammatory Bowel Disease (IBD) pathway*, and *Staphylococcus infection*.

Considering all the genes in the relevant pathways as favorable features, we merge these pathways to get a single set of genes (Fig. 1f). We use this merged set of genes as the selected features for our final subtyping. In our example, the final selected genes are the genes in the six pathways listed above. We then use these genes to construct the final clustering as shown in Fig. 1g.

We note that this feature selection procedure can be used in conjunction with any clustering method. In our experimental studies, we used three clustering methods that belong to different clustering models. The first method is the classical k-means. It is well-known that k-means does not always converge to a global optimal point, it depends on the initialization. To overcome this problem, we ran k-means several times and chose the partitioning that has the smallest residual sum of squares (RSS). In the rest of the manuscript, we refer to this as "RSS k-means". The second method is Similarity Network Fusion (SNF) [4], which is based on spectral clustering. The third one is the traditional hierarchical clustering using cosine similarity as the distance function. We will show that our framework helps to improve the subtyping using any of the three mentioned clustering methods.

3 Results

In this section,we assess the performance of our feature selection for disease subtyping framework using gene expression data (Agilent G4502A-07 platform level 3) generated by the Cancer Genome Atlas (TCGA) (cancergenome.nih.gov). We selected the samples that have miRNA and methylation measurements as were selected in SNF [4]. A copy of the dataset is available in the github repository (http://datad.github.io/disSuptyper). The number of patients is $M = 92$, and the number of genes is $N = 17,814$. For all the performed clusterings, we set the number of clusters as $k = 3$ according to prior knowledge of the number of subtypes of colon cancer [4]. When running our method, we used 184 pathways from the KEGG pathway database [22].

As described in Sect. 2, our framework can be used in conjunction with any unsupervised clustering algorithm. Here we test it using three clustering methods: RSS k-means, SNF [4], hierarchical clustering [2]. For all clustering methods, we first clustered the patients using all the measured genes, then clustered the patients using only the genes selected by our technique. To contrast the difference between the three traditional clustering methods and our pipeline results, we performed survival analysis for all the cases using Kaplan-Meier analysis and Cox p-value.

3.1 Subtyping Using k-means

We clustered the patients from the TCGA colon adenocarcinoma dataset using our pipeline in conjunction with RSS k-means. We used the 184 signaling pathways from the KEGG database [22]. For each pathway P_i, we partitioned the patients using the genes in the pathway P_i as features to get a clustering C_i.

Table 1. List of pathways selected by our approach when using RSS k-means. We first ranked the pathways by FDR adjusted p-value (*p-value.fdr*), then selected the pathways with a nominal *p-value* ≤ 0.05 as relevant pathways.

Pathway	*p-value*	*p-value.fdr*
Complement and coagulation cascades	0.00020	0.03680
AGE-RAGE signaling pathway in diabetic complications	0.00420	0.38640
Peroxisome	0.00670	0.41093
Cytokine-cytokine receptor interaction	0.01040	0.45448
Fc gamma R-mediated phagocytosis	0.01650	0.45448
Phagosome	0.01840	0.45448
Inflammatory bowel disease (IBD)	0.02050	0.45448
Staphylococcus aureus infection	0.02150	0.45448
Leukocyte transendothelial migration	0.02330	0.45448
NF-kappa B signaling pathway	0.03710	0.50048
Renin secretion	0.03850	0.50048
Malaria	0.04780	0.51326
Platelet activation	0.06980	0.54970

After this step, we got a total of 184 clusterings, one per pathway. Also for each pathway, we constructed the empirical distribution and then estimated the *p-value* of how likely the pathway helps to improve disease subtyping. The *p-values* of relevant pathways are shown in Table 1. The horizontal red line represents the significance cutoff at 5%. There are 12 relevant pathways. We then merged the relevant pathways to get a single set of genes that we used as clustering features. This final set of features consists of 851 genes when using RSS k-means algorithm. Finally, we performed RSS k-means clustering using these 851 genes.

Figure 2 shows the survival analysis of the resultant clusterings. Figure 2a shows the resultant clustering when using RSS k-means for all 17,814 genes. The Cox p-value of this clustering is 0.129, which is not significant. Figure 2b shows the resultant clustering using the 851 selected genes. The resultant Cox p-value is 0.0156, which is approximately ten times lower than using all genes.

3.2 Subtyping Using SNF

Similar to the assessment performed for k-means, we clustered the patients from the TCGA colon adenocarcinoma dataset using our pipeline in conjunction with SNF. To perform SNF clustering, we ran the SNFtool Bioconductor package with the parameters suggested by the authors [4]. We used the same input (KEGG pathways), settings (three clusters), and process previously described.

After this step, we obtained 184 clusterings, one per pathway. Then for each pathway, we constructed the empirical distribution and estimated the *p-value* of how likely the pathway helps to improve disease subtyping. The estimated

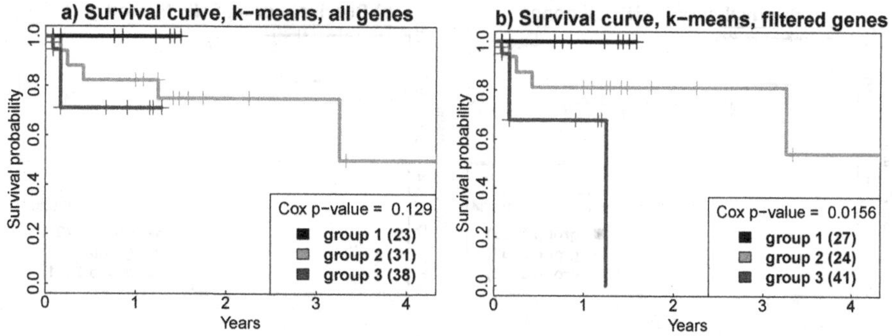

Fig. 2. Kaplan-Meier survival analysis of the obtained subtypes using RSS k-means algorithm. (a) Survival curves using all genes. (b) Survival curves using selected genes.

Table 2. List of pathways that contain relevant genes obtained with our approach when using SNF. We first ranked the pathways by *p-value.fdr*, then selected the pathways with a nominal *p-value* ≤ 0.05.

Pathway	*p-value*	*p-value.fdr*
HTLV-I infection	0.00400	0.37765
Endocrine and other factor-regulated calcium reabsorption	0.00680	0.37765
Complement and coagulation cascades	0.00800	0.37765
Aldosterone-regulated sodium reabsorption	0.00830	0.37765
AMPK signaling pathway	0.01410	0.51324
Phagosome	0.02150	0.54196
Fc epsilon RI signaling pathway	0.02290	0.54196
Cytosolic DNA-sensing pathway	0.02680	0.54196
Peroxisome	0.03900	0.61320
Leishmaniasis	0.04300	0.61320
Non-alcoholic fatty liver disease (NAFLD)	0.05400	0.66544

p-values are shown in Table 2. The horizontal red line represents the significance threshold of 5%. There are 10 relevant pathways. We merged these relevant pathways to get a single set of genes that we used as our final set of selected features. This feature set contains 764 genes for SNF method. Finally, we performed SNF clustering using these 764 genes.

Figure 3 shows the survival analysis of the resultant clusterings. Figure 3a shows the clustering when using SNF for all 17,814 genes. The Cox p-value of this clustering is 0.1836, which is not significant (this resultant is identical to the result reported in [4]). Figure 3b shows the resultant clustering when using the 764 selected genes. The Cox p-value is 0.0207, which is approximately ten times lower than using all genes. Despite this meaningful improvement, none of

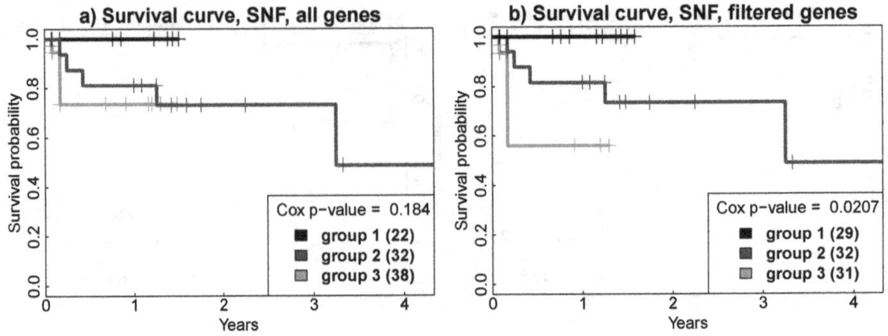

Fig. 3. Kaplan-Meier survival analysis of the obtained subtypes using SNF. (a) Survival curves using all genes. (b) Survival curves using the selected genes.

the pathways has a corrected *p-value.fdr* ≤ 0.05. This shows a lack of statistical power on our approach and an opportunity for improvement.

3.3 Subtyping Using Hierarchical Clustering

Alike the assessment performed previously, we clustered the colon adenocarcinoma patients using our pipeline in conjunction with Hierarchical Clustering (HC) [2]. We used the 184 signaling pathways from KEGG [22]. The estimated *p-values* of the relevant pathways obtained with HC are shown in Table 3. The horizontal red line represents the significance threshold of 5%. We merged these three relevant pathways to get our final set of selected features. This feature set contains 195 genes for HC. Finally, we performed hierarchical clustering using the selected genes only.

Figure 4 shows the survival analysis of the resultant clusterings. Figure 4a shows the clustering when using HC for all 17,814 genes. The Cox p-value of this clustering is 0.799 which is not significant. Figure 4b shows the resultant clustering when using the 195 selected genes. The Cox p-value is 0.151 which is lower than using all genes, but it is still not significant. The subtypes obtained with hierarchical clustering do not separate the patients in clinically meaningful subtypes in any of the cases (neither using all genes nor filtered genes).

Given that our approach requires resampling for computing the *p-values* p_i, this pipeline is more time consuming than traditional approaches. For the computational experiments presented here, we generated 10,000 random samplings and clusterings per each pathway (184 pathways in total). Our pipeline took several hours to subtype the set of patients (about 8 h for k-means, 17 h for SNF, and 46 h for hierarchical clustering) while running any traditional clustering method takes only some minutes (less than 6 min). We ran these experiments on a typical desktop workstation with a 2.6 GHz Intel Core i5, 8 GB of RAM, on a single thread, and the OS X 10.11 operative system.

Table 3. List of pathways selected by our approach when using hierarchical clustering. We first ranked the pathways by FDR adjusted p-value (*p-value.fdr*), then selected the pathways with a nominal *p-value* ≤ 0.05 as relevant pathways.

Pathway	*p-value*	*p-value.fdr*
Cytosolic DNA-sensing pathway	0.01140	0.63874
Peroxisome	0.01200	0.63874
Fc epsilon RI signaling pathway	0.04090	0.63874
Complement and coagulation cascades	0.12390	0.80770

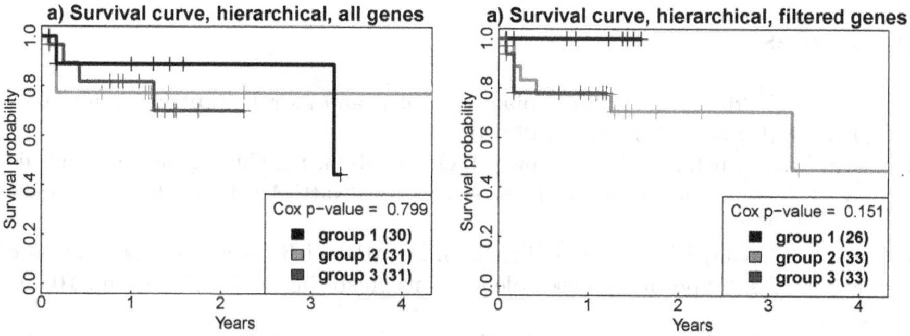

Fig. 4. Kaplan-Meier survival analysis of the obtained subtypes using hierarchical clustering (HC). (a) Survival curves using traditional HC. (b) Survival curves using HC in our pipeline.

4 Conclusions

In this article, we describe a framework to combine gene expression data, survival data, and biological knowledge available in pathway databases for a better disease subtyping. The performance of the new approach was demonstrated on the colon adenocarcinoma data downloaded from TCGA. The described framework was tested in conjunction with k-means, Similarity Network Fusion (SNF) and hierarchical clustering. For these clustering algorithms, our approach greatly improves the subtyping. In all cases, the Cox p-value is folds lower when using the selected features. Cox p-value improved from 0.129 to 0.0156 for k-means, from 0.184 to 0.0207 for SNF, and from 0.799 to 0.151 for hierarchical clustering.

Our contribution is two-folds. First, this framework introduces a way to exploit the additional information available in biological databases. Although the framework was demonstrated on KEGG pathways, it can exploit information available in other databases, such as functional modules available in Gene Ontology database or protein-protein interactions available in the STRING database. Second, this framework is the first one that integrates clinical data, biological pathways, and gene expression data for disease subtyping. For future work, we plan to use other clinical variables besides survival information and integrate multiple datatypes, such us microRNA, for a more comprehensive analysis [30]. Additionally, we plan to analyze the performance of feature selection methods from other contexts into the context of disease subtyping.

Acknowledgments. This study used data generated by the TCGA Research Network; we thank donors and research groups for sharing these valuable data. This research was supported in part by the following grants: NIH R01 DK089167, R42 GM087013 and NSF DBI-0965741, and by the Robert J. Sokol Endowment in Systems Biology. Any opinions, findings, conclusions, or recommendations expressed in this material are those of the authors and do not necessarily reflect the views of any of the funding agencies.

References

1. Saria, S., Goldenberg, A.: Subtyping: what it is and its role in precision medicine. IEEE Intell. Syst. **30**(4), 70–75 (2015)
2. Eisen, M.B., Spellman, P.T., Brown, P.O., Botstein, D.: Cluster analysis and display of genome-wide expression patterns. Proc. Natl. Acad. Sci. **95**(25), 14863–14868 (1998)
3. Kim, E.Y., Kim, S.Y., Ashlock, D., Nam, D.: MULTI-K: accurate classification of microarray subtypes using ensemble k-means clustering. BMC Bioinform. **10**, 260 (2009)
4. Wang, B., Mezlini, A.M., Demir, F., Fiume, M., Tu, Z., Brudno, M., Haibe-Kains, B., Goldenberg, A.: Similarity network fusion for aggregating data types on a genomic scale. Nat. Methods **11**(3), 333–337 (2014)
5. Hsu, J.J., Finkelstein, D.M., Schoenfeld, D.A.: Outcome-driven cluster analysis with application to microarray data. PLoS ONE **10**(11), e0141874 (2015)
6. Shai, R., Shi, T., Kremen, T.J., Horvath, S., Liau, L.M., Cloughesy, T.F., Mischel, P.S., Nelson, S.F.: Gene expression profiling identifies molecular subtypes of gliomas. Oncogene **22**(31), 4918–4923 (2003)
7. Hira, Z.M., Gillies, D.F., Hira, Z.M., Gillies, D.F.: A review of feature selection and feature extraction methods applied on microarray data. Adv. Bioinform. **2015**, e198363 (2015)
8. Huang, G.T., Cunningham, K.I., Benos, P.V., Chennubhotla, C.S.: Spectral clustering strategies for heterogeneous disease expression data. In: Pacific Symposium on Biocomputing, pp. 212–223 (2013)
9. Pyatnitskiy, M., Mazo, I., Shkrob, M., Schwartz, E., Kotelnikova, E.: Clustering gene expression regulators: new approach to disease subtyping. PLoS ONE **9**(1), e84955 (2014)
10. Li, T., Zhang, C., Ogihara, M.: A comparative study of feature selection and multiclass classification methods for tissue classification based on gene expression. Bioinformatics **20**(15), 2429–2437 (2004)
11. Hernández-Torruco, J., Canul-Reich, J., Frausto-Solís, J., Méndez-Castillo, J.J.: Feature selection for better identification of subtypes of Guillain-Barré. Comput. Math. Methods Med. **2014**, e432109 (2014)
12. Sebastiani, F.: Machine learning in automated text categorization. ACM Comput. Surv. **34**(1), 1–47 (2002)
13. Liu, Y., Schumann, M.: Data mining feature selection for credit scoring models. J. Oper. Res. Soc. **56**(9), 1099–1108 (2005)

14. Zheng, Z., Wu, X., Srihari, R.: Feature selection for text categorization on imbalanced data. SIGKDD Explor. Newsl. **6**(1), 80–89 (2004)
15. Hall, M.A.: Correlation-based feature selection for machine learning. Ph.D. thesis, The University of Waikato (1999)
16. Diaz-Uriarte, R., de Andres, S.A.: Gene selection and classification of microarray data using random forest. BMC Bioinform. **7**, 3 (2006)
17. Sharma, A., Imoto, S., Miyano, S., Sharma, V.: Null space based feature selection method for gene expression data. Int. J. Mach. Learn. Cybern. **3**(4), 269–276 (2011)
18. Bair, E., Tibshirani, R.: Semi-supervised methods to predict patient survival from gene expression data. PLOS Biol. **2**(4), e108 (2004)
19. Paoli, S., Jurman, G., Albanese, D., Merler, S., Furlanello, C.: Integrating gene expression profiling and clinical data. Int. J. Approx. Reason. **47**(1), 58–69 (2008)
20. Bushel, P.R., Wolfinger, R.D., Gibson, G.: Simultaneous clustering of gene expression data with clinical chemistry and pathological evaluations reveals phenotypic prototypes. BMC Syst. Biol. **1**, 15 (2007)
21. Chalise, P., Koestler, D.C., Bimali, M., Yu, Q., Fridley, B.L.: Integrative clustering methods for high-dimensional molecular data. Transl. Cancer Res. **3**(3), 202–216 (2014)
22. Kanehisa, M., Goto, S.: KEGG: kyoto encyclopedia of genes and genomes. Nucleic Acids Res. **28**(1), 27–30 (2000)
23. Croft, D., Mundo, A.F., Haw, R., Milacic, M., Weiser, J., Wu, G., Caudy, M., Garapati, P., Gillespie, M., Kamdar, M.R., Jassal, B., Jupe, S., Matthews, L., May, B., Palatnik, S., Rothfels, K., Shamovsky, V., Song, H., Williams, M., Birney, E., Hermjakob, H., Stein, L., D'Eustachio, P.: The Reactome pathway knowledgebase. Nucleic Acids Res. **42**(D1), D472–D477 (2014)
24. Hanisch, D., Zien, A., Zimmer, R., Lengauer, T.: Co-clustering of biological networks and gene expression data. Bioinformatics **18**(suppl. 1), S145–S154 (2002)
25. Huang, D., Pan, W.: Incorporating biological knowledge into distance-based clustering analysis of microarray gene expression data. Bioinformatics **22**(10), 1259–1268 (2006)
26. Rapaport, F., Zinovyev, A., Dutreix, M., Barillot, E., Vert, J.P.: Classification of microarray data using gene networks. BMC Bioinform. **8**, 35 (2007)
27. Pok, G., Liu, J.C.S., Ryu, K.H.: Effective feature selection framework for cluster analysis of microarray data. Bioinformation **4**(8), 385–389 (2010)
28. Prlić, A., Procter, J.B.: Ten Simple rules for the open development of scientific software. PLOS Comput. Biol. **8**(12), e1002802 (2012)
29. Carey, V.J., Stodden, V.: Reproducible research concepts and tools for cancer bioinformatics. In: Ochs, M.F., Casagrande, J.T., Davuluri, R.V. (eds.) Biomedical Informatics for Cancer Research, pp. 149–175. Springer, New York (2010). doi:10.1007/978-1-4419-5714-6_8
30. Diaz, D., Draghici, S.: mirIntegrator: Integrating miRNAs into signaling pathways. R package (2015)

Large-Scale Bandit Recommender System

Frédéric Guillou[1(✉)], Romaric Gaudel[2], and Philippe Preux[2]

[1] Inria, Univ. Lille, CNRS, Villeneuve-d'Ascq, France
frederic.guillou@inria.fr
[2] Univ. Lille, CNRS, Centrale Lille, Inria, UMR 9189 - CRIStAL, Villeneuve-d'Ascq, France
{romaric.gaudel,philippe.preux}@univ-lille3.fr

Abstract. The main target of Recommender Systems (RS) is to propose to users one or several items in which they might be interested. However, as users provide more feedback, the recommendation process has to take these new data into consideration. The necessity of this update phase makes recommendation an intrinsically sequential task. A few approaches were recently proposed to address this issue, but they do not meet the need to scale up to real life applications. In this paper, we present a Collaborative Filtering RS method based on Matrix Factorization and Multi-Armed Bandits. This approach aims at good recommendations with a narrow computation time. Several experiments on large datasets show that the proposed approach performs personalized recommendations in less than a millisecond per recommendation.

1 Introduction

We consider Collaborative Filtering approaches based on Matrix Completion. Such Recommender Systems (RS) recommend items to users and adapt the recommendation to user tastes as inferred from past user behavior. Depending on the application, items can be ads, news, music, videos, movies, etc. This recommendation setting was popularized by the Netflix challenge [4]. Most of approaches model the taste of each user regarding each item as a matrix $\mathbf{R}^*$ [13]. In that matrix, only a few entries are known: the ones corresponding to feedback gathered in the past. In such a context, the RS recovers unknown values in $\mathbf{R}^*$ and the evaluation is done by splitting log data into two parts: the first part (aka. train-set) is used to define the training matrix which is completed by the RS algorithm; the second part (aka. test-set) is used to measure the quality of the matrix returned by the RS. Common measures of that quality are Mean Absolute Error (MAE) and Root Mean Squared Error (RMSE) on the test-set. While such a static batch evaluation makes sense to measure the quality of the matrix-completion step of Collaborative Filtering, it does not evaluate the quality of the final recommendation. A Collaborative Filtering based RS works in reality in a sequential manner and loops through the following steps:

1. Build a model of the users' tastes based on past feedback;
2. Recommend items to users using this model;
3. Gather feedback from users about recommended products.

P.M. Pardalos et al. (Eds.): MOD 2016, LNCS 10122, pp. 204–215, 2016.
DOI: 10.1007/978-3-319-51469-7_17

Note that the model built at step 1 heavily depends on the feedback gathered at previous iterations. This feedback only exists for items which were chosen by the model itself. As such, at step 2, one faces the exploration/exploitation dilemma: either (i) recommend an item which led to the best feedback in the past (aka exploit) or (ii) recommend an item which hopefully brings information on the user's taste (aka explore). This dilemma is the core point of Multi-Armed Bandit Theory [3]. It is already studied in a sub-field of RS which has access to a representation of the context (the user, the webpage ...) [25]. Typical applications are the selection of news or ads to show on a web-page. The corresponding RS builds upon contextual bandits which are supported by strong theoretical results. In contrast with these studies, we focus on the setting where these representations are unknown and have to be inferred solely from users' feedback. In particular, we want to emphasize that we do not use any side information, neither about users, nor items. That field of research is almost empty and the few attempts therein leave out computational complexity constraints [12].

Our paper introduces the first sequential Collaborative Filtering based RS which (i) makes a good trade-off between exploration and exploitation and (ii) is able to properly scale. Extensive experiments are conducted on real word datasets of millions of ratings and convey three main conclusions: first, they highlight the need for a trade-off between exploration and exploitation for a RS to be optimal; second, they demonstrate that the proposed approach brings such optimal trade-off; third, they exhibit that the proposed approach can perform good recommendations in less than a millisecond per recommendation, the time a RS has to make its recommendation in a real, live system.

After introducing the setting in the next section, Sect. 3 recalls the standard matrix factorization approach and introduces the necessary background in bandit theory. In Sect. 4, we introduce an algorithm which fully takes into account the sequential aspect of RS. Section 5 provides an experimental study on real datasets. Finally, Sect. 6 reviews research results related to the proposed approach, and we conclude and draw some future lines of work in Sect. 7.

2 Sequential Recommendation

Let us focus on a particular recommendation scenario, which illustrates more accurately how typical Recommender Systems work. We consider N users, M items, and the unknown matrix $\mathbf{R}^*$ of size $N \times M$ such that $r^*_{i,j}$ is the taste of user i with regards to item j. At each time-step t,

1. a user i_t requests a recommendation from the RS,
2. the RS selects an item j_t among the set of available items,
3. user i_t returns a feedback $r_t \sim \mathcal{D}(r^*_{i_t,j_t})$ for item j_t.

In current paper we assume the mean of distribution $\mathcal{D}(r^*_{i_t,j_t})$ to be $r^*_{i_t,j_t}$. See [14] for an example of a more refined observation/noise model.

We refer to applications where the feedback r_t corresponds to the quantity that has to be optimized, aka. *the reward*. In such a context, the aim of the RS is to maximize the reward accumulated along time-steps $\text{CumRew}_T = \sum_{t=1}^{T} r_t$, or to minimize the *pseudo-regret* $\mathcal{R}_T$ (1) which measures how much the system loses in average by recommending a sub-optimal item:

$$\mathcal{R}_T = \sum_{t=1}^{T} \max_j r^*_{i_t,j} - \mathbb{E}[r_t] = \sum_{t=1}^{T} \max_j r^*_{i_t,j} - r^*_{i_t,j_t}. \tag{1}$$

Along the paper, we use the following notations. We denote $\mathbf{R}_t$ the partially known $N \times M$ matrix such that $r_{i_s,j_s} = r_s$ for any $s \leqslant t$. We note $\mathcal{S}_t$ the set of known entries of $\mathbf{R}_t$ and $\mathcal{I}_t(i)$ (respectively $\mathcal{J}_t(j)$) the set of items j (resp. users i) for which $(i, j) \in \mathcal{S}_t$. For the sake of readability, the subscript t is omitted in the following. Finally, for any matrix $\mathbf{M}$, we denote $\mathbf{M}_i$ the i-th row of $\mathbf{M}$.

3 Building Blocks

We introduce in Sect. 4 a RS which handles sequential recommendations. This RS is composed of two main ingredients: (i) a model to infer an estimate $\widehat{\mathbf{R}}^*$ of the matrix $\mathbf{R}^*$ from known values in $\mathbf{R}$, and (ii) a strategy to choose the item to recommend given $\widehat{\mathbf{R}}^*$. This strategy aims at balancing exploration and exploitation. In this section we go over state of the art approaches for both tasks.

3.1 Matrix Factorization

Since the Netflix challenge [4], many works on RS focus on Matrix Factorization [13]: the unknown matrix $\mathbf{R}^*$ is assumed to be of low rank. Namely, there exist $\mathbf{U}$ and $\mathbf{V}$ such that $\mathbf{R}^* = \mathbf{U}\mathbf{V}^T$, where $\mathbf{U}$ is a matrix of size $N \times k$ representing users features, $\mathbf{V}$ is a matrix of size $M \times k$ representing items features, k is the rank of $\mathbf{R}^*$, and $k \ll \max(N, M)$. Thereafter, the estimator of $\mathbf{R}^*$ is defined as

$$\widehat{\mathbf{R}}^* \stackrel{def}{=} \widehat{\mathbf{U}}\widehat{\mathbf{V}}^T, \text{ s.t. } (\widehat{\mathbf{U}}, \widehat{\mathbf{V}}) = \operatorname*{argmin}_{\mathbf{U},\mathbf{V}} \sum_{\forall (i,j)\in\mathcal{S}} \left(r_{i,j} - \mathbf{U}_i\mathbf{V}_j^T\right)^2 + \lambda \cdot \Omega(\mathbf{U}, \mathbf{V}), \tag{2}$$

in which $\lambda \in \mathbb{R}^+$, and the usual regularization term $\Omega(\mathbf{U}, \mathbf{V})$ is $||\mathbf{U}||^2 + ||\mathbf{V}||^2$. Eq. (2) corresponds to a non-convex optimization problem. The minimization is usually performed either by stochastic gradient descent (SGD), or by alternating least squares (ALS). As an example, ALS-WR [29] regularizes users and items according to their respective importance in the matrix of ratings: $\Omega(\mathbf{U}, \mathbf{V}) = \sum_i \#\mathcal{I}(i)||\mathbf{U}_i||^2 + \sum_j \#\mathcal{J}(j)||\mathbf{V}_j||^2$.

3.2 Multi-armed Bandits

A RS works in a sequential context. As a consequence, while the recommendation made at time-step t aims at collecting a good reward at the present time, it affects the information that is collected, and therefore also the future recommendations and rewards. Specifically, in the context of sequential decision under uncertainty problems, an algorithm which focuses only on short term reward loses w.r.t. expected long term reward. This section recalls standard strategies to handle this short term vs. long term dilemma. For ease of understanding, the setting used in this section is much simpler than the one faced in our paper.

We consider the Multi-Armed Bandits (MAB) setting [3]: we face a bandit machine with M independent arms. At each time-step, we pull an arm j and receive a reward drawn from $[0,1]$ which follows a probability distribution ν_j. Let μ_j denote the mean of ν_j, $j^* = \text{argmax}_j \, \mu_j$ be the best arm and $\mu^* = \max_j \mu_j = \mu_{j^*}$ be the best expected reward. The parameters $\{\nu_j\}$, $\{\mu_j\}$, j^* and μ^* are unknown.

We play T consecutive times and aim at minimizing the *pseudo-regret* $\mathcal{R}_T = \sum_{t=1}^{T} \mu^* - \mu_{j_t}$, where j_t denotes the arm pulled at time-step t. As the parameters are unknown, at each time-step, we face the dilemma: either (i) focus on short-term reward (aka. *exploit*) by pulling the arm which was the best at previous time-steps, or (ii) focus on long-term reward (aka. *explore*) by pulling an arm to improve the estimation of its parameters. Neither of these strategies is optimal. To be optimal, a strategy has to balance exploration and exploitation.

Auer [3] proposes a strategy based on an upper confidence bound (UCB1) to handle this exploration/exploitation dilemma. UCB1 balances exploration and exploitation by playing the arm $j_t = \text{argmax}_j \, \hat{\mu}_j(t) + \sqrt{\frac{2 \ln t}{T_j(t)}}$, where $T_j(t)$ corresponds to the number of pulls of arm j since the first time-step and $\hat{\mu}_j(t)$ denotes the empirical mean reward incurred from arm j up to time t. This equation embodies the exploration/exploitation trade-off: while $\hat{\mu}_j(t)$ promotes exploitation of the arm which looks optimal, the second term of the sum promotes exploration of less played arms. Other flavors of UCB-like algorithms [2,10,18] aim at a strategy closer to the optimal one or at a strategy which benefits from constraints on the reward distribution.

ε_n-greedy is another efficient approach to balance exploration and exploitation [3]. It consists in playing the greedy strategy ($j_t = \text{argmax}_j \, \hat{\mu}_j(t)$) with probability $1 - \varepsilon_t$ and in pulling an arm at random otherwise. Parameter ε_t is set to α/t with α a constant, so that there is more exploration at the beginning of the evaluation and then a decreasing chance to fall on an exploration step.

4 Explore-Exploit Recommender System

This section introduces a RS which handles the sequential aspect of recommendation. More specifically, the proposed approach works in the context presented in Sect. 2 and aims at minimizing the pseudo-regret $\mathcal{R}_T$. As needed, the proposed approach balances exploration and exploitation.

Named SeALS (for *Sequential ALS-WR*), our approach is described in Algorithm 1. It builds upon ALS-WR Matrix Completion approach and ε_n-greedy strategy to tackle the exploration/exploitation dilemma. At time-step t, for a given user i_t, ALS-WR associates an expected reward $\hat{r}_{i_t,j}$ to each item j. Then the item to recommend j_t is chosen by an ε_n-greedy strategy.

Algorithm 1. SeALS: recommend in a sequential context

Input: T_u, p, λ, α **Input/Output:** $\mathbf{R}$, $\mathcal{S}$
$(\widehat{\mathbf{U}}, \widehat{\mathbf{V}}) \leftarrow$ ALS-WR$(\mathbf{R}, \mathcal{S}, \lambda)$
for $t = 1, 2, \ldots$ **do**
  get user i_t and set $\mathcal{A}_t$ of allowed items
  $$j_t \leftarrow \begin{cases} \text{argmax}_{j \in \mathcal{J}_t} \widehat{\mathbf{U}}_{i_t} \widehat{\mathbf{V}}_j^T & \text{, with probability } 1 - \min(\alpha/t, 1) \\ \text{random}(j \in \mathcal{A}_t) & \text{, with probability } \min(\alpha/t, 1) \end{cases}$$
  recommend item j_t and receive rating $r_t = r_{i_t, j_t}$
  update $\mathbf{R}$ and $\mathcal{S}$
  if $t \equiv 0 \mod T_u$ **then** $(\widehat{\mathbf{U}}, \widehat{\mathbf{V}}) \leftarrow$ mBALS-WR$(\widehat{\mathbf{U}}, \widehat{\mathbf{V}}, \mathbf{R}, \mathcal{S}, \lambda, p)$ **end if**
end for

Algorithm 2. mBALS-WR: mini-batch version of ALS-WR

Input: $\mathbf{R}$, $\mathcal{S}$, λ, p, **Input/Output:** $\widehat{\mathbf{U}}$, $\widehat{\mathbf{V}}$
Sample randomly $p\%$ of all users in a list l_{users}
Sample randomly $p\%$ of all items in a list l_{items}
$\forall i \in l_{users}$, $\widehat{\mathbf{U}}_i \leftarrow \text{argmin}_{\mathbf{U}} \sum_{j \in \mathcal{J}_t(i)} \left(r_{i,j} - \mathbf{U}\widehat{\mathbf{V}}_j^T\right)^2 + \lambda \cdot \#\mathcal{I}_t(i) \|\mathbf{U}\|$
$\forall j \in l_{items}$, $\widehat{\mathbf{V}}_j \leftarrow \text{argmin}_{\mathbf{V}} \sum_{i \in \mathcal{I}_t(j)} \left(r_{i,j} - \widehat{\mathbf{U}}_i \mathbf{V}^T\right)^2 + \lambda \cdot \#\mathcal{J}_t(j) \|\mathbf{V}\|$

Obviously, ALS-WR requires too large computation times to be run at each time-step to recompute user and item features. A solution consists in running ALS-WR every T_u time-steps. While such a strategy works well when T_u is small enough, $\mathcal{R}_T$ drastically increases otherwise (see Sect. 5.3). Taking inspiration from stochastic gradient approaches [6], we solve that problem by designing a mini-batch version of ALS-WR, denoted mBALS-WR (see Algorithm 2).

mBALS-WR is designed to work in a sequential context where the matrix decomposition slightly changes between two consecutive calls. As a consequence, there are three main differences between ALS-WR and mBALS-WR. First, instead of computing $\widehat{\mathbf{U}}$ and $\widehat{\mathbf{V}}$ from scratch, mBALS-WR updates both matrices. Second, mBALS-WR performs only one pass on the data. And third, mBALS-WR updates only a fixed percentage of the line of $\widehat{\mathbf{U}}$ and $\widehat{\mathbf{V}}$. When the parameter $p = 100\,\%$, mBALS-WR is a one-pass ALS-WR.

The main advantage of mBALS-WR is in spreading the computing budget along time-steps which means $\widehat{\mathbf{U}}$ and $\widehat{\mathbf{V}}$ are more often up to date. On one hand, ALS-WR consumes a huge computing budget every thousands of time-steps; in between two updates, it selects the items to recommend based on an outdated decomposition. On the other hand, mBALS-WR makes frequent updates of the decomposition. In the extreme case, updates can be done at each time-step.

5 Experimental Investigation

In this section, we empirically evaluate the algorithms in the sequential setting on large real-world datasets. Two series of experiments emphasize two aspects of the sequential RS: Sect. 5.2 shows that exploration improves the quality of the RS model and compares our model with several baselines, while Sect. 5.3 focuses on the influence of the method updating the matrix model on the pseudo-regret and the running time.

5.1 Experimental Setting and Remarks

We use the same setting as the one used in the paper by Kawale et al. [12]. For each dataset, we start with an empty matrix $\mathbf{R}$ to simulate an extreme cold-start scenario where no information is available at all. Then, for a given number of time-steps, we loop on the following procedure:

1. we select a user i_t uniformly at random,
2. the algorithm chooses an item j_t to recommend,
3. we reveal the value of $r_t = r^*_{i_t,j_t}$ and increment the pseudo-regret score (1).

We assume that $\mathbf{R}^*$ corresponds to the values in the dataset. To compute the regret, the maximization term $\max_j r^*_{i_t,j}$ is taken w.r.t. the known values. Note that it is allowed to play an arm several times: once an item has been recommended, it is not discarded from the set of future possible recommendations.

We consider four real-world datasets for our experiments: Movielens1M/20M [11] and Douban [19] for datasets on movies, and Yahoo! Music user ratings of musical artists[1]. Characteristics of these datasets are reported in Table 1. For the Yahoo! dataset, we remove users with less than 20 ratings.

Some difficulties arise when using real datasets: in most cases, the ground truth is unknown, and only a very small fraction of ratings is known since users gave ratings only to items they have purchased/listened/watched. This makes the evaluation of algorithms uneasy considering we need in advance the reward of items we include in the list of possible recommendations. This is the case in our experiments, as we do not have access to the full matrix $\mathbf{R}$ in all datasets.

This issue is solved in the case of contextual bandits by using reject sampling [17]: the algorithm chooses an arm (item), and if the arm does not appear in logged data, the choice is discarded, as if the algorithm had not been called at all. For a well collected dataset, this estimator has no bias and has a known bound on the decrease of the error rate [15]. With our setting, we need no more

Table 1. Dataset characteristics.

	Movielens1M	Movielens20M	Douban	Yahoo!
Number of users	6,040	138,493	129,490	1,065,258
Number of items	3,706	26,744	58,541	98,209
Number of ratings	1,000,209	20,000,263	16,830,839	109,485,914

[1] https://webscope.sandbox.yahoo.com/.

to rely on reject sampling: we restrict the possible choices for a user at time-step t to the items with a known rating in the dataset.

The SeALS algorithm is compared to the following baselines:

- Random: at each iteration, a random item is recommended to the user.
- Popular: this approach assumes we know the most popular items based on the ground truth matrix. At each iteration, the most popular item (restricted to the items rated by the user on the dataset) is recommended. This is a strong baseline as it knows beforehand which items have the highest average ratings.
- UCB1: this bandit approach considers each reward r_{i_t,j_t} as an independent realization of a distribution ν_{j_t}. In other words, it recommends an item without taking into account the identity of the user requesting the recommendation.
- PTS [12]: this approach builds upon a statistical model of the matrix $\mathbf{R}$. The recommendations are done after a Thompson Sampling strategy [7] which is implemented with a Particle Filter. We present the results obtained with the non-Bayesian version, as it obtains very similar results with the Bayesian one.

5.2 Impact of Exploration

The first set of experiments compares two strategies to recommend an item: SeALS with $\alpha > 0$, and SeALS with $\alpha = 0$ (denoted Greedy) which corresponds to the greedy strategy. Both strategies use the maximum possible value for p ($p = 100\,\%$), which means we update all the users and items every T_u time-steps. The value of T_u used is the same for SeALS and Greedy, and α is set to 2,000 for Movielens1M, 10,000 for Douban and Movielens20M and 250,000 for Yahoo!. We set $\lambda = 0.1$ for Greedy and $\lambda = 0.15$ for SeALS. Parameter k is set to 15.

We also compare these two approaches with PTS. By fixing the value of the parameters of PTS as mentioned in [12] (30 particles and $k = 2$), the experimental results we obtain are not as good as the ones presented in that paper. However, we recover results similar to the ones presented in [12] by setting k to 15. So, we use that value of k for the results of the PTS approach we are displaying.

Figure 1 displays the pseudo-regret $\mathcal{R}_T$ obtained by the Recommender System after a given number of iterations (all results are averaged over 50 runs).

Results on all datasets demonstrate the need of exploration during the recommendation process: by properly fixing α, SeALS gets lower pseudo-regret value than Greedy. SeALS also obtains the best results on all datasets.

The PTS method which also tackles the exploration/exploitation dilemma appears only on the Movielens1M evaluation as this method does not scale well on large datasets (see Sect. 5.3 for the running time of PTS on Movielens1M). However, it is important to note that on the original PTS paper, this approach performs only comparably or slightly better than the Popular baseline on the evaluation provided on all small datasets, while our approach consistently performs much better than this baseline. One can reasonably assume that SeALS would perform better than PTS even if the latter one didn't have a scaling issue.

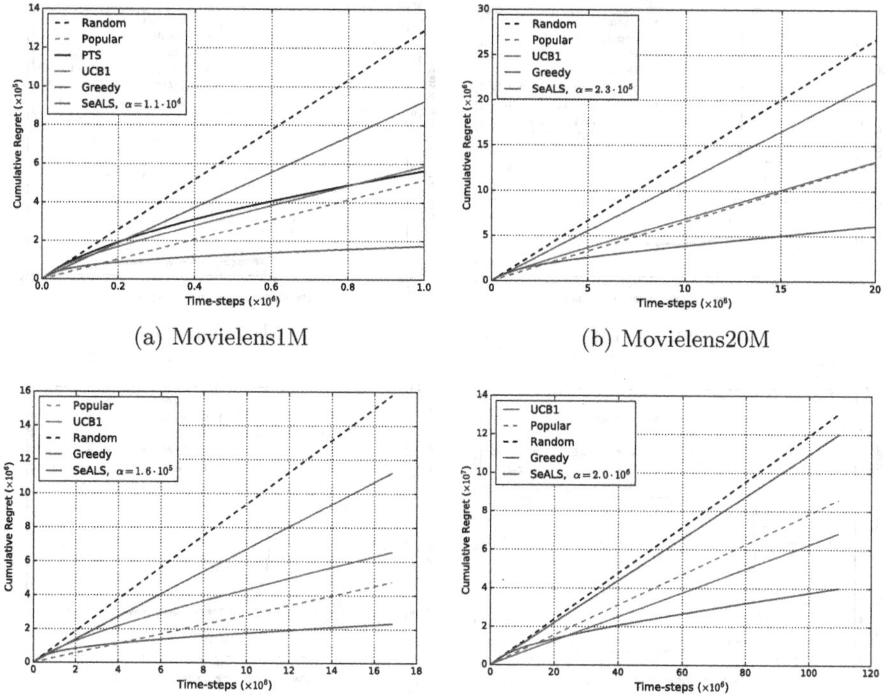

(a) Movielens1M (b) Movielens20M

(c) Douban (d) Yahoo

Fig. 1. Impact of exploration on four datasets.

5.3 Impact of the Update Strategy

To evaluate the impact of the method used to update the model as well as the period at which updates take place, we set up a different evaluation display for this second experiment. For each RS, we run experiments as presented in Sect. 5.1, and we store the final running time as well as the pseudo-regret at the end of all iterations. Such evaluation methodology allows finding which size of mini-batch and update period leads to the best trade-off between the final pseudo-regret score and the running time. This is an important point as a recommendation should both (i) be accurate and (ii) be quickly provided to the user in a real-word RS.

Figure 2 displays the results of this experiment. Each curve corresponds to a fixed size of mini-batch p, and every point of a same curve represents a specific value of the update period T_u. A point located at a high running time results from a small value of T_u (meaning the model was updated very often). For SeALS with $p = 100\,\%$, the period T_u of the updates varies in the range $[2\,10^3; 2\,10^5]$ for Movielens1M, $[10^4; 5\,10^6]$ for Douban and Movielens20M, and $[2.5\,10^5; 10^7]$ for Yahoo!. For $p = 10\,\%$ and $p = 0.1\,\%$, the considered periods are the same ones as for $p = 100\,\%$, but respectively divided by 10 and 10^3 in order to obtain comparable running times. Indeed, since we update a smaller portion of the

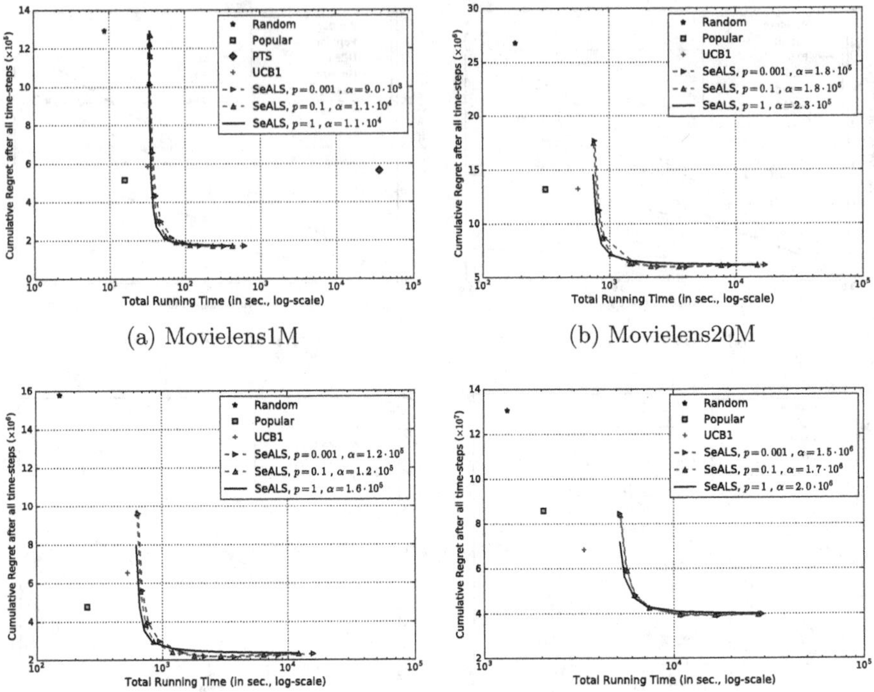

(a) Movielens1M (b) Movielens20M

(c) Douban (d) Yahoo

Fig. 2. Impact of the update strategy on four datasets.

matrix, it is possible to run this update more often and choose a small period T_u. For each value of p, we display the curve with the value of α (for the exploration) which reaches the lowest pseudo-regret.

Three main conclusions are drawn from this experiment: first, the results on Movielens1M highlight the non-scalability of the PTS algorithm, which takes several hours to complete 1 million iterations while SeALS only takes a few minutes. PTS does not seem to be an appropriate algorithm to provide quick recommendations as it takes too long updating the model. Second, on each dataset, each curve concerning SeALS quickly decreases: there is a rapid transition from a poor score to a good pseudo-regret. This means finding the appropriate period of update is sufficient to obtain a good RS. Third, on the large datasets, decreasing the portion of the users and items to update with a smaller p results in a worse score when using large update periods, but leads to a slightly better trade-off between the pseudo-regret and the running time at some point, when the updates are happening more often. One has to notice the best results for smaller values of p are obtained with a smaller value of α, which implies less steps of exploration has been done. We guess some sort of exploration is added in the system by not updating each user and each item at each time-step as it happens when $p = 100\,\%$ is chosen: while the whole model is not updated, it does not play optimally, which means it explores.

6 Related Work

To the best of our knowledge, only few papers consider a RS setting similar to the one presented in the current paper, where exploration/exploitation dilemma is tackled with ratings only [12,20,21,27,28]. [20] focus on a different recommendation setting where an item can be recommended only one time per user. Their approach builds upon ALS Matrix Factorization framework and extends linear bandits [16] to handle the exploration/exploitation dilemma. The use of linear bandit framework prevents this approach from scaling.

Other papers propose a solution based on a Bayesian model. The approach PTS, introduced in [12], tackles the exact same problem as our paper. PTS is based on Thompson Sampling [7] and Particle Filtering. However, their approach requires a huge computing time which scales with k^3. As a consequence, [12] only provides experiments on "small" datasets, with $k = 2$. SeALS is the first approach which both tackles the exploration/exploitation dilemma and scales up well to large datasets and high number of recommendations, while building an accurate representation of users and items ($k \gg 2$).

Contrary to the non-contextual case, the exploration/exploitation dilemma is already well-studied in the field of RS which has access to a representation of the context [25]. Compared to them, we focus on the setting where these representations are unknown and have to be inferred from users feedback.

Some papers focusing on the cold-start setting also focuses on the need for exploration [1,5]: the goal in this case is to deal with new users or new items. While some approaches look at external information on the user [1], some papers rewrite the problem as an Active Learning problem [5]: perform recommendation in order to gather information on the user as fast as possible. Targeted applications would first "explore" the user and then "exploit" him. Unlike Active Learning strategies, (i) we spread the cost of exploration along time, and (ii) we handle the need for a never-ending exploration to reach optimality [3].

Finally, some researches focus on a ranking algorithm instead of trying to target a good RMSE score. Cremonesi et al. [8] compare state of the art RS algorithms with respect to a rank-based scoring and shows that winning algorithms are not the ones which reach the smallest RMSE score. Following the same guideline, [22,24,26] propose RS which directly target a good ranking of the top items instead of a full-completion of the matrix. During the training phase, they replace the L2 loss of Eq. (2) by rank-based losses (AUC, MRR, NDCG...). While targeting a rank-based measure during the training phase could increase the accuracy of the RS, [9] and [23] show that the cumulative reward/regret is the unique good metric to evaluate the RS algorithm.

7 Conclusion and Future Work

In this paper we handle Recommender Systems based on Matrix Completion in the suitable context: an endless loop which alternates (i) learning of the model and (ii) recommendations given the model. Our proposed approach, SeALS, meets both challenges which arise in such a context. First, SeALS handles both short-term and long-term reward by balancing exploration and exploitation. Second, SeALS handles constraints on computing budget by adapting mini-batch strategy to alternating least square optimization. Experiments on real-life datasets show that (i) exploration is a necessary evil to acquire information and eventually improve the performance of the RS, and (ii) SeALS runs in an acceptable amount of time (less than a millisecond per iteration). SeALS paves the way to many extensions. SeALS builds upon ALS-WR which is intrinsically parallel; implementations of SeALS in real-life systems should benefit from parallel computing. SeALS could also be extended to mix user feedback and contextual information. All in all, we hope SeALS is revealing a new playground for other bandit algorithms beyond ε_n-greedy.

Aknowledgments. The authors would like to acknowledge the stimulating environment provided by SequeL research group, Inria and CRIStAL. This work was supported by French Ministry of Higher Education and Research, by CPER Nord-Pas de Calais/FEDER DATA Advanced data science and technologies 2015–2020, and by FUI Hermès. Experiments were carried out using Grid'5000 testbed, supported by Inria, CNRS, RENATER and several universities as well as other organizations.

References

1. Agarwal, D., Chen, B.C., Elango, P., Motgi, N., Park, S.T., Ramakrishnan, R., Roy, S., Zachariah, J.: Online models for content optimization. In: Proceedings of NIPS 2008, pp. 17–24 (2008)
2. Audibert, J.Y., Munos, R., Szepesvári, C.: Exploration-exploitation tradeoff using variance estimates in multi-armed bandits. Theoret. Comput. Sci. **410**(19), 1876–1902 (2009)
3. Auer, P., Cesa-Bianchi, N., Fischer, P.: Finite-time analysis of the multiarmed bandit problem. Mach. Learn. **47**, 235–256 (2002)
4. Bennett, J., Lanning, S.: Netflix: the Netflix prize. In: KDD Cup and Workshop (2007)
5. Bhagat, S., Weinsberg, U., Ioannidis, S., Taft, N.: Recommending with an agenda: active learning of private attributes using matrix factorization. In: Proceedings of RecSys 2014, pp. 65–72 (2014)
6. Bottou, L., Bousquet, O.: The tradeoffs of large scale learning. In: Proceedings of NIPS, pp. 161–168 (2007)
7. Chapelle, O., Li, L.: An empirical evaluation of thompson sampling. In: Proceedings of NIPS 2011, pp. 2249–2257 (2011)
8. Cremonesi, P., Koren, Y., Turrin, R.: Performance of recommender algorithms on top-N recommendation tasks. In: Proceedings of RecSys 2010, pp. 39–46 (2010)
9. Garcin, F., Faltings, B., Donatsch, O., Alazzawi, A., Bruttin, C., Huber, A.: Offline and online evaluation of news recommender systems at swissinfo.ch. In: Proceedings of RecSys 2014, pp. 169–176. ACM, New York (2014)
10. Garivier, A., Cappé, O.: The KL-UCB algorithm for bounded stochastic bandits and beyond. In: Proceedings of COLT 2011, pp. 359–376 (2011)
11. Harper, F.M., Konstan, J.A.: The movielens datasets: history and context. ACM Trans. Interact. Intell. Syst. (TiiS) **5**(4), 19 (2015)

12. Kawale, J., Bui, H., Kveton, B., Thanh, L.T., Chawla, S.: Efficient thompson sampling for online matrix-factorization recommendation. In: NIPS 2015 (2015)
13. Koren, Y., Bell, R., Volinsky, C.: Matrix factorization techniques for recommender systems. Computer **42**(8), 30–37 (2009)
14. Koren, Y., Sill, J.: Ordrec: an ordinal model for predicting personalized item rating distributions. In: Proceedings of RecSys 2011, pp. 117–124 (2011)
15. Langford, J., Strehl, A., Wortman, J.: Exploration scavenging. In: Proceedings of ICML, pp. 528–535 (2008)
16. Li, L., Chu, W., Langford, J., Schapire, R.E.: A contextual-bandit approach to personalized news article recommendation. In: Proceedings of the 19th International Conference on World Wide Web, pp. 661–670. ACM, New York (2010)
17. Li, L., Chu, W., Langford, J., Wang, X.: Unbiased offline evaluation of contextual-bandit-based news article recommendation algorithms. In: Proceedings of WSDM 2011, pp. 297–306 (2011)
18. Li, L., Chu, W., Langford, J., Schapire, R.E.: A contextual-bandit approach to personalized news article recommendation. In: Proceedings of World Wide Web (WWW 2010), pp. 661–670 (2010)
19. Ma, H., Zhou, D., Liu, C., Lyu, M.R., King, I.: Recommender systems with social regularization. In: Proceedings of the Fourth ACM International Conference on Web Search and Data Mining, pp. 287–296. ACM (2011)
20. Mary, J., Gaudel, R., Preux, P.: Bandits and Recommender Systems. In: Pardalos, P., Pavone, M., Farinella, G.M., Cutello, V. (eds.) MOD 2015. LNCS, vol. 9432, pp. 325–336. Springer, Cham (2015). doi:10.1007/978-3-319-27926-8_29
21. Nakamura, A.: A UCB-like strategy of collaborative filtering. In: Proceedings of ACML 2014 (2014)
22. Rendle, S., Freudenthaler, C., Gantner, Z., Schmidt-Thieme, L.: BPR: bayesian personalized ranking from implicit feedback. In: Proceedings of UAI 2009, pp. 452–461 (2009)
23. Said, A., Bellogín, A.: Comparative recommender system evaluation: benchmarking recommendation frameworks. In: Proceedings of RecSys 2014, pp. 129–136 (2014)
24. Shi, Y., Karatzoglou, A., Baltrunas, L., Larson, M., Oliver, N., Hanjalic, A.: CLiMF: learning to maximize reciprocal rank with collaborative less-is-more filtering. In: Proceedings of RecSys 2012, pp. 139–146 (2012)
25. Tang, L., Jiang, Y., Li, L., Li, T.: Ensemble contextual bandits for personalized recommendation. In: Proceedings of RecSys 2014 (2014)
26. Weston, J., Yee, H., Weiss, R.J.: Learning to rank recommendations with the k-order statistic loss. In: Proc. of RecSys 2013, pp. 245–248 (2013)
27. Xing, Z., Wang, X., Wang, Y.: Enhancing collaborative filtering music recommendation by balancing exploration and exploitation. In: Proceedings of International Society of Music Information Retrieval (ISMIR), pp. 445–450 (2014)
28. Zhao, X., Zhang, W., Wang, J.: Interactive collaborative filtering. In: CKIM 2013, pp. 1411–1420 (2013)
29. Zhou, Y., Wilkinson, D., Schreiber, R., Pan, R.: Large-Scale Parallel Collaborative Filtering for the Netflix Prize. In: Fleischer, R., Xu, J. (eds.) AAIM 2008. LNCS, vol. 5034, pp. 337–348. Springer, Heidelberg (2008). doi:10.1007/978-3-540-68880-8_32

Automatic Generation of Sitemaps Based on Navigation Systems

Pasqua Fabiana Lanotte[1(✉)], Fabio Fumarola[1], Donato Malerba[1,2], and Michelangelo Ceci[1,2]

[1] Dipartimento di Informatica, Università Degli Studi di Bari Aldo Modo, via Orabona 4, 70125 Bari, Italy
{pasqua.lanotte,fabio.fumarola,donato.malerba, michelangelo.ceci}@uniba.it

[2] CINI: National Interuniversity Consortium for Informatics, Bari, Italy

Abstract. In this paper we present a method to automatically discover sitemaps from websites. Given a website, existing automatic solutions extract only a flat list of urls that do not show the hierarchical structure of its content. Manual approaches, performed by web-masters, extract deeper sitemaps (with respect to automatic methods). However, in many cases, also because of the natural evolution of the websites' content, generated sitemaps do not reflect the actual content becoming soon helpless and confusing for users. We propose a different approach that is both automatic and effective. Our solution combines an algorithm to extract frequent patterns from navigation systems (e.g. menu, nav-bar, content list, etc.) contained in a website, with a hierarchy extraction algorithm able to discover rich hierarchies that unveil relationships among web pages (e.g. relationships of super/sub topic). Experimental results, show how our approach discovers high quality sitemaps that have a deep hierarchy and are complete in the extracted urls.

Keywords: Sitemaps · Web mining · Sequential pattern mining · Optimization

1 Introduction

As described by Jakob Nielsen [5], "one of the oldest hypertext usability principles is to offer a visual representation of the information space in order to help users understand where they can go". Sitemaps can provide such a visualization, offering a useful supplement to the navigation system on a website. A sitemap represents an explicit specification of the design concept and knowledge organization of a website. They help *users* and *search engine bots* by providing a hierarchical overview of a website's content: (i) to find pages on the site, and (ii) to understand, at a glance, contextual information such as relationships between web pages (e.g. relationships of super/sub topic). For example, on the contrary of the *keyword-based search* a sitemap can be useful in a university website to look for all professors, or the research areas; or it can be used to discover

P.M. Pardalos et al. (Eds.): MOD 2016, LNCS 10122, pp. 216–223, 2016.
DOI: 10.1007/978-3-319-51469-7_18

organization's information assets, such as affiliated companies to the website's organization. This kind of queries are hard to answer by a search engine, but really simple to extract from a well designed sitemap.

Having clarified why sitemaps are important, we can concentrate on how sitemaps are generated. Before the introduction of the Google Sitemap Protocol (2005), sitemaps were mostly manually generated. However, as websites got bigger, it became difficult for web-masters to keep the sitemap page updated, as well as list all the pages and all the contents in community-building tools like forums, blogs and message boards. As regards search engines, they cannot keep track of all this material, skipping information as they crawl through changing websites. To solve this issue several automatic tools were proposed on the Web[1]. These services generate an XML sitemap, used by *search bots*, which enumerate a flat list of urls and do not output the hierarchical structure of websites. Other solutions extract the hierarchical organization of a website by using the content and the hyperlinks structure of web pages. In particular, Weninger et al. proposed the HDTM algorithm [6] which uses *random walks with restart from homepage* to sample a distribution to assign to each website's page. Then, an iterative algorithm based on Gibbs sampling discovers hierarchies. Final hierarchies are obtained selecting, for each node, the most probable parent, in the way that higher nodes in the hierarchy contain more general terms, and lower nodes in the hierarchy contain terms which are more specific to their location.

The algorithm we propose, named *SMAP*, operates in a different way and, specifically, takes advantage of navigation systems of websites (e.g. menu, navbar, content list, etc.) organized as *web lists* [3]. Given a website's homepage, *SMAP* alternates a four-steps strategy that: (1) generates the web graph using web-lists, (2) extracts sequences which represent navigation path by exploiting random walk theory, (3) mines frequent closed sequential patterns through the application of an algorithm for sequential pattern mining, and (4) prunes discovered patterns to extract the sitemap in the form of an hierarchy. Experimental results show the validity of the proposed approach with respect to HDTM.

2 Extraction of Sitemaps: Preliminary Definitions

Before describing the proposed methodology for automatic sitemap extraction, we provide some formal definitions.

A web page is characterized by multiple representations, such as a textual representation composed by web page terms, a visual representation and a structural representation. Our method takes into account both the visual and the structural representation. For this reason, we provide the following formal definitions which identify the sources of information we exploit.

Definition 1. *A web page is characterized by a* ***Structural Representation*** *composed by web elements inscribed in HTML tags and organized in a tree-based structure. HTML tags can be applied to pieces of text, hyperlinks and multimedia data to give them different meaning and rendering in the web page.*

[1] http://slickplan.com/, http://www.screamingfrog.co.uk, https://www.xml-sitemaps.com/.

Definition 2. ***Web Page Visual Representation.*** *When a web page is rendered in a web browser, the CSS2 visual formatting model [4] represents the web page's elements by rectangular boxes that are laid out one after the other or nested inside each other by forming a tree, called* ***Rendered Box Tree****. By associating the web page with a coordinate system whose origin is at the top-left corner, the spatial position of each web page's element is fully determined by the tuple* (x, y, h, w)*, where* (x, y) *are the coordinates of the top-left corner of its corresponding box, and* (h, w) *are the box's height and width respectively. Therefore, the* ***Visual Representation*** *of a web page is given by its Rendered Box Tree.*

Definition 3. *A* ***Web List*** *is a collection of two or more web elements codified as rendered boxes having a similar HTML structure, and visually adjacent and aligned. This alignment can occur via the x-axis (i.e. a vertical list), the y-axis (i.e. horizontal list), or in a tiled manner (i.e. aligned vertically and horizontally) [3].*

To check whether two web elements e_i, e_j have similar HTML structure, we first codify the HTML tree having as root the web element e_i (e_j) in a string composed by HTML tags ordered applying a breadth search on the rooted tree; then, we apply to the generated strings the *Normalized Edit Distance*. Moreover, two web elements e_i, e_j are adjacent if they are sibling in the Rendered Box Tree and there is no a web element e_w, with $e_w \neq e_i \neq e_j$, which is rendered between them. Finally, two web elements e_i, e_j are aligned on: *(i)* the x-axis if the have same x coordinate, that is $e_i.x = e_j.x$; *(ii)* the y-axis if they share the same y value, that is $e_i.y = e_j.y$. Additional details about web list definition and extraction see [2].

Definition 4. *A* ***Website*** *is a directed graph* $G = (V, E)$*, where* V *is the set of web pages and E is the set of hyperlinks. In most cases, the homepage* h *of a website represents the website's entry page and,. thus, allows the website to be viewed as a rooted directed graph.*

As clarified before, our algorithm for the automatic identification of sitemaps exploits a sequential pattern mining step. The main rationale behind this choice is to see a navigation path in a website as a sequence of urls. This aspect is formalized in the following definitions:

Definition 5. ***Sequence:*** *Let* $G = (V, E)$ *be a website, a sequence* S *or equivalently a navigation path is defined as* $S = \langle t_1, t_2, \ldots, t_m \rangle$*, where each* $t_j \in V$ *denotes the web page at the* j*-th step in the navigation path* S*.*

Given a database of sequences SDB and a single sequence S the *absolute support* of S in SDB is the number of sequences in SDB which contain S, while its *relative support* is the absolute support divided by the size of database (i.e., $|SDB|$). With the term support will refer to the relative support, unless otherwise specified. A sequence is frequent if its support is greater than a user defined threshold.

The task of sequential pattern mining indicates the task of discovering frequent (sub-)sequences from SDB. This task is considered computationally challenging because algorithms have to generate and/or test a combinatorially explosive number of intermediate subsequences. For this reason, we consider the simplest problem (in terms of time complexity) of identifying *closed* sequential patterns which are compact and provide a lossless representation of all sequential patterns.

Definition 6. ***Closed Sequence:*** *Given two sequences S_i and S_j, if S_i is a supersequence of S_j and their support in SDB is the same, we say that S_i absorbs S_j. A sequential pattern S_j is* **closed** *if no sequence S_i that absorbs S_j exists.*

3 Extraction of Sitemaps: Problem Definition and Proposed Solution

After these preliminary definitions which are useful to introduce the approach from a methodological viewpoint, we can now formally define the properties of the extracted sitemaps:

Definition 7. ***Sitemap:*** *Given a web graph $G = (V, E)$ where $h \in V$ is the homepage, given a user-defined threshold t and a weight function $w : E \rightarrow \mathbb{R}$, then $T = \arg\max_{T_i} \Phi(T_i)$ is a sitemap if:*

- *$T_i = (V_i', E_i')$ is a tree rooted in h, where $V_i' \subseteq V$ and $E_i' \subseteq E$;*
- *$\Phi(T_i) = \sum_{e \in E_i'} w(e)$;*
- *$\forall\ e = (j_1, j_2) \in E_i', j_2 \in webList(j_1)$, that is, the url of the web page j_2 is contained in some web list of the web page j_1 (See Def. 3);*
- *$\forall e \in E_i',\ w(e) \geq t$.*

Intuitively, a naïve solution for the definition of the weight function $w(e)$ can be the reciprocal of the minimum number of edges, starting from the homepage, that it is necessary to follow to reach e by means of a breadth search of the web graph. Using this approach, links which are closer to the homepage have higher weights. Therefore, the extracted hierarchy is determined by the shortest paths from the homepage to each website's page. However, this solution has the drawback that the shortest paths not necessarily define the website's organization, as designed by web master and codified through the navigation systems. In fact, short paths can represent noisy links, such as short-cuts used to highlight web pages (e.g. links in the University website's homepage to some professor who won a prize), which may not be relevant to the sitemap generation process. In this work we adopt a more informative solution to define $w(\cdot)$. This solution, described below, exploits the support of a sequence.

Our solution does not need to enumerate all possible solutions T_i to optimize $\Phi(\cdot)$, yet it is able to extract directly T analyzing a sub-graph $G' \subseteq G$. This is done by exploiting properties of closed sequential patterns in a four steps algorithm:

1. **Web Graph Generation:** Given a website, we extract the web graph $G' = (V', E')$ ($V' \subseteq V$ and $E' \subseteq E$), such that $\exists (j_1, j_2) \in E' \Leftrightarrow j_2 \in webList(j_1)$;
2. **Sequence Dataset Generation:** Given the web graph G' extracted in the previous step, and two numbers $L_1, L_2 \in \mathbb{N}$, the random walk theory is used to navigate the web graph G' from the homepage h. The output of this phase is a sequence database composed by L_1 random walks having length L_2 and starting from h. In our experiments, we set $L_1 = 500,000$ and $L_2 = 10$;
3. **Closed Sequential Pattern Mining:** Given the sequence database, extracted at the step 2, and a user defined threshold t (i.e. minimum support), we apply a closed sequential pattern mining algorithm to extract all the frequent sequences. In this phase we used the algorithm CloFAST [1] as closed sequential pattern mining algorithm. CloFAST showed to provide compact results, saving computational time and space if compared with other closed sequential pattern mining algorithms. CloFAST returns a tree T and a function $s : E' \to \mathbb{R}$ which associates each node $e = (j_1, j_2) \in E'$ with the *relative* support of the sequence $\langle h, \ldots, j_1, j_2 \rangle$ in T. We use this function to define the function $w(\cdot)$ used in Definition 7. In this way, the constructed sitemap is constructed by taking into account the most frequent navigation paths in the web site.
4. **Tree Pruning:** The last task of the sitemap generation is the pruning process. In particular, since in the hierarchy we intend to extract, there is a unique frequent path from the root h to each page, the goal of this step is to select for each web page in T the best path to reach it. In particular, from T we prune all the sequences $S = \langle u_1, \ldots, u_i, u_{i+1}, \ldots, u_n \rangle$, with $u_j \in V'$, such that $\nexists\ (u_i, u_{i+1}) \in E$. In this way, frequent paths that do not exist in the real web graph are removed. Moreover, to obtain the tree that maximizes the function $\Phi(\cdot)$, for each node j which is present more than once in the extracted tree (this means that multiple frequent sequences exist starting from h and ending with j), we compare all the sub-trees of T having j as root. From them, we only keep the one whose sum of weights is maximum.

4 Experiments and Discussion

The empirical evaluation of the effectiveness of our approach (SMAP) was not a trivial task because, at the best of our knowledge, there is no dataset generated for the specific task of sitemap extraction. Thus, two solutions were possible: *(i)* involving human experts to extract the hierarchical organization of websites and generating a ground truth dataset for each considered website; *(ii)* using existing sitemap pages, which are manually generated by web masters and provided by some website, as ground truth.

In the first case, the involvement of experts requires extensive human effort and working time. This is not feasible in the context of the Web especially for websites having a great amount of web pages strongly connected and websites having deep hierarchies. In the second case, sitemap pages are in general manually created by web designers. Therefore, the risk is that sitemap pages are not updated (i.e. they can contain links to non-existent pages or they ignore the existence of new sections) or are very abstract (i.e. contain shallow hierarchies composed by few pages). For this reason, to empirically evaluate SMAP, we performed experiments on following websites which provide updated sitemap pages: cs.illinois.edu, www.cs.ox.ac.uk, www.cs.princeton.edu, and www.enel.it.

The evaluation has been performed in terms of Precision, Recall and F-measure of edges. In particular, the Precision measures how many of the extracted edges belong to the real sitemap. The Recall measures how many edges, belonging to the real sitemap are extracted. We also included the F-Measure which is the weighted harmonic means of Precision and Recall. These measures are evaluated counting how many edges of the real sitemap are found by SMAP. More formally:

$$Precision = \frac{|\{e|e \in SMAP(G) \wedge e \in GT\_Sitemap\}|}{|\{e|e \in SMAP(G)\}|} \tag{1}$$

$$Recall = \frac{|\{e|e \in SMAP(G) \wedge e \in GT\_Sitemap\}|}{|\{e|e \in GT\_Sitemap\}|} \tag{2}$$

$$F = \frac{2(precision \times recall)}{precision + recall} \tag{3}$$

In these formulas $SMAP(G)$ represents the set of edges extracted by $SMAP$, whereas $GT\_Sitemap$ represents the set of edges in the real sitemap (ground truth).

The results are compared with those obtained by HDTM [6]. For HDTM we set $\gamma = 0.25$ (to avoid too shallow or too deep hierarchies) and $5,000$ Gibbs iterations, as suggested by the authors in their paper.

Table 1 shows the effectiveness of SMAP varying the minimum support. It is interesting to note that when the minimum support threshold decreases, we are able to obtain deeper and wider sitemaps. While, as expected, by reducing the minimum support threshold, precision increases and recall decreases. This is due to the fact that, by decreasing the support, the number of generated sequences increases and the extracted hierarchy becomes deeper and wider, including website sections which are not included in the sitemap page. This behaviour can be observed by analyzing the F-measure which, from website to website, shows different trends for different values of minimum support. Interestingly, our algorithm (SMAP) significantly outperforms HDTM, independently of the website and of the minimum support threshold. This can be motivated by the different nature of the two algorithms: on the contrary of SMAP, HDTM organizes website's pages in a hierarchy using the distribution of web pages' terms. Then, it can happen that, for example, for a Computer Science department website HDTM organizes the web page of a *professor* as a child of its *research area* web page rather than as a child of the *professors* web page. In this way, web pages clustered together as siblings in the hierarchy by web masters are split in different parts of the extracted hierarchy.

Table 1. Experimental results of SMAP and HDTM.

Website	Algorithm	Min. supp	Precision	Recall	F-Measure
www.cs.illinois.edu	SMAP	0.005	0.38	0.78	0.51
www.cs.illinois.edu	SMAP	0.001	0.66	0.48	0.56
www.cs.illinois.edu	SMAP	0.0005	0.66	0.33	0.44
www.cs.illinois.edu	HDTM	–	0.12	0.1	0.11
www.cs.ox.ac.uk	SMAP	0.005	0.72	0.3	0.42
www.cs.ox.ac.uk	SMAP	0.001	0.72	0.21	0.33
www.cs.ox.ac.uk	SMAP	0.0005	0.72	0.21	0.33
www.cs.ox.ac.uk	HDTM	–	0.37	0.15	0.21
www.cs.princeton.edu	SMAP	0.005	0.61	0.55	0.58
www.cs.princeton.edu	SMAP	0.001	0.89	0.23	0.36
www.cs.princeton.edu	SMAP	0.0005	0.89	0.2	0.33
www.cs.princeton.edu	HDTM	–	0.36	0.08	0.13
www.enel.it	SMAP	0.005	0.31	0.74	0.43
www.enel.it	SMAP	0.001	0.36	0.72	0.48
www.enel.it	SMAP	0.0005	0.8	0.73	0.76
www.enel.it	HDTM	–	0.35	0.75	0.48

5 Conclusions

In this paper, we presented a new method for the automatic generation of sitemaps. SMAP successfully addresses the open issue of the extraction of websites' hierarchical organization as understood and codified by web masters through websites' navigation systems (e.g. menu, navbar, content list, etc.). Moreover, it provides to web masters a tool for automatic sitemap generation which can be helpful for providing a explicit specification of the design concept and knowledge organization of websites. The experimental results prove the effectiveness of SMAP compared to HDTM, a state of art algorithm, which generates sitemaps based on the distribution of web pages' terms. As future work we will analyze the effect of the input parameters of SMAP (e.g. the dimensions L_1 and L_2 used in the Sequence Dataset Generation step) on the extracted hierarchies. Moreover, we will study how to combine sitemaps which describe the topical organization of a website (e.g. sitemaps extracted by HDTM) with sitemaps which describe the website's structural organization.

Acknowledgment. This project has received funding from the European Commission through the project MAESTRA - Learning from Massive, Incompletely annotated, and Structured Data (Grant number ICT-2013-612944).

References

1. Fumarola, F., Lanotte, P.F., Ceci, M., Malerba, D.: CloFAST: closed sequential pattern mining using sparse and vertical id-lists. Know. Inf. Syst **48**(2), 429–463 (2016)
2. Fumarola, F., Weninger, T., Barber, R., Malerba, D., Han, J.: Hylien: A hybrid approach to general list extraction on the web. In: Proceedings of the 20th International Conference Companion on World Wide Web, WWW 2011, pp. 35–36. ACM, New York (2011)
3. Lanotte, P.F., Fumarola, F., Ceci, M., Scarpino, A., Torelli, M.D., Malerba, D.: Automatic extraction of logical web lists. In: Andreasen, T., Christiansen, H., Cubero, J.-C., Raś, Z.W. (eds.) ISMIS 2014. LNCS (LNAI), vol. 8502, pp. 365–374. Springer, Heidelberg (2014). doi:10.1007/978-3-319-08326-1_37
4. Lie, H.W., Bos, B., Sheets, C.S.: Designing for the Web, 2nd edn. Addison-Wesley Professional, Reading (1999).
5. Nielsen, J., Loranger, H.: Prioritizing Web Usability. New Riders Publishing, Thousand Oaks (2006)
6. Weninger, T., Bisk, Y., Han, J.: Document-topic hierarchies from document graphs. In: Proceedings of the 21st ACM International Conference on Information and Knowledge Management, CIKM 2012, pp. 635–644. ACM, New York (2012)

A Customer Relationship Management Case Study Based on Banking Data

Ivan Luciano Danesi(✉) and Cristina Rea

D&A Data Science, UniCredit Business Integrated Solutions S.C.p.A., via Livio Cambi 1, 20151 Milan, Italy
{ivanluciano.danesi,cristina.rea}@unicredit.eu

Abstract. This work aims to show a product recommender construction approach within the banking industry. Such a model costruction should respect several methodological and business constraints. In particular, analysis' outcome should be a model which must be easily interpretable when shown to business people. We start from a Customer Relationship Management data set collected in Banking industry. Formerly, data is prepared by managing missing values and keeping only the most relevant variables. Latterly, we apply some algorithms and evaluate them using diagnostic tools.

Keywords: Customer Relationship Management · Machine learning · Missing values · Variables selection

1 Introduction

Customer Relationship Management (CRM) comprises the set of processes and techniques used by a company in order to manage and improve interactions with current and future customers. Data mining techniques applied in the CRM environment are evolving due to new technology developments in Big Data management and analytics [6,12]. This development requires promptly dealing with a great amount of etherogenous data. For a review of data mining techniques in CRM please refer to [2,8].

Aim of this work is to outline a methodology for building a predictive analytical tool to improve CRM capability to promptly predict customers financial service needs and behaviour. In this study case we focus on customer identification, since we want to construct a recommender system to determine a favourable target population: the built model will describe customer's buying propensity for a specific financial product. Most used techniques for CRM recommenders can be found in [1,9]. In these works, a series of applications are presented for constructing a CRM recommender model. Here, we face problems that arises when data from a specific domain (Banking and Finance) are considered. Predictive models developed under this initiative combine the existing set of static data (monthly variables, social and demographical information) with dynamic daily data in order to use all the available information. Nevertheless, at the end of the

P.M. Pardalos et al. (Eds.): MOD 2016, LNCS 10122, pp. 224–235, 2016.
DOI: 10.1007/978-3-319-51469-7_19

analysis, the resulting model is intended to be interpreted by business parties for a practical application of the results. At this level, we favour the possibility of analysing the model in its details. Furthermore, the resulting model should be easily interpretable and explainable. In order to obtain results comprehensible by all involved parties, a compromise is needed between the chosen model's complexity and its predictive power.

This work analyzes a real dataset that includes a series of issues as etherogeneous data, high number of missing values and a few number of relevant features (in the order of tens) in a great number of observed fetures (some hundreds). Moreover, the dataset is strongly umbalanced. For dealing with all of these issues the Authors define a procedure composed by different steps that can be replicated for other problems in the same field. This procedure considers dataset characteristics that are common in CRM.

This paper is organized as follows. In Sect. 2 we present the data set and we deal with the missing data, also performing a first feature selection. In Sect. 3 we apply some algorithms for model construction, then we present the results. In Sect. 4 we draw the relative conclusions starting from a comparison between all models built.

2 The Dataset

In a Big Data environment, data considered for CRM purposes are described in terms of a great number of variables both direct or synthesized. Clearly, these variables are often raw data that can present some percentage of missing values or a high level of correlation: a preliminary treatment is therefore mandatory.

We have collected customers' data[1] referred to seven subsequent months. We have chosen to use six months for model construction (the train set, composed by the first six months) then we keep the last month as test set for model evaluation. The hypothesis is that the observed phenomenon remains stable across the time, at least during the observed months. The response variable is known for all of the 7 considered month since we are observing past data (from May 2015 to November 2015).
For each month, the available data sources are

- monthly customer data (composed by raw customers data and some features computed by business experts, from now on called artificial features);
- daily customer balances;
- daily transactional data.

Monthly customer data consists of approximately nine hundred variables referred to the time same interval used for observing the outcome variable. Since daily customer balances and daily transactional data is collected at a different time

[1] Considered data is referred to a subset of a leading Bank's clients. More detailed informations and exploratory analysis are not here reported due to Compliance issues.

interval, it is mandatory to artificially synthesize descriptive variables in the same reference frame of the outcome variable. We have constructed some tens of synthetic descriptive features starting by the latter two dataset. For example the mean of the daily customer balance over the month, the number of transactions in the month and the maximum time interval between two consecutive transactions. The final dimension of the dataset after the creation of these features is approximately 1100 features.

2.1 Balancing the Dataset

Descriptive univariate analysis on the outcome variable was performed: this variable represents the target event. This variable is collected with a monthly delay due to the essential time to develop CRM campaigns and obtain customers feedback.

The original data set is deeply unbalanced: indeed, with respect to the dichotomous outcome, only 0.04% of the contacted potential customers was observed to provide a positive response.

In order to obtain non-biased model results, it is necessary to balance the dataset used for model construction (the train set) or to modify the predictive models.

We have decided to balance the dataset stratifying with respect to the months. This means that, for each month, we considered the positive outcomes (the group less numerous) and a fixed percentage of the negative outcomes (the group most numerous). The percentage is decided arbitrary at this stage (in this case 30% of positive outcomes), then this decision is refined after model definition.

2.2 Missing Values Management

Statistical models and Machine Learning algorithms are typically biased when missing values are present. In order to deal with missing values, it is important to understand their reason of being. Missing values can derive not only by generic errors (*e.g.* wrong separator in a flat file), but also (i) unanswered questions in a form; (ii) the customer does not have the product, then the related mesures (*e.g.* opening date) are missing; (iii) wrong codification of zero values; (iv) data structure modification across the time.

Missing values should be treated in different ways according to their cause and percentage to improve the batch data quality. In the observed data we can observe that approximately the 60% of the features present between 0 and 5% of missing values. These missing values are typically due to cause (i) or generic errors and are here filled with mean/median values. Regarding the other features, three actions are put in practice. Firstly, the pattern in the missing values is observed and, when possible is corrected. For example, all of the features that measure transactions or count some quantities, when the phenomena has not been registered are coded as missing but should be considered as 0 instead. Secondly, the variables with more than 95% of missing values are removed. Thirdly, the other missing values are imputed by assigning the mean/median value.

2.3 Feature Selection

The feature selection which we are going to describe follows two steps. Firstly, the dataset is simplified by removing textual data: CRM model refinement including textual information is beyond the scope of this work. Secondly the only the most relevant variables for the object of study are selected through the following considerations.

1. Identification of correlated information. If we have, for example, different features with minimum, average and maximum value of the same numeric explanatory variable stored in separate fields, and such variables are highly correlated, only a single value is preserved.
2. Frequency and cumulative value of a numeric specific explanatory variable can be presented for many different temporal periods (last month, last three months, third month and so on). In this case, the variables regarding the most recent period are considered (last month or those referred to the last three months). Where appropriate, average information are preferred.
3. Ratios between explanatory variables, computed following the previously described points, are kept only if strictly necessary.
4. Principal Component Analysis (PCA) for determining the variables explaining variance's higher percentage.
5. Analysis of the correlation between the variables of the data set and with respect to the outcome variable (also considering descriptive statistics and graphical visualizations).
6. Hypothesis testing, in order to observe whether the observed variable is significantly different in the populations characterized by positive and negative value of the outcome.

Exploiting the approach previously described in order to improve data quality and remove unnecessary variables (removing features with high percentage of missing values, text data and redundances), 101 features are selected, representing the starting point for the analysis development. More precisely

- 74 most relevant variables are considered coming from monthly records (13 of them are artificial features);
- 27 indicators are synthesized from daily data, 2 of them are obtained from credit and debit card transactions;

are kept.

At the conclusion of all the statistical procedures for variable reduction (also PCA and hypothesis testing), the original data set has been reduced to 41 explanatory variables, which can be considered in the process of statistical modelling.

3 The Predictive Model Construction

3.1 Selected Algorithms

As mentioned before, the algorithms considered for models creation (listed below) are selected following two main goals: predictive power and possibility to be explained to business counterparties.

The benchmark. In the recommender system domain, heuristic models are still the most diffused ones [9]. We consider an expert-judgement-driven classification tree based on business knowledge of the problem. The three classify the customers in groups with a determined level of buying propensity. This model is created by domain experts with a deep knowledge of the market and the products. The splits of the classification trees are made mainly with respect to artificial features. This approach is here considered as the benchmark with respect to evaluate and compare the other algorithms.

Regression tree. The goal of prediction trees is to generate a response or a class Y using a set of inputs $X_1, \ldots, X_p$ using a binary tree. Since in this work we are using prediction trees as an approximation to an unknown regression function, we are dealing with regression trees. The resulting tree is similar in the interpretation to the benchmark model, but is constructed using a data-driven approach.
After the estimation step, the resulting regression tree presents homogeneous observations buckets in the end nodes. Regarding the predictive process, binary tree is used as starting point by a root node and then the procedure hierarchically chooses at every node of the tree the path accordingly to the observation's characteristics. The tree construction is usually performed finding the best partition optimising the homogeneity of the nodes. For further details see [4].

Random forest. Random Forests usually consider a number of k bootstrapped training samples. On each of these k samples, a prediction tree is constructed considering a random sample of $m < p$ predictors at each split. By sampling at each split m predictors, the strong predictor is considered $(p-m)/p$ times. This results in a "forest" composed by de-correlated trees. Therefore, averaging uncorrelated quantities leads to a reduction in variance which is higher than averaging high correlated quantities. In Random Forests trees are usually fully grown. Random Forests improve predictive accuracy by generating a large number of random trees; the process of classification of each record is performed by using each tree of this new forest and deciding a final predicted outcome through a combination of the results across all of the trees [5]. The impact of single variables on the outcome can be explained to the business counterpart using the partial dependence plots as presented in [10,11].

Regression tree on less variables. The regression tree is performed only on the most influential variables in order to reduce eventual noise and not necessary variables.

Random forest on less variables. The random forest is performed only on the most influential variables.

Logistic regression with stepwise algorithm. The logistic regression model is a Generalized Linear Model (GLM) widely used in risk and in propensity contexts [7]. When the number of variables is in the order of some tens, to estimate directly a GLM can not be the best solution. We implemented a stepwise approach instead, for the inclusion and the evaluation of the explanatory variables. This allows to proceed iteratively in forward or backward direction, adding and removing variables. The evaluation of the variables is done by observing the value of Akaike Information Criteria (AIC). The business interpretation of the resulting model can be easily explained in terms of odds ratios.

The first model (benchmark) is performed on all of the initial available monthly customer raw data. The second and the third models (regression tree and random forest) are constructed with the 41 selected features. The fourth and the fifth models (regression tree and random forest on less variables) consider a subsample of features (the most 10 relevant ones). This additional variable selection is performed using the random forest algorithm (for an example of random forest used for this purpose see [3]) and sorting the features observing an aritmetic mean of impurity and Gini measures [4]. The sixth algorithm (stepwise GLM) considers the list of 41 features, using as starting point the 10 most relevant and using the forward direction. The trees are constructed using impurity measures for splitting rules and for determining the correct dimension.

3.2 Observed Performance Indices

The model listed in Subsect. 3.1 are evaluated and compared by considering the following quantities.

1. Threshold (THR): discriminant value of the predicted probability above which an individual is classified as positive, namely if $\hat{p}_i > THR$, i^{th} customer is classified as a positive.
2. Number of True Negatives (TN): number of negative outcomes who are correctly classified.
3. Number of False Negatives (FN): number of positive outcomes who are misclassified as negative outcomes.
4. Number of False Positives (FP): number of negative outcomes who are misclassified as positive outcomes.
5. Number of True Positives (TP): number of positive outcomes who are correctly classified.
6. Specificity (SPEC): proportion of negative outcomes who are correctly classified, namely SPEC = TN/(TN + FP).
7. Recall (Sensitivity) (SENS): proportion of positive outcomes who are correctly classified, namely SENS = TP/(TP + FN).
8. Precision (Positive Predictive Value) (PPV): proportion of positives who are effectively positive outcomes, namely PPV = TP/(TP + FP).

9. F1: harmonic mean between Recall and Precision, namely

$$\text{F1} = \frac{2}{\frac{1}{\text{RECALL}} + \frac{1}{\text{PREC}}} = \frac{2\text{TP}}{2\text{TP} + \text{FP} + \text{FN}}.$$

10. Receiver Operating Characteristic (ROC) curve: graphical plot of the True Positive Rate (*i.e.* Recall) against the False Positive Rate (*i.e.* 1 - Specificity) as the threshold varies between 0 and 1.
11. Area Under Curve (AUC): integral of ROC curve. AUC ranges from 0.5 (model which predicts randomly) to 1 (perfect model).
12. UPLIFT: graphical plot where first customers are ordered decreasingly by their predicted propensity probability and divided into bins; then, for each bin the percentage of correctly classified positive outcomes (*i.e.* Recall increase) is reported. In this report each bin contains 5% of the customers.

3.3 Results

Algorithms introduced in Subsect. 3.1 are applied to the train set prepared by following procedure outlined in Sect. 2. The estimated models are then used for evaluate the outcome variable on the test set.

The random forests are constructed with 1000 trees and with 6 variables considered in each node when launched with 41 variables, and 3 when launched with 20 variables (approximately the square root of the number of considered features). These algorithm parameters, as well as the percentage of data balance, are then modified in order to maximize the value of the AUC.

The output of the GLM is reported in Table 1. The algorithm confirm the presence of 10 relevant features. An increasing value of the explanatory variables `VARIABLE_10` and `VARIABLE_1` could decrease the predicted probability to have a

Table 1. Logistic regression model with stepwise selection method

Variable	Estimate	Standard error	z-value	p-value	Significance
Intercept	−0.542	0.225	−2.41	0.016	*
VARIABLE_1	−0.023	0.005	−4.81	<0.001	***
VARIABLE_2	1.320	0.335	3.93	<0.001	***
VARIABLE_3	−0.000000197	<0.001	−4.49	<0.001	***
VARIABLE_4	0.0000802	<0.001	2.03	0.042	*
VARIABLE_5	−0.00000183	<0.001	−6.22	<0.001	***
VARIABLE_6	0.00000415	<0.001	3.37	<0.001	***
VARIABLE_7	−0.0000202	<0.001	−2.48	0.013	*
VARIABLE_8	−1.660	0.425	−3.91	<0.001	***
VARIABLE_9	0.00000128	<0.001	4.88	<0.001	***
VARIABLE_10	−0.865	0.325	−2.66	0.008	***

positive outcome and therefore to become an effective buying customer. On the other hand, an increasing value of the explanatory variable VARIABLE_3 could increase the predicted probability to have a positive outcome. These most statistically significant variables in terms of prediction can be chosen according to the last column of Table 1, where the following legend is adopted:

- High statistical significance of the estimates: *** (p-value less than 0.001)
- Mid-high statistical significance of the estimates: ** (p-value less than 0.01)
- Medium statistical significance of the estimates: * (p-value less than 0.05)
- Low statistical significance of the estimates: *blank* (p-value greater than 0.05)

In Tables 2 and 3, main confusion matrix statistics for different thresholds for the five models (benchmark is excluded) are reported.

Table 2. Performances of tree model (upper sub table) and random forest model (lower sub table) for the first 100, 200, 500, 1000, 2000, 5000, 10000, 20000, 50000 customers (N CUST), ordered by decreasing estimated customer's buying propensity. Thresholds (THR) and proportions of customers (PROP CUST) are also reported. The reported performances are number of True Negatives (TN), number of False Negatives (FN), number of False Positives (FP), number of True Positives (TP), Specificity (SPEC), Sensitivity (SENS), Predictive Positive Value (PPV), F1 (F1).

THR	N CUST	PROP CUST	TN	FN	FP	TP	SPEC	SENS	PPV	F1
1.000	100	0.000	209,848	139	0	0	1.000	0.000		0.000
1.000	200	0.001	209,848	139	0	0	1.000	0.000		0.000
1.000	500	0.002	209,848	139	0	0	1.000	0.000		0.000
1.000	1,000	0.005	209,848	139	0	0	1.000	0.000		0.000
1.000	2,000	0.010	209,848	139	0	0	1.000	0.000		0.000
1.000	5,000	0.024	209,848	139	0	0	1.000	0.000		0.000
1.000	10,000	0.048	209,848	139	0	0	1.000	0.000		0.000
0.623	20,000	0.095	198,294	110	11,554	29	0.945	0.209	0.003	0.005
0.418	50,000	0.238	172,354	48	37,494	91	0.821	0.655	0.002	0.005
THR	N CUST	PROP CUST	TN	FN	FP	TP	SPEC	SENS	PPV	F1
0.865	100	0.000	209,749	139	99	0	1.000	0.000	0.000	0.000
0.845	200	0.001	209,651	137	197	2	0.999	0.014	0.010	0.012
0.815	500	0.002	209,354	133	494	6	0.998	0.043	0.012	0.019
0.790	1,000	0.005	208,862	130	986	9	0.995	0.065	0.009	0.016
0.760	2,000	0.010	207,877	124	1,971	15	0.991	0.108	0.008	0.014
0.706	5,000	0.024	204,914	106	4,934	33	0.976	0.237	0.007	0.013
0.647	10,000	0.048	199,945	78	9,903	61	0.953	0.439	0.006	0.012
0.560	20,000	0.095	190,065	46	19,783	93	0.906	0.669	0.005	0.009
0.360	50,000	0.238	160,138	18	49,710	121	0.763	0.871	0.002	0.005

Table 3. Performances of the tree on selected variables (upper sub table), random forest on selected variables (middle sub table) and logistic model stepwise on selected variables (lower sub table) for the first 100, 200, 500, 1000, 2000, 5000, 10000, 20000, 50000 customers (N CUST), ordered by decreasing estimated probability of customer's buying propensity. Thresholds (THR) and proportions of customers (PROP CUST) are also reported. The reported performances are number of True Negatives (TN), number of False Negatives (FN), number of False Positives (FP), number of True Positives (TP), Specificity (SPEC), Sensitivity (SENS), Predictive Positive Value (PPV), F1 (F1).

THR	N CUST	PROP CUST	TN	FN	FP	TP	SPEC	SENS	PPV	F1
1.000	100	0.000	209,848	139	0	0	1.000	0.000		0.000
1.000	200	0.001	209,848	139	0	0	1.000	0.000		0.000
1.000	500	0.002	209,848	139	0	0	1.000	0.000		0.000
1.000	1,000	0.005	209,848	139	0	0	1.000	0.000		0.000
1.000	2,000	0.010	209,848	139	0	0	1.000	0.000		0.000
1.000	5,000	0.024	209,848	139	0	0	1.000	0.000		0.000
1.000	10,000	0.048	209,848	139	0	0	1.000	0.000		0.000
0.524	20,000	0.095	190,019	80	19,829	59	0.906	0.424	0.003	0.006
0.418	50,000	0.238	163,248	43	46,600	96	0.778	0.691	0.002	0.004
THR	N CUST	PROP CUST	TN	FN	FP	TP	SPEC	SENS	PPV	F1
0.886	100	0.000	209,750	139	98	0	1.000	0.000	0.000	0.000
0.870	200	0.001	209,655	138	193	1	0.999	0.007	0.005	0.006
0.838	500	0.002	209,357	135	491	4	0.998	0.029	0.008	0.013
0.813	1,000	0.005	208,867	132	981	7	0.995	0.050	0.007	0.012
0.780	2,000	0.010	207,874	125	1,974	14	0.991	0.101	0.007	0.013
0.723	5,000	0.024	204,910	109	4,938	30	0.976	0.216	0.006	0.012
0.663	10,000	0.048	199,949	87	9,899	52	0.953	0.374	0.005	0.010
0.572	20,000	0.095	189,980	49	19,868	90	0.905	0.647	0.005	0.009
0.360	50,000	0.238	160,003	20	49,845	119	0.762	0.856	0.002	0.005
THR	N CUST	PROP CUST	TN	FN	FP	TP	SPEC	SENS	PPV	F1
1.000	100	0.000	209,749	138	99	1	1.000	0.007	0.010	0.008
0.997	200	0.001	209,649	138	199	1	0.999	0.007	0.005	0.006
0.983	500	0.002	209,351	136	497	3	0.998	0.022	0.006	0.009
0.956	1,000	0.005	208,851	136	997	3	0.995	0.022	0.003	0.005
0.899	2,000	0.010	207,851	136	1,997	3	0.990	0.022	0.002	0.003
0.770	5,000	0.024	204,865	122	4,983	17	0.976	0.122	0.003	0.007
0.646	10,000	0.048	199,887	100	9,961	39	0.953	0.281	0.004	0.008
0.508	20,000	0.095	189,914	73	19,934	66	0.905	0.475	0.003	0.007
0.321	50,000	0.238	159,954	33	49,894	106	0.762	0.763	0.002	0.004

A model's performance improvement can be observed for all of the possible thresholds by the increasing rectangularization of relative ROC curve, presented in Fig. 1.

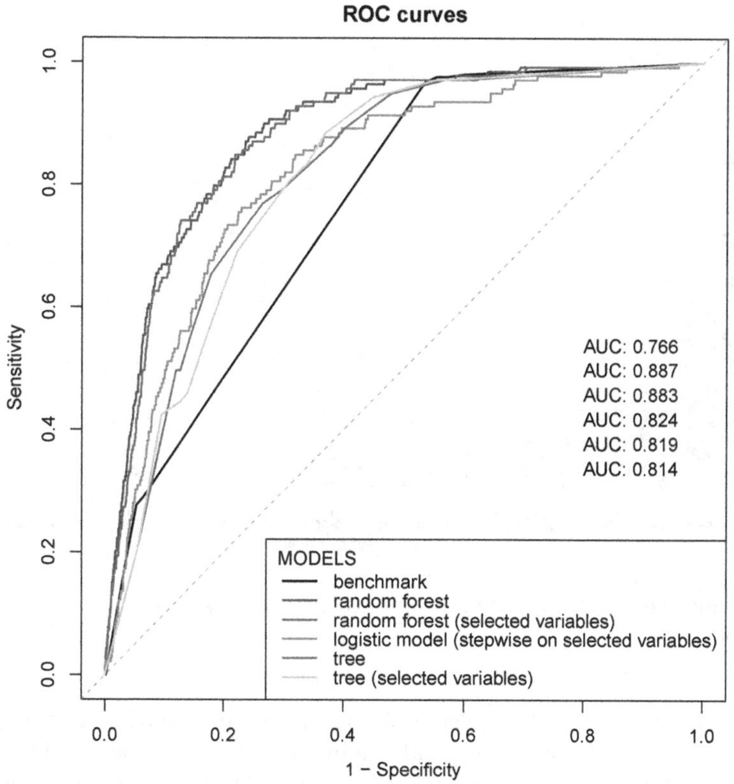

Fig. 1. ROC curves for all the fitted statistical models and respective AUC values

All the fitted statistical models (particularly, Random Forests) present better performances with respect to the benchmark model, the black curve in Fig. 1. The random forest on the 41 variables is the best performing model considering all of the diagnostic presented so far: it has the higher value of AUC and has the best results in terms of the confusion matrices reported in Tables 2 and 3. The scope of these models is to catch the positive outcomes in the first percentiles of customers ordered by propensity. This can be easily observed in the uplift. For Random Forest model the uplift is reported in Fig. 2.

4 Discussion

We have analyzed data from different sources with the objective of creating a CRM recommender. Part of all available data relates to a different time interval than the outcome variable. Due to this reason we have artificially synthesized descriptive variables in the same reference frame of the outcome variable. The resulting variables were reduced in number in two steps. Formerly, the variables are removed in a data cleaning process. Latterly, only the most predictive variables are considered.

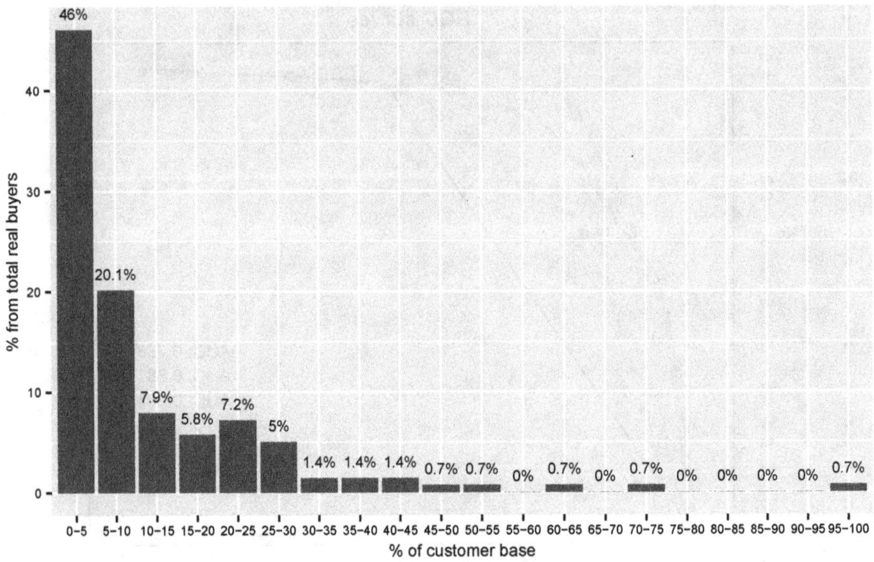

Fig. 2. Captured clients according to ordered propensities predicted by the random forest model

Some algorithms are then selected for the purposes of having a great predictive power and to be easy understandable by different interlocutors. These algorithms are regression trees, random forests and GLM. As benchmark, a heuristic model is considered.

Random forest algorithm performed on all the variables survived to the variable selection procedure results in the best performing model. Captured clients according to random forest ordered predicted propensities shows how such an approach allows a good results in terms of CRM scope. The heuristic model has lower performances than the other models considering all of the diagnostic tools. Due to the aims of the model, order/rank measures (ROC curve and AUC) are the most relevant. Such diagnostic tools show how well the positive cases are ordered before negative cases. The thresholds measures (*e.g.* F1) are relatively low since the data are highly unbalanced. For this reason these metrics should be evaluated using a comparative perspective.

This case study does not evidence that expert-judgement-driven analysis are not useful. At the contrary, the business-driven artificial features has been evidenced as relevant for the best performing model.

However, the presented approach shows some drawbacks. The analyst should make several decisions in the processes of (i) missing values (thresholds definition), (ii) variables selection (how to compare the different diagnostics) and (iii) algorithm evaluation (which characteristics have to be prioritized). This may introduce subjectivity in model creation: for instance, two analysts could create two different models with the same data and aims.

As future developments, we think to include text mining procedures to prevent data loss in the information included into text data. Furthermore, we want to proceed into refining data cleaning procedure and algorithm selection for different cases, in order to obtain a more objective approach.

Declaration of Intent. The views and opinions expressed in this paper are those of the authors only, and do not necessarily represent the views and opinions of UniCredit Business Integrated Solutions S.C.p.A. or any other organization. All the computation have been conducted on anonimized data on UniCredit servers by UniCredit employee. The results have been observed only in aggregated form.

Acknowledgments. We want to acknowledge Raffaele Brevetti, Luca Cilumbriello, Federica Perugini, Nicolò Russo and Dr. Enrico Tonini for their contribute in this work.

References

1. Agarwal, D.K., Chen, B.C.: Statistical Methods for Recommender Systems. Cambridge University Press, Cambridge (2015)
2. Baser, N.C., Thakar, D.G.: A literature review on customer relationship management in banks. Int. J. Customer Relat. Mark. Manage. (IJCRMM) **6**(4), 43–56 (2015)
3. Genuer, R., Poggi, J.M., Tuleau-Malot, C.: Variable selection using random forests. Pattern Recogn. Lett. **31**(14), 2225–2236 (2010)
4. Hardle, W., Simar, L.: Applied Multivariate Statistical Analysis. Springer, Heidelberg (2007)
5. James, G., Witten, D., Hastie, T., Tibshirani, R.: An Introduction to Statistical Learning. Springer, Heidelberg (2013)
6. Kevork, E.K., Vrechopoulos, A.P.: CRM literature: conceptual and functional insights by keyword analysis. Mark. Intell. Plann. **27**(1), 48–85 (2009)
7. McCullagh, P., Nelder, J.A.: Generalized Linear Models, vol. 37. CRC Press, Boca Raton (1989)
8. Ngai, E.W.T., Xiu, L., Chau, D.C.K.: Application of data mining techniques in customer relationship management: a literature review and classification. Expert Syst. Appl. **36**, 2592–2602 (2009)
9. Park, D.H., Kim, H.K., Choi, I.Y., Kim, J.K.: A literature review and classification of recommender systems research. Expert Syst. Appl. **39**, 10059–10072 (2012)
10. Strobl, C., Boulesteix, A.L., Kneib, T., Augustin, T., Zeileis, A.: Conditional variable importance for random forests. BMC Bioinf. **9**(1), 307 (2008)
11. Welling, S.H., Refsgaard, H.H.F., Brockhoff, P.B., Clemmensen, L.H.: Forest floor visualizations of random forests. arXiv preprint. arXiv:1605.09196 (2016)
12. Zineldin, M., Vasicheva, V.: Banking and financial sector in the cloud: knowledge, quality and innovation management. In: Cloud Systems in Supply Chains, pp. 178–194 (2015)

Lagrangian Relaxation Bounds for a Production-Inventory-Routing Problem

Agostinho Agra[1(✉)], Adelaide Cerveira[2,3], and Cristina Requejo[1]

[1] University of Aveiro and CIDMA, 3810-193 Aveiro, Portugal
{aagra,crequejo}@ua.pt
[2] University of Trás-os-Montes e Alto-Douro, 5001-801 Vila Real, Portugal
cerveira@utad.pt
[3] INESC TEC, Porto, Portugal

Abstract. We consider a single item Production-Inventory-Routing problem with a single producer/supplier and multiple retailers. Inventory management constraints are considered both at the producer and at the retailers, following a vendor managed inventory approach, where the supplier monitors the inventory at retailers and decides on the replenishment policy for each retailer. We assume a constant production capacity. Based on the mathematical formulation we discuss a classical Lagrangian relaxation which allows to decompose the problem into four subproblems, and a new Lagrangian decomposition which decomposes the problem into just a production-inventory subproblem and a routing subproblem. The new decomposition is enhanced with valid inequalities. A computational study is reported to compare the bounds from the two approaches.

Keywords: Inventory routing · Lagrangian relaxation · Lagrangian decomposition · Lower bounds

1 Introduction

We consider a single item Production-Inventory-Routing (PIR) problem with a single producer/supplier and multiple clients/retailers. Such complex problems combining production, inventory and routing decisions have been receiving a great attention in recent years with a large increase of the number of publications. Applications to complex supply chain problems can be found, for instance, in [3, 4, 6, 10, 12, 13, 16, 19]. For surveys on complex inventory routing problems covering theoretical and industrial aspects see [1, 7, 11]. The problem considered in this paper fits in the land transportation mode since for each time period a vehicle routing problem is solved in order to find the low cost distribution plan for the deliveries assigned to that period. This contrasts with maritime transportation where, in general, a route may take several time periods [3, 4, 6, 10].

In the past, such complex problems were typically decomposed and solved separately. Lagrangian relaxations for related problems have been used in [9, 15, 23] where the problem is decomposed into several simpler subproblems.

P.M. Pardalos et al. (Eds.): MOD 2016, LNCS 10122, pp. 236–245, 2016.
DOI: 10.1007/978-3-319-51469-7_20

Recent works have shown that there are gains in considering the interrelated problems simultaneously in order to coordinate the different decisions [5,20,23]. There are a few papers reporting exact algorithms to solve these problems. Such exact algorithms are based on branch-and-cut procedures. Some examples can be found in [3,4,8,20,22]. In order to solve efficiently such problems by branch-and-cut it is important to use good formulations [3,4]. Therefore, some relevant work has been done on the modeling of complex problems, either by including valid inequalities or by using extended formulations, see [2–5,13,20,22,24].

We consider a classical model for the PIR problem and discuss a Lagrangian relaxation and a Lagrangian decomposition. First we adapt the Lagrangian relaxation given by Fumero and Vercellis in [15], that decomposes the problem into (i) a trivial production subproblem at the supplier, (ii) a trivial lotsizing problem at the supplier, (iii) a lotsizing problem at each client, and (iv) a capacitated vehicle routing problem in each time period. Then we propose a new Lagrangian decomposition that uses a duplication of variables. The problem is decomposed into (i) a subproblem that considers the production at the supplier and the inventory management at the supplier and clients, and (ii) several vehicle routing subproblems, one for each time period. The Lagrangian decomposition is enhanced with the inclusion of valid inequalities. Although the subproblems are more complex than the subproblems resulting from the classical Lagrangian relaxation, computational tests show that by considering such more complex subproblems (all of them are NP-hard) one obtains better bounds when the subproblems can be solved to optimality.

The outline of the paper is as follows. In Sect. 2 we introduce a mixed integer formulation to the PIR problem. In Sect. 3 we present a Lagrangian relaxation that is similar to the one given in [15]. Then, in Sect. 4 we introduce a new Lagrangian decomposition and discuss enhancements to tighten each subproblem resulting from the decomposition. In Sect. 5 the details of benchmark instances generation are given, and computational results to test the model and compare the Lagrangian relaxation with the Lagrangian decomposition are presented. Finally, the conclusions are stated in Sect. 6.

2 Mathematical Model

In this section we introduce a mixed integer formulation to the PIR problem.

Consider parameters n, nt, and m representing the number of clients, the number of periods, and the number of available vehicles, respectively. Define the sets $N_c = \{1, \ldots, n\}$, and $T = \{1, \ldots, nt\}$ representing the set of clients and the set of periods, respectively. Considering node 0 as the producer, $N = \{0\} \cup N_c$ is the set of nodes (producer and clients), and $A = \{(i, j) : i, j \in N, i \neq j\}$ is the set of arcs. Additionally, consider the following parameters: d_{it} is the demand of client $i \in N_c$ in period $t \in T$; S_i^0 is the initial stock at node $i \in N$; $\overline{I}_i$ is the inventory limit at node $i \in N$; $\overline{P}_t$ is the production capacity in period $t \in T$; Q is the vehicles capacity; $\overline{Q}_{it} = \min\{Q, \sum_{\ell \in t \ldots nt} d_{i\ell}, d_{it} + \overline{I}_i\}$ is the delivery quantity limit in period $t \in T$ at client $i \in N_c$. For cost parameters, SC is the set up cost for producing in a period, PC is the unit production cost, VC is the cost for using a vehicle, H_i is the unit holding cost at node $i \in N$, and C_{ij} is the traveling cost from node i to node j, $(i, j) \in A$.

For the production and inventory decisions we define the following variables: y_t is a binary variable that indicates whether there is production in period $t \in T$ or not; p_t is the production level in period $t \in T$; s_{it} is the stock level at node $i \in N$ at the end of period $t \in T$; q_{it} indicates the quantity delivered at node $i \in N_c$ in period $t \in T$. For the routing decisions we define the variables: x_{ijt} indicates whether a vehicle travels in arc $(i, j) \in A$ in period $t \in T$ or not; z_{it} is a binary variable that is one if node $i \in N_c$ is visited in period $t \in T$; v_t indicates the number of vehicles leaving the producer in period $t \in T$, and f_{ijt} is the quantity of product transported in arc $(i, j) \in A$ in period $t \in T$.

The PIR model is as follows:

$$\min \sum_{t\in T} SC\, y_t + \sum_{t\in T} PC\, p_t + \sum_{t\in T}\sum_{i\in N} H_i s_{it} + \sum_{t\in T}\sum_{(i,j)\in A} C_{ij}x_{ijt} + \sum_{t\in T} VC v_t \quad (1)$$

$$\text{subject to} \quad S_0^0 + p_t = \sum_{i\in N_c} q_{it} + s_{0t} \qquad t = 1 \quad (2)$$

$$s_{0,t-1} + p_t = \sum_{i\in N_c} q_{it} + s_{0t} \qquad \forall t \in T,\ t > 1 \quad (3)$$

$$S_i^0 + q_{it} = d_{it} + s_{it} \qquad \forall i \in N_c, t = 1 \quad (4)$$

$$s_{i,t-1} + q_{it} = d_{it} + s_{it} \qquad \forall i \in N_c, \forall t \in T, t > 1 \quad (5)$$

$$s_{it} \le \overline{I}_i \qquad \forall i \in N, \forall t \in T \quad (6)$$

$$p_t \le \overline{P}_t\, y_t \qquad \forall t \in T \quad (7)$$

$$q_{it} \le \overline{Q}_{it}\, z_{it} \qquad \forall i \in N_c, \forall t \in T \quad (8)$$

$$\sum_{j\in N} x_{ijt} = z_{it} \qquad \forall i \in N_c, \forall t \in T \quad (9)$$

$$\sum_{j\in N} x_{jit} + \sum_{j\in N} x_{ijt} = 2z_{it} \qquad \forall i \in N, \forall t \in T \quad (10)$$

$$\sum_{j\in N} x_{0jt} = v_t \qquad \forall t \in T \quad (11)$$

$$v_t \le m \qquad \forall t \in T \quad (12)$$

$$\sum_{i\in N} f_{ijt} - \sum_{i\in N_c} f_{jit} = q_{jt} \qquad \forall j \in N_c, \forall t \in T \quad (13)$$

$$f_{ijt} \le Q\, x_{ijt} \qquad \forall (i, j) \in A, \forall t \in T \quad (14)$$

$$y_t, z_{it}, x_{ijt} \in \{0, 1\} \qquad \forall i, j \in N, \forall t \in T \quad (15)$$

$$v_t, f_{ijt} \in \mathbb{Z}^+ \qquad \forall i, j \in N, \forall t \in T \quad (16)$$

$$s_{it}, p_t, q_{it} \ge 0 \qquad \forall i \in N, \forall t \in T \quad (17)$$

The objective function (1) is to minimize the total cost which includes the production set-up, the production, the holding, the traveling, and the vehicle usage costs. Constraints (2) and (3) are the inventory conservation constraints at the producer, and (4) and (5) are the inventory conservation constraints at clients. Constraints (6) impose a storage capacity at the producer and at each client. Constraints (7) impose limits on the production at each period, and constraints (8) impose limits on the delivery quantity at each client for each period. Constraints (9) and (10) are the routing constraints. Constraints (11) together with constraints (12) guarantee that the number of vehicles leaving the producer does not exceeds the available number of vehicles. Constraints (13) are the flow balance constraints at clients and constraints (14) guarantee that the capacity of each vehicle is not exceeded. Constraints (15), (16), and (17) are the variables domain constraints.

3 Lagrangian Relaxation

Here we present a Lagrangian relaxation that follows the one from Fumero and Vercelis [15], however in [15] constraints (8) are not considered. In order to derive the Lagrangian relaxation we associate multipliers λ_t^S to constraints (2) (for $t = 1$) and (3) (for $t > 1$), nonnegative multipliers λ_{it}^D to constraints (8) and nonnegative multipliers λ_{ijt}^F to constraints (14) and dualize these constraints in the usual Lagrangian way.

This leads to the following relaxation

$$\begin{aligned}\mathscr{L}(\lambda^S,\lambda^D,\lambda^F) = \min \sum_{t\in T} SCy_t + \sum_{t\in T} PCp_t + \sum_{t\in T}\sum_{i\in N} H_i s_{it} + \sum_{t\in T}\sum_{(i,j)\in A} C_{ij}x_{ijt} + \sum_{t\in T} VCv_t + \\ + \lambda_1^S(S_0^0 + p_1 - \sum_{i\in N} q_{i1} - s_{01}) + \sum_{t\in T,\ t>1} \lambda_t^S(s_{0,t-1} + p_t - \sum_{i\in N} q_{it} - s_{0t}) + \\ + \sum_{t\in T}\sum_{i\in N_c} \lambda_{ijt}^D(q_{it} - \overline{Q}_{it}) + \sum_{t\in T}\sum_{(i,j)\in A} \lambda_{ijt}^F(f_{ijt} - Q\, x_{ijt})\end{aligned}$$

subject to $(4)-(13),(15)-(17)$.

This problem can be separated into four subproblems: a production subproblem (on variables y_t, p_t), an inventory subproblem at the supplier (on variables s_{0t}), an inventory subproblem at the retailers (on variables s_{it}, q_{it}, f_{ijt}) and a routing subproblem (on variables x_{ijt}, z_{it}, v_t). The first two subproblems can be solved by inspection.

For each set of multipliers $(\lambda^S, \lambda^D, \lambda^F)$, with $\lambda^S \in \mathbb{R}$, $\lambda^D \in \mathbb{R}_0^+$, and $\lambda^F \in \mathbb{R}_0^+$, the value of $\mathscr{L}(\lambda^S, \lambda^D, \lambda^F)$ gives a lower bound on the optimum value of the PIR problem. To obtain the best lower bound the following dual Lagrangian problem: $\max_{\lambda^S\in\mathbb{R},\lambda^D\geq 0,\lambda^F\geq 0} \mathscr{L}(\lambda^S, \lambda^D, \lambda^F)$, has to be solved.

A subgradient optimization procedure [17] is used to solve the dual Lagrangian problem. This procedure starts by initializing the Lagrangian multipliers $(\lambda^S, \lambda^D, \lambda^F)$ to $(\lambda_0^S, \lambda_0^D, \lambda_0^F)$. We set $(\lambda_0^S)_t = S_0^0, t \in T$, $(\lambda_0^S)_{nt+1} = 0$, $(\lambda_0^D)_{it} = (S_0^0 + n)/\overline{Q}_{it}, i \in N_c, t \in T$ and $(\lambda_0^F)_{ijt} = 1/Q, (i,j) \in A, t \in T$. Then, iteratively, at each iteration k, the relaxed problem $\mathscr{L}(\lambda_k^S, \lambda_k^D, \lambda_k^F)$ is solved and the Lagrangian multipliers are updated using a direction (d_k^S, d_k^D, d_k^F) and a step-size (s_k^S, s_k^D, s_k^F) as follows:

$$\lambda_{k+1}^S = \max\{\lambda_k^S + s_k^S d_k^S\},\ \lambda_{k+1}^D = \max\{\lambda_k^D + s_k^D d_k^D\},\ \text{and } \lambda_{k+1}^F = \max\{\lambda_k^F + s_k^F d_k^F\}.$$

The direction, following Held, Wolfe and Crowder [17], is updated as follows:

$$d_k^S = \nabla_k^S,\ d_k^D = \nabla_k^D,\ \text{ and } d_k^F = \nabla_k^F,$$

considering the following subgradients: $(\nabla_k^S)_t = S_0^0 + p_t - \sum_{i \in N_c} q_{it} - s_{0t},\ t = 1$, $(\nabla_k^S)_t = s_{0,t-1} + p_t - \sum_{i \in N_c} q_{it} - s_{0t},\ t \in T,\ t > 1$, $(\nabla_k^D)_{it} = q_{it} - \overline{Q}_{it},\ i \in N_c, t \in T$ and $(\nabla_k^F)_{ijt} = f_{ijt} - Q\ x_{ijt},\ (i,j) \in A, t \in T$. For the step-size, following Shor [21], we consider:

$$s_k^S = \rho \frac{UB - \mathscr{L}(\lambda_k^S, \lambda_k^D, \lambda_k^F)}{\nabla_k^S\ d_k^S},\ s_k^D = \rho \frac{UB - \mathscr{L}(\lambda_k^S, \lambda_k^D, \lambda_k^F)}{\nabla_k^D\ d_k^D},\ s_k^F = \rho \frac{UB - \mathscr{L}(\lambda_k^S, \lambda_k^D, \lambda_k^F)}{\nabla_k^F\ d_k^F}$$

with $\rho \in]0, 2[$, and where UB is the best upper bound known.

4 Lagrangian Decomposition

In this section a Lagrangian decomposition of the PIR model is considered, where variables q_{it} and z_{it} are duplicated. New variables q_{it}^1 replace q_{it} in constraints (2)–(5), variables z_{it}^2 replace z_{it} in constraints (9)–(10), and variables q_{it}^2 replace q_{it} in constraints (13). Constraints (8) are also duplicated and replaced by the following two sets of constraints

$$q_{it}^1 \le \overline{Q}_{it}\ z_{it}^1,\quad \forall i \in N_c, \forall t \in T, \tag{18}$$

$$q_{it}^2 \le \overline{Q}_{it}\ z_{it}^2,\quad \forall i \in N_c, \forall t \in T. \tag{19}$$

Additionally the following equalities are added

$$q_{it}^1 = q_{it}^2 \quad \forall i \in N_c, \forall t \in T \tag{20}$$

$$z_{it}^1 = z_{it}^2 \quad \forall i \in N_c, \forall t \in T \tag{21}$$

Associating multipliers λ^q to constraints (20) and multipliers λ^z to constraints (21), and dualizing these constraints on the Lagrangian way, two separated subproblems are obtained: a production-inventory subproblem, and a routing subproblem.

The production-inventory subproblem corresponds to the relaxation

$$\mathscr{L}^{PI}(\lambda^q, \lambda^z) = \min \sum_{t \in T} SCy_t + \sum_{t \in T} PCp_t + \sum_{t \in T} \sum_{i \in N} H_i s_{it} + \sum_{t \in T} \sum_{i \in N_c} (\lambda_{it}^q q_{it}^1 + \lambda_{it}^z z_{it}^1)$$

subject to the modified constraints (2)–(7) and (18), with variables z_{it} and q_{it} replaced by z_{it}^1 and q_{it}^1, respectively. While the routing subproblem corresponds to the relaxation

$$\mathscr{L}^{R}(\lambda^q, \lambda^z) = \min \sum_{t \in T} \sum_{(i,j) \in A} C_{ij} x_{ijt} + \sum_{t \in T} VCv_t - \sum_{t \in T} \sum_{i \in N_c} (\lambda_{it}^q q_{it}^2 + \lambda_{it}^z z_{it}^2)$$

subject to the modified constraints (9)–(14) and (19), with variables z_{it} and q_{it} replaced by z^2_{it} and q^2_{it}, respectively. This routing subproblem can be further separated into nt routing subproblems, one for each period.

All these subproblems are NP-hard but can be solved in general for reasonable size instances. For each set of multipliers (λ^q, λ^z) the value of $\mathscr{L}^{PI}(\lambda^q, \lambda^z) + \mathscr{L}^{R}(\lambda^q, \lambda^z)$ gives a lower bound on the optimum value of the PIR problem. To obtain the best lower bound the corresponding dual Lagrangian problem has to be solved. A subgradient optimization procedure [17] as described above is used. The multipliers are initialized as follows: $(\lambda^z_0)_{it} = C_{0i}, i \in N_c, t \in T$, $(\lambda^q_0)_{it} = (S^0_i + n)/\overline{Q}_{it}, i \in N_c, t \in T$.

To improve the Lagrangian decomposition each subproblem is tightened. For the production-inventory subproblem the following valid inequalities are considered.

$$\sum_{\ell=1}^{t} y_\ell \geq \left\lceil \frac{1}{\overline{P}_t} \Big(\sum_{i \in N_c} \sum_{\ell=1}^{t} d_{i\ell} - \sum_{i \in N} S^0_i \Big) \right\rceil \qquad \forall t \in T,\ t > 1 \tag{22}$$

$$\sum_{\ell=1}^{t} z^1_{i\ell} \geq \left\lceil \frac{1}{\min\{\overline{I}_i, Q\}} \sum_{\ell=1}^{t} (d_{i\ell} - S^0_i) \right\rceil \qquad \forall i \in N_c, \forall t \in T \tag{23}$$

Inequalities (22) impose a minimum number of production periods, while inequalities (23) impose a minimum number of visits to each client i.

For the routing subproblems we consider the Miller-Tucker-Zemlin [18] reformulation. We introduce new variables w_{it} indicating the load on the vehicle when client $i \in N_c$ is visited in period $t \in T$ (before delivering), and include the classical set of constraints,

$$w_{it} \geq w_{jt} + q_{it} - Q(1 - x_{ijt}), \qquad \forall (i,j) \in A, \forall t \in T, \tag{24}$$

$$w_{it} \leq Q z_{it}, \qquad \forall i \in N_c, \forall t \in T, \tag{25}$$

$$w_{it} \geq 0, \qquad \forall i \in N_c, \forall t \in T. \tag{26}$$

Inequalities (24)–(26) are added to each routing subproblem.

5 Computational Results

A computational experimentation was conducted to test the bounds provided by the two Lagrangian approaches. The computational tests were conducted using the Xpress-Optimizer 28.01.04 [14] solver with the default options.

First we provide details on the generation of benchmark instances. The coordinates of the clients are randomly generated in a 100 by 100 square grid, and the producer is located in the center of the grid. For the number of clients two values are considered: $n = 10$, and $n = 20$. For the number of periods three values are considered: $nt = 5$, $nt = 10$, and $nt = 15$. For each pair of values n, nt, three instances are generated, giving a total of 18 instances.

For each value of n, a complete graph with a symmetric traveling cost matrix associated to the set of arcs is considered. The traveling costs C_{ij} and C_{ji} are the Euclidean distance between the nodes i and j in the grid. The demand values,

d_{it}, are randomly generate between 40 and 80 units. The initial stock at producer S_0^0 is zero, and the initial stock S_i^0 at client i is randomly generated between 0 and three times the average demand of client i. The holding cost H_i is one for all $i \in N$. The maximum inventory level $\overline{I}_i$ is 500 for all $i \in N$. The production capacity $\overline{P}_t$ is 50 % of the average demand. The production set up cost and the unit production cost are given by $SC = 100$ and $PC = 1$, respectively. The number of available vehicles is $m = 3$. The vehicle usage cost is $VC = 50$, and the vehicle capacity is $Q = 500$.

Table 1 reports the instance data and the bounds obtained through the computational experiments. First column displays the number n of clients, the second column displays the number nt of periods, and the third column is the instance identifier number for each pair n, nt. The next columns report the results obtained with the linear programming relaxation of model PIR (columns *LP*), the Lagrangian relaxation given in Sect. 3 (columns *LR*), the Lagrangian decomposition given in Sect. 4 with all enhancements (columns *ILD*) and the best upper bound obtained using the branch and bound procedure from the solver Xpress to solve the PIR model given in Sect. 2 with a time limit of 2 h (column *UB*). None of the tested instances could be solved to optimality within the given running time limit. For columns *LP* we report the value of the lower bound (*LB*) and the running time (*Time*). For columns *LR* and *ILD* we report

Table 1. Comparison of bounds obtained via Lagrangian approaches.

			LP		*LR*			*IDL*			*UB*
n	nt	Inst.	*LB*	*Time*	*LB*	*Iter*	*Time*	*LB*	*Iter*	*Time*	
10	5	1	1736.5	0	1496.0	500	19.5	2338.8	41	56.3	2754
10	5	2	1491.6	0	1236.4	500	19.4	1690.2	16	37.9	2637
10	5	3	1491.9	0	1240.5	500	19.2	1910.8	25	34.7	2835
10	10	1	2864.2	0	2531.9	500	72.2	4604.7	33	167.8	5603
10	10	2	2730.3	0	2247.6	500	74.0	3618.9	13	89.0	5464
10	10	3	2673.0	0	2215.5	500	71.9	3678.9	20	111.0	5692
10	15	1	3452.2	0.1	3070.6	500	251.3	6640.9	27	202.8	8261
10	15	2	3603.6	0.1	2855.9	500	250.5	5294.8	17	169.5	7828
10	15	3	3928.7	0.1	3262.2	500	251.2	5588.3	18	151.3	8729
20	5	1	2504.5	0.1	1630.8	500	122.5	2805.6	14	451.6	3919
20	5	2	2670.6	0.1	1952.7	500	124.4	2618.4	15	550.5	4252
20	5	3	2917.4	0.1	2123.5	500	118.0	2970.2	17	336.6	4438
20	10	1	4535.9	0.3	2809.9	500	239.4	5665.7	17	1213.3	7914
20	10	2	4116.7	0.3	2514.2	500	238.7	4791.9	15	1283.2	7644
20	10	3	4571.8	0.3	2788.3	500	241.7	5676.0	22	683.2	8805
20	15	1	6230.2	0.4	3749.2	500	493.8	7578.9	8	1253.5	12067
20	15	2	5917.3	0.5	3403.5	500	498.3	6890.4	10	1235.4	11785
20	15	3	6739.8	0.4	4091.3	500	494.9	8464.0	18	1066.3	13335

the value of the lower bound (*LB*), the number of iterations of the subgradient algorithm (*Iter*), and the running time (*Time*).

We can see that the linear relaxation is fast. The lower bounds obtained using the Lagrangian relaxation are lower than the lower bounds obtained using the linear relaxation. In theory one should observe the opposite relation between bounds when the optimal multipliers are used. This means that we could not identify near optimal multipliers. The Lagrangian decomposition provides always the best lower bounds. The Lagrangian relaxation terminates when the maximum number of iterations (500) was attained, while the Lagrangian decomposition terminates, in most of the cases, when the time limit of 20 min was attained. Thus, one may conclude that solving the Lagrangian decomposition at each iteration of the subgradient algorithm requires high running times (as expected, since more complex subproblems are solved in each iteration) but provides better lower bounds.

6 Conclusions

We consider a complex production-inventory-routing problem. For this problem two Lagrangian approaches are tested. One is a classical Lagrangian relaxation that allows to split the problem into several small subproblems, where two of them can be solved by inspection. The other approach decomposes the problem into two main subproblems (one is a production-inventory problem and the other is a routing problem). Such subproblems are still complex and NP-hard. Nevertheless, they can be solved for reasonable size instances using a commercial solver. The computational tests indicate that keeping a higher degree of complexity in the subproblems leads to harder subproblems, requiring larger running times to solve to optimality, but allow us to derive better lower bounds. Also, as the quality of the lower bounds obtained through Lagrangian approaches depend greatly on the quality of the Lagrangian multipliers, the computational results show that finding near optimal multipliers is a challenging task when we consider the classical approach that uses many multipliers. Overall we may conclude that, when the subproblems can be solved to optimality, then it may be preferable to consider such complex subproblems and perform less iterations of a subgradient method using a few number of Lagrangian multipliers, than to consider simpler subproblems which allow to perform more iterations, but require the tuning of an high number of Lagrangian multipliers. As future research we aim to investigate decomposition approaches that are based on similar decomposition ideas as the one we tested here, but where no computation of Lagrangian multipliers is required.

Acknowledgements. The research of the first and third authors was supported through CIDMA and FCT, the Portuguese Foundation for Science and Technology, within project UID/MAT/ 04106/2013. The research of the second author was financed by the ERDF - European Regional Development Fund through the Operational Programme for Competitiveness and Internationalisation - COMPETE 2020 Programme within project "POCI-01-0145-FEDER-006961", and by FCT within project UID/EEA/50014/2013.

References

1. Adulyasak, Y., Cordeau, J., Jans, R.: The production routing problem: a review of formulations and solution algorithms. Comput. Oper. Res. **55**, 141–152 (2015)
2. Adulyasak, Y., Cordeau, J., Jans, R.: Formulations and branch and cut algorithms for multi-vehicle production and inventory routing problems. Inf. J. Comput. **26**(1), 103–120 (2014)
3. Agra, A., Andersson, H., Christiansen, M., Wolsey, L.: A maritime inventory routing problem: discrete time formulations and valid inequalities. Networks **62**, 297–314 (2013)
4. Agra, A., Christiansen, M., Delgado, A.: Mixed integer formulations for a short sea fuel oil distribution problem. Transp. Sci. **47**, 108–124 (2013)
5. Agra, A., Christiansen, M., Delgado, A., Simonetti, L.: Hybrid heuristics for a short sea inventory routing problem. Eur. J. Oper. Res. **236**, 924–935 (2014)
6. Agra, A., Christiansen, M., Ivarsoy, K., Solhaug, I., Tomasgard, A.: Combined ship routing and inventory management in the salmon farming industry. Ann. Oper. Res. (in press)
7. Andersson, H., Hoff, A., Christiansen, M., Hasle, G., Løkketangen, A.: Industrial aspects and literature survey: combined inventory management and routing. Comput. Oper. Res. **37**(9), 1515–1536 (2010)
8. Archetti, C., Bertazzi, L., Laporte, G., Speranza, M.G.: A branch-and-cut algorithm for a vendor-managed inventory-routing problem. Transp. Sci. **41**(3), 382–391 (2007)
9. Bell, W.J., Dalberto, L.M., Fisher, M.L., Greenfield, A.J., Jaikumar, R., Kedia, P.: Improving the distribution of industrial gases with an on-line computerized routing and scheduling optimizer. Interfaces **13**(6), 4–23 (1983)
10. Christiansen, M.: Decomposition of a combined inventory and time constrained ship routing problem. Transp. Sci. **33**(1), 3–16 (1999)
11. Christiansen, M., Fagerholt, K.: Maritime Inventory Routing Problems. In: Floudas, C., Pardalos, P. (eds.) Encyclopedia of Optimization, 2nd edn, pp. 1947–1955. Springer, New York (2009)
12. Eksioglu, S.D., Romeijn, H.E., Pardalos, P.M.: Cross-facility management of production and trasportation planning problem. Comput. Oper. Res. **33**(11), 3231–3251 (2006)
13. Eppen, G.D., Martin, R.K.: Solving multi-item capacitated lot-sizing problems using variable redefinition. Oper. Res. **35**, 832–848 (1997)
14. FICO Xpress Optimization Suite
15. Fumero, F., Vercellis, C.: Synchronized development of production, inventory, and distribution schedules. Transp. Sci. **33**(3), 330–340 (1999)
16. Geunes, J., Pardalos, P.M.: Network optimization in supply chain management and financial engineering: an annotated bibliography. Networks **42**(2), 66–84 (2003)
17. Held, M., Wolfe, P., Crowder, H.P.: Validation of subgradient optimization. Math. Program. **6**, 62–88 (1974)
18. Miller, C., Tucker, A., Zemlin, R.: Integer programming formulations and travelling salesman problems. J. Assoc. Comput. Mach. **7**(4), 326–329 (1960)
19. Pei, J., Pardalos, P.M., Liu, X., Fan, W., Yang, S., Wang, L.: Coordination of production and transportation in supply chain scheduling. J. Ind. Manage. Optim. **11**(2), 399–419 (2015)
20. Ruokokoski, M., Solyali, O., Cordeau, J.-F., Jans, R., Süral, H.: Efficient formulations and a branch-and-cut algorithm for a production-routing problem. GERAD Technical report G-2010-66. HEC Montréal, Canada (2010)

21. Shor, N.Z.: Minimization Methods for Non-Differentiable Functions. Springer, Heidelberg (1985)
22. Solyali, O., Süral, H.: A branch-and-cut algorithm using a strong formulation and an a priori tour-based heuristic for an inventory-routing problem. Transp. Sci. **45**(3), 335–345 (2011)
23. Solyali, O., Süral, H.: A relaxation based solution approach for the inventory control and vehicle routing problem in vendor managed systems. In: Neogy, S.K., Das, A.K., Bapat, R.B. (eds.) Modeling, computation and Optimization, pp. 171–189. World Scientific, Singapore (2009)
24. Solyali, O., Süral, H.: The one-warehouse multi-retailer problem: reformulation, classification and computational results. Ann. Oper. Res. **196**(1), 517–541 (2012)

Convergence Rate Evaluation of Derivative-Free Optimization Techniques

Thomas Lux(✉)

Roanoke College, Salem, VA 24153, USA
thlux@mail.roanoke.edu
http://cs.roanoke.edu/~thlux/

Abstract. This paper presents a convergence rate comparison of five different derivative-free numerical optimization techniques across a set of 50 benchmark objective functions. Results suggest that Adaptive Memory Programming for constrained Global Optimization, and a variant of Simulated Annealing are two of the fastest-converging numerical optimization techniques in this set. Lastly, there is a mechanism for expanding the set of optimization algorithms provided.

Keywords: Optimization · Convergence · Metaheuristic · Derivative-free

1 Introduction

Many well defined engineering problems in the real world do not allow for the timely computation of an optimal solution. The task of *Optimization* is to find desirable solutions to problems in spaces that prohibit exhaustive search. In order to find desirable solutions while only sampling a relatively small subset of possible solutions, optimization algorithms make assumptions about the search space for the problem at hand. The goal of an optimization algorithm can vary, but for the purpose of this paper we consider algorithms attempting to find "good" solutions with the fewest computations possible.

The problems that optimization algorithms solve are often formulated as a function which takes a set of parameters as input and then returns the performance of that set of parameters as output. We define an *objective function* as a deterministic function of multiple real-valued parameters from $\mathbb{R}^n$ to $\mathbb{R}$. The computational cost of exhaustively searching for a minimum return value of one of these objective functions grows exponentially with respect to the number of input parameters and their acceptable values.

The class of objective functions that we focus on are those for which each parameter is bounded, and no derivative information is known about the objective function. In mathematical terms we define this as: given derivative-free objective function $f : \mathbb{R}^n \rightarrow \mathbb{R}$ minimize $f(\boldsymbol{x})$ subject to $a_i \leq x_i \leq b_i(a_i, b_i \in \mathbb{R})$.

In this paper we analyze five different numerical optimization techniques, four of which are relatively new, and compare their *rates of convergence* on a set

P.M. Pardalos et al. (Eds.): MOD 2016, LNCS 10122, pp. 246–256, 2016.
DOI: 10.1007/978-3-319-51469-7_21

of well-understood objective functions with varying dimensionality. The purpose of this study is to propose a ranking of these algorithms in terms of their likelihood to converge to an optimal solution in a restricted number of objective function evaluations. The five different numerical optimization techniques that we compare, in order of their creation, are: Simulated Annealing (SA) [1], Adaptive Memory Programming for constrained Global Optimization (AMPGO) [2], Cuckoo Search (CK) [3], Backtracking Search optimization Algorithm (BSA) [4], and quick Artificial Bee Colony (qABC) [5]. In Sect. 3 we list the objective functions, their bounds, the metrics for convergence evaluation, and the general formulation of each optimization algorithm. In Sect. 4 we provide figures and tables demonstrating the performance of each algorithm across our set of objective functions.

1.1 Metaheuristic Optimization Algorithms

Metaheuristic algorithms are useful for searching through objective function spaces for which there is no known derivative because they rely strictly on the objective values obtained and the internal heuristics of the algorithm. In [6] the general form of a metaheuristic optimization algorithm and the behavior of a random walk is defined. It is also mentioned that the mechanisms by which optimization algorithms achieve this random walk behavior is quite different. All of the numerical optimization algorithms that we analyze in this paper, with the exception of AMPGO, are metaheuristic algorithms. Each of the metaheuristic algorithms can be generalized to a random walk search model while AMPGO utilizes a memoized approximation of the objective function gradient for convergence.

2 Related Works

Numerical optimization is currently a rapidly expanding field of research. Every year there are more numerical optimization techniques and algorithms introduced and the five being compared in this paper provide only a small subset of all algorithms. This paper focuses specifically on derivative-free optimization techniques because they are useful for optimizing complex objective functions [7,8]. These five derivative-free algorithms are only a few of large existing set and have been selected because they each have independent assertions claiming them to be the "best" optimization algorithms by various sources within recent years [3–5,9,10].

Due to the sheer quantity of optimization research papers being published, even the most thorough review papers [6,11] do not compare all algorithms. Many optimization algorithms that have been proposed and asserted as comparatively better than others are compared strictly to older less powerful algorithms. Compounding this difficulty, much of the source code is written in languages for proprietary software such as Matlab and is not readily usable by the public. For these reasons, more research comparing modern numerical optimization algorithms as well as providing open sourced implementations of the algorithms in a freely available language needs to be done.

3 Methods

The 50 objective functions chosen for this evaluation are a sampling of functions used on other optimization algorithms as well as in some popular benchmark data sets. The function definitions can be found in the Appendix Table 1, while the number of dimensions used in testing and acceptable ranges for values are all listed in the Appendix Table 2. Plots of a representative subset of these functions in their two dimensional variants are seen in Fig. 1.

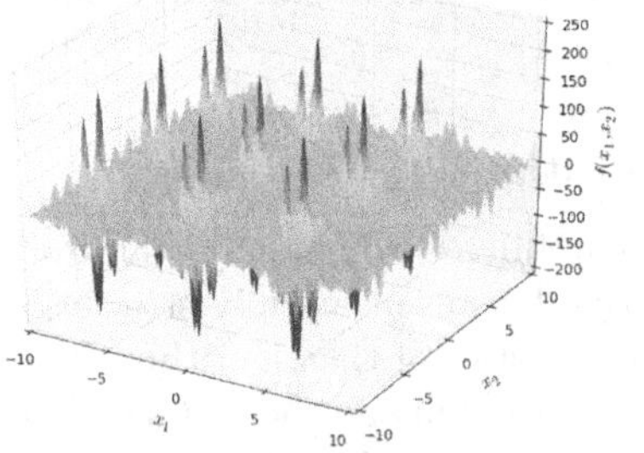

(a) Shubert Function

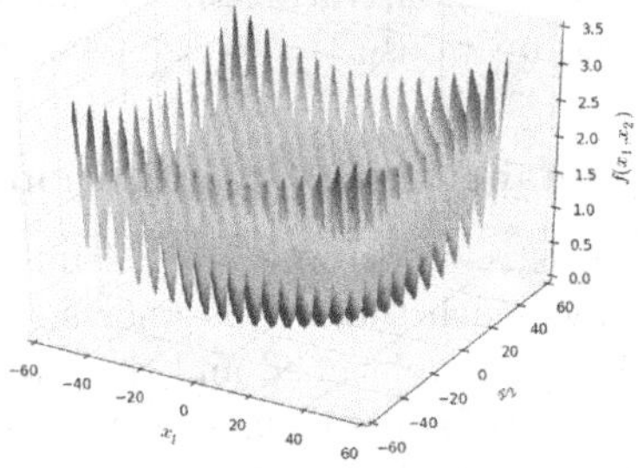

(b) Griewank Function

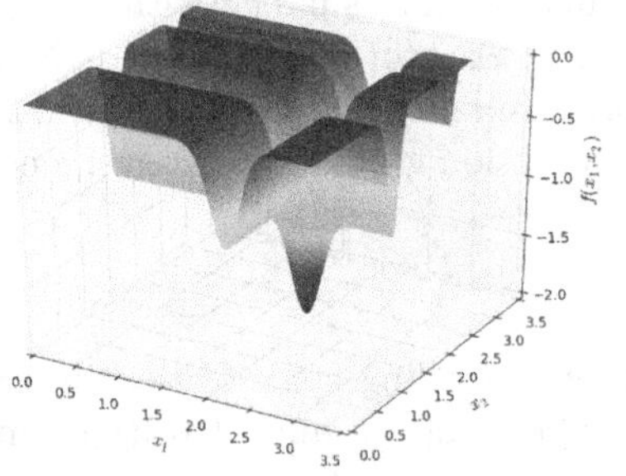

(c) Michalewicz Function

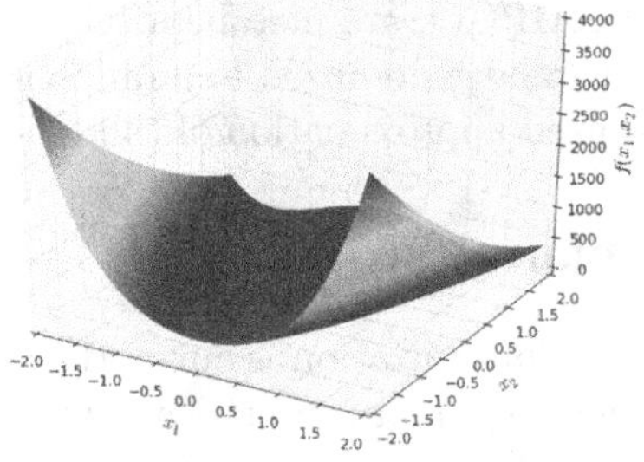

(d) Rosenbrock Function

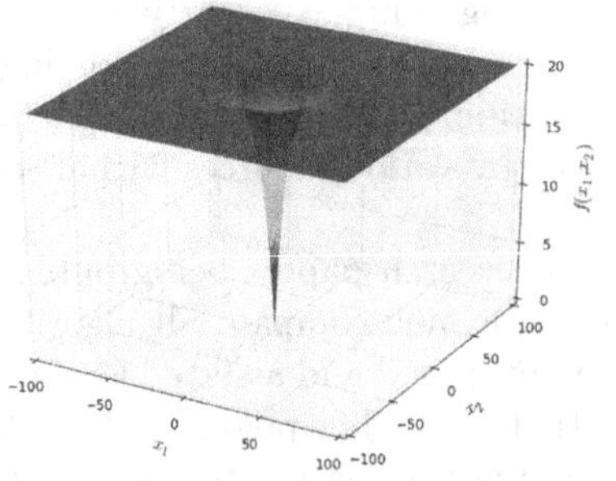

(e) Easom Function

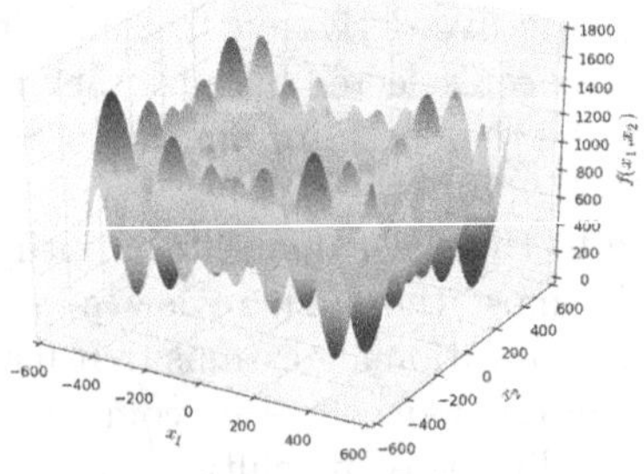

(f) Schwefel Function

Fig. 1. The 2 dimensional variants of some of the objective functions, provided as a sampling of the types of spaces used in testing. These plots were generated by the code freely available at [9].

In order to compare the relative performance of the different optimization algorithms across multiple objective functions, we use *Data Profiles*, *rank 0 probability*, and *best solution probability.* The remaining statistics that are normally included in a comparative study of optimization algorithms (maximum, minimum, average, and standard deviation of objective function values obtained) are available through electronic supplementary materials.

The parameters we use for each of the optimization algorithms are the default suggested values provided by the authors of the algorithms. The one exception to this is the simulated annealing algorithm, for which we use an *acceptance probability* of 0, and add a modification that increases the *temperature* each time a better solution is discovered. For exact parameters and implementation details, please see the referenced source code. Each optimization algorithm is allowed 5000 executions of each objective function. 5000 is selected under the assumption that these optimization algorithms would be used in the future on moderately difficult-to-compute objective functions which require at least 10 s to evaluate per iteration under 12–24 h time constraints.

We use a *convergence in mean* stopping criteria over the objective function value obtained by an optimization algorithm after 5000 executions to determine an appropriate number of repeated trials. By using convergence in mean, we attempt to equally weight the results in terms of randomness. We consider the final objective function value obtained by an optimization algorithm to have converged in mean after ($t > 500$) trials if: for 100 sequential trials, the mean of the cumulative final objective function values shifts by less than 0.01%, in terms of the range of mean values encountered. For this set of objective functions and optimization algorithms the average number of trials needed to pass this convergence test was 1106.

Data Profile. The *Performance Profile*, introduced by Dolan in [12], compares the *performance ratio* of multiple optimization algorithms given a stopping criteria such as a time restriction or maximum number of executions of the objective function. Performance Profiles can be used to compare optimization algorithms with the only drawback being the need for a singular stopping criteria. *Data Profiles*, created by Moré in [7], address the need for establishing a singular stopping criteria by measuring the convergence of an optimization algorithm after successive executions of the objective function. The primary measurement performed in a Data Profile is at each execution of the objective function. Supposing an optimization algorithm has executed an objective function f, for n iterations, starting with initial solution $\mathbf{x}_0$, where the best solution obtained by any optimization algorithm in the comparison set at the nth execution of the objective function is f_L, the performance of that optimization algorithm can be measured by the percentage of solutions that pass the following criteria:

$$f(x_0) - f(x) \geq (1 - \tau)(f(x_0) - f_L) \tag{1}$$

$\tau > 0$ is the convergence tolerance, and represents how close an optimization algorithm should be after n executions of the objective function to f_L given a specific f_0. We provide plots for each $\tau \in \{10^{-1}, 10^{-3}, 10^{-5}, 10^{-7}\}$.

Rank 0 Probability. It is often desirable to rank optimization algorithms in terms of their performance. This becomes difficult when the algorithms incorporate randomness, but given some number of independent repeated trials we can still use simple counting techniques to calculate the *probability* that a selected algorithm will be ranked in a certain position. Particularly the position of interest for comparative studies is often that of the best algorithm, which we refer to as rank 0. Consider the following explanation.

Given some set of optimization algorithms A and an optimization algorithm of interest $\beta \in A$, consider objective function f where each optimization algorithm has been allowed n executions of f. At the nth execution, β along with each other algorithm in A, has some set of objective function values obtained from repeated trials T_β, where $n(T_\beta)$ is the number of repeated trials. Now, assuming that the values in T_β are independent and hence equally likely, we can count the number of ways that a trial value from T_β could be the smallest of trial values selected for each algorithm. This counts the number of ways that β could be the best algorithm in A after n executions of the objective function. Using a simple counting technique, we have that the rank 0 probability R_0 of β, can be computed as:

$$R_0(\beta) = \frac{1}{n(T_\beta)} \sum_{v \in T_\beta} \frac{\prod_{\alpha \in A \setminus \{\beta\}} n(\{i \in T_\alpha \text{ s.t. } v \leq i\})}{\prod_{\alpha \in A \setminus \{\beta\}} n(T_\alpha)} \tag{2}$$

Note that the entire computation of rank 0 probability can be done in $alog(a)$ with $a = n_{avg}(T_\alpha)$[1] for each execution of the objective function. This is not prohibitive with any reasonable number of trials.

Best Solution Probability. It could also be interesting to know which optimization algorithms achieved the best solutions and how often each algorithm was the one to achieve the best solution in the benchmark objective function set. We consider an algorithm to have achieved the best possible solution if it is within 1% of the best objective function value obtained by any other algorithm after the same number of executions of the objective function. For objective function f, the 1% distance, $f_{1\%}$ is defined from the range $max_f - min_f$, where max_f is the maximum value of f that any optimization algorithm achieved and min_f is the minimum. The best solution probability bsp of optimization algorithm α after n executions of each objective function f, in the set of benchmark objective functions F, is:

$$bsp(\alpha) = \frac{n(\{f \in F \text{ s.t. } min(T_\alpha) - min_{f,n} < f_{1\%}\})}{n(F)} \tag{3}$$

where T_α is the set of trials for α after n executions of f and $min_{f,n}$ is the minimum value achieved for f after n iterations by any optimization algorithm.

[1] The computation of $n(\{i \in T_\alpha \text{ s.t. } v \leq i\})$ in the upper product sum can be done in $log(n(T_\alpha))$ if each T_α is sorted. n_{avg} represents the average number of trials for all $\alpha \in A$.

For verification and reuse, the code for each of these algorithms is available at the following web address implemented in python3 with Numpy and SciPy: https://github.com/thlux/Convergence_Rate_Evaluation.

4 Results

The four different tolerance values for data profiles display important nuances related to this set of optimization algorithms. As can be seen in Fig. 2, Annealing converges at least weakly more often than any other optimization algorithm, qABC is close behind. As we shift from weak convergence ($\tau = 10^{-1}$) toward stronger convergence ($\tau = 10^{-7}$), the results transition to show that AMPGO is the optimization algorithm most capable of converging quickly. It is clear from the four different tolerance values that Annealing and AMPGO are the two contenders for fastest converging optimization algorithms. Annealing converges more consistently, but not as quickly as AMPGO.

The rank 0 probability results in Fig. 3 also suggest that Annealing and AMPGO are close competitors. For the first 20 executions of the objective function, the random sampling provided by the populations of qABC, BSA, and Cuckoo Search allow for the fastest convergence. Beyond 20 and until 500 executions of the objective function, it is unclear which of Annealing or AMPGO

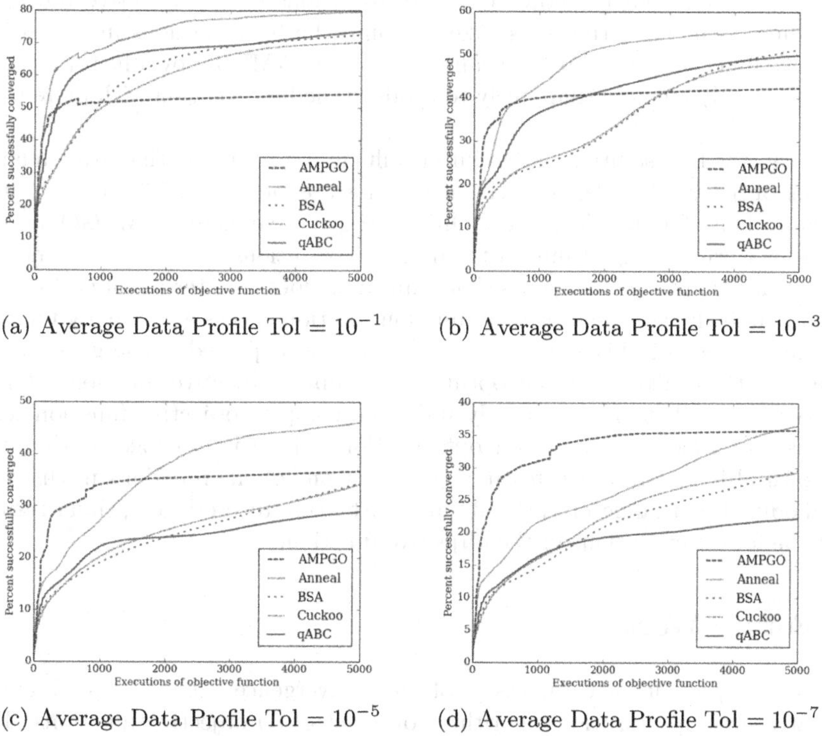

(a) Average Data Profile Tol = 10^{-1}

(b) Average Data Profile Tol = 10^{-3}

(c) Average Data Profile Tol = 10^{-5}

(d) Average Data Profile Tol = 10^{-7}

Fig. 2. The average data profile results across all 50 objective functions

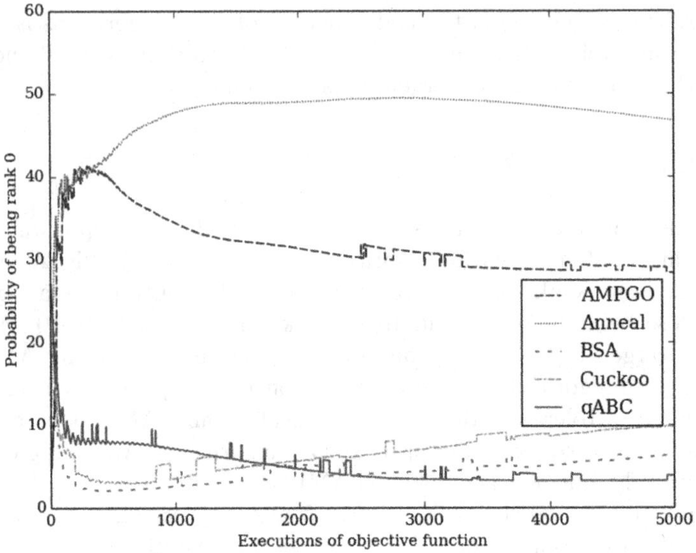

Fig. 3. The average rank 0 probability across all 50 objective functions

is a better choice. Given more than 500 executions of the objective function, Annealing is the algorithm most likely to find the best solutions in this set of objective functions. Notably, Annealing only beats AMPGO by a factor of 50%, and would only be more than likely to produce the best solution with more than two trials.

Lastly, the best solution probability results in Fig. 4 suggest that given a large number of repeated trials, all of the algorithms are more than 50% likely to find the equivalent of a "best" solution. Once again, Annealing and AMPGO are the highest performing algorithms for the first 500 executions. The high performance of Annealing in this plot suggests that Annealing does a better job of performing global search though the objective function solution spaces given many trials.

The purpose of this study is to compare the expected convergence rates of each of these algorithms for optimizing complex objective functions. These results suggest that given a mostly unknown complex objective function with bounds and time constraints on optimization, where it is expected that the unknown objective function resembles at least one of the functions in this test set, modified Annealing or AMPGO are most likely to produce optimal results with the fewest executions of the objective function.

5 Future Work

This paper presents a comparison of the convergence rates of five different derivative-free optimization algorithms on a set of 50 objective functions. The goal of this work is to provide insight as to which optimization algorithm will

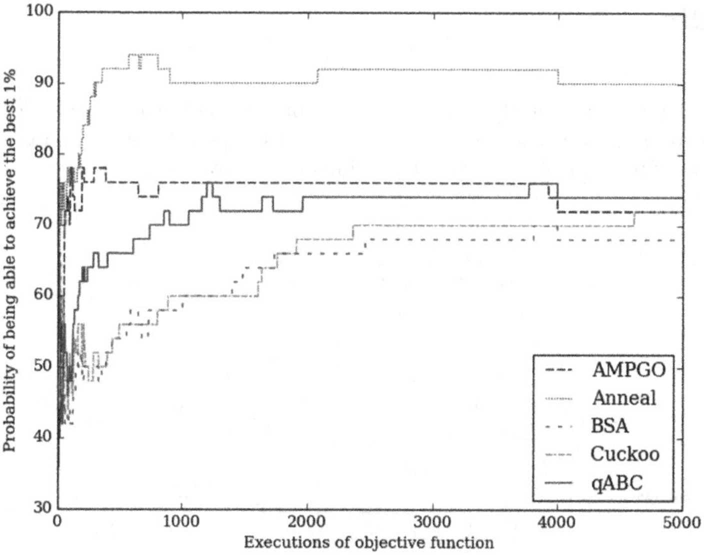

Fig. 4. The average probability of being able to achieve within 1% of the best solution across all 50 objective functions

produce the best results on real-world problems. There are many potential directions this study could take with that goal in mind.

It is necessary to continue introducing new derivative-free optimization algorithms to this test set. In order to encourage the global comparison of new optimization algorithms, the source code for all of this research is readily available in Python. Any new algorithm can be introduced to this comparison by porting existing code to a python function header with an example provided.

This study could be repeated with real-world complex optimization problems to see if the projected results hold under the noise of different objective functions. Finally, the modified Annealing algorithm used in this paper may be improved by the techniques introduced in [13].

6 Conclusion

Comparative study suggests that AMPGO and modified Simulated Annealing converge more reliably on objective functions in our set than do qABC, BSA and Cuckoo search. More optimization algorithms need to be incorporated into this form of comparison in order to expand the search for the fastest-converging general optimization algorithms for bounded derivative-free optimization on difficult-to-compute objective functions.

Acknowledgments. This research project was funded by the Roanoke College Mathematics Computer Science and Physics Department. The python code for AMPGO and the benchmarking library were a result of the freely available work done by Andrea Gavana at [9].

Appendix

Table 1. These are the mathematical formulations of the functions used to compare the optimization algorithms. For the definitions of the Needle Eye, Penalty02, Rana, and Zero Sum functions please see code provided in electronic supplementary materials.

Name	Definition
Ackley	$-20e^{-0.2\sqrt{\frac{1}{n}\sum_{i=1}^{n} x_i^2}} - e^{\frac{1}{n}\sum_{i=1}^{n}\cos(2\pi x_i)} + 20 + e$
Adjiman	$\cos(x_1)\sin(x_2) - \frac{x_1}{(x_2^2+1)}$
Alpine	$\sum_{i=1}^{n} \lvert x_i \sin(x_i) + 0.1x_i \rvert$
Beale	$(x_1x_2 - x_1 + 1.5)^2 + \left(x_1x_2^2 - x_1 + 2.25\right)^2 + \left(x_1x_2^3 - x_1 + 2.625\right)^2$
Bohachevsky	$\sum_{i=1}^{n-1}\left[x_i^2 + 2x_{i+1}^2 - 0.3\cos(3\pi x_i) - 0.4\cos(4\pi x_{i+1}) + 0.7\right]$
Cosine Mixture	$-0.1\sum_{i=1}^{n}\cos(5\pi x_i) - \sum_{i=1}^{n} x_i^2$
Deceptive	$-\left[\frac{1}{n}\sum_{i=1}^{n} g_i(x_i)\right]^{\beta}$
Deflected Corrugated Spring	$0.1\sum_{i=1}^{n}\left[(x_i-\alpha)^2 - \cos\left(K\sqrt{\sum_{i=1}^{n}(x_i-\alpha)^2}\right)\right]$
Drop Wave	$-\frac{1+\cos\left(12\sqrt{\sum_{i=1}^{n} x_i^2}\right)}{2+0.5\sum_{i=1}^{n} x_i^2}$
Easom	$a - \frac{a}{e^{b\sqrt{\frac{c}{n}}}} + e - e^{\frac{d}{n}}, c = \sum_{i=1}^{n} x_i^2, d = \sum_{i=1}^{n}\cos(cx_i)$
Egg Holder	$-x_1\sin\left(\sqrt{\lvert x_1 - x_2 - 47\rvert}\right) - (x_2+47)\sin\left(\sqrt{\left\lvert \frac{1}{2}x_1 + x_2 + 47\right\rvert}\right)$
Exponential	$-e^{-0.5\sum_{i=1}^{n} x_i^2}$
Giunta	$0.6 + \sum_{i=1}^{n}\left[\sin^2\left(1 - \frac{16}{15}x_i\right) - \frac{1}{50}\sin\left(4 - \frac{64}{15}x_i\right) - \sin\left(1 - \frac{16}{15}x_i\right)\right]$
Goldstein Price	$\left[1 + (x_1+x_2+1)^2(19 - 14x_1 + 3x_1^2 - 14x_2 + 6x_1x_2 + 3x_2^2)\right] +$ $\left[30 + (2x_1 - 3x_2)^2(18 - 32x_1 + 12x_1^2 + 48x_2 - 36x_1x_2 + 27x_2^2)\right]$
Griewank	$\frac{1}{4000}\sum_{i=1}^{n} x_i^2 - \prod_{i=1}^{n}\cos\left(\frac{x_i}{\sqrt{i}}\right) + 1$
Katsuura	$\prod_{i=0}^{n-1}\left[1 + (i+1)\sum_{k=1}^{d}\lfloor(2^k x_i)\rfloor 2^{-k}\right]$
Langermann	$-\sum_{i=1}^{5}\frac{c_i\cos\left\{\pi\left[(x_1-a_i)^2+(x_2-b_i)^2\right]\right\}}{e^{\frac{(x_1-a_i)^2+(x_2-b_i)^2}{\pi}}}$
Levy	$\sin^2(\pi y_1) + \sum_{i=1}^{n-1}(y_i-1)^2[1 + 10\sin^2(\pi y_{i+1})] + (y_n-1)^2$
Michalewicz	$-\sum_{i=1}^{2}\sin(x_i)\sin^{2m}\left(\frac{ix_i^2}{\pi}\right)$
Miele Cantrell	$(e^{-x_1} - x_2)^4 + 100(x_2 - x_3)^6 + \tan^4(x_3 - x_4) + x_1^8$
Mishra01	$(1+x_n)^{x_n} \quad ; \quad x_n = n - \sum_{i=1}^{n-1}\frac{(x_i+x_{i+1})}{2}$
Mishra02	$\left[\frac{1}{n}\sum_{i=1}^{n}\lvert x_i\rvert - \left(\prod_{i=1}^{n}\lvert x_i\rvert\right)^{\frac{1}{n}}\right]^2$
Multi Modal	$-20e^{-0.2\sqrt{\frac{1}{n}\sum_{i=1}^{n} x_i^2}} - e^{\frac{1}{n}\sum_{i=1}^{n}\cos(2\pi x_i)} + 20 + e$
Odd Square	$-e^{-\frac{d}{2\pi}}\cos(\pi d)\left(1 + \frac{0.02h}{d+0.01}\right)$
Pathological	$\sum_{i=1}^{n-1}\frac{\sin^2\left(\sqrt{100x_{i+1}^2+x_i^2}\right)-0.5}{0.001\left(x_i-x_{i+1}\right)^4+0.50}$
Penalty01	$\frac{\pi}{30}\left\{10\sin^2(\pi y_1) + \sum_{i=1}^{n-1}(y_i-1)^2\left[1 + 10\sin^2(\pi y_{i+1})\right] + (y_n-1)^2\right\} +$ $\sum_{i=1}^{n} u(x_i, 10, 100, 4)$
Qing	$\sum_{i=1}^{n}(x_i^2 - i)^2$
Quintic	$\sum_{i=1}^{n}\left\lvert x_i^5 - 3x_i^4 + 4x_i^3 + 2x_i^2 - 10x_i - 4\right\rvert$
Rastrigin	$10n + \sum_{i=1}^{n}\left[x_i^2 - 10\cos(2\pi x_i)\right]$
Ripple	$\sum_{i=1}^{2} -e^{-2\log 2(\frac{x_i-0.1}{0.8})^2}\left[\sin^6(5\pi x_i) + 0.1\cos^2(500\pi x_i)\right]$
Rosenbrock	$\sum_{i=1}^{n-1}[100(x_i^2 - x_{i+1})^2 + (x_i-1)^2]$
Paviani	$\sum_{i=1}^{10}\left[\log^2(10 - x_i) + \log^2(x_i - 2)\right] - \left(\prod_{i=1}^{10} x_i^{10}\right)^{0.2}$
Plateau	$30 + \sum_{i=1}^{n}\lfloor x_i\rfloor$
Salomon	$1 - \cos\left(2\pi\sqrt{\sum_{i=1}^{n} x_i^2}\right) + 0.1\sqrt{\sum_{i=1}^{n} x_i^2}$

Table 1. (*continued*)

Name	Definition
Sargan	$\sum_{i=1}^{n} n \left(x_i^2 + 0.4 \sum_{i \neq j}^{n} x_i x_j \right)$
Schwefel01	$418.9829n - \sum_{i=1}^{n} x_i \sin(\sqrt{\lvert x_i \rvert})$
Schwefel02	$\sum_{i=1}^{n} \lvert x_i \rvert + \prod_{i=1}^{n} \lvert x_i \rvert$
Shubert	$\left(\sum_{i=1}^{5} i \cos[(i+1)x_1 + i] \right) \left(\sum_{i=1}^{5} i \cos[(i+1)x_2 + i] \right)$
Sine Envelope	$- \sum_{i=1}^{n-1} \left[\frac{\sin^2(\sqrt{x_{i+1}^2 + x_i^2} - 0.5)}{(0.001(x_{i+1}^2 + x_i^2) + 1)^2} + 0.5 \right]$
Six Hump Camel	$4x_1^2 + x_1 x_2 - 4x_2^2 - 2.1x_1^4 + 4x_2^4 + \frac{1}{3} x_1^6$
Trigonometric	$1 + \sum_{i=1}^{n} 8 \sin^2 \left[7(x_i - 0.9)^2 \right] + 6 \sin^2 \left[14(x_i - 0.9)^2 \right] + (x_i - 0.9)^2$
Ursem Waves	$-0.9x_1^2 + (x_2^2 - 4.5x_2^2)x_1 x_2 + 4.7 \cos \left[2x_1 - x_2^2(2 + x_1) \right] \sin(2.5\pi x_1)$
Vincent	$- \sum_{i=1}^{n} \sin(10 \log(x))$
Wavy	$1 - \frac{1}{n} \sum_{i=1}^{n} \cos(k x_i) e^{-\frac{x_i^2}{2}}$
Weierstrass	$\sum_{i=1}^{n} \left[\sum_{k=0}^{kmax} a^k \cos \left(2\pi b^k (x_i + 0.5) \right) - n \sum_{k=0}^{kmax} a^k \cos(\pi b^k) \right]$
Whitley	$\sum_{i=1}^{n} \sum_{j=1}^{n} \left[\frac{(100(x_i^2 - x_j)^2 + (1 - x_j)^2)^2}{4000} - \cos(100(x_i^2 - x_j)^2 + (1 - x_j)^2) + 1 \right]$

Table 2. Objective functions used for evaluating the five optimization algorithms.

Name	Dimensions	Bounds	Name	Dimensions	Bounds
Adjiman	2	$[-1, 2]_x$ $[-1, 1]_y$	Beale	2	$[-4.5, 4.5]$
Egg Holder	2	$[-512, 512]$	Goldstein Price	2	$[-2, 2]$
Langermann	2	$[0, 10]$	Shubert	2	$[-10, 10]$
Six Hump Camel	2	$[-5, 5]$	Ursem Waves	2	$[-0.9, 1.2]_x$ $[-1.2, 1.2]_y$
Drop Wave	4	$[-5.12, 5.12]$	Whitley	4	$[-10.24, 10.24]$
Miele Cantrell	4	$[-1, 1]$	Weierstrass	4	$[-0.5, 0.5]$
Rastrigin	4	$[-5.12, 5.12]$	Katsuura	4	$[0, 10]$
Salomon	4	$[-100, 100]$	Deceptive	8	$[0, 1]$
Giunta	8	$[-1, 1]$	Griewank	8	$[-600, 600]$
Trigonometric	8	$[-500, 500]$	Paviani	8	$[2.001, 9.999]$
Sargan	8	$[-100, 100]$	Zero Sum	8	$[-10, 10]$
Plateau	16	$[-5.12, 5.12]$	Michalewicz	16	$[0, \pi]$
Mishra11	16	$[-10, 10]$	Odd Square	16	[−5*pi, 5*pi]
Qing	16	$[-500, 500]$	Rosenbrock	16	$[-5, 10]$
Alpine01	16	$[-10, 10]$	Bohachevsky	24	$[-15, 15]$
Easom	24	$[-100, 100]$	Levy03	24	$[-10, 10]$
Multi Modal	24	$[-32, 32]$	Penalty02	24	$[-50, 50]$
Quintic	24	$[-10, 10]$	Vincent	24	$[0.25, 10]$
Ackley	48	$[-32, 32]$	Cosine Mixture	48	$[-1, 1]$
Wavy	48	[−pi, pi]	Pathological	48	$[-100, 100]$
Schwefel22	48	$[-100, 100]$	Deflected Corrugated Spring	96	[0, 2*alpha]
Mishra02	96	$[0, 1 + 1e - 9]$	Penalty01	96	$[-50, 50]$
Exponential	96	$[-1, 1]$	Ripple01	96	$[0, 1]$
Schwefel	96	$[-512, 512]$	Sine Envelope	96	$[-100, 100]$

References

1. Kirkpatrick, S.: Optimization by simulated annealing: quantitative studies. J. Stat. Phys. **34**(5–6), 975–986 (1984)
2. Lasdon, L., Duarte, A., Glover, F., Laguna, M., Martí, R.: Adaptive memory programming for constrained global optimization. Comput. Oper. Res. **37**(8), 1500–1509 (2010)
3. Yang, X.S., Deb, S.: Cuckoo search via lévy flights. In: World Congress on Nature and Biologically Inspired Computing, NaBIC 2009, pp. 210–214. IEEE (2009)
4. Civicioglu, P.: Backtracking search optimization algorithm for numerical optimization problems. Appl. Math. Comput. **219**(15), 8121–8144 (2013)
5. Karaboga, D., Gorkemli, B.: A quick artificial bee colony (qABC) algorithm and its performance on optimization problems. Appl. Soft Comput. **23**, 227–238 (2014)
6. Civicioglu, P., Besdok, E.: A conceptual comparison of the cuckoo-search, particle swarm optimization, differential evolution and artificial bee colony algorithms. Artif. Intell. Rev. **39**(4), 315–346 (2013)
7. Moré, J.J., Wild, S.M.: Benchmarking derivative-free optimization algorithms. SIAM J. Optim. **20**(1), 172–191 (2009)
8. Floudas, C.A., Pardalos, P.M.: Encyclopedia of Optimization. Springer Science and Business Media, Heidelberg (2009)
9. Gavana, A.: Global optimization benchmarks and AMPGO. Accessed Apr 2016 (2014)
10. Karaboga, D., Basturk, B.: A powerful and efficient algorithm for numerical function optimization: artificial bee colony (ABC) algorithm. J. Glob. Optim. **39**(3), 459–471 (2007)
11. Rios, L.M., Sahinidis, N.V.: Derivative-free optimization: a review of algorithms and comparison of software implementations. J. Glob. Optim. **56**(3), 1247–1293 (2013)
12. Dolan, E.D., Moré, J.J.: Benchmarking optimization software with performance profiles. Math. Program. **91**(2), 201–213 (2002)
13. Alizamir, S., Pardalos, P.M., Rebennack, S.: Improving the neighborhood selection strategy in simulated annealing using the optimal stopping problem. INTECH Open Access Publisher (2008)

The Learnability of Business Rules

Olivier Wang[1,2], Changhai Ke[1], Leo Liberti[2(✉)], and Christian de Sainte Marie[1]

[1] IBM France, 9 Rue de Verdun, 94250 Gentilly, France
[2] CNRS LIX, Ecole Polytechnique, 91128 Palaiseau, France
olivier.wang@polytechnique.edu, liberti@lix.polytechnique.fr

Abstract. Among programming languages, a popular one in corporate environments is Business Rules. These are conditional statements which can be seen as a sort of "programming for non-programmers", since they remove loops and function calls, which are typically the most difficult programming constructs to master by laypeople. A Business Rules program consists of a sequence of "IF *condition* THEN *actions*" statements. Conditions are verified over a set of variables, and actions assign new values to the variables. Medium-sized to large corporations often enforce, document and define their business processes by means of Business Rules programs. Such programs are executed in a special purpose virtual machine which verifies conditions and executes actions in an implicit loop. A problem of extreme interest in business environments is enforcing high-level strategic decisions by configuring the parameters of Business Rules programs so that they behave in a certain prescribed way on average. In this paper we show that Business Rules are Turing-complete. As a consequence, we argue that there can exist no algorithm for configuring the average behavior of all possible Business Rules programs.

1 Introduction

Business Rules (BR) are used in corporate environments to define business processes. Since not every employee is a computer programmer, the BR language is conceived as a "programming language for non-programmers". Typically, laypeople understand conditions and assignments much better than function calls and loops. Therefore, BR programs consist of sequences of conditional statements of the form

```
IF
        condition
THEN
        actions
```

where the *condition* is enforced on a vector of BR program variables $x = (x_1, \ldots, x_n)$, and the *actions* are assignments of the form

$$x \leftarrow f(x),$$

for some function f specified with the usual arithmetic operators and transcendental functions found in any standard mathematical library. Currently, there

P.M. Pardalos et al. (Eds.): MOD 2016, LNCS 10122, pp. 257–268, 2016.
DOI: 10.1007/978-3-319-51469-7_22

exist software packages called business rules management systems for creating, verifying, storing, retrieving, managing and executing BR programs, such as IBM's Operational Decision Management (ODM) [7], previously ILOG's JRules.

A typical example is provided by a bank which defines the process of deciding whether to grant a loan to a customer. The conditions will verify anagraphical, work-related and credit ranking type information about the customer: since there may be dependency relationship in such information, it makes sense to break down this verification as a set of BRs, some of which might depend on the preceding ones. The action triggered by a BR might, through the assignment of a binary value to a Boolean variable or a scoring value to a scalar variable, switch on or off the activation of subsequent BRs. Finally, the output of this BR program will be an assignment of a binary value to a Boolean variable linked to whether the loan will be granted or not. The BRs will be initially written by employees on the basis of policies, regulations and personal work experience, and then modified to take the evolution of the business process into account.

Banks, however, come under political scrutiny. They might be required to prove to the legislators that they grant a certain percentage of all demanded loans. Accordingly, banks ask their technical staff that their business processes, encoded as BR programs, be configured so as to behave as required on average. The average mentioned here refers to all possible requests to come; banks would probably interpret it over "average over all requests received so far", and their technical staff would likely pick an appropriate sample and run a supervised Machine Learning (ML) algorithm to learn the BR program configuration parameter values on this sample used as a training set.

This begs the question: does there exist an algorithm for the task? More precisely, is there an algorithm such that, given a configurable BR program and a training set as input, would provide parameter values as output that conform to an accepted learning paradigm? This sounds like a philosophical question, but by fixing the learning paradigm to some concrete example, such as *Probably Approximately Correct* (PAC) learning [14], then the question can be answered.

The objective of this paper is precisely to answer this question. Unfortunately, it turns out that the answer is negative. We prove that the BR language is universal, meaning that interpreting BR programs in their natural computational environment is Turing-complete [10,15]. This means that any program can be written in the BR language, including, e.g. the implementation of certain types of pseudorandom functions. Such functions are known to be *PAC unlearnable* [3,5], which proves our result. Of course, our general negative result does not prevent the existence of learning algorithms which work on restricted classes of instances. Although such algorithms will not be discussed in this paper, some can be devised by means of mathematical programming [16].

2 How Business Rules Work

As already mentioned, BR consist of IF ... THEN constructs. For simplicity, they do away with function calls and loops. On the other hand, this removal is only apparent, as BR replace function calls with *meta-variables* and loops with their execution environment.

Meta-variables and variables are stored in different symbol tables. Whereas the variables' symbol table simply matches variable symbols to variables, the meta-variables' symbol table matches meta-variable symbols to variable symbols. Every time a condition or an action in a BR refers to a meta-variable, the BR is instantiated for every variable symbol corresponding to the meta-variable, creating a matching number of IF ... THEN code fragments. This is akin to calling a function with a given parameter.

BR programs consist of a sequence of BRs. The execution environment is set up as an implicit loop which keeps scanning and executing rules until a termination condition is achieved, at which point the loop ends and the return variable – a Boolean in our example – is passed back to the calling process. Such a condition is usually that every rule condition is evaluated to FALSE.

2.1 Definitions and Notation

A BR program consists in a set of type declarations and a set of rules. A type declaration consists of either the creation of a type (*create_new_type*(type)) or the assignment of a type to a variable (*type*(var) ← type), where var can be either a variable or a meta-variable. A rule is defined as follows. Given α the typed meta-variables and x the typed variables, a rule is written:

if $T(\alpha, x)$ **then**
  $\alpha \leftarrow A(\alpha, x)$
  $x \leftarrow B(\alpha, x)$
end if

where T is the condition and the couple (A, B) describes the action. The action can be modifying the value of some fixed variable (function B) or the value of the variable matched to a meta-variable (function A), or both. In the first case, that action will be the same across all rule instances made from this rule; while in the second case the action will be different depending on the instance. For example, with variables x_1 and x_2 and meta-variable α_1, all of type int we might have the following rule:

if $\alpha_1 = 13$ **then**
  $\alpha_1 \leftarrow \alpha_1 + 3$
end if

in which case we have an action that has only A: $A(\alpha, x) = \alpha + 3$; $B(\alpha, x) = x$. The rule instance matching α_1 to x_1 would modify x_1, and the rule instance matching α_1 to x_2 would modify x_2. If the rule was instead:

if $\alpha_1 = 13$ **then**
  $x_1 \leftarrow x_1 + 2$
  $x_2 \leftarrow 30$
end if

we would have an action with only B: $A(\alpha, x) = \alpha$; $B(\alpha, x) = (x_1 + 2, 30)$. The rule instance matching α_1 to x_1 and the rule instance matching α_1 to x_2 would execute the same action under different conditions. At least one of the variables, say x_1 without loss of generality, is selected to be the *output* of the BR program.

The first part of a BR program interpreter is to compile the rule instances derived from each rule. At compile time, α is replaced by every type-feasible reordering of the x input variable vector. For $x \in X \subseteq \mathbb{R}^n$, the explicit set of rule instances compiled from this rule is the type-feasible part of the following code fragments, using $(\sigma_j \mid j \in \{1, \ldots, n!\})$ the permutations of $\{1, \ldots, n\}$:

if $T((x_{\sigma_1(1)}, \ldots, x_{\sigma_1(n)}), (x_1, \ldots, x_n))$ **then**
$(x_{\sigma_1(1)}, \ldots, x_{\sigma_1(n)}) \leftarrow A((x_{\sigma_1(1)}, \ldots, x_{\sigma_1(n)}), (x_1, \ldots, x_n))$
$(x_1, \ldots, x_n) \leftarrow B((x_{\sigma_1(1)}, \ldots, x_{\sigma_1(n)}), (x_1, \ldots, x_n))$
end if

...

if $T((x_{\sigma_n(1)}, \ldots, x_{\sigma_n(n)}), (x_1, \ldots, x_n))$ **then**
$(x_{\sigma_n(1)}, \ldots, x_{\sigma_n(n)}) \leftarrow A((x_{\sigma_n(1)}, \ldots, x_{\sigma_n(n)}), (x_1, \ldots, x_n))$
$(x_1, \ldots, x_n) \leftarrow B((x_{\sigma_n(1)}, \ldots, x_{\sigma_n(n)}), (x_1, \ldots, x_n))$
end if

The size of this set varies. The typing of the variables and meta-variables matters, and depending on T, A and B some of these operations might also be computationally equivalent. The number of rule instances compiled from a given rule can be 0 for an invalid rule ($T(\alpha, x) = \mathsf{False}$), 1 for a static rule ($T(\alpha, x)$ and $B(\alpha, x)$ do not vary with α, and $A(\alpha, x) = \alpha$), and up to $n!$ for some rules if every variable has the same typing.

2.2 The BR Interpreter

The "implicit loop" referred to above can be considered as a *BR interpreter*: it takes a BR program and turns it into a computational process which provides an output corresponding to a given input.

Any common interpreter for a BR program is algorithmically equivalent to creating all rule instances as a first step, and as a second step continuously verifying the given conditions, executing one assignment action corresponding to a condition which is True at each iteration of the loop, until either every condition evaluates to False or a user-defined termination condition becomes active. Although there exist many such interpreters, such as [4], for the purpose of this paper we only consider the most basic interpreter $\mathscr{I}_0$ which follows the simple algorithm below.

1. Match variables to type-appropriate meta-variables in rules to create all possible rule instances
2. Select the rule instances for which the condition is true, using the current values of the variables
3. Execute the action of the first rule instance in the current selection or stop if there is no such rule instance
4. Restart from step 2.

The Rules (in order)
R_1 :
if $(\alpha_1 \geq 1 \wedge \alpha_2 = 2)$
then $(\alpha_1 \leftarrow 0, \alpha_2 \leftarrow 0)$
R_2 :
if $(\alpha_3 = 1)$
then $(\alpha_3 \leftarrow \alpha_3 + 1)$

The typed Variables (in order)
int x $\leftarrow$ 1
int age $\leftarrow$ 90

The typed Meta-variables
int α_1
int α_2
int α_3

The Rule Instances
In the total order considered by the algorithm, they are:
r_1 :
if (x $\geq$ 1$\wedge$age = 2)
then (x $\leftarrow$ 0,age $\leftarrow$ 0)
r_1' :
if (age $\geq$ 1$\wedge$x = 2)
then (age $\leftarrow$ 0,x $\leftarrow$ 0)
r_2 :
if (x = 1)
then (x $\leftarrow$x+1)
r_2' :
if (age = 1)
then (age $\leftarrow$age+1)

The Execution
Iteration 1:
Truth value of conditions:
$t(r_1) = \mathsf{False}$
$t(r_1') = \mathsf{False}$
$t(r_2) = \mathsf{True}$
$t(r_2') = \mathsf{False}$
Rule instances selected (in order): r_2
Rule executed: r_2
Variable values:
x = 2
age = 90

Iteration 2:
$t(r_1) = \mathsf{False}$
$t(r_1') = \mathsf{True}$
$t(r_2) = \mathsf{False}$
$t(r_2') = \mathsf{False}$
Rules selected: r_1'
Rule executed: r_1'
Variable values:
x = 0
age = 0

Iteration 3:
$t(r_1) = \mathsf{False}$
$t(r_1') = \mathsf{False}$
$t(r_2) = \mathsf{False}$
$t(r_2') = \mathsf{False}$
Rules selected: None
Rule executed: None

END

Fig. 1. Example illustrating the execution algorithm

The order of rule instances used in Step 3 is defined by the order induced by a predefined order on the rules in the BR program and on the input variables.

We expect any more complicated ones to be able to simulate the most basic, so that BR executed using any interpreter are at least as powerful as what our study concludes. We also expect that $\mathscr{I}_0$ is able to simulate other interpreters, by using additional variables. This expectation comes from the idea that the difference with our interpreter will come from Step 3. The conflict resolution strategy, i.e. the strategy for selecting which rule among the ones obtained in Step 2 is to be executed, is what characterizes BR interpreters. Our algorithm has a simple conflict resolution strategy: whenever more than one rule instance could be executed, the choice is obtained from a given (fixed) total order on rule instances. An example of the execution of such an algorithm is described in Fig. 1.

The most common conflict resolution strategies combine one or more of the following three elements [4]:

- *Refraction* which prevents a rule instance from firing (being selected by the conflict resolution algorithm) again unless its condition clause has been reset.
- *Priority* which is a kind of partial order on rules, leading of course to a partial order on rule instances.

- *Recency* which orders rule instances in decreasing order of continued validity duration (when rule instances are created at run time, it is often expressed as increasing order of rule instance creation time).

These elements can be simulated by the basic interpreter by adding more types, variables or rules, in broad strokes:

- *Refraction* results in the use of an additional boolean variable $needsReset$ per rule per variable permutation (i.e. per rule instance), an additional test clause in each rule and an additional rule (**if** $needsReset_{x,r} = true \wedge T_r(\alpha, x) = false$ **then** $needsReset_{x,r} = false$) per rule r.
- *Priority* results in one additional integer variable p, an additional test clause in each rule ($p = \pi_r$), an additional action clause in each rule ($p \leftarrow p_{\max}$), and two additional rules that come dead last in the predefined order on rules (**if** $p > 0$ **then** $p = p - 1$; **if** $p = 0$ **then** Stop).
- *Recency* is the most complicated. A possible simulation could involve an additional integer variable $validityStart$ per rule instance, an additional integer variable $timer$, an additional action clause in each rule ($timer \leftarrow timer + 1$) and a similar setup to the one suggested for *priority*, using an integer variable $priorValidity$ which would this time start at 0 and end at $timer$.

3 Turing-Completeness of Business Rules

Since a BR program is executed in a loop construct as discussed in Sect. 2.2, which only terminates when all conditions evaluate to False, a fundamental question is whether we can decide if the execution terminates at all. Asking the same question of any programming language amounts to asking whether the HALTING PROBLEM (HALT) can be solved on the class of Turing Machines (TM) that the programming language is able to describe. Since it is well known that HALT cannot be solved for Universal TMs (UTMs), the question is whether BRs can describe a UTM. In this section prove that this is indeed the case. In other words, we prove that BRs are Turing-complete.

For our original question, i.e. whether there is an algorithm $\mathcal{A}$ for arbitrarily changing a BR program so it statistically behaves according to a given target, Turing-completeness shows that we cannot hope to ever find an algorithm $\mathcal{A}$ which works on *all* BR programs. At the very least, we shall have to limit our attention to all *terminating* BR programs.

Proofs of Turing-completeness are sometimes obtained by using μ-recursive functions [9] such as in [8]. Our proof of Turing-completeness for BR programs consists in spelling out a BR program $\mathcal{U}$ which takes as input any TM description with its own input. When executed $\mathcal{U}$, simulates the TM acting on its input.

A UTM is a TM which can simulate any other TM on arbitrary input [12,13]. It does that by taking as input a description T of any TM as well as its input x. We use the usual definition of a Turing Machine [6]. We note the states of a TM $q_1, \dots, q_Q$, its tape symbols $s_1, \dots, s_S$, its blank symbol s_b, its transition function $(q^i, s^i) \rightarrow (q^f, s^f, \mathsf{act})$ where $\mathsf{act} \in \{$"left", "right", "stay"$\}$, its initial state q^0, and its accepting states $T_e r$. An initial tape T_0 is said to be accepted by a TM if the TM reading this tape stops and has a final state in $T_e r$.

A BR program which simulates a UTM by being able to simulate any TM is described in Fig. 2. It uses the same notations, with initial values of $q = q^0$; of T a truncated T^0 containing a finite number of symbols, containing T_0^0 and containing all non-blank symbols of T^0; of $l = \text{size}(T)$; and of $p = 0$.

We suppose the variables include the following:
- many (static) state objects of type "state": $q_1, \ldots, q_Q$
- many (static) symbol objects of type "symbol": $s_1, \ldots, s_S$
- a (static) finite set of terminal states of type "terminal": T_{er}
- a (static) blank symbol of type "symbol": s_b
- a (static) set of Turing rules of type "rules", of the form $(\text{state}_{\text{initial}}, \text{symbol}_{\text{initial}}, \text{right}|\text{left}|\text{stay}, \text{state}_{\text{next}}, \text{symbol}_{\text{written}})$:
  $\mathcal{R} = \{(q_r^i, s_r^i, \mathsf{act}_r, q_r^f, s_r^f) \mid \mathsf{act}_r \in \{\text{"left"}, \text{"right"}, \text{"stay"}\}\}_r$
- the current state of type "state": q
- the length of the visible tape data, of type "length": l
- the current visible tape data of type "tape": $T = \{(i, s_i) \mid i \in \mathbb{N}, 0 \leq i \leq l-1\}$ where l is the length of the visible tape data
- the current place on the tape of type "position": p

We use the following meta-variables in the BR program that simulates a UTM:
- α_{qf} of type "state"
- α_{sf} of type "symbol"

The rule set to simulate a UTM is then written in a compact form:

R_1:
if
$(q, T(p), \mathsf{act}, \alpha_{qf}, \alpha_{sf}) \in \mathcal{R}$
then
$q \leftarrow \alpha_{qf}$
$T \leftarrow (T \setminus \{(p, T(p))\}) \cup \{(p, \alpha_{sf})\}$
$p \leftarrow p \pm 1$(Depending on the value of act)
$l \leftarrow l \pm 1$(Depending on the respective values of act, p and l)

R_2:
if
$(q \in T_{er})$
then
Stop;

Fig. 2. A BR program which describes a UTM.

Some simplifications have been made for the sake of clarity: R_1 should clearly be at least three different rules each replacing act with one of "left", "right", "stay". The complete formally correct form would in fact have two more rules, in order to increase the length of the tape as needed, using the variable s_b.

Theorem 1. *The BR program described in Fig. 2 simulates any TM given an accepted tape. The final value of the tape and the final symbol of the TM will be identical to the final values of T and q.*

Proof. We prove that the simulation is correct by induction over the number of steps n taken by the TM.

Before the TM takes any step ($n = 0$), its tape is identical to the value of T in the BR program before it executes any rule, as that tape is given to the BR as an input value. Similarly, the place on the tape at that point is the value of p and the state of the TM is equal to the value of q.

Assume that the tape, the position on it, and the state of the TM after step n are accurately represented by the BR program after n rule executions. We call T^{TM} the sequence representing the tape, p^{TM} the current place on the tape, and q^{TM} the current state of the TM.

If the TM halts, that means $\forall(\mathsf{act}, q^f, s^f), (q^{TM}, T^{TM}_{p^{TM}}, \mathsf{act}, q^f, s^f) \notin \mathcal{R}$. Thus, the BR does not execute R_1. Further, as the initial tape is accepted by the TM, we have $q^{TM} \in T_e r$, which fulfills the condition for R_2. The BR program terminates at the same time as the TM, and its output is correct as R_2 does not modify values.

Otherwise, the TM will follow a rule in $\mathcal{R}$. Let us call it

$$r = (q^{TM}, T^{TM}_{p^{TM}}, \mathsf{act}, q^f, s^f).$$

In this case, the next BR executed will be R_1, and the only member of $\mathcal{R}$ to match will be r. In other words, the only relevant rule instance is:

R_1:

if $(q, T(p), \mathsf{act}, q^f, s^f) \in \mathcal{R}$

then $q \leftarrow q^f$

$T \leftarrow (T \setminus \{(p, T(p))\}) \cup \{(p, s^f)\}$

$p \leftarrow p \pm 1$(Depending on the value of act)

$l \leftarrow l \pm 1$(Depending on the respective values of act, p and l)

because $q = q^{TM}$ and $T(p) = T^{TM}_{p^{TM}}$. As the action on this BR instance corresponds exactly to the modifications to the state and tape of the TM, the values stored on the tape of the TM after $n + 1$ steps will again be the same as the values in the BR program. □

4 Unlearnability

Let us go back once more to the original question: is there an algorithm for arbitrarily changing a BR program so it statistically behaves according to a given target? By Sect. 3, we know that this algorithm cannot exist in the most general terms, since BR programs might not even terminate in finite time. But what if we just look at those BR programs which *do* terminate? The question can be considered as a learnability problem: Does there exist an algorithm $\mathcal{A}$ which taking a class $\mathscr{P}_p$ of terminating BR programs parametrized over p, a data distribution $\mathcal{D}$ over its input domain X and a goal g for the value of the average output $\mathbb{E}_{\mathcal{D}}(\mathscr{P}_p)$, efficiently and weakly learns p? In other words, is p learnable in this context?

Other Turing-complete languages have looked at the learning problem. Algorithms for learning some restricted classes of programs exist for inductive logic programming [2] or nonmonotonic inductive logic programming [11]. However, we are asking about the general case of learning in all programs.

We limit ourselves to the well-studied and powerful family of algorithms known as Probably Approximately Correct (PAC) learning algorithms introduced in [14]. In this section we use a class of pseudorandom functions to provide a negative answer to the question, when posed in these very general terms. Of course, it may still be possible to achieve our stated purpose for less general, but still useful classes of BR programs.

4.1 Background

Pseudorandom functions (PRF), introduced by Goldreich et al. [5], are indexed families of functions F_p for which there exists a polynomial-time algorithm to evaluate $F_p(x)$, but no probabilistic polynomial-time algorithm can distinguish the function from a truly random function F_{rand} without knowing p, even if allowed access to an oracle.

A PAC learning algorithm identifies a concept (i.e. a function $X \to \{0, 1\}$) among a concept class $\mathscr{C}$ (i.e. a family of concepts). For a concept $f \in \mathscr{C}$ and a list S of data points in X of length λ, an algorithm $\mathcal{A}$ is an (ϵ, δ)-PAC learning algorithm for $\mathscr{C}$ if for all sufficiently large λ:

$$\mathbb{P}[\mathcal{A}(f) = h \mid h \text{ is an } \epsilon\text{-approximation to } f] \geq 1 - \delta$$

where $\mathcal{A}$ has access to an oracle for f.

- $\mathcal{A}$ is said to be efficient if the time complexity of $\mathcal{A}$ and h are polynomial in $1/\epsilon$; $1/\delta$; and λ.
- $\mathcal{A}$ is said to weakly learn $\mathscr{C}$ if there exist some polynomials $p_\epsilon(\lambda)$; $p_\delta(\lambda)$ for which $\epsilon \leq \frac{1}{2} - \frac{1}{p_\epsilon(\lambda)}$ and $\delta \leq 1 - \frac{1}{p_\delta(\lambda)}$.
- We say a concept class is PAC learnable if it is both efficiently and weakly learnable. Otherwise, it is unlearnable.

It is known that PRF are unlearnable using PAC algorithms [3,5]. In the rest of this section, we consider F_p such a PRF, and note $\text{Eval}_{p,x}(F_p(x))$ the complexity of evaluating $F_p(x)$.

4.2 Unlearnability Result

We call $(\mathscr{P}_p)_{p \in \pi}$ a class of terminating BR programs indexed by p, S a list of items from the input domain X with $|S| = \lambda$, and g a goal for the value of the average output $\mathbb{E}_S(\mathscr{P}_p)$. We consider C the concept class whose members are $f : (\mathscr{P}_p)_{p \in \pi} \to \{0, 1\}$.

Theorem 2. *The concept class C is unlearnable: specifically, the concept $h \in C$ defined as $h(p) = 1$ iff $\mathbb{E}_S(\mathscr{P}_p) = g$ cannot be learned using a PAC learning algorithm in the general case.*

In other words, there is no practically viable algorithm that can learn a BR program out of a class of BR programs in the general case, even with access to a perfect oracle. This is a consequence of both the Turing-completeness of BR programs and the unlearnability of PRF.

Proof. As BR programs are Turing-complete, we choose the family $(\mathscr{P}_p)_{p\in\pi}$ to be a PRF. Any algorithm that learns C also learns $(\mathbf{1}_f(p))_p \subset C$, where $\mathbf{1}_f(p) = 1$ iff $\mathscr{P}_p = f$. Learning the latter is trivially the same as learning a PRF, which is proven to be impossible. □

The specific example mentioned in the theorem answers our original question. Even if the concept we wish to learn is described more broadly than by providing an oracle for a specific BR program, it is impossible to adjust the statistical behavior of BR programs according to a predefined goal.

4.3 Complete Unlearnability

We have used the fact that PRFs are not PAC learnable in the sense that no PAC algorithm can efficiently and weakly learn a PRF. We now demonstrate an example of a concept class that cannot be learned by PAC algorithms at all. This example is based on the intuition that chaos cannot be predicted, and so cannot be learned.

We use a known chaotic map, the logistic map $f_{n+1}(x) = af_n(x)(1 - f_n(x))$, $f_0(x) = x$, with the parameter $a = 4$. Some of its properties are presented by Berliner [1]. We call $C_n(x)$ the concept class such that $C_n(x) = 1$ iff $f_n(x) \geq 0.5$ and $C_x(n) = 0$ otherwise, where $x \in [0, 1]$ follows the arcsine distribution, i.e. the probability density function is $p(x) = \frac{1}{\pi\sqrt{x(1-x)}}$, and with $n \in \mathbb{N}$ following the uniform distribution.

Theorem 3. *The concept class $C_n(x)$ cannot be learned with any accuracy. To be precise, for all algorithms $\mathcal{A}$ calling the oracle $C_n(x)$ a finite number of times, we have:*

$$\mathbb{P}_{n\in\mathbb{N}}(\mathbb{P}_{x\in X}(\mathcal{A}(C_n)(x) \neq C_n(x)) = 0.5) = 1$$

Proof. The proof relies heavily on Berliner's paper [1]. From it, we know that as the logistic map is chaotic, each sequence $(f_n(x))_n$ is either eventually periodic or is dense in $[0, 1]$. We also know that as X follows the arcsine distribution, the $C_n(X)$ are i.i.d. Bernoulli random variables, such that $\mathbb{P}_{x\in X}(C_n(x) = 1) = 0.5$.

Suppose $\mathcal{A}$ calls $C_n(x)$ for values of $x \in \{x_1, \ldots, x_k\}$. We call n^0 the value such that $\mathcal{A}(C_n) = C_{n^0}$. As $\mathbb{P}_{x\in X}(C_{n^0}(x) = 1) = 0.5$ does not depend on n^0, and the $C_n(X)$ are i.i.d., we have $\mathbb{P}_{x\in X}(C_{n^0}(x) \neq C_n(x)) = 0.5$ iff $n^0 \neq n$ and $\mathbb{P}_{x\in X}(C_{n^0}(x) \neq C_n(x)) = 0$ otherwise. The theorem is thus the same as saying that $\mathcal{A}$ almost certainly (in the probabilistic sense) cannot match n^0 to the exact value of n. We now prove that there almost always exists $n^1 \neq n$ which is indistinguishable from n by $\mathcal{A}$, i.e. $C_{n^1}(x_1) = C_n(x_1), \ldots, C_{n^1}(x_k) = C_n(x_k)$.

Let us call $Y_i^1 = C_i(x_1), \ldots, Y_i^k = C_i(x_l)$ with $i \in \mathbb{N}$. Some of the sequences Y^j are periodic after some rank, and some are not. Without loss of generality, we assume $Y^1, \ldots, Y^{k_1}$ are periodic, and $Y^{k_1+1}, \ldots, Y^k$ are not. Almost certainly, $(Y^1)_{i \geq n}, \ldots, (Y^{k_1})_{i \geq n}$ are periodic (n is big enough). Using $P \in \mathbb{N}$ to denote the smaller common multiple of those sequences' periods, we notice that $C_{n+Pi}(x^1) = C_n(x^1), \ldots, C_{n+Pi}(x^{k_1}) = C_n(x^{k_1})$. We note $y_i = n + Pi$.

As the sequences $Y^{k_1+1}, \ldots, Y^k$ are not eventually periodic, we know that each sequence $(f_n(x^{k_1+1}))_n, \ldots, (f_n(x^k))_n$ is dense in $[0,1]$. Consequently, for any sequence of $k - k_1$ bits, there exists a countable number of $n^1 \in (Y_i)_{i \in \mathbb{N}}$ such that it is equal to $Y^{k_1+1}, \ldots, Y^k$. In particular, if this sequence is $C_n(x^{k_1+1}), \ldots, C_n(x^k)$, any of those n^1 different from n proves the theorem. □

It must be noted that no practical application would ever try to learn this type of concept class. A key part of the proof is allowing the concept class to be infinite and indexed by a natural number, without bounding that index at all. This is unlikely to happen for computational reasons, the usual way to represent a natural number being with integer or long typed variables. Another difficulty is representing and computing real numbers, which can be done using Real RAM machines, to compute $f_n(x^1), \ldots, f_n(x^l)$. In the case of the logistic map with parameter $a = 4$, the task is made slightly easier by the existence of an exact solution, but other chaotic maps would require expensive recursive computations.

The existence of such extreme cases of unlearnability is nevertheless something to be careful of. It must be noted that none of the aforementioned computational difficulties are impossibilities, and that such unlearnable concepts are thus possible problems for BR programs, among other Turing-Complete programming languages.

5 Conclusion

Business Rules seem simple enough, repeatedly treating data according to a simple algorithm. The complexity of BR programs actually comes from the interpreters. In particular, almost any interpreter that uses a looping algorithm can make BR programs Turing-complete, as is the case with the simplistic algorithm we have presented in this paper. The proof of such is simple, yet it is a result that has been overlooked so far (to the best of our knowledge). The Turing completeness of BR programs can have important theoretical implications: it links the usual Rules research on Inference Rules and ontologies with more traditional research on programming languages and computability.

The second part of our paper studies shows another theoretical implication of this result. As a Turing-complete language, no PAC algorithm can adjust the parameters in a BR program for a specific statistical behavior, even with a perfect oracle. This impossibility has practical implications. In particular, BR programs are often used to model business processes, which companies might want to optimize over an average output. That there exists no such algorithm in the general case means that algorithms working on a specific class of BR programs are their only way of automatically modifying a BR program with such an aim. Furthermore, we demonstrate that learning at all is not always possible, and so general learning algorithms for Turing-complete programming languages cannot exist as such. In cases that treat the full generality of the learning problem, learning heuristics are thus not only computationally efficient, but also absolutely necessary.

Acknowledgments. The first author (OW) is supported by an IBM France/ANRT CIFRE Ph.D. thesis award.

References

1. Berliner, L.M.: Statistics, probability and chaos. Stat. Sci. **7**, 69–90 (1992)
2. Blockeel, H., De Raedt, L.: Top-down induction of first-order logical decision trees. Artif. Intell. **101**(1), 285–297 (1998)
3. Cohen, A., Goldwasser, S., Vaikuntanathan, V.: Aggregate pseudorandom functions and connections to learning. In: Dodis, Y., Nielsen, J.B. (eds.) TCC 2015. LNCS, vol. 9015, pp. 61–89. Springer, Heidelberg (2015). doi:10.1007/978-3-662-46497-7_3
4. de Sainte Marie, C., Hallmark, G., Paschke, A.: RIF Production Rule Dialect, 2nd edn. W3C Recommendation (2013)
5. Goldreich, O., Goldwasser, S., Micali, S.: How to construct random functions. J. ACM **4**(33), 792–807 (1986)
6. Hopcroft, J.E., Ullman, J.D.: Introduction to Automata Theory, Languages, and Computation. Addison-Wesley, Boston (1979)
7. IBM. Operational Decision Manager 8.8 (2015)
8. Kepser, S.: A simple proof of the turing-completeness of XSLT and XQuery. In: Usdin, T. (ed.) Extreme Markup Languages 2004 (2004)
9. Kleene, S.C.: Introduction to Metamathematics. North-Holland Publishing Co., Amsterdam (1952)
10. Liberti, L., Marinelli, F.: Mathematical programming: turing completeness and applications to software analysis. J. Comb. Optim. **28**(1), 82–104 (2014)
11. De Raedt, L., Džeroski, S.: First-order jk-clausal theories are PAC-learnable. Artif. Intell. **104**(1), 375–392 (1994)
12. Shannon, C.: A Universal Turing machine with two internal states. In: Shannon, C., McCarthy, J. (eds.) Automata Studies. Annals of Mathematics Studies, vol. 34, pp. 157–165. Princeton University Press, Princeton (1956)
13. Turing, A.: On computable numbers, with an application to the Entscheidungsproblem. Proc. Lond. Math. Soc. **42**(1), 230–265 (1937)
14. Valiant, L.G.: A theory of the learnable. Commun. ACM **11**(27), 1134–1142 (1984)
15. Wang, O., Kai, C., Liberti, L., De Sainte Marie, C.: Business rule sets as programs: turing-completeness and structural operational semantics. In: Treizièmes Rencontres des Jeunes Chercheurs en Intelligence Artificielle (RJCIA 2015) (2015)
16. Wang, O., Liberti, L., D'Ambrosio, C., Sainte Marie, C., Ke, C.: Controlling the average behavior of business rules programs. In: Alferes, J.J.J., Bertossi, L., Governatori, G., Fodor, P., Roman, D. (eds.) RuleML 2016. LNCS, vol. 9718, pp. 83–96. Springer, Heidelberg (2016). doi:10.1007/978-3-319-42019-6_6

Dynamic Programming with Approximation Function for Nurse Scheduling

Peng Shi(✉) and Dario Landa-Silva

School of Computer Science, ASAP Research Group, The University of Nottingham, Nottingham, UK
{psxps4,dario.landasilva}@nottingham.ac.uk

Abstract. Although dynamic programming could ideally solve any combinatorial optimization problem, the curse of dimensionality of the search space seriously limits its application to large optimization problems. For example, only few papers in the literature have reported the application of dynamic programming to workforce scheduling problems. This paper investigates approximate dynamic programming to tackle nurse scheduling problems of size that dynamic programming cannot tackle in practice. Nurse scheduling is one of the problems within workforce scheduling that has been tackled with a considerable number of algorithms particularly meta-heuristics. Experimental results indicate that approximate dynamic programming is a suitable method to solve this problem effectively.

Keywords: Markov decision process · Approximate dynamic programming · Nurse scheduling problem

1 Introduction

The 1956 paper by Richard Bellman [1] described the *principle of optimality* and *dynamic programming*, the algorithm based on this principle. Basically, dynamic programming breaks an optimization problem into simpler sub-problems that can be solved recursively and it has been applied to a variety of optimization problems [2]. The nurse scheduling problem is a complex combinatorial optimization problem that consists in assigning shifts to nurses in each day of a given planning period (usually a number of weeks). This problem has been investigated to a considerable extent and many algorithms have been proposed to solve it [3]. Constructing high-quality schedules for their nurses are crucial for many hospitals because minimizing salary costs and maximizing the overall satisfaction of nurses with their working patterns are key aims. In the context of workforce scheduling, it appears that dynamic programming algorithms have been applied only on small problem instances [4].

The limitation of dynamic programming to tackle larger workforce scheduling problems lies on the curse of dimensionality, which means that the search space grows exponentially as the input size increases. This means that implementations

P.M. Pardalos et al. (Eds.): MOD 2016, LNCS 10122, pp. 269–280, 2016.
DOI: 10.1007/978-3-319-51469-7_23

of dynamic programming require too much memory to store the search space and too much computation time for evaluating all states of the search. To make the search more efficiently, approximate dynamic programming selects only a small part of the search space based on approximation functions [2]. The solution obtained by approximate dynamic programming is close to the optimal solution and the computational time is decreased considerably.

This paper investigates the ability of approximate dynamic programming to solve nurse scheduling problems. There seems to be no previous papers conducting such investigation. The intended contribution of this work is to propose a methodology to solve larger workforce scheduling problems to near-optimality in practical computation time using approximate dynamic programming. The structure of this paper is organized as follows. Section 2 reviews the literature on approximate dynamic programming and nurse scheduling. Section 3 describes how the nurse scheduling problem can be seen as a typical Markov decision model. Section 4 describes the proposed methodology and Sect. 5 discusses the experimental results. Conclusions are given in Sect. 6.

2 Literature Review

This section consists of two parts. Some background on the nurse scheduling problem is given firstly and then followed by an overview of the approximate dynamic programming technique.

2.1 Nurse Scheduling Problem

The Nurse Scheduling Problem (NSP) can be grouped into two categories: cyclic scheduling problem and non-cyclic scheduling problem. In cyclic scheduling, nurse shift preferences are not considered so assigned shift patterns can be replicated across planning periods. In non-cyclic scheduling, nurse shift preferences are considered which limits such replication of shift patters. Surveys such as [5,6] provide detailed reviews of terminology, constraints and objectives. In this paper we tackle the non-cyclic NSP. In this type of NSP, hard constraints are the shift requirements for the whole solution, such as maximum or minimum working shifts, while soft constraints are related to daily requirements on the number of nurses.

There are two widely-used benchmark data sets for non-cyclic nurse scheduling maintained by researchers at the University of Nottingham [7] and the University of Ghent [8] respectively. Benchmark in [7] consists of nurse scheduling problems collected from various real-world scenarios. The number of instances is limited but the problem size ranges from small with 2 weeks, 8 nurses and 1 shift type to larger with 52 weeks, 150 nurses and 32 shifts. On the other hand, benchmark in [8] is artificially generated based on 9 indicators. But overall there are 6 sets of instances according to the number of nurses and the planning period ranges from one week to one month. Both benchmarks also keep a record of the best known solutions.

A wide range of methodologies, such as constraint programming and meta-heuristics have been applied to solve nurse scheduling problems and achieving very good solutions [5,6]. However, it seems that no paper has applied dynamic programming directly to solve this type of problem due to the curse of dimensionality. Only one publication describes the use of dynamic programming as part of a column generation method to solve NSP [7].

2.2 Approximate Dynamic Programming

The Bellman Equation (1) conveys the strength of dynamic programming because of the principle of optimality, but this equation also reflects the issue that prevents applying dynamic programming broadly.

$$V(S_t) = max\{C(A(S_t, S_{t+1})) + V(S_{t+1})\} \tag{1}$$

In this equation, which refers to a maximisation problem, t indicates a given stage, S_t and S_{t+1} represents the space of sub-problem combinations at stages t and $t+1$ respectively, $A(*)$ is an action space that records all connections from stage t to stage $t+1$, $C(*)$ is the action cost function, $V(*)$ is the value function that being in a particular stage. The purpose of dynamic programming is to evaluate all the combinations terminating in each stage and select the optimal one. However, for a problem with a large action space, it is infeasible for dynamic programming to calculate the optimal solution because of either time or memory consumption. This is known as the *curse of dimensionality* and this issue makes dynamic programming algorithms non-practical to be applied widely to large combinatorial optimization problems.

Hence, developing algorithms that satisfy the principle of optimality but are also practical to solve large combinatorial problems has attracted the attention from researchers in recent decades. Hence, research into Approximate Dynamic Programming (ADP) has increased. Basically, ADP algorithms are divided into two categories. In one category, there are algorithms using linear programming based approaches with dynamic programming to resolve approximations. These approaches have obtained impressive results on solving problems such as backgammon, job shop scheduling and elevator scheduling [9]. The other category is to construct the approximate solution with simulation based procedures [10]. Instead of evaluating the whole action space, the Monte-Carlo simulation mechanism is employed to rank the importance of stage links by processing a number of iterations. The final solution is constructed by selecting the links with minimum penalty cost.

The whole idea of Approximate Dynamic Programming is also known as Reinforcement Learning [11] in Artificial Intelligence or Neuro-Dynamic Programming [10] in control theory. ADP has achieved impressive progress on solving Markov Decision Process (MDP) problems. MDP is a branch of mathematics based on probability theory, optimal control and mathematical analysis [12]. Any MDP model can be summarized in the underlying object collection level with four important factors $\{S, A, Pr(s,a), R(s,a)\}$ where S is a state search space, A is an action search space that moves to the next possible state, $Pr(s, a)$ is the probability of selecting a particular action a from the current state s to move to the next state, and $R(s, a)$ is the reward function that calculates the reward of taking the action a.

The work presented by Colorini and Dorigo [13] is similar to the idea of simulation based ADP in combinatorial optimisation. The aim of their algorithm is to search an optimal solution based on simulation outcomes of a parallel clients set. This algorithm is called Ant Colony System (ACS) and mimics the behaviour of ants seeking food. They applied ACS to solve a travelling salesman problem. However, the simulation behaviours of ADP and ACS are distinguishable, especially with the new ACS extensions developed.

Approximate dynamic programming algorithms are also being considered for solving stochastic optimization problems. Schuetz and Kolisch [9] applied ADP to solve capacity allocation problems in the service industry. Their algorithm, called λ-SMART, is a simulation based ADP. Their algorithm performed well in terms of solution quality, computational time and memory usage. Koole and Pot [14] applied ADP to solve a stochastic workforce scheduling problem with multi-skills in a call center. Their paper suggested a general structure to apply ADP to this type of problem. It also presented some mathematical proofs on the accuracy of ADP. Instead of comparing with other methods, Koole and Pot investigated the relationship between parameters setting and the quality of the obtained solutions.

Given the successful applications of ADP in the literature, we believe this mechanism can also be applicable to solve the nurse scheduling problem and hence the motivation for the work in this paper as a new research direction on tackling nurse scheduling. This paper uses Q-learning as an ADP approach to solve the nurse scheduling problem. The aim of Q-learning [2] is to solve any problem with a representation based on environmental states S, possible actions A from states, and the value of state-action pairs, called Q value. Basically, Q-learning evaluates state-action pairs and increases or reduces the corresponding Q value depending on the outcome (i.e. state that the state-action led to).

3 Nurse Scheduling as a Markov Decision Process

Every MDP model is summarized as $\{S, A, Pr(s, a), R(s, a)\}$ and Fig. 1 depicts the decision process for NSP and explained in what follows. The state space S is represented as **STATE** in the figure. There two types of states: pre-condition state and post-condition state. In every MDP iteration, a state is selected as a starting point. This state is called pre-condition state while the post-condition state is the outcome state after making a decision. In the NSP, an example of pre-condition state is the nurse schedule completed until day 4 and the post-condition state is the schedule completed until day 5 as depicted in the figure.

The action space A is represented as **ACTION** in the figure. The daily requirement in NSP is to select w nurses out of the total W nurses with various shift assignment for each nurse. In the consequence, the action space for NSP

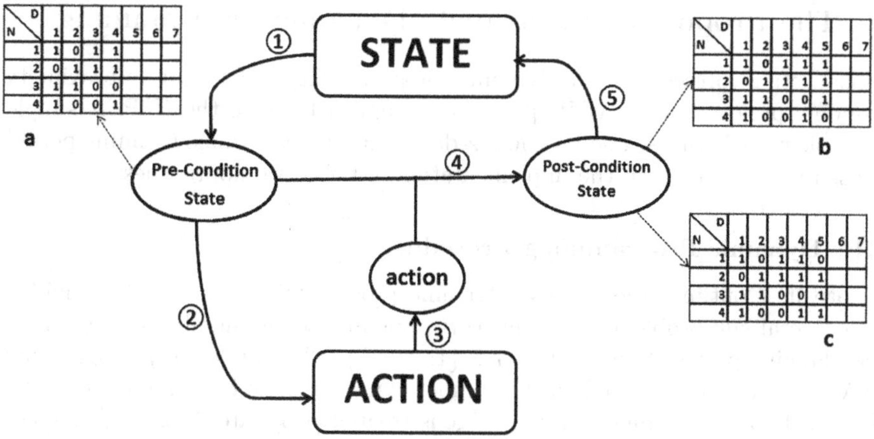

Fig. 1. Transition process between state and action in Markov decision process

is modelled as any possible assignment in the next day for any given scheduling period. In order to cover the curse of dimensionality, the action space is decomposed with two parts. The first part is to select a subset of w required nurses (with constraints) or any number (without constraints). With these selected nurses there will be w^{SH} various shift assignment among these nurses in total. In the consequence, the search space is decreased because only a subset need to be evaluated in the iterations. Finally, the selected action will be allocated to different nurses. As shown in Fig. 1, the allocation process is the outcome to produce a post-condition state represented by schedule b or c.

$Pr(s, a)$ is the transition probability from the state in the current stage with a selected action to the next stage. In most of the MDP models, this value will be updated within the algorithm no matter what is the initial value. In the NSP, the action is the possible shift assignment in the next day. So, the probability value is unchanged in the current implementation.

$R(s, a)$ is the reward function for taking an action from the current stage. The reward should be considered in two parts. In each scheduling day, there are hard constraints and/or soft constraints to be considered. For a produced schedule it can be evaluated if there are any violations of nurse preferences hard constraints. The number of nurses scheduled or on duty is a kind of soft constraint that is evaluated according to the daily nurse requirements. For the next day in the scheduling we should consider the previous working patterns. For example, in most NSP models, there is a hard constraint on the maximum number of working days over a certain period of time. Even though the pre-condition state is feasible, the post-condition state should be tested after the selected action. The value of the reward will be fully described in the next section.

4 The Proposed Approximate Dynamic Programming

This section presents both dynamic programming (DP) and approximate dynamic programming (ADP) procedures applied to solve the NSP. A simple example with 2 nurses, 2 shifts and a 2-days section of the whole planning period is used below to explain the implementation of these two approaches.

4.1 Dynamic Programming Procedure

Algorithm 1 outlines the steps of dynamic programming. The NSP is divided into several sub-problems or stages corresponding to the number of days T in the planning period. In each stage $t \in \{1, \cdots, T\}$, all possible daily assignments (NA) are constructed based on the number of nurses W and the number of shifts SH. Each daily assignment in the stage is treated as a state. In a single step of dynamic programming, every state in the current stage t is added to the input. This input is a partially-built solution until stage $t-1$. Solutions are constructed by recursively repeating this single step from an initial stage until the final stage.

Algorithm 1. Dynamic Programming Procedure

```
begin
    if t = T then
        return Sol;
    else
        NA ← ConstructAssignment(t + 1);
        for na ∈ NA do
            ta ← Combine(Sol, na);
            DP(ta, t + 1);
```

A NSP example is presented in Fig. 2. The initial input in this example is $\{D, N\}$ in stage T. D is a day shift and N is a night shift. The circle in each stage represents one possible assignment for the two nurses and the set of circles is the value of NA in the algorithm. For instance, the circle with pair $\{D, N\}$ in stage $t+1$ means that a day shift is assigned to the first nurse a night shift is assigned to the second nurse in that day. The solution with the optimal value will be selected based on the objective function (2).

$$f = \min(V + \sum_{i=1}^{W} x_{it} \times c_{it}), t \in [1, \cdots, T] \tag{2}$$

W is the number of nurses and t represents different stages. x_{it} is an individual nurse shift assignment and the cost is c_{it}. V is the value of soft and hard constraints violations that is calculated in (3). V_{hc} is the number of hard constraint violations and c_h is the penalty of hard constraint violations. V_{sc} is the number of soft constraint violations and c_s is the penalty of soft constraint violations.

$$V = c_h \times V_{hc} + c_s \times V_{sc} \tag{3}$$

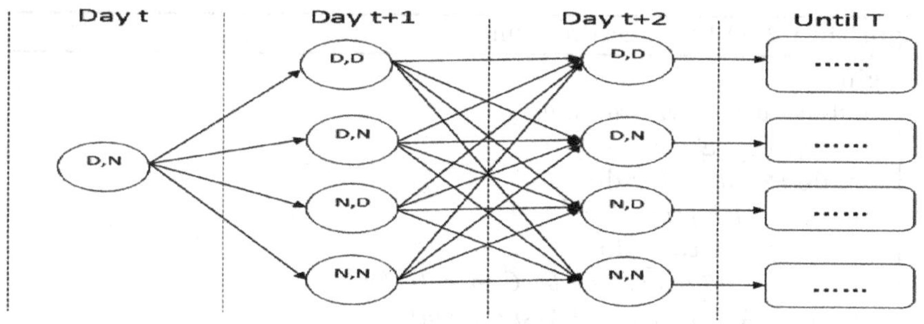

Fig. 2. Structure of NSP for dynamic Programming

4.2 Simulation Based Approximate Dynamic Programming

The approximate dynamic programming implementation in this paper is a modification from Algorithm 1 and it uses a simulation process. The objective function and the recursive process is kept as in Algorithm 1. Different to dynamic programming, the essence of ADP is to only evaluate one solution per process. With this purpose, the storage space of ADP is changed as recording the importance of a link that connected stages t and $t+1$. Our proposed ADP is an extension of Q-Learning based on the following equations.

$$
\begin{aligned}
a_t^n &= \begin{cases} RandomSelection, & c < \epsilon \\ \min_{a \in A} Q_{t-1}(s_t^n, a), & otherwise \end{cases} \\
Q_t(S_t^n, a_t^n) &= \alpha_n q_t^n + (1-\alpha_n) Q_{t-1}(S_t^n, a_t^n) \\
q_t^n &= c(S_t^n, a_t^n) + \alpha min_{a' \in A}(Q_{t-1}(S_t^{n+1}, a'))
\end{aligned}
\tag{4}
$$

Instead of updating V as described in the previous section, the whole process of Q-learning is to update the Q value, $Q(s,a)$, which is the value of a pair state s and action a. In Eq. (4), t is the iteration value, n is the process value, s is a single state. One iteration in ADP is finished by completing the input to a fully constructed solution with a number of processes. In this paper, one iteration is finished until a feasible solution is explored. In every process, only one action will be selected between two stages. The action selection could be random or follow some designed rules. In our proposed algorithm, the selection method follows the idea of ϵ-greedy [2]. The random and greedy selection method will be processed based on the value of ϵ. This selection method is able to guarantee that Q-learning explores a wide search space and converges into a optimum. In the nurse scheduling problem s_t^n is the nurse assignment combination from the initial stage 1 to stage n in current the iteration t, a is the action space, a_t^n is the link between stage $n-1$ and n in iteration t. A is the set of all links between any two continuous stages. The purpose of calculating a_t^n is to select the next action or links from the current stage to the next stage. The structure of this approximate dynamic programming is outlined in Algorithm 2.

Algorithm 2. Q-Learning Procedure

```
1  begin
2      Initial value of α, ε and max_iter;
3      i ← 0, M ← Empty;
4      while i < max_iter do
5          Sol ← Empty;
6          for j ← 1 to T do
7              c ← RandomNumberGenerator();
8              NA ← ConstructAssignment(t);
9              if c < ε then
10                 a ← RandomSelection(NA);
11             else
12                 a ← GreedySelection(NA);
13             q(Sol, a) = FitnessFunction(Sol, a);
14             if (Sol, a) ∈ M then
15                 Q(Sol, a) = GetQValue(Sol, a);
16             else
17                 Q(Sol, a) = 0;
18             Q(Sol, a) ← α × q(Sol, a) + (1 − α) × Q(Sol, a);
19             Insert(M, (Sol, a));
20             Sol ← Combine(Sol, a);
21             j ← j + 1;
22         Update(α, ε);
23         i ← i + 1;
```

Algorithm 2 outlines the steps of Q-learning on solving NSP instance. These steps are explained with the same previous example. A set of parameters need to be initialized: α is the learning rate and ϵ is the searching rate. Larger values for these parameters guarantee that Q-learning explores a larger portion of the search space. M is a lookup table that record the q value of a link between two stages.

The number of processes per iteration in this example is 2, and its output is moved to the next stage. For any given iteration, the initial input is $\{D, N\}$ at day t. In the first process, there are four actions leading to the next stage. The method to select an action is based on the random value c. If c is less than ϵ then the action will be randomly selected from the four actions. Otherwise, the action with the lowest Q value will be selected. Once an action is selected, for example $\{D, D\}$, then there is a state-action pair $\{\{D, N\}, \{D, D\}\}$. This selection will be evaluated through both the daily preference cost of $\{D, D\}$ and the validity of this pair. In the next step, the Q value of the pair will be updated based on line 18 and added to M. Before starting a new process, part of a temporal solution will be constructed based on the selected action. All the rest processes are repeated until a solution is fully explored. Q-learning will start a new iteration until this solution is feasible.

One possible solution in the last iteration could be $\{\{D, N\}, \{D, D\}, \{N, N\}\}$. With a higher value of ϵ this solution will not be repeated in the next few iterations. Moreover, a number of state-action pairs will be evaluated until a feasible solution is found. These two points will guarantee the Q-learning process to explore more of the search space.

After a number of iterations, the greedy selection method will be considered for evaluating all selected state-action pairs. With this method, not all new pairs will be added to the lookup table M, but only the pairs that appear promising according to the previous step, i.e. random selection, in the algorithm. At the end of the whole algorithm, there is a temporal solution that records the best solution found during the iterations. While there is also the storage space M that indicates which pair is more important (with lower value) for constructing the optimal solution.

5 Experiments and Results

The Q-learning and dynamic programming (DP) algorithm presented in the previous section were implemented in Java and all computations were performed on an Intel (R) Core (TM) i7 CPU with 3.2 GHz and RAM 6 GB. A sub-group of instances from the NSPLib benchmark was considered in these experiments. There are 7290 instances in this sub-group, each with 25 nurses, 7 scheduling days and 4 shift types. The set of hard and soft constraints were selected from the *Case* 1 file. The detailed descriptions of the instances is given in [8, 15]. Maenhout and Vanhoucke [16] published their mechanisms to solve the NSPLib instances. Here, results from the proposed ADP are compared to the best known solutions from the literature.

5.1 Comparing with Dynamic Programming Procedure

The parameter values for the Q-learning method to produce the results shown in the table are the best after tuning. The initial value of α and ϵ are both set to 0.9 and the maximum number of iterations is set upto 1000. Table 1 shows a summary of the experimental results from applying Q-learning and DP to solve the selected NSP instances. The performance of the proposed Q-learning and DP methods is compared using four aspects. *No. Solutions* is the number of solutions that each algorithm found. *Avg. Cost* is the average objective cost of the solutions. *Avg. Violation* is the average soft constraint violations in the solutions. *Avg. Time* is the average time to generate the solutions.

The contribution of approximate dynamic programming is to extend the usage of dynamic programming on larger instances. To give an illustration, none of the selected instances could be solved by dynamic programming due to the state space exceeding the memory. These experimental results serve as evidence that approximate dynamic programming is an improvement from the standard

Table 1. Comparison of Q-learning with dynamic programming

	No. solutions	Avg. cost	Avg. violation	Avg. time
Q-learning	7290	306.0798	0.471056241	9.946227709
DP	Could not be solved			

dynamic programming approach that is able to tackle the nurse scheduling problem. Not only the computational speed but also the algorithm effectiveness is improved by considering only the useful partial state space.

In our experiments, the outcome of one iteration is to construct and evaluate a feasible solution, as described in the previous section. Constructing such feasible solution might cause a large number of processes because of the infeasible solutions. The value state-action pairs that construct an infeasible solution are updated within the same iteration. In the consequence, the number of evaluations within one iteration is larger than those required for a single feasible solution. This is also the reason why the average computation time is longer when comparing to other mechanisms in the next section. Figure 3 shows the solution convergence for a problem instance. The number of iterations is set to 2000 but the objective function value settles at around 817 in iteration 453 for this problem instance. Experimental results show similar results for other instances.

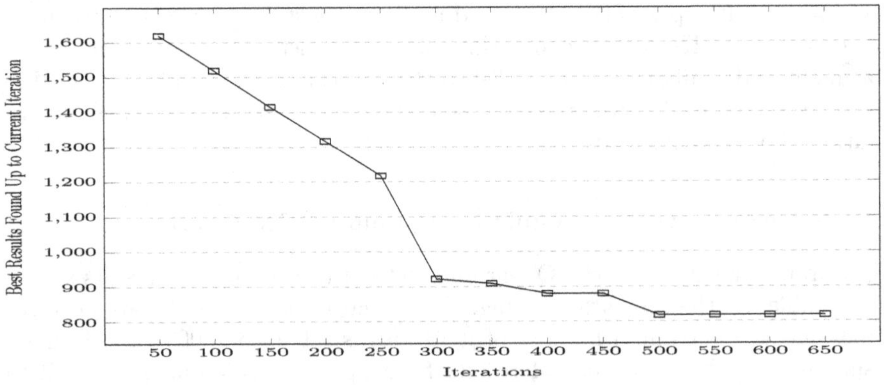

Fig. 3. The convergence of objective value updating for sample instance

5.2 Comparing with Meta-Heuristic Approaches

A number of published papers describe approaches to solve the NSPLib instances. For comparison in this paper we use the results from [16] because they provide detailed results for every instance. That paper combined two

Table 2. Comparison of Q-learning with Electromagnetism (EM) and Scatter Search (SS) [16]

	No. solutions	Avg. cost	Avg. violation	Avg. time	Achievements
Q-learning	7290	306.0798	0.471056241	9.946227709	84.79%
EM and SS	7290	305.776	0.53	0.532	88.27%

meta-heuristics, an Electromagnetism (EM) Approach and Scatter Search (SS). Comparison is carried out using the 5 aspects shown in Table 2.

The achievement is calculated using Eq. (5). $V(Approach)$ is the number of instances for which the given approach achieves a solution that is not worse than the best solution reported [16]. The average objective cost, violation and achievements for both the proposed Q-learning and meta-heuristics are very close, a difference within 5%. From this point of view, the quality of our solutions is comparable to published results. Hence, we argue that approximate dynamic programming could make contributions to the state of art for solving the Nurse Scheduling Problem. It should be noted that the computation time of solving instances by meta-heuristics approaches are much faster than the proposed algorithm. This shortcoming means that future research can be diverted at improving Q-learning or approximate dynamic programming to make it more efficient.

$$Achievement(\%) = V(Approach)/7290 \times 100\% \quad (5)$$

6 Conclusion

This paper proposes an approximate dynamic programming (ADP) algorithm to solve the nurse scheduling problem (NSP). For this, the NSP is modelled as a Markov Decision Process. Then, a typical ADP algorithm, Q-Learning, is applied to generate solutions. Instead of evaluating the whole state-action space like in standard dynamic programming, ADP only works on a subset of the space, the most promising state-actions according to the objective function. The experimental results here support the idea of modelling NSP as a Markov Decision Process and provide evidence that ADP exhibits very good performance by achieving as good solutions as heuristics proposed in the literature.

This paper has shown that ADP is able to solve effectively a range of NSP instances that dynamic programming is not able to solve. Although the instances solved are of realistic size, it is still a challenge to solve larger instances with ADP. Hence, future work will focus on speeding up ADP and make improvements on the procedure so that applying it to solve more challenging problem instances becomes practical in terms of computational time. This paper demonstrates that applying ADP to solve difficult combinatorial optimization problems like workforce scheduling is a viable and interesting research direction.

References

1. Bellman, R.: Dynamic programming and lagrange multipliers. Proc. Nat. Acad. Sci. **42**(10), 767–769 (1956)
2. Powell, W.B.: Approximate Dynamic Programming: Solving the Curses of Dimensionality, vol. 703. Wiley, Hoboken (2007)
3. Van den Bergh, J., Beliën, J., De Bruecker, P., Demeulemeester, E., De Boeck, L.: Personnel scheduling: a literature review. Eur. J. Oper. Res. **226**(3), 367–385 (2013)
4. Elshafei, M., Alfares, H.K.: A dynamic programming algorithm for days-off scheduling with sequence dependent labor costs. J. Sched. **11**(2), 85–93 (2008)
5. Cheang, B., Li, H., Lim, A., Rodrigues, B.: Nurse rostering problems–a bibliographic survey. Eur. J. Oper. Res. **151**(3), 447–460 (2003)
6. De Causmaecker, P., Berghe, G.V.: A categorisation of nurse rostering problems. J. Sched. **14**(1), 3–16 (2011)
7. Curtois, T.: Employee shift scheduling benchmark data sets. Technical report, School of Computer Science, The University of Nottingham, Nottingham, UK, September 2014
8. Vanhoucke, M., Maenhout, B.: Characterisation and generation of nurse scheduling problem instances. Technical report, Ghent University, Faculty of Economics and Business Administration (2005)
9. Schuetz, H.-J., Kolisch, R.: Approximate dynamic programming for capacity allocation in the service industry. Eur. J. Oper. Res. **218**(1), 239–250 (2012)
10. Bertsekas, D.P.: Dynamic Programming and Optimal Control, vol. 1. Athena Scientific, Belmont (1995). (No. 2)
11. Sutton, R.S., Barto, A.G.: Reinforcement Learning: An Introduction, vol. 1. MIT Press, Cambridge (1998). (No. 1)
12. Puterman, M.L.: Markov Decision Processes: Discrete Stochastic Dynamic Programming. Wiley, Hoboken (2014)
13. Dorigo, M., Gambardella, L.M.: Ant colony system: a cooperative learning approach to the traveling salesman problem. IEEE Trans. Evol. Comput. **1**(1), 53–66 (1997)
14. Koole, G., Pot, A.: Approximate dynamic programming in multi-skill call centers. In: 2005 Proceedings of the Winter Simulation Conference, pp. 576–583. IEEE (2005)
15. Maenhout, B., Vanhoucke, M.: NSPLib - a nurse scheduling problem library: a tool to evaluate (meta-)heuristic procedures. In: OR in Health, pp. 151–165. Elsevier (2005)
16. Maenhout, B., Vanhoucke, M.: New computational results for the nurse scheduling problem: a scatter search algorithm. In: Gottlieb, J., Raidl, G.R. (eds.) EvoCOP 2006. LNCS, vol. 3906, pp. 159–170. Springer, Heidelberg (2006). doi:10.1007/11730095_14

Breast Cancer's Microarray Data: Pattern Discovery Using Nonnegative Matrix Factorizations

Nicoletta Del Buono[1(✉)], Flavia Esposito[1], Fabio Fumarola[2], Angelina Boccarelli[3], and Mauro Coluccia[4]

[1] Department of Mathematics, University of Bari Aldo Moro, 70125 Bari, Italy
nicoletta.delbuono@uniba.it
[2] Department of Informatics, University of Bari Aldo Moro, 70125 Bari, Italy
[3] Department of Biomedical Sciences and Human Oncology, University of Bari Aldo Moro, 70125 Bari, Italy
[4] Department of Pharmacy – Drug Sciences, University of Bari Aldo Moro, 70125 Bari, Italy

Abstract. One challenge in microarray analysis is to discover and capture valuable knowledge to understand biological processes and human disease mechanisms. Nonnegative Matrix Factorization (NMF) – a constrained optimization mechanism which decomposes a data matrix in terms of additive combination of non-negative factors– has been demonstrated to be a useful tool to reduce the dimension of gene expression data and to identify potentially interesting genes which explain latent structure hidden in microarray data.

In this paper, we detail how to use Nonnegative Matrix Factorization based on generalized Kullback-Leibler divergence to analyze gene expression profile data related to the cell line of mammary cancer MCF-7 and to pharmaceutical compounds connected to the metabolism of arachidonic acid. NMF technique is able to reduce the dimension of the considered genes-compounds matrix from thousands of genes to few metagenes and to extract information about the drugs that more affect these genes. We provide an experimental framework illustrating the technical steps one has to perform to use NMF to discover useful patterns from microarray data. In fact, the results obtained by NMF method could be used to select and characterize therapies that can be effective on biological functions involved in the neoplastic transformation process and to perform further biological investigations.

Keywords: Nonnegative matrix factorization · Microarray data · Metagenes · Breast cancer

1 Introduction

Gene expression data plays an important role in understanding life. Recent advances in microarrays technology make possible to measure the expression levels of large numbers of genes and proteins simultaneously. The amount of this

P.M. Pardalos et al. (Eds.): MOD 2016, LNCS 10122, pp. 281–292, 2016.
DOI: 10.1007/978-3-319-51469-7_24

type of biological data is growing exponentially as microarrays become larger so the main challenge would be to discover and capture valuable knowledge to understand biological processes and human disease mechanisms [14]. Clearly human capabilities are unsuitable to process big amounts of data, hence automatic mechanisms – such as clustering algorithms and dimensionality reductions techniques – are indispensable tools to extract useful information and knowledge from biological data. Dimensionality reduction techniques can be categorized in two classes: feature selection and feature extraction. A feature selection method is a process that selects a subset of k original (and supposed relevant) features for spanning a reduced space that may better describe the phenomena of interest. On the other hand, feature extraction methods try to capture hidden properties of data and discover the minimum number of uncorrelated or lowly correlated factors that can be used to better describe the phenomena of interest. This can be accomplished by the creation of new features obtained as functions of the original data. Reduction of the computational complexity of data, both in time (for elaboration) and in space (for storage), and discovery of latent structure hidden in data, (meaningful structures and/or unexpected relationships among variables) are some of the advantages resulting from feature extraction methods [3]. Singular Value Decomposition and Principal Component Analysis (PCA), Independent Component Analysis, Network Component Analysis have been widely used to effectively analyzing and interpreting microarray data [11].

In this paper we adopt Nonnegative Matrix Factorization (NMF) –a constrained optimization mechanism– to analyze a data set of mammary cancer (MCF-7) and then to select and characterize therapies that can be effective on biological functions involved in the neoplastic transformation process. The aim is twofold: firstly (i) to present an experimental framework which details the steps one has to perform when NMF is used on microarray data and then (ii) to emphasize the advantages that NMF provides in terms of interpretability of its factors in a sensible context. Particularly, we describe in some details how NMF techniques are able to extract information about a subset of genes which represents the whole data set and about the drugs that more affect these genes. The proposed experimental framework –based on standard NMF technique– reduces the dimensionality of the considered microarrays and discovers useful patterns that can be adopted to perform biological analysis. Moreover, the nonnegativity of the factors extracted by NMF allows an interpretation in terms of the probability that a gene is relevant in each metagene and a metagenes expression pattern is important in the corresponding sample. This is an advantage that NMF can achieved over an approach such as PCA, which is more popular into a biology context.

The paper is organized as follows. A brief introduction of NMF learning mechanism, the related constrained optimization problem and the properties of NMF factors in the context of microarray analysis is firstly presented. Then the results obtained applying the NMF algorithm based on the generalized Kullback-Leibler divergence to a cancer microarray data set are details. A final section highlighting the biological meaning of the obtained results and sketching some future works and open problems closes the paper.

2 The NMF Approach for Microarrays

NMF is an unsupervised technique for linear dimensionality reduction of a given data matrix which is able to explain data in terms of additive combination of non-negative factors that represent realistic building blocks for the original data (provided that data are non-negative too) [13]. The non-negativity constraint is useful for learning part-based representations and finds its justifications both by the rules of physics has (in many applications one knows that the quantities involved cannot be negative) and by the intuitively and physiological principles assuming that parts are generally combined additively (and not subtracted) to form a whole. Humans learn objects as part-based physiological principles assuming that parts are generally combined additively allowing meaningful interpretations of information extracted from a given data matrix.

In the context of microarray experiments, gene expression data are collected in a numerical non-negative matrix $X \in \mathbb{R}^{n \times m}$, in which rows correspond to different genes, columns correspond to samples (which may represent distinct tissues, experiments, conditions or time points) and elements x_{ij} indicate the expression level of the gene i in the sample j [5]. Typically, the number of genes n is much larger than the number m of samples. The NMF can be used to decompose the microarray data matrix X into two non-negative matrices, such as $X \approx WH$, where $W \in \mathbb{R}^{n \times r}$ and $H \in \mathbb{R}^{r \times m}$. The columns of the matrix W – called the *basis* or *metagenes matrix* – are the metagenes, while the rows in the *encoding* or *metagene expression profiles matrix* H represent the metagenes expression pattern of the corresponding sample. Hence each element h_{ij} reveals the effect that the metagene i has on the sample j: if the value h_{ij} is very small, then the corresponding metagene is useless in approximating that particular sample. Under some hypotheses, the columns of W can be interpreted as prototypes of data clusters so that the coefficients h_{ij} can be easily interpreter as membership degrees of each sample to each cluster [1,2,5,19]. It is worthy to observe that the decomposition $X \approx WH$ could be dually viewed as individuating metasamples (rather than metagenes) and groups of genes (rather than of samples) when the entries of W are taken into account [4]. The rank r of the factorization – that is the number of metagenes (or metasamples) – represents the number of latent factors in the decomposition and is generally chosen so that $(n+m)r < nm$, hence a compression of the given dataset is obtained. This parameter is problem dependent and user-specified and plays a fundamental role in the factorization process: in fact, different values of r lead to different factorization results.

From a computational point of view, NMF may be re-written as a non-linear constrained optimization problem over a specified divergence measure D which evaluates how well the low dimensional matrix WH approximates X. In the microarray context, the most commonly used divergence D are the Least Squared (LS) error $Div(X, WH) = \frac{1}{2} \|X - WH\|_F^2$ (being $\| \cdot \|_F$ the matrix Frobenius norm), designed under the assumption of normal independently identically distributed noise, and the generalized Kullback-Leibler (KL) divergence

$Div(X, WH) = \sum_{ij}\left(X_{ij}log\left(\frac{X_{ij}}{(WH)_{ij}}\right) - X_{ij} + (WH)_{ij}\right)$, which corresponds to maximum likelihood estimation under an independent Poisson assumption. The factorization problem reduces to the minimization of the selected divergence measure with non-negativity constraint. The constrained optimization problem is NP-hard and non convex in both the unknowns W and H. However, the problem is convex with respect to each variable taken separately: this makes alternating optimization techniques, i.e., solving at each iteration two separate convex problems, very adapted: first fixing H to estimate W, and then fixing W to estimate H. Multiplicative update algorithms early proposed in [13] for both LS and KL divergences follow this approach. Several other algorithms appeared in literature: some recent and comprehensive reviews can be found in [7,18]. For the KL divergence function, the multiplicative rules which update W and H are given by:

$$W_{ij} \leftarrow W_{ij}\frac{\sum_k (H_{jk}X_{ik})/(WH)_{ik}}{\sum_k H_{jk}} \qquad H_{ij} \leftarrow H_{ij}\frac{\sum_k (W_{ki}X_{kj})/(WH)_{kj}}{\sum_k W_{ki}} \tag{1}$$

The complexity per iteration is $O(rmn)$ and it has been proved that the generalized KL-divergence is invariant under these updates if and only if W and H are at a stationary point at the divergence [13].

3 Methods and Results

In this section we detail the use of KL-divergence based NMF algorithm to identify subsets of genes (in a cancer dataset) which affect drugs influencing biological functions involved in the neoplastic process. We provide a standardization of the technical steps (which lacks in the existing literature) that one has to perform when NMF mechanisms are used to treat microarray datasets. In this sense, we describe an experimental framework based on standard NMF for pattern discovery and we apply it to a real cancer dataset in order to highlight the advantages –in terms of interpretability– that the factors extracted by NMF show.

3.1 The Breast Cancer Dataset and NMF Algorithm

We considered a dataset composed by 10331 genes of mammary cancer (related only to the cell line MCF-7) and 434 pharmaceutical compounds related to the metabolism of arachidonic acid, a fat acid that makes some substances involved in inflammatory processes [8,15,17] extracted from dataset present in Connectivity Map (CMap) [12]. The extracted gene expression levels and the pharmaceutical compounds were preprocessed via MAS 5.0 statistical algorithm in the Affymetrix Microarray Suite version 5.0 (http://www.affymetrix.com/support/technical/technotesmain.affx) to derive the microarray input matrix to be factorized via NMF. Algorithms in NMF package in R [6] were then adopted to analyze the obtained dataset via NMF. The used updates rules are mainly based on multiplicative updates in (1); an additional stabilization step has been added (every 10 iterations) to shifting up all entries from zero to a very small positive value.

3.2 Rank Estimation

The choice of the rank parameter r is critical in NMF and no automatic mechanisms are know to select it. In a microarray data analysis context r defines the number of metagenes used to approximate the microarray data matrix. Several approaches have been proposed to select r in a given interval of values. In the illustrated experimental framework, we restrict the evaluation onto the interval $[2, 50]$ and we select r accordingly to: (i) the first values for which the cophenetic correlation coefficient (CCC) starts a significantly decrease [1] and (ii) the first values at which the variation in the Residual Sum of Squares curve (RSS) –between the target matrix and its NMF estimate– presents an inflection point [9]. The CCC quantifies the stability of the dimensionality reduction parameter r and of the hierarchical clusters derived from the NMF factors [1,6]. For stable value it will be close to 1, while for unstable value it is near to 0. Peaks in the plot represent stable values. However, one has to look for consistency, that is a peak followed by a slow drop off.

Figure 1(a) plots the CCC against different r in the range $[2, 50]$. Based upon the previous considerations, we chose $r = 4, 5, 6$ as optimal values, while Fig. 1(b)

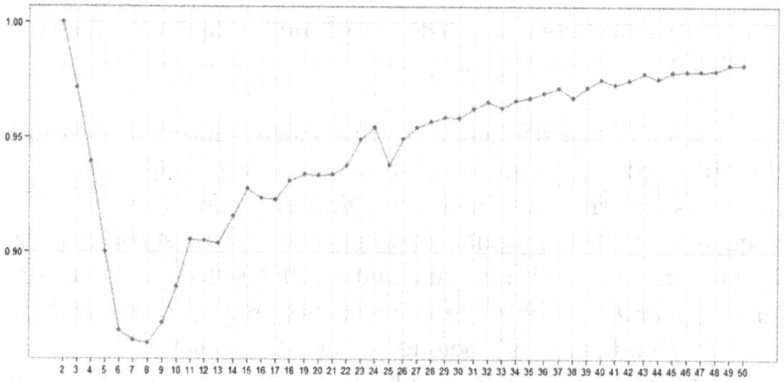

(a) Cophenetic Correlation Coefficient

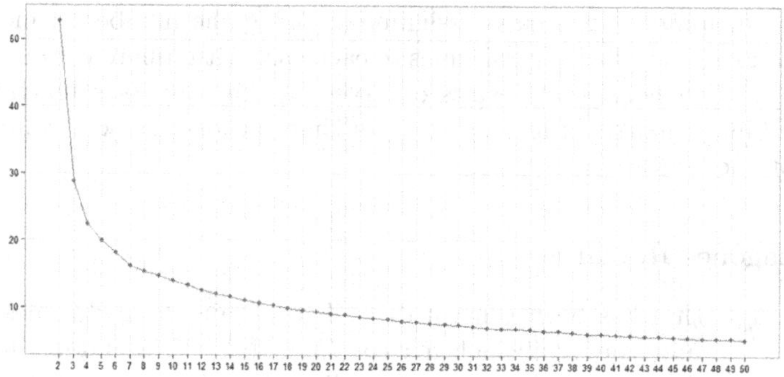

(b) Residual Sum of Squares curve

Fig. 1. (a) Behavior of the CCC in the interval $[2, 50]$. A decreasing behavior followed by a slow drop off can be observed in the sub-interval $[2, 8]$. (b) Behavior of the RSS curve in the interval $[2, 50]$. A local inflection point is reached for $r = 4$.

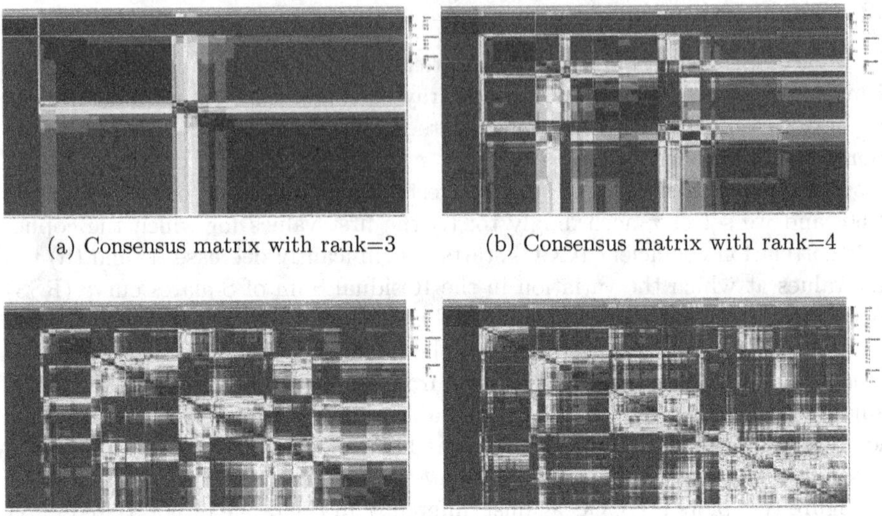

(a) Consensus matrix with rank=3 (b) Consensus matrix with rank=4

(c) Consensus matrix with rank=5 (d) Consensus matrix with rank=6

Fig. 2. Consensus matrices computed for each rank value in the interval [3, 6].

depicts the behavior of the RSS curve in the selected interval: a local inflection point can be observed for the same values of r (i.e. $r \in [4, 6]$).

The consensus matrix (obtained averaging the connectivity matrices) has been also used as an additional qualitative measure of the stability of the factorization. Connectivity matrix is a binary matrix that reflects cluster relationships between samples, for k samples the connectivity matrix $C \in \mathbb{R}^{k \times k}$ has values $c_{ij} = 1$ if sample i and sample j belong to the same cluster, $c_{ij} = 0$ otherwise. From each run of NMF, in order to simply the cluster interpretation, the connectivity matrix is reordered to form a block diagonal matrix. Then elements of a consensus matrix range in $[0, 1]$ and they can be interpreted as the probabilities of the reproducibility of the class assignment. Being the number of metagenes know, NMF can be applied several times to each case with different value of rank to choose the best consistent results observed in the consensus matrix which tends to a binary matrix. Figure 2 suggests that the best rank decomposition value is obtained with $r = 4$.

3.3 Metagenes Research

The NMF algorithm has been run with $r = 4$ to obtain the metagenes matrix $W \in \mathbb{R}^{10331 \times 4}$ and the coefficient matrix $H \in \mathbb{R}^{4 \times 434}$. Random initialization mechanism has been adopted to initialize W_0: the reported results are the average value obtained on different runs. A normalization step (to sum up to one) has been performed on the metagenes matrix and rows have been ordered using hierarchical clustering. Metagenes matrix W is illustrated in

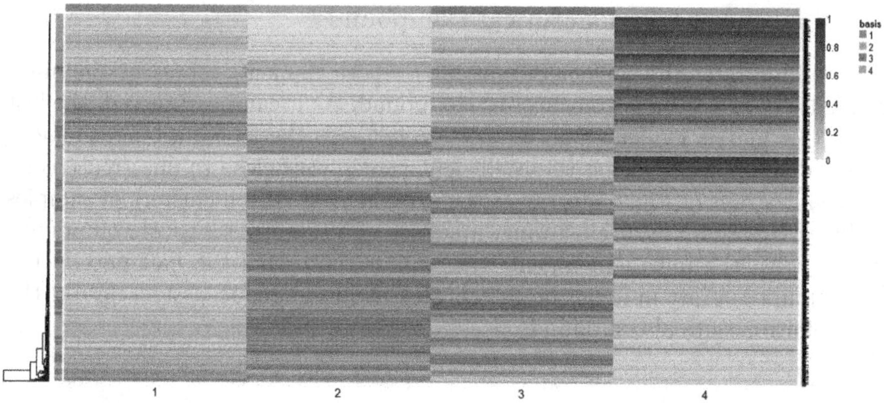

Fig. 3. Graphical illustration of the basis matrix W obtained for rank value $r = 4$.

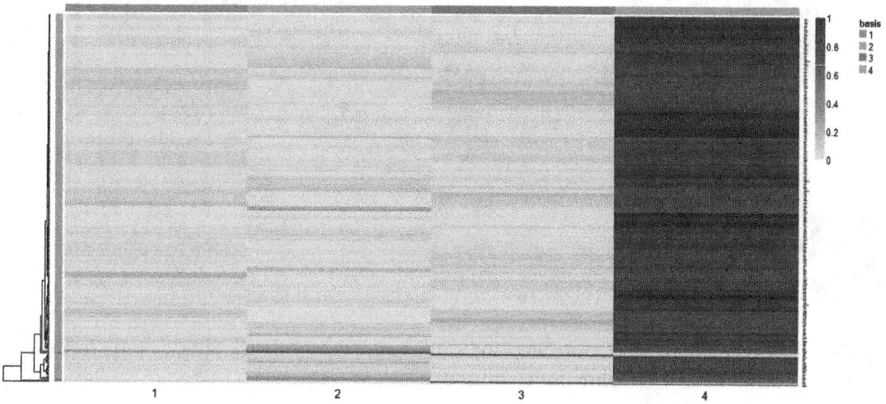

Fig. 4. Basis matrix W with rank value $r = 4$ obtained with gene extraction and gene ranking

Fig. 3: as it can be noted the fourth mategene provides a good representation of the whole set of 10331 genes (in fact, its elements are close to the maximum value 1). The *gene.score* method [10] has been used to select genes which assume relatively large importance in each biological process. Particularly, genes with *gene.score* value higher than $\hat{\mu} + 3\hat{\sigma}$ were selected (being $gene.score(i) = 1 + \frac{1}{\log_2(r)} \sum_{q=1}^{r} p(i,q) \log_2(p(i,q))$, where $p(i,q)$ is the probability that the i-th gene contributes to cluster q, $\hat{\mu}$ and $\hat{\sigma}$ the median and the median absolute deviation of *gene.scores*, respectively) and genes with the maximal values in the corresponding rows of W larger than the median of all elements in W. The high bearing of the fourth metagene, showed in Fig. 3, can be confirmed by genes extraction and visual analysis drawn in Fig. 4. Particularly, Fig. 4 depicts the matrix W in which the basis specific features are kept in row subsets obtained using the method previously described.

3.4 Analysis of Pharmaceutical Compounds

Figure 5 illustrates the coefficient matrix H whose elements represent the influence of the k-th metagene on the j-th compound. As it can be observed, the last row of this matrix evidences an interaction between the fourth metagene and a subset of compounds, being the coefficient values very close to one (red color). The dendrogram in Fig. 6 illustrates the arrangement of the clusters of samples produced by hierarchical clustering: a subset of 39 pharmaceutical compounds interact with the fourth metagene: this subset represents the half part of the entire compound set and can be interpreted as latent reduced structure in the entire pharmaceutical compounds.

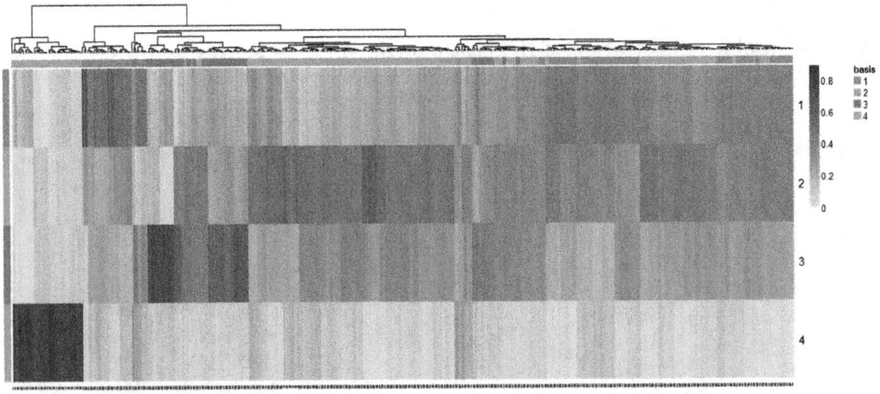

Fig. 5. Coefficient matrix H obtained for rank value $r = 4$. Dark shade (red) indicates coefficient value near 1. (Color figure online)

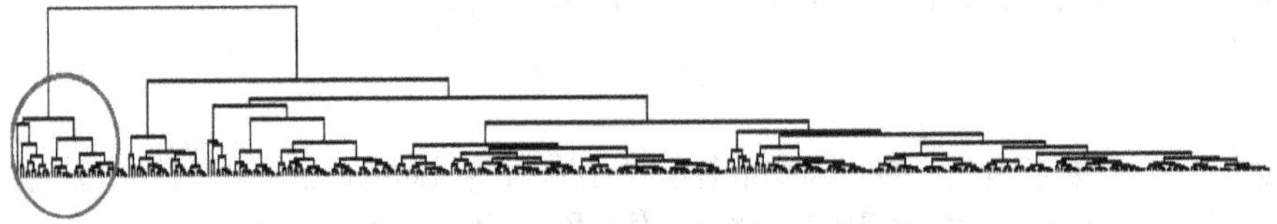

Fig. 6. Dendrogram obtained from NMF with rank $r = 4$. The circle highlights the subset of 39 compounds which interacts with the fourth metagene (i.e. the last columns of the basis matrix W with $r = 4$.)

3.5 Analysis with Different Ranks Decomposition

The results obtained with $r = 4$ proved to fulfill the preservation of most the information of the original dataset which can be compressed in reduced basis and coefficient matrices. To be guaranteed of this, we performed the same analysis also for $r = 5, 6$ (these values also satisfy criteria described in Sect. 3.2). We focus on genes extracted by these values of factorization ranks to prove that the

Table 1. Gene comparison

Rank value	Number of relevant genes	Percentage of equal genes
4	305	100%
5	399	96.1%
6	401	87.5%

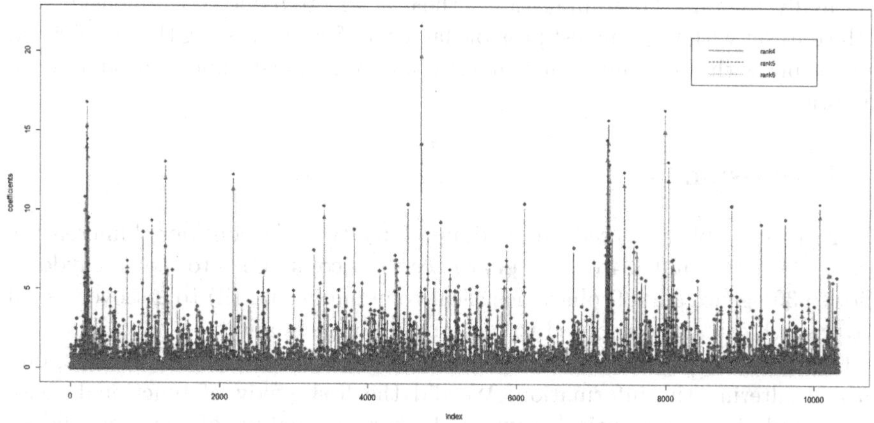

Fig. 7. Density of genes coefficients in the relevant metagenes computed by the different rank factorizations $r = 4, 5, 6$

choosing of a decomposition with rank $r = 4$ ensured the lesser dimensionality while maintaining the original information of the dataset. Table 1 compares the results of the different genes outcome: the first column indicates the rank value for each NMF computation, the second column provides for the number of relevant genes extracted with the *gene.score* method for each rank value and the last column reports the percentage of genes extracted with $r = 4$ which also belong to the factorizations obtained with the other rank values. As the rank value increases it can be observed a growing number of relevant genes and a decreasing percentage of included genes. Additionally, Fig. 7 illustrates the density of the relevant metagenes for the three different factorizations computed with $r = 4, 5, 6$, and show that the coefficients of the metagene computed with the decomposition rank $r = 4$ are almost entirely overlying on the coefficients of the relevant metagene computed with the decomposition rank $r = 5$ and the most part is included in the coefficients obtained with $r = 6$.

Moreover, for the compounds extraction with rank $r = 5, 6$, it is worthy to note that both cases present the same results of the previous analysis. In fact, the coefficients matrices H, obtained from the two different rank factorizations, represent the influence of the each metagene on each of the 434 compounds. It should be highlight that r rank NMF is able to identify a particular subset of compounds interacting with the selected metagenes, in both cases, and that this subset is composed by the same 39 pharmaceutical compounds found with the analysis performed with $r = 4$. Figure 2(b) upholds this observation: the most stable cluster on the left corner on the top of the consensus matrix, is formed by 40 pharmaceutical compounds, 39 of which are the same 39 compounds selected by NMF and illustrated in the dendrogram reported in Fig. 6. These 39 compounds represent the half part of the entire compound set and can be interpreted as a particular and reduced stable structure in the entire pharmaceutical compounds.

Finally, the rank 4 decomposition is the best way to reduce the dimensionality of the dataset with the smallest possible factor. In fact, increasing the factor rank, compatibly with the estimation rank criteria, no significant information loss can be noticed.

4 Discussion

The NMF was able (i) to reduce the dimensionality of the considered microarray from 10331 genes and 434 related pharmaceutical compounds to the arachydonic acid, to 301 genes and 39 pharmaceutical compounds and (ii) to discover useful patterns which can be adopted to perform a biological analysis.

Our goal was to find out if we could change the size of our initial group without altering the information. We did the first study of functional genes and second of pharmaceutical compounds. For the analysis of genes, we did an analysis tool on line GENECODICS. This software allows analysis of gene lists GENECODIS integrating separate databases; GENECODIS is a tool that integrates multiple sources of information to search for records that are often present in a set of genes [16]. The analysis can provide useful information for the interpretation of the genes and molecular biological selected. Biological processes selected were the following: the development of multicellular organisms, ion transport, signal transduction, and the immune response. Using the web tool PubChem, from the selected 39 pharmaceutical compounds we have grouped those that have relations with anti-neoplastic agents, anti-inflammatory agents such as cyclooxygenase inhibitors as shown in Fig. 8.

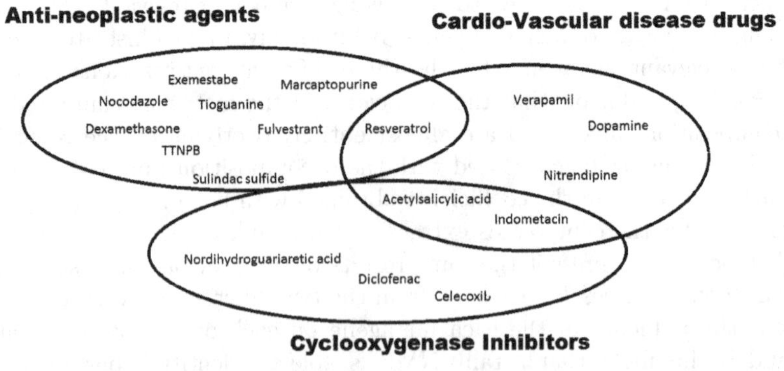

Fig. 8. Groups of pharmaceutical compounds obtained with PubChem web tool.

In this paper we have shown that the biological events related to the selected pharmaceutical compounds are representative of the two initial studies. This method provides a useful endpoint for screening and activity of drugs. In summary, it shows a promising chemical genomic strategy to select modulators of complex cancer phenotypes and establish their mechanisms of action.

To demonstrate the power of NMF techniques and investigate their limits, future works aim to compare the presented results with those obtained using bi-clustering techniques, such as the 3-NMF factorization with constraints [4]. Moreover, we are applying the proposed experimental framework on a blood cancer dataset in order to validate the procedure and develop additional hypothesis, but these new results will appear on a future paper of ours.

References

1. Brunet, J.P., Tamayo, P., Golub, T.R., Mesirov, J.P.: Metagenes and molecular pattern discovery using matrix factorization. Proc. Nat. Acad. Sci. **101**(12), 4164–4169 (2004)
2. Carmona-Saez, P., Pascual-Marqui, R.D., Tirado, F., Carazo, J.M., Pascual-Montano, A.: Biclustering of gene expression data by non-smooth non-negative matrix factorization. BMC Bioinform. **7**(1), 1 (2006)
3. Casalino, G., Del Buono, N., Mencar, C.: Non negative matrix factorizations for intelligent data analysis. In: Naik, G.R. (ed.) Non-negative Matrix Factorization Techniques: Advances in Theory and Applications, pp. 49–74. Springer, Heidelberg (2016)
4. Del Buono, N., Pio, G.: Non-negative matrix tri-factorization for co-clustering. Inf. Sci. **301**(C), 13–26 (2015)
5. Devarajan, K.: Nonnegative matrix factorization: an analytical and interpretive tool in computational biology. PLoS Comput. Biol. **4**(7), e1000029 (2008)
6. Gaujoux, R., Seoighe, C.: A flexible R package for nonnegative matrix factorization. BMC Bioinform. **11**(1), 1 (2010)
7. Gillis, N.: The why and how of nonnegative matrix factorization (2014). http://arxiv.org/pdf/1401.5226v2.pdf
8. Howe, L.R.: Inflammation and breast cancer. Cyclooxygenase/prostaglandin signaling and breast cancer. Breast Cancer Res. **9**(4), 210 (2007)
9. Hutchins, L.N., Murphy, S.M., Singh, P., Graber, J.H.: Position-dependent motif characterization using non-negative matrix factorization. Bioinformatics **24**, 2684–2690 (2008)
10. Kim, H., Park, H.: Sparse non-negative matrix factorizations via alternating non-negativity-constrained least squares for microarray data analysis. Bioinformatics **23**(12), 1495–1502 (2007)
11. Kossenkov, A.V., Ochs, M.F.: Matrix factorisation methods applied in microarray data analysis. Int. J. Data Min. Bioinform. **4**(1), 72–90 (2010)
12. Lamb, J., Crawford, E.D., Peck, D., Modell, J.W., Blat, I.C., Wrobel, M.J., Lerner, J., Brunet, J.P., Subramanian, A., Ross, K.N., et al.: The connectivity map: using gene-expression signatures to connect small molecules, genes, and disease. Science **313**(5795), 1929–1935 (2006)
13. Lee, D.D., Seung, H.S.: Algorithms for non-negative matrix factorization. In: Proceedings of the Advances in Neural Information Processing Systems Conference, vol. 13, pp. 556–562. MIT Press (2000)

14. Moschetta, M., Basile, A., Ferrucci, A., Frassanito, M., Rao, L., Ria, R., Solimando, A., Giuliani, N., Boccarelli, A., Fumarola, F., Coluccia, M., Rossini, B., Ruggieri, S., Nico, B., Maiorano, E., Ribatti, D., Roccaro, A., Vacca, A.: Novel targeting of phospho-cMET overcomes drug resistance and induces antitumor activity in multiple myeloma. Clin. Cancer Res. **19**(26), 4371–4382 (2013)
15. Muller, A., Homey, B., Soto, H., Ge, N., Catron, D., Buchanan, M.E., McClanahan, T., Murphy, E., Yuan, W., Wagner, S.N., Barrera, J.L., Mohar, A., Verastegui, E., Zlotnik, A.: Involvement of chemokine receptors in breast cancer metastasis. Nature **410**, 50–56 (2001)
16. Nogales-Cadenas, R., Carmona-Saez, P., Vazquez, M., Vicente, C., Yang, X., Tirado, F., Carazo, J.M., Pascual-Montano, A.: GeneCodis: interpreting gene lists through enrichment analysis and integration of diverse biological information. Nucleic Acids Res. **37**(suppl. 2), W317–W322 (2009). http://nar.oxfordjournals.org/content/37/suppl_2/W317.abstract
17. Harris, R.E., Casto, B.C., Harris, Z.H.: Cyclooxygenase-2 and the inflammogenesis of breast cancer. World J. Clin. Oncol. **5**(4), 677–692 (2014)
18. Wang, Y.X., Zhang, Y.J.: Nonnegative matrix factorization: a comprehensive review. IEEE Trans. Knowl. Data Eng. **25**(6), 1336–1353 (2013)
19. Yoo, J., Choi, S.: Orthogonal nonnegative matrix tri-factorization for co-clustering: multiplicative updates on Stiefel manifolds. Inf. Process. Manag. **46**(5), 559–570 (2010)

Optimizing the Location of Helicopter Emergency Medical Service Operating Sites

Maurizio Bruglieri[1(✉)], Cesare Cardani[2], and Matteo Putzu[2]

[1] Department of Design, Politecnico di Milano, Via Durando 38/a, 20158 Milano, Italy
maurizio.bruglieri@polimi.it

[2] Aerospace Science and Technology Department, Politecnico di Milano, Via La Masa 34, 20156 Milano, Italy

Abstract. The European Commission Regulation (EU) No 965/2012, now completely operative in all the European countries, allows helicopter night landings for emergency medical service in dedicated spaces, provided with a minimum amount of facilities, called HEMS Operating Sites. This possibility opens new scenarios for the mixed, ambulance/helicopter, rescue procedure, today not fully exploited. The paper studies the problem of optimal positioning for HEMS sites, where the transfer of the patient from ambulance to helicopter takes place, through the use of Geographic Information System (GIS) and optimization algorithms integrated in the software *ArcGIS for Desktop*. The optimum is defined in terms of the minimum intervention time. The solution approach has been applied to the area of competence of "SOREU dei Laghi", in Lombardia region, with a catchment area of almost two million people.

Keywords: Helicopter Emergency Medical Service · Maximal covering location · Minimum facility location · Orographic constraints · ArcGIS

1 Introduction

In Italy, as well as in the whole Europe, the Helicopter Emergency Medical Service (HEMS) is a spread activity whose growth is consolidating. Indeed the uniqueness of the capabilities of rotorcrafts in the emergency services is today thoroughly appreciated, nevertheless the costs connected to the management of such complex activities are growing too and there is a strong push to increase not only the effectiveness but also the efficiency of the HEMS.

A good opportunity in this sense is represented by the European Commission Regulation (EU) No 965/2012 of Oct. 5th, 2012, that from Oct. 28th, 2014 is definitely entered in force in all European countries allowing night landings of helicopters for emergency medical service in dedicated spaces called HEMS *Operating Site*. These new landing areas must meet certain operating conditions reported in the mentioned regulation as well as in the *Acceptable Means of Compliance* and in the *Guidance Material* of the EASA[1] of Nov. 25th, 2012. These requirements may be satisfied by sports fields, which are flat and usually already

[1] European Aviation Safety Agency.

P.M. Pardalos et al. (Eds.): MOD 2016, LNCS 10122, pp. 293–305, 2016.
DOI: 10.1007/978-3-319-51469-7_25

provided with a lighting system and for which the installation of further equipment (e.g. windsock) has a low cost. For these reasons in a country like Italy, the sites eligible to become HEMS operating sites are numerous and well distributed. Furthermore they allow the introduction of the mixed rescue procedure - the so called rendez-vous - between the ambulance and the helicopter, that today is the only feasible way to take advantage of helicopter flight, even at night.

The paper studies the problem of optimal positioning for HEMS operating sites, where the transfer of the patient from the ambulance to the helicopter takes place. In particular we solve two optimization problems related to this. In the *Maximum Coverage of Demand Points* (MCDP) problem we want to open a fixed number of HEMS maximizing the number of demand points (i.e. injured people) that can be served within a specified threshold time. While, in the *Minimum HEMS Sites Opening* (MHSO) problem we want to cover all the demand points, within their threshold times, minimizing the total number of HEMS sites opened.

Both of the optimization problems are NP-hard and therefore cannot be solved in exact way through a Mixed Integer Linear Program (MILP) in real applications like that faced by us, with about 100 candidate HEMS sites and about 10,000 demand points (Sect. 5). Indeed, the MILP would require the introduction of about 1 million of binary variables to decide which demand points assign to each open candidate HEMS site and no MILP commercial solver (e.g., CPLEX, GUROBI) is able to solve problems of this size (usually they can only solve MILP up to a few thousands of binary variables).

For this reason, both of the optimization problems are solved in heuristic way through the location-allocation solver present in the software *ArcGIS for Desktop 10*. *ArcGIS* is a Geographic Information System (GIS) produced by ESRI [13] that allows creating, sharing, managing and analyzing maps and geographic information stored in a database. Thanks to *ArcGIS*, our study overcomes some limits met in previous works. In particular the orography, that, to the best of our knowledge, has always been neglected in all previous studies on this subject. Moreover the ambulance travel times are computed with more accuracy considering different speeds according to the road classification.

1.1 Original Contributions and Paper Organization

A first original contribution of our work is to identify for the first time two optimization problems of practical interest in the context of Emergency Medical Service, taking into account the aforementioned European Commission Regulation 965/2012. This aspect is not trivial, since the commission regulation does not directly deal with these optimization problems. The latter have been identified only after a deep analysis of the Emergency Medical Service actual needs and of the EU regulation.

A second original contribution consists in defining a method to collect all the input data necessary to solve the optimization problems. Also this aspect is not trivial since it requires to do reasonable hypotheses (e.g., identifying the demand points with the road intersections, etc.) and to analyze the data base available (Sects. 4.1 and 5.1) including also georeferenced information (as explained before).

A third original contribution is the development of a pre-processing algorithm to preventively eliminate the demand points for which the ambulance intervention mode is faster than the rendez-vous intervention (Sect. 4.2).

Finally, the solution approaches are tested to a real case, under the responsibility of "SOREU[2] dei Laghi", with a catchment area of almost two million people in the provinces of Varese, Como and Lecco in Lombardia region (Italy). We perform various sensitivity analysis, through the changes of some parameters such as the average road speeds, flight speed, operational transfer times or the number of HEMS operating sites opened, obtaining useful insights.

The outline of the paper is the following: in Sect. 2 we review the related literature; in Sect. 3 we formally define the optimization problems addressed; in Sect. 4 we describe our solution approaches; in Sect. 5 we show the numerical results for a case study; finally, in Sect. 6 we draw some conclusions.

2 Literature

In the literature there are several studies that prove the effectiveness of using helicopters in emergency interventions. For example, [7] shows via a statistical analysis that there are no complications for children related with their emergency transport by helicopter. Other studies, e.g., [8,11] focus on demonstrating through a historical and statistical analysis that helicopters are faster than ambulances both in inter-hospital transport and from the accident site to the hospital. However, only very few works address the problem of the optimal location of HEMS operating sites, although location problems are widely investigated in other domains, e.g. for antenna positioning in cellular networks [10].

Paper [1] offers two models of population coverage to solve the problem of simultaneous localization of ambulances, HEMS bases, and HEMS sites: *Set Cover with Backup Model* and *Maximal Cover for a Given Budget Model.* In such study, however, the specific orography area that heavily affects the choice of a candidate HEMS site is not taken into account. Moreover, the evaluation of transport times of both ambulance and helicopter are too rough since are computed simply considering a constant average speed and the Euclidean distance.

Two location problems concerning HEMS services are addressed in [2]: *minisum* and *minimax.* Given the required rescue points (origins) and hospitals (destinations), the localization of the points for rendez-vous sites and helicopters bases are selected in order to: minimize the average transport time to the hospital of the total rescue points weighted according to the number of people present in each origins (objective minisum); minimize the maximum transport time from rescue point to the hospital (objective minimax). However, also in this work there are the same limitations of [1] in defining the real response time of ambulances and helicopters, since no network displacement has been defined.

[2] SOREU is the Italian acronym of the regional emergency-urgency operations room.

In conclusion, to the best of our knowledge, no paper in the literature solves the problem of the optimal location of HEMS sites computing accurately the transport time of both ambulance and helicopter by using suitable tools, like GIS, to model properly their networks and to take into account the orography constraints in the helicopter flight. Our work aims to overcome this lack.

3 Statement of the Problems

Let us consider a fleet of helicopters, each one located in a base $b \in B$, a set S of candidate HEMS sites for rendez-vous interventions, a set D of demand points, a set H of hospitals specialized in emergency treatment and provided with helipad, a threshold time τ_d within which the injured person $d \in D$ has to be hospitalized and a number $n << |S|$ of HEMS sites that can be opened. Let us suppose that for each pair of locations i, j we know a priori the minimum times t'_{ij} and t''_{ij} necessary to go from i to j by ambulance and helicopter, respectively.

The first problem that we want to solve consists in deciding which n HEMS sites are to open to maximize the number of demand points that can be served through a rendez-vous intervention within their threshold times. The latter condition imposes that a demand point $d \in D$ is served if there is an open HEMS site s, a base $b \in B$ and a hospital $h \in H$ such that (see Fig. 1):

$$max\{t'_{ds}, t''_{bs}\} + t''_{sh} \leq \tau_d \quad (1)$$

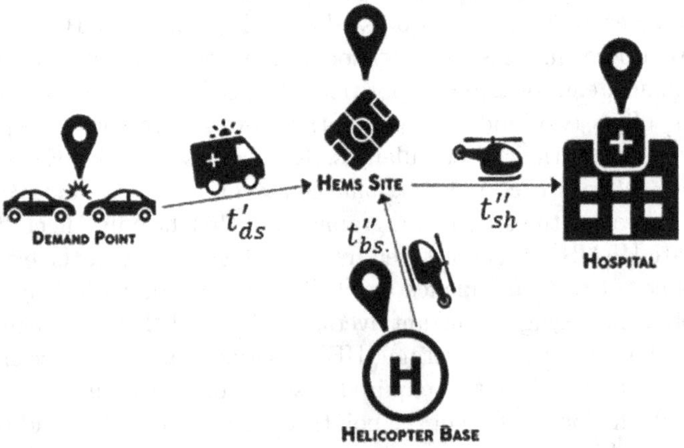

Fig. 1. Rendez-vous intervention through ambulance and helicopter.

We call this problem the *Maximum Coverage of Demand Points* (MCDP). MCDP is an NP-hard combinatorial optimization problem since it is a *maximal covering location* problem [3,6]. Roughly speaking, this means that the number of feasible solutions of the problem grows quickly as the input size increases, so that finding the optimal solution in a reasonable computational time becomes prohibitive. Therefore it is necessary to consider a heuristic solution algorithm, that although cannot ensure to find an optimal solution, returns an acceptable feasible solution within a definitely shorter time.

The second optimization problem that we want to solve consists in covering all the demand points, within their threshold times, minimizing the total number of HEMS sites opened. We call this second problem the *Minimum HEMS Sites Opening* (MHSO). This problem is also NP-hard since can be viewed as a *minimum facility location* problem [3,4].

4 Modeling and Solution Algorithms

In Sect. 4.1 we explain in detail how the input of the problems MCDP and MHSO have been properly built to model the real situation; in Sect. 4.2 we describe a pre-processing algorithm to reduce the number of demand points; finally, in Sect. 4.3 we describe the algorithms used to solve both the problems.

4.1 Modeling the Input Data

HEMS Bases (set B)
A HEMS base is the place where the helicopter is hosted on a pad and/or in a hangar with line maintenance facilities and where resting flying crews and medical personnel are ready to intervene. Their locations are chosen on the basis of a wide variety of considerations, sometimes not directly connected to the service. In our work the set B of HEMS bases is considered a given data.

Candidate HEMS Sites (set S)
The EASA definition of HEMS operating site constrains only its size and the lighting system it is provided with. Sports fields appear excellent candidates because: (i) they are very wide-spread, being located in almost all Italian municipalities; (ii) they enjoy a lighting system already present; (iii) they generally have size that meets the constraints imposed by EASA, taking into account also the helicopter size.

Demand Points (set D)
Like for every location allocation problem it is required to choose the potential rescue points in need of coverage, the so called demand points. To be sure that they represent real places, we decided to use all the intersections in the road network. Although this is a somehow arbitrary choice, it is straight and easily implemented and it is coherent with the observation that intersections are the places where road accidents are statistically more frequent.

Hospitals Specialized in Emergency Treatment (sets $\tilde{H}$ and H)
With set $\tilde{H}$ we indicate in general all facilities provided with first aid service able to grant acceptance to people in need of emergency treatments (e.g., in Italy, the so called DEA = Emergency Acceptance Department). Within set $\tilde{H}$ we consider the subset H including only the hospitals provided with helipad. Such hospitals are the only ones suitable for rendez-vous intervention mode.

Number of HEMS Sites to Open (n)
This parameter strictly depends on the budget. By simulating different investment amounts it has been possible to analyze the relative number of covered demand points. This value can range from 1 to the maximum number of sports fields present in the interest area, with dimension compliant with EASA rules.

Threshold Time (τ_d)
We model the threshold time of rescue intervention by considering different values for each demand point since the success of an intervention depends on the kind of trauma to be addressed and that can have a different level of urgency. It is also interesting to investigate the following value of threshold time for each $d \in D$:

$$\hat{\tau}_d = min_{s \in S, b \in B, h \in H}\{max\{t'_{ds}, t''_{bs}\} + t''_{sh}\} \tag{2}$$

since it represents the minimum time to serve each demand point with the rendez-vous intervention. Through this threshold time value a demand point is considered to be covered only if candidate HEMS site, which grants the minimum intervention time, is open. Since this threshold time is very restrictive, in the experimentation we perform also a sensitivity analysis around it.

Ambulance Travel Times (t'_{ij})
First of all, to model ambulance travel times we need the information on the road network of the area of interest. This can be obtained through a GIS file. Moreover it is necessary to define a speed for each kind of road arc of the network, in order to obtain the time necessary to travel it by ambulance. Then $\forall i \in D, j \in S \cup \tilde{H}$ the ambulance travel times, t'_{ij}, can be computed by applying the Dijkstra's algorithm contained in ArcGIS Network Analyst, on such a network.

Helicopter Travel Times (t''_{ij})
We can create a network to approximate the free movement of the helicopter using shape file containing the height surveying of the area of interest by a spot height map representation (helicopter network). We build such a network following the altimetric profile land just over 300 m of altitude - from sea level - to avoid unnecessary irregularities over the flat area. The helicopter minimum travel times, t''_{ij}, can be computed by applying again the Dijkstra's algorithm of ArcGIS Network Analyst, on such a network.

4.2 Pre-processing Algorithm

Before solving both the MCDP and the MHSO problems, we run a pre-processing algorithm to eliminate from the set D the demand points for which the ambulance intervention mode is faster than a rendez-vous intervention. For this purpose, from set D we eliminate the demand points d such that (see again Fig. 1):

$$\min_{h \in \tilde{H}} t'_{dh} \leq \hat{\tau}_d \tag{3}$$

where $\hat{\tau}_d$ is defined by (2). In this way set D will contain only the demand points that can really guarantee a reduction of travel time by the rendez-vous intervention mode.

4.3 Solving the MCDP and the MHSO

The MCDP problem has been solved by exploiting the high potentiality of the location-allocation solver present in ArcGIS for Desktop 10. In particular, the tool of ArcGIS Network Analyst [14] for solving the *maximize coverage* problem has been used. Such a tool requires the definition of the following parameters.
- *Facilities to choose*: it is the number of facilities that must be located. In our case it is the number n of HEMS sites to be opened.
- *Impedance cutoff*: this parameter represents the maximum time beyond which a demand point results not allocated, i.e. not-covered. For each demand point $d \in D$ we set the cut-off time equal to the threshold time τ_d.

In similar way, the MHSO problem has been solved by using the ArcGIS tool created to address the *minimize facility* problem. This tool requires the definition of the same parameters of MCDP except for the *Facilities to choose* parameter since it will be known through the solution of the MHSO itself.

Both the *maximize coverage* and the *minimize facility* problems are two variants of the more general *facility location* problem. Given a set F of candidate facilities and a set D of demand points with a weight, the latter consists in choosing a subset of the facilities, P, of fixed cardinality, such that the sum of the weighted distances from each element of D to the closest element in P is minimized. Since this is an NP-hard combinatorial optimization, heuristics are necessary to solve it in reasonable time. To this aim, the location-allocation solver starts by generating an origin-destination matrix of shortest-path costs between all the facilities and demand point locations along the network. It then builds an edited version of the cost matrix by a process known as Hillsman editing [5] to solve different problem types. Then the location-allocation solver generates a set of semi-randomized solutions and applies a vertex substitution heuristic [9] to refine them in order to create a set of good solutions. A metaheuristic then combines the latter to build better solutions. When no additional improvement is possible, the metaheuristic returns the best solution found. The combination of an edited matrix, semi-randomized initial solutions, a vertex substitution heuristic and a refining metaheuristic quickly yields near-optimal results.

5 Computational Experiments

5.1 Construction of the Data Set

The study has been focused on Como, Lecco and Varese provinces, an area subjected to the responsibility of "SOREU dei Laghi". In this area of interest there is only one HEMS base located in Villa Guardia (Como province) that currently uses a helicopter AW139 whose largest dimension is equal to 16.66 m. Therefore, in this application, the candidate HEMS sites must have an extension of at least 3,200 m^2, according to EASA constraints.

Ambulance Network

Thanks to free download of GIS files from the Lombardia's Geoportal [12] website it has been possible to model the real road network of the three provinces. In this model the average speeds along four different kinds of road, have been considered according to Table 1. The resulting ambulance network is defined by more than 20,000 arcs and 8,313 intersections.

Table 1. Classification of the road arcs in the ambulance network.

Type of road arc	Average speed [Km/h]
Highway	100
State road	60
Town road	45
Intersection	30

Helicopter Network

The network created to approximate the free movement of the helicopter has been obtained from the shape file (Point Listed 10000 CT10, [12]) containing the height surveying of the Lombardia Region by a spot height map representation. The HEMS network that has been created follows the altimetric profile land just over 300 m of altitude - from sea level - as stated before. The rescue flight mission has been approximated as follows: (i) take-off from Villa Guardia HEMS base, with a Rate of Climb (ROC) of 1,000 ft/min and a True Air Speed (TAS) of 80 kt; (ii) cruise constrained along the network border at a speed of about 124 kt; (iii) landing, with a descent rate equal to 385 ft/min and a TAS of 70 kt.

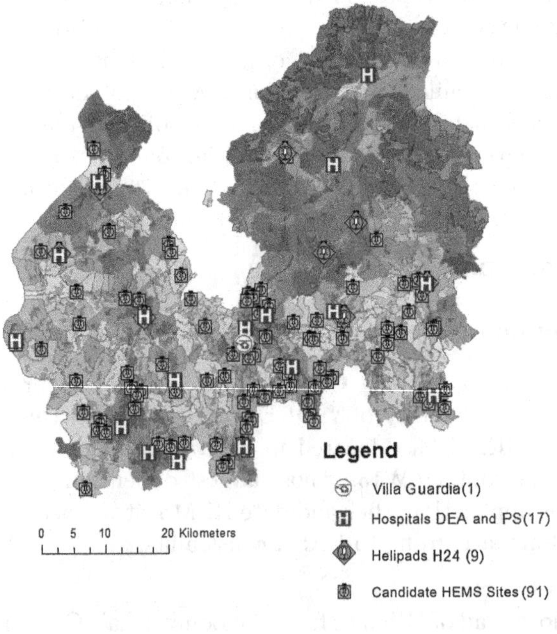

Fig. 2. Location of the HEMS base "Villa Guardia", candidate HEMS sites and Hospitals (DEA) in the case study. (Color figure online)

More than 170,000 arcs and 56,991 nodes have been created to simulate the helicopter flight. The huge number of arcs and nodes of the helicopter network compared to the ambulance network, is due to the fact that the former is more discretized to better simulate the HEMS flight freedom.

The HEMS candidate sites have been determined through a conditional query relative to the minimum size (at least 3,200 m^2, as explained before) of all sports fields, in the area of interest: it results that 91 of them are suitable. Their positions are indicated in Fig. 2 by the green squares. In the covering of the demand points, we take also into account existing helipads already dedicated to helicopter rescue H24: they are 9, as indicated by the red diamonds. In the same figure the locations of the 17 hospitals specialized in emergency treatment and equipped by helipad (set H) are also indicated. The demand points are given by all the intersections in the road network, therefore $|D| = 8,313$.

Hereafter, we consider also the H24 helipads as candidate HEMS sites to verify their usefulness in the rendez-vous interventions. Five of them are excluded being too close to the hospitals. In this way the candidate HEMS sites considered become 95.

5.2 Computational Results

Solution of the Pre-processing Algorithm

The first numerical result is the application of the pre-processing algorithm (Sect. 4.2) to the data set described in Sect. 5.1, in order to consider only the demand points that can actually benefit of a travel time saving by a rendez-vous intervention. Thanks to the pre-processing, a 73.8% reduction of all the demand points has been obtained. As represented by the light-blue circles, in Fig. 3, the 2,180 demand points resulting are located far from hospitals, according to the fact that if a demand point is close to a hospital, the ambulance transport mode is faster. In addition it has been evaluated the sensitivity of the number of demand points, originated from the pre-processing algorithm, to ambulance network speed changes. The outcome of this evaluation is shown in Fig. 4.

The result obtained shows how a less number of demand points for which the rendez-vous transport mode is more convenient (exclusively in terms of time) corresponds to an increase of ambulance average speed. This evaluation makes evident that it is fundamental a statistical analysis on a real ambulance average speeds, considering different time slots, for a correct representation of the model. We also performed a sensitivity analysis of the pre-processing algorithm to the operational transfer times (e.g., the time spent to slow down and lock the rotors, to transfer the patient from the ambulance to the helicopter, etc.). The decrease of the demand points, given by the pre-processing, becomes considerable by increasing the transfer time: e.g. for a transfer time of 5 min, $|D|$ slumps to about 200.

Solution of the MCDP

This problem has been solved for different values of n, while, for each demand point $d \in D$, the threshold time has been initially fixed to the value $\hat{\tau}_d$ given by

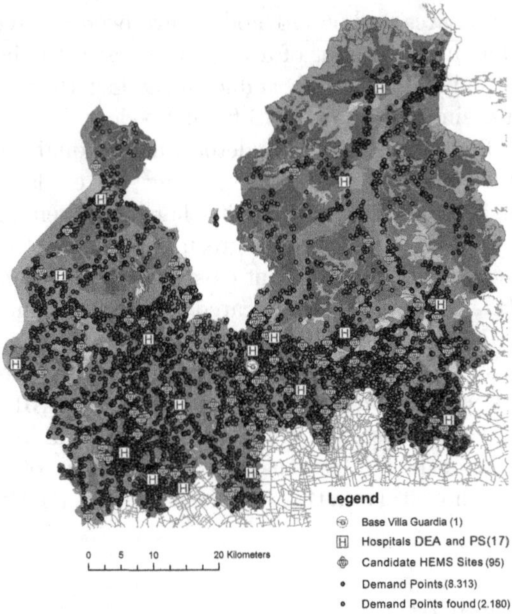

Fig. 3. Reduction of the candidate HEMS sites through the pre-processing algorithm. (Color figure online)

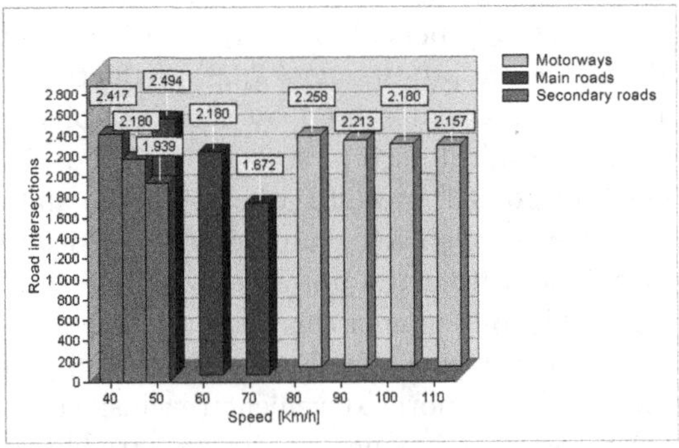

Fig. 4. Sensitivity of pre-processing to ambulance network speed changes.

formula (2). As shown in Fig. 5, the number of demand points covered increases with the number n of candidate HEMS site opened. For instance, from this graphic, it can be found that opening 12 HEMS sites, of the 95 candidates, it is possible to ensure a coverage of the 60% of all the demand points (obtained after the preprocessing). Their location can be seen in Fig. 6.

Solution of the MHSO

The solution of the MHSO shows the minimum number of HEMS sites necessary to guarantee the complete coverage of the demand points d, within their

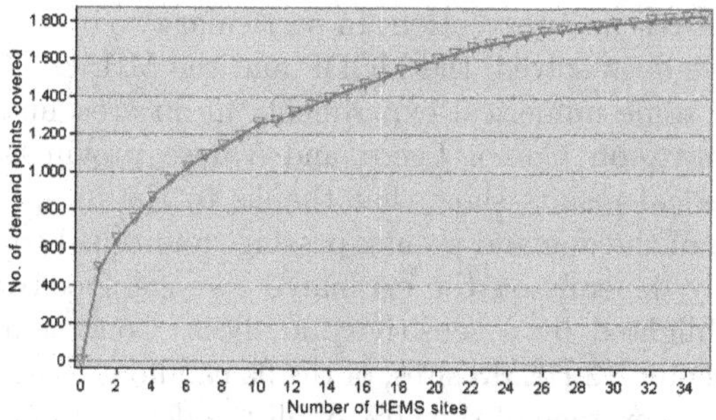

Fig. 5. Number of demand points covered varying the number of HEMS sites opened

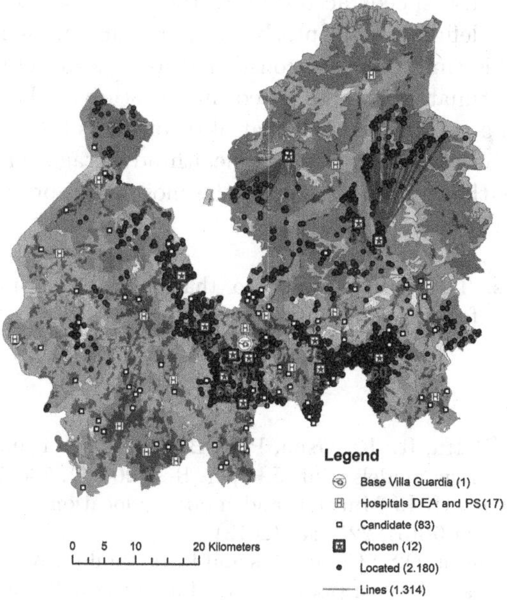

Fig. 6. Location of the HEMS sites in the solution of the MCDP with $n = 12$.

threshold times τ_d. We experimented different values of τ_d, starting from the more restrictive value $\hat{\tau}_d$ and increasing it by 1 min until $\hat{\tau}_d + 10$. The number of necessary HEMS sites to be opened for covering all demand points within threshold times $\hat{\tau}_d$ is equal to 52. While when $\tau_d = \hat{\tau}_d + 10$, all demand points are covered opening only 12 HEMS sites.

6 Conclusions

In this paper we addressed the strategic positioning of HEMS sites, where the transfer of injured patient from the ambulance to the helicopter takes place, for the rendez-vous rescue intervention. In particular two optimization problems related to this have been solved: the MCDP and the MHSO.

We performed some numerical experiments in an area under "SOREU dei Laghi" responsibility, on Como, Lecco and Varese provinces, in Lombardia region. The numerical results show that thanks to the pre-processing there is a 73.8% reduction of the demand points, passing from 8,313 to 2,180. Moreover this result strongly depends on the estimated average speeds used in network definition, as highlighted by a sensitivity analysis. Concerning the MCDP it results that by opening 12 HEMS sites, of the 95 candidates, it would be already possible to ensure a coverage of the 60% of all the demand points.

Thanks to GIS, this work overcomes some limits met in previous studies. Indeed, not only the exact layout of the road network has been evaluated, but also the different cruise speeds according to the road classification. Even more has been done for defining helicopter's coverage timing. In fact, while in all previous studies this time was just considered as a ratio between the traveled distance and the estimated average speed, in our study, it has been accurately computed through a special network that also considers the territory altimetry.

Future works could consider also the health advantages given by the possibility to transport the injured person to the most appropriate hospital rather than to the nearest one.

Acknowledgments. The authors want to thank Eng. Giovanna Marchionni of Politecnico di Milano for her help in using ArcGIS.

References

1. Erdermir, E.T., Batta, R., Rogerson, P.A., Blatt, A., Flanigan, M.: Joint ground and air EMS coverage models. Eur. J. Oper. Res. **207**(2), 736–749 (2010)
2. Furuta, T., Tanaka, K.I.: Minisum and minimax location models for HEMS. J. Oper. Res. Soc. Jpn **56**(3), 221–242 (2012)
3. Garey, M.R., Johnson, D.S.: Computers and Intractability: A Guide to the Theory of NP-Completeness. Series of Books in the Mathematical Sciences. W.H. Freeman & Co., San Francisco (1979)
4. Guha, S., Khuller, S.: Greedy strikes back: improved facility location algorithms. J. Alg. **31**, 228–248 (1999)
5. Hillsman, E.L.: The p-median structure as a unified linear model for location-allocation analysis. Environ. Plann. A **16**(3), 305–318 (1984)
6. Luiz, A.N.L., Pereira, M.A.: A lagrangean/surrogate heuristic for the maximal covering location problem using Hillsman's edition. Int. J. Ind. Eng. **9**(1), 57–67 (2002)
7. Peters, J., Beekers, C., Eijk, R., Edwards, M., Hoogerwerf, N.: Evaluation of Dutch helicopter emergency medical services in transporting children. Air Med. J. **33**(3), 112–114 (2014)
8. Svenson, J.E., O'Connor, J.E., Lindsay, M.B.: Is air transport faster? a comparison of air versus ground transport times for interfacility transfers in a regional referral system. Air Med. J. **24**(5), 170–172 (2005)
9. Teitz, M.B., Bart, P.: Heuristic methods for estimating the generalized vertex median of a weighted graph. Operations Res. **16**, 955–961 (1968)

10. Vasquez, M., Hao, J.K.: A heuristic approach for antenna positioning in cellular networks. J. Heuristics **7**(5), 443–472 (2001)
11. Weerheijm, D.V., Wieringa, M.H., Biert, J., Hoogerwerf, N.: Optimizing transport time from accident to hospital: when to drive and when to fly? ISRN Emerg. Med. **2012**, Article ID 508579, 1–5 (2012)
12. http://www.cartografia.regione.lombardia.it/geoportale
13. http://www.esri.com/software/arcgis/arcgis-for-desktop
14. http://desktop.arcgis.com/en/arcmap/latest/extensions/network-analyst/algorithms-used-by-network-analyst.htm

An Enhanced Infra-Chromatic Bound for the Maximum Clique Problem

Pablo San Segundo(✉), Jorge Artieda, Rafael Leon, and Cristobal Tapia

Center for Automation and Robotics (CAR), Polytechnic University of Madrid (UPM), C/ Jose Gutiérrez Abascal, 2, 28006 Madrid, Spain
pablo.sansegundo@upm.es

Abstract. There has been a rising interest in experimental exact algorithms for the maximum clique problem because the gap between the expected theoretical performance and the reported results in practice is becoming surprisingly large. One reason for this is the family of bounding functions denoted as *infra-chromatic* because they produce bounds which can be lower than the chromatic number of the bounded subgraph. In this paper we describe a way to enhance exact solvers with an additional infra-chromatic bounding function and report performance over a number of graphs from well known data sets. Moreover, the reported results show that the new enhanced procedure significantly outperforms state-of-the-art.

Keywords: Infra-chromatic · Clique · Approximate-coloring · Branch-and-bound · Combinatorial optimization · Search

1 Introduction

Given a simple undirected graph $G = (V, E)$, with n vertices and m edges, $N(v)$ refers to the neighbor set of vertices adjacent to a vertex $v \in V$ and $G[U]$ to the subgraph induced by a vertex set $U \subseteq V$. A complete subgraph (or clique) is an induced subgraph such that all its vertices are pairwise adjacent, and the maximum clique problem (MCP) consists in finding a clique of maximum cardinality. The MCP is a theoretical deeply studied NP-hard problem which has also found many applications [1]. Examples of clique search appear in network analysis, coding theory, fingerprint matching, bioinformatics [2], robotics [3, 4] and computer vision [5].

In the last decade there has been rising interest in the exact solving of the MCP. At present, almost all successful algorithms are branch-and-bound and use a greedy vertex coloring procedure as bound [6–15]. Interestingly, the reports provided by these practical algorithms average a much better performance than is to be expected by the theoretical algorithms alone. A brief overview of theoretical results now follows:

P.M. Pardalos et al. (Eds.): MOD 2016, LNCS 10122, pp. 306–316, 2016.
DOI: 10.1007/978-3-319-51469-7_26

- The MCP is *NP*-hard and can not even be approximated in polynomial time within a factor of $|V|^{1/3-\varepsilon}$, for any $\varepsilon > 0$(unless $P = NP$) [16].
- There can be as much as $3^{n/3}$ distinct maximal cliques to enumerate in a graph of size n [17], which is an upper bound for all enumerative solvers.
- Branch-and-bound solvers do not need to explore all the maximal cliques. An important number of them may be discarded when, during construction, it is shown that they cannot improve the current incumbent clique. In [18], the worst case running time was set to $O(2^{n/3})$ and, at present, the value of this bound is $O(2^{n/4})$ [19], in both cases with exponential space consumption. We note that these algorithms have not been implemented in practice and literature provides no experimental evaluation.
- An interesting recent work is [20] which is concerned with the family of branch-and-bound solvers which use coloring as bounding function; these are also the reference algorithms for this work. [20] sets $\Omega(2^{n/5})$ as lower bound for this family of solvers.

The latter result is especially relevant because it is related to state-of-the-art exact algorithms and, as far as we are aware, the gap between the good performances shown by the experimental algorithms and the theoretical result remains to be explained. Motivated by this fact, this work contributes with an empirical study of cutting-edge *infra-chromatic* bounding functions which have recently been described for the MCP. These functions are able to bind the clique number of a given subproblem below its chromatic number by additional computational effort. More precisely, this implies that the theoretical $\Omega(2^{n/5})$ bound does not hold for these algorithms.

Leading infra-chromatic experimental algorithms for the MCP at present are MaxCLQ [14], and its improved variant IncMaxCLQ [15], as well as bit-parallel BBMCX [11] — see recent comparison surveys [21, 22]. The contribution of this work is experimental. First we describe a way to enhance BBMCX with an additional infra-chromatic bounding function closely related to the one described in MaxCLQ. This bounding function can also be applied to earlier BBMC variants [8–10]. The paper reports performances of two enhanced variants (including BBMCX), together with IncMaxCLQ.

The remaining part of this work is structured in the following way: Sects. 2 and 3 cover prior related algorithms. The new bounding function for exact maximum clique is described in Sect. 4 and validated empirically in Sect. 5. Finally, Sect. 6 briefly discusses the reported results and Sect. 7 presents some conclusions.

2 Related Bounding Functions for the MCP

In the last decade, greedy sequential vertex coloring (SEQ) has proved to be a good bounding function for exact branch-and-bound solvers for the MCP. SEQ is an $O(n^2)$ worst-case bounded procedure which iteratively assigns the lowest possible color number to each vertex that is consistent with the current partial coloring. The first maximum clique algorithm of this type described in literature is Fahle's [6]. Since then, much research has been devoted to improve the basic SEQ bounding function.

We denote by $C(G) = \{C_1, C_2, \ldots, C_k\}$ a k-coloring of graph G, where C_i refers to the (color) set which contains vertices assigned color number i. For a given k-coloring $C(G)$, maximum clique solvers use a *recoloring* bounding function to lower the color number k of a vertex $v_1 \in C_k$ by a swap movement with two already colored vertices $v_2 \in C_i$ and $v_3 \in C_j$ where $i<j<k$. A successful recoloring leads to the new coloring $v_1 \in C_i$ and v_2, $v_3 \in C_j$. In practice, recoloring can be applied to every vertex of the SEQ coloring, as described originally for algorithm MCS [12], or relaxed to a particular subset, as in BBMCR [9].

A bounding function that could produce bounds below the chromatic number of the bounded subproblem was first described in algorithm MaxCLQ [14], and later improved in IncMaxCLQ [15]. It was defined in terms of a reduction of the MCP over a colored graph to a partial maximum satisfiability problem (PMAX-SAT) as follows: (I) Each vertex of the graph maps to a logical variable. (II) Each PMAX-SAT hard clause consists of two negated literals and represents a non-adjacency relation between the pair of corresponding vertices. (III) Each PMAX-SAT soft clause represents a distinct color set and contains the corresponding positive literals that map to vertices of the set.

Figure 1 provides a simple example of the reduction when applied to an odd-cycle:

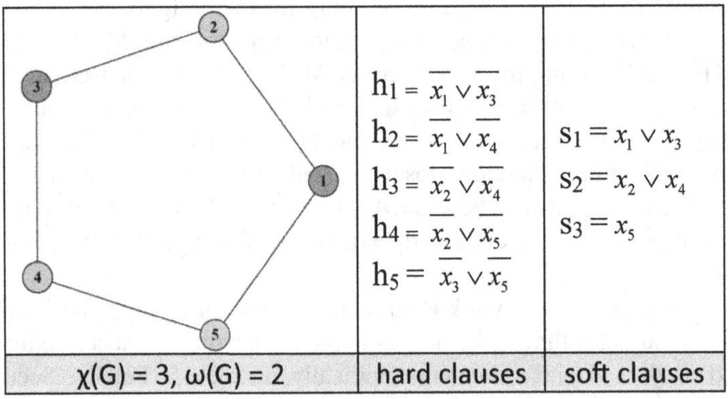

Fig. 1. An example of an MCP reduction to PMAX-SAT. (Color figure online)

The PMAX-SAT problem's equivalence to the MCP consists in finding an assignment of literals which satisfy the maximum number of soft clauses (the color sets) and all hard clauses (the graph structure) — see [14, 15] for a more detailed description. The cleverness of this reduction for branch-and-bound exact MCP solvers is that well-known pruning strategies for PMAX_SAT may be used as bounding functions for the MCP. The main inference is based on propagating literal assignments of *unit* soft clauses (singleton color sets) until an empty clause is reached. When this occurs, the soft clauses (color sets) that make part of the reasoning are said to be *inconsistent*. Intuitively, the subgraph induced by an inconsistent set of k colors cannot contain a clique of size k or bigger, so the bound provided by the number of colors in the original coloring may be decremented by one *for each inconsistent set of colors found*.

In the example of Fig. 1, one such inconsistent set of soft clauses is $\{s_1, s_2, s_3\}$, which can be obtained by the following inference: (I) $x_5 \leftarrow TRUE$ since s_3 is a unit clause. (II) Propagating $x_5 \leftarrow TRUE$ value leads to $s_2 = x_4$, according to hard clause h_4, and $s_1 = x_1$, according to hard clause h_5. (III) $x_4 \leftarrow TRUE$, since s_2 is a unit clause. IV) Propagating $x_4 \leftarrow TRUE$ leads to the empty clause $s_1 = \emptyset$, according to the hard clause h_2. Consequently, the original color bound of the graph, 3, can be reduced to 2. Note that the chromatic number of the graph $\chi(G) = 3$ is higher than the new bound.

The term *infra-chromatic* describing bounding functions for the MCP was first employed in the bit-parallel BBMCX algorithm. There, the authors propose an inference procedure in terms of color sets which integrates well with the bitstring encoding and branching of the family of BBMC algorithms [8–11]. More precisely, BBMCX looks for triplets of color sets $\{C_x, C_y, C_z\}$ such that $|C_x| = 1$ and no triangles exist in the induced graph $G[C_x \cup C_y \cup C_z]$. The reported results show that this particular case of inconsistency can be computed very efficiently achieving a very good tradeoff between pruning ability and computational effort.

3 The Prior Infra-Chromatic Procedure

Rising interest in exact maximum clique search combines exhaustive enumeration of maximal cliques with a bounding function based on approximate coloring. A step of the search (typically a recursive call to the algorithm) branches on a vertex to enlarge a clique, and a branch of the search tree corresponds to a maximal clique. The solution to the MCP is any leaf node of maximum cardinality.

In this work we borrow the notation for sets employed in BBMCX [11]. More specifically:

- S denotes the clique to be enlarged at any point during search.
- S_{max} is the largest clique found at any point during search.
- U denotes the set of vertices of the current subproblem.
- U_v denotes the set of vertices of the child subproblem resulting from branching on vertex $v \in U$.

- $G_v = G[U_v]$ refers to the induced graph subproblem which results from branching on vertex v.
- F is a set of (forbidden) color numbers employed in the infra-chromatic bounding function described in BBMCX.

Of interest to this work is the bounding function called for each subproblem U during search. Algorithm 1 outlines the bounding procedure which roughly corresponds with the one described in BBMCX. Procedure UPPERBOUND computes the SEQ coloring of the subproblem $G[U_v]$ in the main loop (steps 1 to 12), by enlarging one color set at a time. Once all color sets below a certain pruning threshold k_{min} have been constructed, each selected vertex goes through the infra-chromatic filter INFRACHROM_I described in Algorithm 2.

Algorithm 1. Outline of the BBMCX [11] bounding function.

INPUT: A graph $G[U_v]$ and a color threshold k_{min}

UPPERBOUND (U_v, $k_{min} = |S_{max}| - |S| + 1$)

Initial step: $U \leftarrow U_v$, $F \leftarrow \varnothing$, $k \leftarrow 1$, $C \leftarrow \varnothing$

1. **repeat until** $U = \varnothing$
2. $C_k \leftarrow U$
3. **repeat until** all vertices in C_k have been selected
4. select the first possible vertex v from C_k
5. result ← FAIL
6. **if** $k \geq k_{min}$ **then** result ← INFRACHROM_I (v, k_{min}, $C_1, C_2, \cdots, C_{k_{min}-1}, C_k$, F)
7. **if** result = FAIL **then** $C_k \leftarrow C_k \setminus N(v)$
8. $U \leftarrow U \setminus \{v\}$
9. **endrepeat**
10. $C \leftarrow C \cup C_k$
11. $k := k + 1$
12. **endrepeat**
13. **return** $|C|$

The filter function looks for pairs of distinct color classes (i, j) such that vertex v is adjacent to exactly just one vertex w in color i and color j contains no common neighbors to both v and w. If the filter succeeds (and returns SUCCESS) the vertex is removed from the original coloring and the other two inconsistent color sets are tagged as forbidden, so that they do not take part in further inferences. We refer the reader to [11] for additional design and implementation details.

Algorithm 2. Outline of the infra-chromatic bounding function called by Algorithm 1.

INFRACHROM_I (v, k_{min}, $C_1, C_2, \cdots, C_{k_{min}-1}, C_k$, F) //F is a list of forbidden colors, $v \in C_k$

1. **find** any color class i not in F such that it has only one neighbor w in common with v
2. **find** any color class $j \neq i$ not in F such that it contains no common neighbors to both w and v
3. **if** steps 1 and 2 are successful
4. add color labels i and j to F
5. remove v from C_k
6. **return** SUCCESS
7. **repeat** steps 1 to 6 with all feasible colors
8. **return** FAIL

4 The Enhanced Infra-Chromatic Procedure

We enhance the previously described bounding function with another infra-chromatic bound which employs unit clause propagation described in PMAX-SAT based solvers. Algorithms 3 and 4 describe the general outline of the new enhanced function. The coloring C obtained after applying INFRACHROM_I to SEQ is now filtered again by new INFRACHROM_II. The function searches for inconsistent color sets using unit clause inferences and returns the number of conflicts found. As explained in the introductory section, the resulting new bound is the difference between the previous bound (the size of the coloring) and the number of conflicts outputted by INFRACHROM_II.

Algorithm 3. Outline of the enhanced Algorithm 1.

UPPERBOUND (U_v, $k_{min} = |S_{max}| - |S| + 1$) //$k_{min}$ is a color number threshold

//... same steps as Algorithm 1

12. **endrepeat**
13. **if** $|C| \geq k_{\min}$ **then**
14. *num_conf*←INFRACHROM_II(C , F)
15. **endif**
16. **return** $|C|$ - *num_conf*

In the current implementation of INFRACHROM_II, the forbidden color set F and the color and neighbor sets are encoded as bitsets. The main inference is the set intersection operation in step 7 which benefits from bit-masking, as well as the initial step 1. The procedure runs in $O(|V|^2)$ in the worst-case. In the current implementation, we obtain the best performance when color sets are examined in non-increasing order. This strategy tends to find conflicts earlier because the highest color sets of the SEQ coloring are expected to have a small size. We note that it is easy to envisage more greedy variants of Algorithm 4 by filtering out color sets from the reasoning with additional constraints. It is also worth noting that the integration of INFRACHROM_II in the prior bounding function (see Algorithm 3) is not restricted to BBMCX, and may also make part of other exact approximate-color exact solvers such as MCS, BBMC, BBMCR and other published BBMC variants.

Algorithm 4. Outline of the infra-chromatic procedure called by Algorithm 3.

```
INPUT: A vertex coloring C and a list F of color numbers
INFRACHROM_II (C, F)
Initial step 1: Remove all color sets with labels in F from C
Initial step 2: UC ← all singleton color sets in C, num_conf ← 0
repeat until UC = ∅
UC' ← ∅
move the first singleton color set in UC to UC'
    repeat until UC' = ∅
        select (and remove) a singleton color set C_k = {v} from UC'
        tag C_k as inconsistent
        for every C_{j≠k} do C_j ← C_j ∩ N(v)
        for every C_{j≠k}
            if |C_{j≠k}| = 1 then UC' ← UC' ∪ C_j
            else if |C_{j≠k}| = ∅ then
                num_conf ← num_conf + 1
                remove C_j and all sets tagged as inconsistent from UC
                restore all color sets not tagged as inconsistent
                goto step 1
            endif
        endfor
    endrepeat
endrepeat
return num_conf
```

5 Experiments

To evaluate the new proposed infra-chromatic bounding function we have considered the following algorithms in this research:

1. BBMC [8, 9]: The original bit-parallel algorithm for the MCP, which uses only the SEQ color bound.
2. BBMC+*I*: BBMC enhanced with the new PMAX-SAT based infra-chromatic bound described in Algorithm 3.

3. BBMCR [9]+*I:* The BBMC recoloring variant enhanced with the new infra-chromatic bound.
4. BBMCX [11]+*I:* The BBMC infra-chromatic variant enhanced with the new infra-chromatic bound.
5. IncMaxCLQ [15]: The latest PMAX-SAT based algorithm, as provided by its main developer.

All the algorithms were tested over a big number of structured graphs from public datasets which may be found in other maximum clique reports elsewhere. The majority of structured instances were presented at the Second DIMACS Implementation Challenge[1]. The rest are taken from the BHOSHLIB benchmark[2] (more specifically, the *frb30-15* collection). The algorithms were also run against uniform random graphs, with sizes ranging from 150 to 15000 vertices and varying densities.

The hardware used in the experiments was a 20 core XEON with 64 GB of RAM running a Linux OS and all algorithms considered were run on a single core of this machine. With respect to initial configurations, the enhanced BBMC infra-chromatic variants use the more recent preprocessing ideas, that is, the initial sorting of vertices described in IncMaxCLQ, and a strong initial solution computed by a state-of-the-art heuristic, as recommended in [13].

Due to space constraints we report in this manuscript the performance of algorithms BBMCR+*I,* BBMCX+*I* and IncMaxCLQ against 37 representative structured graphs — solved in more than 0.01 s by at least one algorithm.

6 Discussion

Table 1 reports the performance of the enhanced algorithms BBMCR+*I,* BBMCX+*I* as well as IncMaxCLQ. Interestingly, out of the 37 instances reported, BBMCR+*I* is faster than BBMCX+*I* in 25 of them. This would indicate that recoloring (BBMCR), followed by a single infra-chromatic filter, is more appropriate than the double infra-chromatic filter of BBMCX+*I.* This is corroborated by the number of steps taken by both algorithms.

In relation to IncMaxCLQ, the enhanced algorithms perform better in 27 of the 37 instances. Worth noting is that IncMaxCLQ clearly outperforms the latter in *keller5* and *C250.9*. Overall, IncMaxCLQ produces smaller search trees but with higher computational cost, the tradeoff being favorable only in very dense graphs (that is, $p \geq 0.9$) with the exception of *keller5*. It is also worth noting that the number of steps taken by both enhanced algorithms are much less, on average, than those reported for BBMCX in [11].

1 http://cs.hbg.psu.edu/txn131/clique.html.

2 http://www.nlsde.buaa.edu.cn/~kexu/benchmarks/graph-benchmarks.htm.

Table 1. A comparison of different infra-chromatic exact maximum clique algorithms. Time is measured in seconds with precision of milliseconds. Header *Inc/BBMC* compares performance as a ratio. Header *p* refers to uniform density and header ω is the size of the maximum clique. Cells in bold show the best times in each row. A *step* is a recursive call in the BBMC enhanced algorithms.

			IncMaxCLQ [15]		BBMCR [9]+/		BBMCX [11]+/		Inc/BBMC
	p	ω	steps	time	steps	time	steps	time	time
C250.9	0.90	44	734186	**228**	40940184	832	34457025	1144	0.3
MANN_a27	0.99	126	290	**0.266**	3947	0.330	2749	0.318	0.8
MANN_a45	0.99	345	21595	101	40024	**39.6**	39944	41.4	2.5
brock200_1	0.75	21	2700	0.482	23006	**0.250**	29960	0.357	1.9
brock400_1	0.75	27	890529	**182**	11695876	233	16761126	358	0.8
brock400_2	0.75	29	711295	139	3594552	**84.4**	5372259	132	1.7
brock400_3	0.75	31	887764	170	803747	**25.0**	1339642	40.9	6.8
brock400_4	0.75	33	640279	129	575792	**19.9**	575787	20.2	6.5
brock800_1	0.65	23	39290054	7846	199341746	6381	134188081	**4295**	1.8
brock800_2	0.65	24	49803966	10011	160679838	5380	107980185	**3586**	2.8
brock800_3	0.65	25	17522630	3759	52854672	**2012**	79775569	3037	1.9
brock800_4	0.65	26	24600360	4768	45605176	**1669**	68040462	2532	2.9
dsjc1000.5	0.50	15	1562003	197	5323240	**157**	7226392	211	1.3
dsjc500.5	0.50	13	19303	2.63	93413	**1.26**	123963	1.66	2.1
frb30-15-4	0.82	30	22	0.213	1	**<0.001**	1	**<0.001**	213
frb30-15-5	0.82	30	9	0.163	1	**<0.001**	1	**<0.001**	163
gen400_p0.9_55	0.90	55	320	1.05	1	**<0.001**	1	**<0.001**	1052
gen400_p0.9_65	0.90	65	75	0.285	2	**<0.001**	2	**<0.001**	285
gen400_p0.9_75	0.90	75	160	0.238	1333	**0.063**	1314	0.073	3.8
hamming10-2	0.99	512	0	25.0	1	**<0.001**	1	**<0.001**	24992
keller5	0.75	27	249409	**127**	90179996	2122	95938710	2667	0.1
p_hat1000-1	0.25	10	1700	0.503	17102	**0.301**	20948	0.361	1.7
p_hat1000-2	0.49	46	51658	**43.2**	843212	54.1	1237187	93.2	0.8
p_hat1500-1	0.25	12	56404	4.64	76509	**2.37**	88973	3.18	2.0
p_hat300-3	0.74	36	506	**0.313**	11205	0.359	15797	0.531	0.9
p_hat500-2	0.51	36	254	0.165	2007	**0.093**	2945	0.148	1.8
p_hat500-3	0.75	50	28338	**19.6**	605339	27.7	833148	44.7	0.7
p_hat700-2	0.50	44	613	0.826	16688	**0.815**	22152	1.42	1.0
p_hat700-3	0.75	62	247906	**241**	5474929	408	8241315	718	0.6
san1000	0.50	15	646	0.421	15210	**0.412**	9981	0.750	1.0
san400_0.7_2	0.70	30	736	0.357	1	**<0.001**	1	**<0.001**	357
san400_0.7_3	0.70	22	7364	2.12	1	**<0.001**	1	**<0.001**	2121
san400_0.9_1	0.90	100	99	0.223	1	**<0.001**	2	**<0.001**	223
sanr200_0.7	0.70	18	1366	**0.193**	15210	1.003	9981	0.750	0.3
sanr200_0.9	0.90	42	8029	**2.28**	320666	6.33	305083	9.19	0.4
sanr400_0.5	0.50	13	5041	0.491	13947	**0.265**	18246	0.336	1.9
sanr400_0.7	0.70	21	477635	90.2	4749393	**73.6**	6450260	107	1.2

7 Conclusions

This work describes a simple way to enhance a leading bit-parallel infra-chromatic solver for the maximum clique problem with a PMAX-SAT based infra-chromatic filter, and compares the performance of two enhanced published solvers with another state-of-the-art algorithm. The contribution is a step forward in the search of an adequate explanation for the gap between theoretical results and the excellent performance shown by experimental algorithms for this problem. Moreover, the enhanced algorithms show some of the best times published, to the best of our knowledge, for a number of structured graphs from well known public data sets.

Acknowledgments. This work is funded by the Spanish Ministry of Economy and Competitiveness (grant NAVEGASE: DPI 2014-53525-C3-1-R).

References

1. Bomze, I., Budinich, M., Pardalos, P., Pelillo, M.: The maximum clique problem. In: Du, D.-Z., Pardalos, P.M. (eds.) Handbook of Combinatorial Optimization, vol. 4, pp. 1–74. Springer, New York (1999)
2. Butenko, S., Chaovalitwongse, W., Pardalos, P. (eds.): Clustering Challenges in Biological Networks. World Scientific, Singapore (2009)
3. San Segundo, P., Rodriguez-Losada, P., Matia, D., Galan, R.: Fast exact feature based data correspondence search with an efficient bit-parallel MCP solver. Appl. Intell. **32**(3), 311–329 (2010)
4. San Segundo, P., Rodriguez-Losada, D.: Robust global feature based data association with a sparse bit optimized maximum clique algorithm. IEEE Trans. Robot. **99**, 1–7 (2013)
5. San Segundo, P., Artieda, J.: A novel clique formulation for the visual feature matching problem. Appl. Intell. **43**(2), 325–342 (2015)
6. Fahle, T.: Simple and fast: improving a branch-and-bound algorithm for maximum clique. In: Möhring, R., Raman, R. (eds.) ESA 2002. LNCS, vol. 2461, pp. 485–498. Springer, Heidelberg (2002). doi:10.1007/3-540-45749-6_44
7. Tomita, E., Seki, T.: An efficient branch-and-bound algorithm for finding a maximum clique. In: Calude, C.S., Dinneen, M.J., Vajnovszki, V. (eds.) DMTCS 2003. LNCS, vol. 2731, pp. 278–289. Springer, Heidelberg (2003). doi:10.1007/3-540-45066-1_22
8. Segundo, S., Rodriguez-Losada, D, Jimenez, A.: An exact bit-parallel algorithm for the maximum clique problem. Comput. Oper. Res. **38**(2), 571–581 (2011)
9. San Segundo, P., Matia, F., Rodriguez-Losada, D., Hernando, M.: An improved bit parallel exact maximum clique algorithm. Optim. Lett. **7**(3), 467–479 (2013)
10. San Segundo, P., Tapia, C.: Relaxed approximate coloring in exact maximum clique search. Comput. Oper. Res. **44**, 185–192 (2014)
11. San Segundo, P., Nikolaev, A., Batsyn, M.: Infra-chromatic bound for exact maximum clique search. Comput. Oper. Res. **64**, 293–303 (2015)
12. Tomita, E., Sutani, Y., Higashi, T., Takahashi, S., Wakatsuki, M.: A simple and faster branch-and-bound algorithm for finding a maximum clique. In: Rahman, Md.S, Fujita, S. (eds.) WALCOM 2010. LNCS, vol. 5942, pp. 191–203. Springer, Heidelberg (2010). doi:10.1007/978-3-642-11440-3_18
13. Batsyn, M., Goldengorin, B., Maslov, E., Pardalos, P.: Improvements to MCS algorithm for the maximum clique problem. J. Comb. Optim. **27**, 397–416 (2014)
14. Li, C.-M., Quan, Z.: Combining graph structure exploitation and propositional reasoning for the maximum clique problem. In: Proceedings of ICTAI, pp. 344–351 (2010)
15. Li, C.-M., Fang, Z., Xu, K.: Combining MaxSAT reasoning and incremental upper bound for the maximum clique problem. In: Proceedings of ICTAI, pp. 939–946 (2013)
16. Bellare, M., Goldreich, O., Sudan, M.: Free bits, PCPs and nonapproximability — towards tight results. In: 1995 Proceedings of 36th Annual Symposium on Foundations of Computer Science, pp. 422–431. IEEE (1995)
17. Moon, J., Moser, L.: On cliques in graphs. Isr. J. Math. **3**(1), 23–28 (1965)
18. Tarjan, R.E., Trojanowski, A.E.: Finding a maximum independent set. Technical report, Computer Science Department, School of Humanities and Sciences, Stanford University, Stanford, CA, USA (1976)

19. Robson, J.M.: Finding a maximum independent set in time $O(2^{n/4})$. Technical report 1251-01, LaBRI, Université de Bordeaux I (2001)
20. Lavnikevich, N.: On the complexity of maximum clique algorithms: usage of coloring heuristics leads to the $\Omega(2^{n/5})$ algorithm running time lower bound (2013)
21. Prosser, P.: Exact algorithms for maximum clique: a computational study. Algorithms **5**(4), 545–587 (2012)
22. Wu, Q., Hao, J.K.: A review on algorithms for maximum clique problems. Eur. J. Oper. Res. **242**(3), 693–709 (2015)

Cultural Ant Colony Optimization on GPUs for Travelling Salesman Problem

Olgierd Unold(✉) and Radosław Tarnawski

Department of Computer Engineering, Faculty of Electronics,
Wroclaw University of Science and Technology,
Wyb. Wyspianskiego 25, 50-370 Wroclaw, Poland
olgierd.unold@pwr.edu.pl

Abstract. Ant Colony Optimization (ACO) is a well-established metaheuristic successfully applied to solve hard combinatorial optimization problems, including Travelling Salesman Problem (TSP). However, ACO algorithm as many population-based approaches has some disadvantages, such as lower solution quality and longer computational time. To overcome these issues, parallel Cultural Ant Colony Optimization (pCACO) is introduced in this paper. The proposed approach hybridises Cultural Algorithm with ACO-based $\mathcal{MAX}$-$\mathcal{MIN}$ Ant System. pCACO has been implemented on Graphics Processing Units (GPUs) using CUDA programming model. Through testing nine benchmark asymmetric TSP problems, the experimental results show the new method enhances the solution quality when compared to sequential and parallel ACO, yielding comparable computational time to parallel ACO.

Keywords: Travelling Salesman Problem · Ant Colony Optimization · Cultural Algorithm · GPU computing · CUDA architecture

1 Introduction

Ant Colony Optimization is a population-based metaheuristic proposed by M. Dorigo and colleagues in the early 1990s [10,12]. The approach relies on the idea of modelling ants group foraging. In ACO the indirect communication of a colony of agents called (artificial) ants is implemented. This communication between ants is based on so-called pheromone trails, which is a kind of distributed numeric information. The ants adapt pheromone trails while searching a solution to the problem being solved.

From the very beginning, ACO has been successfully applied to solve a variety of $\mathcal{NP}$-Problems, including Travelling Salesman Problem (TSP) [11,28]. TSP is a typical $\mathcal{NP}$-hard optimization problem, where a travelling salesman wants to travel all cities but each city is supposed to be visited only once [20]. Note that many others $\mathcal{NP}$-problems can be attributed to TSP, such a postman problem, product assembly line, clustering of data arrays, or even DNA sequencing [16].

P.M. Pardalos et al. (Eds.): MOD 2016, LNCS 10122, pp. 317–329, 2016.
DOI: 10.1007/978-3-319-51469-7_27

Starting from the first ACO algorithm, called Ant System [12], all the later improvements have been tested on TSP, such as Ant Colony System [13], $\mathcal{MAX}$-$\mathcal{MIN}$ Ant System ($\mathcal{MMAS}$) [29], Ant-Q system [14], rank-based Ant System [2]. Despite its utility, there are some drawbacks with ACO. A common problem addressed in literature is that its performance suffers from often longer iteration time than others swarm approaches. It is due to slow accumulation of pheromones needed to influence on the algorithm. Another drawback with ACO is that ACO may be susceptible to local minima.

To address these two issues, firstly, we propose to speed-up ACO in the $\mathcal{MMAS}$ version by using Graphics Processing Unit (GPU) computing, and secondly, to hybridise $\mathcal{MMAS}$ and Cultural Algorithm (CA) [25] to avoid premature convergence. CA was proved to be an efficient approach for optimize ACO [30,34], mainly due to so-called its *dual inheritance mechanism*. CA works using two evolution spaces communicating through accept-influence protocol. This double evolutionary mechanism allows converging faster with a better solution.

The proposed parallel Cultural ACO algorithm has been implemented on GPUs using Compute Unified Device Architecture (CUDA) [23] and experimentally compared with a sequential and a parallel ACO (pACO) on a speed-up and solution quality over asymmetric TSP problems taken from TSPLIB library.

The remainder of this paper is organised as follows. In Sect. 2, we briefly review the concepts of Travelling Salesman Problem, (sequential) Ant Colony Optimization, (sequential) Cultural Algorithm and General-Purpose Computing on GPUs. Section 3 introduces some related works on GPU-based ACO and ACO extended by CA. The proposed algorithm is discussed in Sect. 4. Empirical experiments of evaluating the pCACO are conducted in Sect. 5. At last, Sect. 6 gives some concluding remarks and points out the further research directions.

2 Preliminaries

2.1 Travelling Salesman Problem

Travelling Salesman Problem is a well-known combinatorial optimization task [20], proved to be $\mathcal{NP}$-hard. In TSP a set of cities to be visited and distances between them are given. TSP is to find the shortest way of visiting all the cities and returning to the starting point. Regarding a set of cities as vertices connected in pairs by weighted edges, the goal of TSP is to find a Hamiltonian cycle with the least weight in a complete weighted graph. There are many types of TSP, including symmetric euclidean and non-euclidean, asymmetric, dynamic, and special cases, like multiple TSP [36]. In this paper, the asymmetric TSP (ATSP) is considered, which is the more general version, where the distances between the cities are dependent on the direction of travelling of the edges.

2.2 Sequential Ant Colony Optimization

Ant colony optimization has been inspired by real ant colony's foraging behaviour, where ants can often find the shortest path between a food source and their nest. For obvious reasons, ACO has been first applied to Travelling Salesman Problem [10]. The transition probability from city i to city j for the k-th ant is defined as follows:

$$P_k(i,j) = \begin{cases} \dfrac{[\tau(i,j)]^\alpha \cdot [\eta(i,j)]^\beta}{\sum_{u \in J_k^i} [\tau(i,u)]^\alpha \cdot [\eta(i,u)]^\beta}, & \text{if } j \in J_k^i, \\ 0, & \text{otherwise.} \end{cases} \tag{1}$$

where J_k^i is the set of cities allowed to be visited by ant k from the city i, $\tau(i,j)$ is the amount of pheromone trail on the edge i to j, $\eta(i,j)$ denotes inverse of path length from i to j, α and β are parameters that control the relative importance of trail versus visibility.

After the ants end all their tours, the pheromone trails $\tau(i,j)$ are updated according to the formula:

$$\tau(i,j) \leftarrow (1-\rho) \cdot \tau(i,j) + \sum_{k=1}^{m} \Delta\tau_k(i,j) \tag{2}$$

where m is the number of ants, $(1-\rho)$ is the evaporation rate such that $(0 \leq \rho \leq 1)$, $\Delta\tau_k(i,j)$ is the amount of pheromones remaining on the path at current iteration for ant k. This amount is calculated as:

$$\Delta\tau_k(i,j) = \begin{cases} \dfrac{Q}{L_k}, & \text{if } (i,j) \in k\text{-th ant tour}, \\ 0, & \text{otherwise.} \end{cases} \tag{3}$$

where Q is a constant, and L_k is the total length of k's ant tour.

2.3 Sequential Cultural Algorithm

In this paper, Cultural Algorithm supports ACO to find better results by avoiding premature convergence. CA [25] depicts cultural evolution as a process of dual inheritance from both a micro-evolutionary level (so-called population space, PS) and a macro-evolutionary level (belief space, BS). From the perspective of evolution, any computational framework according to the requirement of CA can be used to represent or describe both of the spaces. The population space is composed of the individuals representing the search space of possible solutions, whereas the belief space consists of the experienced knowledge acquired during evolution process. These two spaces interact with each other through operations *accept*() and *influence*(). The PS conducts the evolution and periodically contributes to the BS using *accept*() operation to update BS. BS performs evolution as well and calls *influence*() operation to direct the evolution process in PS. When *accept*() operation is invoked, the solution is updated by the local best (i.e. the shortest path in TSP case) from PS, and when *influence*() is fired, the global best solution in BS is transferred to PS. The fulfilled termination condition ends the algorithm.

In this study, the population space is supported by ACO in $\mathcal{MMAS}$ version and the belief space, responsible for updating the knowledge, is evolved by genetic algorithm (GA) [15]. GA inspired by nature and proposed by Holland in 1975, encodes the parameters of a solution into the chromosome consisted of genes. GA randomly generates a set of chromosomes as the initial population. Then, it randomly selects two chromosomes from the population to perform some genetic operators, like the crossover and the mutation operations, repeatedly until the result satisfies the termination condition. After performing the genetic operators, the system calculates the fitness value of each chromosome. Chromosomes with higher fitness values are to be selected into the gene pool for reproduction in the next generation. The application of GA as a robust population-based metaheuristic for TSP is already intensively studied since the 1980s, and many special dedicated representations and genetic operators were proposed [19].

2.4 General-Purpose Computing on GPUs

In the CUDA architecture a program to be executed is divided into CPU and GPU part (so-called Heterogeneous Programming). In most cases the GPU (known as a device) works as a coprocessor of CPU (known as a host). The host and the device maintain their own separate memory spaces. The host has got access to devices global, texture and constant memory respectively. CUDA program is based on functions executed in parallel by a given number of CUDA threads called kernels. Threads are grouped together into blocks, which are executed independently to each other. Threads within a block can communicate by sharing data through the shared memory, which is fast in access time but relatively small in size. Shared memory is divided into memory banks. Only one thread can access one memory bank at a time. Otherwise, an access has to be serialised. In a single block, it is also possible to synchronise threads execution to coordinate memory accesses. To avoid data hazard, synchronisation and atomic (also related to global memory) functions are available. Threads in different blocks are independent. Blocks of threads are, in turn, grouped into grids. When a kernel is invoked it needs to know the number of blocks in a grid and the number of threads in a block. Both numbers can be of integer or dim3 type (an integer vector type used in CUDA code). During kernel execution, blocks of the grid are independently scheduled among the GPUs Streaming Multiprocessors (SM). It is also possible to run multiple kernels concurrently and overlap data transfer using different CUDA streams. Each SM is composed of CUDA cores. SM executes on successive clock cycles a single warp of threads. A warp is a group of 32 related threads. Flow control instructions may force threads of the same warp to serialise their execution paths. Each thread has an access to local memory too, which can be registers or a specific region of the global memory. Registers are the fastest available GPUs memories, but the data cannot be stored in directly there. In the GPU, tens of thousands threads can be executed concurrently, which can significantly improve execution time.

3 Related Works

ACO algorithm as a population-based approach is very easy to be parallelized since the ants do not depend on each other while moving in the search space. Delévacq et al. [7] divides parallelization strategies for the Ant Colony Optimization into 3 groups: 1. parallel ants, 2. multiple ant colonies, 3. hardware-oriented parallel ACO.

In 1998 Bullnheimer et al. [3] proposed the first parallel ACO based on parallel ants approach. They examined two parallelization strategies which used a message passing and distributed-memory architecture. Similar approach was proposed, among others, in [9,17]. Delise et al. [8] implemented this scheme of parallelization on shared-memory architecture and multi-core processors.

In 1998 Stützle [27] introduced the parallel execution of multiple ant colonies. This approach based on a message passing and distributed memory architecture. Four different information exchange strategies between ant colonies were tested in [22]. An important conclusion has been drawn not to share too much information between parallel colonies. The other exchange strategies were proposed in [1,5,21].

In 2004 Scheuermann et al. [26] showed working parallel ACO on Field Programmable Gate Arrays. However, the most used parallel hardware today is undoubtedly Graphics Processing Unit. In 2007 Catala et al. [4] solved the Orienteering Problem on GPUs. In [18] 30 city TSP was attacked by $\mathcal{MMAS}$. The number of TSP cities was extended to 800 in [35], and to 2103 in [7]. Speed-ups of GPU-based ACO algorithms on different combinatorial problems were reported, among others, in [31,32].

The algorithm we propose hybridise Cultural Algorithm with ACO-based $\mathcal{MMAS}$. So far very limited works have been reported on Cultural ACO. The work from 2013 [34] solves continuous optimization problem by so-called CACO algorithm. What is interesting, the belief space is composed of two parts: individual and population BS. In [33] authors employed multiple ant colonies as population spaces supported by 3-OPT local search algorithm in belief space. This modified Ant Colony Algorithm was tested on TSP. In 2015 the improved culture algorithm-ant colony algorithm (CA-ACA) was proposed to solve the problem of nodes deployment [30]. Both spaces of CA use the same ant individual representation, but applying different evolutionary schemes.

This paper proposes parallel Cultural Ant Colony Optimization (pCACO) for Travelling Salesman Problem. Parallelization of pCACO is obtained by computing on GPUs using CUDA framework. The parallel ACO works in population space, whereas parallel genetic algorithm evolves the belief population.

Next section presents the proposed pCACO algorithm on GPU architectures.

4 Implementation of Proposed Algorithm on GPUs

Proposed algorithm applies Cultural Algorithm framework [25] in order to improve the performance of $\mathcal{MAX}$-$\mathcal{MIN}$ Ant System. $\mathcal{MMAS}$ is one of the most efficient ACO-based algorithm [29], in which the amount of pheromone over the edge between two vertices i and j is restricted $\tau_{min} \leq \tau(i,j) \leq \tau_{max}$ to avoid search stagnation.

In our approach, the belief and population space evolve independently starting from the randomly initiated pools. Both spaces communicate each other using protocol of Cultural Algorithm. The pseudo-code is shown as follows:

```
In CPU:
1. Allocate instance and pheromone array in GPU global memory.
2. Generate random population for belief space.
3. Allocate parents and offsprings arrays in GPU global memory and copy generated population
   into parents array.
4. Allocate data exchange arrays (of proper size) for population and belief space in GPU global memory.
5. Create CUDA streams for population and belief space.
6. Run population space in a single block and chosen number of threads/ants
   (and other earlier mentioned parameters).
7. Run belief space in a single block and chosen population size (threads).

In GPU (population space):
1. Allocate local variables and parameters in shared memory.
2. Initialize pheromon array (single thread).
3. For each iteration:
3.1. For each turn:
3.1.1. Create tour for an ant (thread).
3.1.2.  <accept> If an ant has created new best tour (for itself) save it to exchange array in global memory (thread).
3.1.3.  Update pheromon values.
3.2. <influence> If it is possible to get solution from belief space (data available), increase pheromon values
     based on the route info.

In GPU (belief space):
1. Randomly choose one parent and generate first offspring (thread).
2. Calculate fitness value of the parent and the offspring (thread).
3. For each iteration:
3.1. For each generation:
3.1.1. Compare the parent with the offspring and swap the parent if an offspring codes a better solution (thread).
3.1.2. Sort parents based on their fitness values (thread).
3.1.3. <influence> Save chosen number of best solutions to exchange array in global memory.
3.1.4. Randomly choose one parent and generate an offspring (thread).
3.1.5. Apply mutation operator on the offspring if drawn (thread).
3.1.6. Calculate fitness value of the offspring (thread).
3.2. <accept> If it is possible to get solution from population space (data available), exchange one of the
     worst parents (bounded to a thread) with the best solution provided by the ant from population space (thread).

In CPU (after population space last iteration):
1. Terminate belief space.
2. Copy the population and belief space data back to main memory.
3. Get the best solution from results.
```

4.1 Population Space

Population space is based on extended pACO by added CA *accept* and *influence* operations. Algorithm starts on a host side with global memory allocation and data initialization. Then on the device side so-called *reference_value* is determined. The value is a product of the longest instance edge and the instance size. Additionally, the *pheromon_multiplier* value is determined, as a quotient of instance size and used ants number. Both parameters are saved in a shared memory due to broadcast mechanism usage. They are both used for pheromone update amount computation.

The route selection step performed by each ant is parallelized. Each ant is associated with a single thread in a thread block. A single thread block is used too. It is recommended to use a number of ants/threads being multiple of 32, as it is a size of warp in CUDA architecture. Global GPU memory is used for both instance and pheromone array.

PS is responsible for generating solutions. Every turn each ant creates solution based on pheromone heuristic information known as acquired experience. If the tour related to the specific ant is the best one created by that ant so far (shortest length) it is saved at an ant-related index to earlier mentioned array. Every ant has got corresponding array index. Described process is the first part of an accept operation. After all turns end, PS is trying to indirectly execute *influence* operation. In the current version, only the best solution provided by BS at a certain point of algorithm influences pheromone array. It is made by adding base pheromone value (τ_{baz}, but limited by $\mathcal{MAX}$-$\mathcal{MIN}$ rule) to every best-BS-tour edge. Next all turns are repeated or the algorithm ends.

Considering the usage of a global memory it is important to remember that the CUDA programming model assumes a device with a *weakly-ordered memory* model. To make sure that all memory changes are visible to other threads in a thread block memory fence functions are used. The *cuRAND* library is used. Each thread has got access to a generator initialized with a different seed. The part of an algorithm responsible for route selection uses intrinsic functions. They are less accurate but much faster versions of standard mathematical functions. Algorithm run faster and the quality of solutions remains the same. Due to a costly modulo operation during edge selection, it is replaced with a bitwise operation.

4.2 Belief Space

The simple genetic algorithm was employed to evolve belief space. It is implemented entirely on the CUDA architecture, parameterized by a number of generations and constant population size. BS makes use of routes created by ants and tries to improve them. The route may be associated with an individual. Cyclically the best individual influences PS's pheromone array which directly translates into further decisions made by ants. Once again all parents and offsprings are stored in a global memory. Additionally, the structure of arrays is used as it is preferred in *SIMD* architectures. The initial population is generated randomly by CPU. Each thread in a thread block is related to a single parent and offspring. The order crossover operator and inversion mutation are used. Each offspring is created on the basis of the related parent and the randomly chosen individual. Only one offspring is created after crossover operation. When newly created offspring after optional mutation is better (based on a standard fitness function) than the predecessor, it takes parent place in memory. Otherwise, parent remains unchanged. Then all individuals are sorted according to their fitness function value. *Bitonic mergesort algorithm* is used. According to a number of PS ants, the same quantity of best individuals is saved to a BS related array, it is the first part of the *influence* cultural operation. After defined number of generations, BS tries to substitute the same amount of worst individuals to new ones based on available PS solutions. It has been assumed that the number of ants in PS is generally smaller than the number of individuals in BS, because of individuals substitution during the accept operation. The whole pCACO algorithm ends with the last PS iteration.

5 Experimental Results

In this study, we used a PC with one Intel Core i5-4670K (4 cores, 3.4 GHz) processor, a single Asus GeForce GTX770 2048 MB 256bit DirectCU II OC, and Kingston 8192 MB 1600 MHz HyperX Blu Red CL10. The OS was Windows 7 Professional 64bit. For CUDA program compilation, Microsoft Visual Studio 2012 and CUDA Toolkit 7.5 were used.

The instances on which we tested our algorithm were taken from the TSPLIB benchmark library [24]. In the experiments, we used 9 instances which were classified as asymmetric Travelling Salesman Problem. 50 runs were performed for each instance.

The parameters settings are as follows, for ACO: $\tau_{baz} = 1000$, $\tau_{min} = 1$, $\tau_{max} = 5 \cdot \tau_{baz}$, $\alpha = 1$, $\beta = 3$, $Q = reference\_value$, $\rho = 0.1$, $m = 64$, $t_{end} = 100$; for pACO: $m = threads = 64$; and for pCACO: $size\_of\_population = 256$, $no\_of\_generation = 1000$, $probability\_of\_inversion\_mutation = 0.02$, $probability\_of\_crossover = 1.0$.

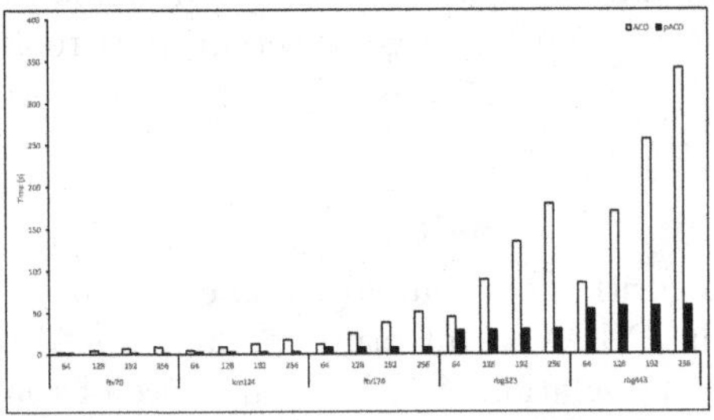

Fig. 1. Observed execution speeds of sequential ACO and parallel ACO (pACO) against benchmarks sets and with different ants/threads (64, 128, 192, 256)

In the preliminary experiments sequential and parallel version of ACO were compared over a different size of problems and with a variable number of ants or threads (in the parallel ACO each ant corresponds to one thread). Figure 1 shows numerical results of observed execution speeds of compared algorithms. The results show a speed-up of up to ca. 2 × faster than the sequential implementation for 64 ants/treads, to ca. 8 × faster for 256 ants/threads. Note that only the route selection step was parallelized. What is interesting, this proportion of speed-ups is observed for each tested dataset.

In Table 1 we report on experimental results obtained with proposed algorithm pCACO, compared to results gained by a sequential and parallel ACO. Friedman post-hoc test was used to check if the differences were statistically significant at the significance level $\alpha = 0.01$. The computational results show that parallel Cultural ACO is able to find very good solutions for all tested instances and in most cases pCACO achieves a significantly better average solution quality than ACO and pACO (except the simplest dataset ftv33, on which all algorithms

Table 1. Performance comparison of sequential Ant Colony Optimization (ACO) with parallel Ant Colony Optimization (pACO), and parallel Cultural Ant Colony Optimization (pCACO) applied to nine asymmetric TSP instances, available in TSPLIB. Given are the instance name (Instance), the best known solution (Best known), the algorithm used (Algorithm), the best solution obtained (Best), the average solution (Average) with standard deviation in parentheses, the average time in seconds to find the best solution in a run (t_{avg}). Averages are taken over 50 trials, 64 ants were used. Values in bold refer to significant differences to the other algorithms.

Instance	Best known	Algorithm	Best	Average	t_{avg}
ftv33	1286	ACO	1286	1286.00 (0.00)	2.5057
		pACO	1286	1286.00 (0.00)	1.2537
		pCACO	1286	1286.00 (0.00)	1.2622
ftv44	1613	ACO	1624	1683.92 (13.87)	4.3463
		pACO	1650	1684.68 (9.88)	2.3485
		pCACO	1623	**1666.72 (15.42)**	2.3554
ftv55	1608	ACO	1635	1696.84 (17.57)	7.1225
		pACO	1674	1691.08 (17.84)	3.7326
		pCACO	1635	**1676.50 (18.80)**	3.7357
ftv64	1839	ACO	1905	1941.32 (11.80)	9.4212
		pACO	1902	1930.50 (14.80)	5.1313
		pCACO	1879	**1911.72 (13.16)**	5.1454
ftv70	1950	ACO	2093	2148.74 (18.87)	10.7068
		pACO	2068	2143.90 (24.31)	6.1712
		pCACO	2003	**2102.38 (33.97)**	6.1960
kro124	36230	ACO	38682	39937.98 (417.49)	21.0575
		pACO	39397	39968.78 (324.13)	12.9537
		pCACO	38450	**39362.12 (377.94)**	12.9688
ftv170	2755	ACO	3130	3269.88 (53.41)	59.1272
		pACO	3150	3274.62 (51.29)	38.7157
		pCACO	3083	3268.30 (63.11)	39.4822
rbg323	1326	ACO	1466	1485.82 (6.03)	211.2016
		pACO	1467	1486.40 (7.25)	140.3757
		pCACO	1468	1483.76 (7.49)	147.4379
rbg443	2720	ACO	3271	3304.88 (13.26)	390.3430
		pACO	3273	3306.24 (12.90)	268.2758
		pCACO	3212	**3262.18 (17.10)**	283.8038

found the best route, and ftv170, and rbg323 instances). What is interesting, our parallel Cultural-based algorithm, which evolves two populations, yields comparable results to parallel ACO in terms of time performing. It is because PS, as well as BS, uses different streams, therefore both spaces are invisible to themselves. The device resources are independently distributed between PS and BS. Parallel CACO runtime depends only on the PS. With the last iteration of PS, BS makes its last iteration too. Minor differences in observed execution speed of pACO and pCACO appear due to CA *accept* and *influence* operations when there is a small chance some data is not ready yet (both algorithms try to use the same data structure).

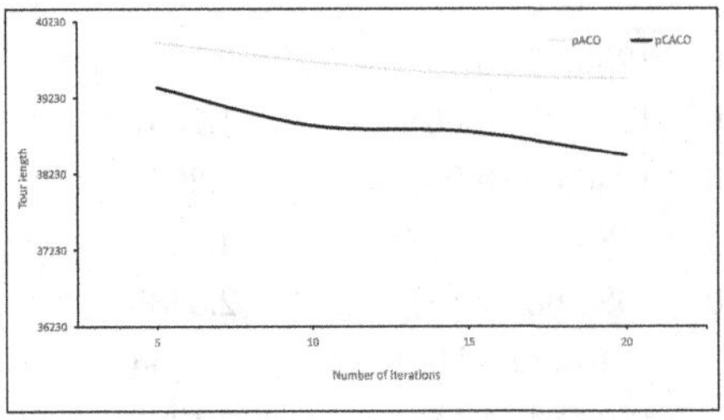

Fig. 2. Evolution of average tour length of kro124 (size of 100 vertices) dataset for pACO and pCACO

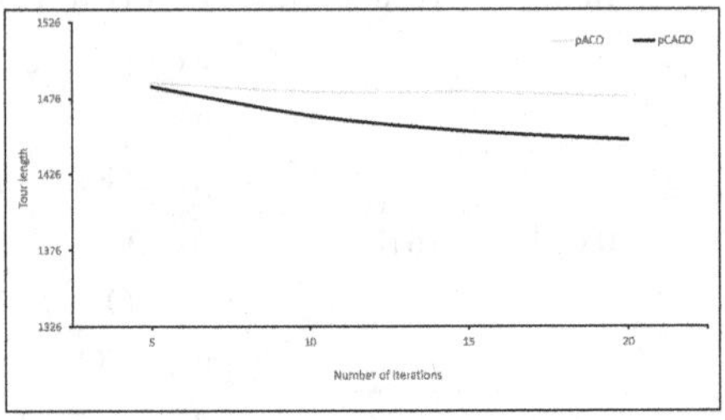

Fig. 3. Evolution of average tour length of rbg323 (size of 323 vertices) dataset for pACO and pCACO

In Figs. 2 and 3 we present the results of comparing the average tour and convergence speed between our approach and parallel ACO for two exemplary datasets kro124 and rbg323. Both figures show that the proposed Cultural ACO achieves average shorter tour than parallel ACO with similar or better speed of convergence.

6 Conclusion

In this paper, we propose a $\mathcal{MAX}$-$\mathcal{MIN}$ Ant System for solving asymmetric Travelling Salesman Problem on a GPU by combining Cultural Algorithm in CUDA. To the best of our knowledge, it is the first use of general-purpose computing on GPUs for speeding Cultural Ant Colony Optimization in a TSP task.

The experimental results show that parallel Cultural ACO outperforms both sequential and parallel ACO in terms of solution quality. Moreover, the average time needed to deliver the solution is comparable to the time performing of parallel ACO, and what is obvious, much better than the sequential version of ACO.

In the future, we will further investigate the potential of other evolutionary algorithms for evolving belief space. Much still can be done for the effective use of GPUs. For example, instead of one population space, we can perform simultaneously many population spaces, and transfer to BS current best solution of each population space. It would be also interesting and we hope useful to implement proposed algorithm in a Google-like Map-Reduce paradigm [6].

Author Contribution. OU initiated and designed the study, supervised the work, made statistical tests. RT performed the experiments. Both authors wrote and approved the final manuscript.

Acknowledgments. The work was supported by statutory grant of the Wroclaw University of Science and Technology, Poland.

References

1. Alba, E., Leguizamon, G., Ordonez, G.: Two models of parallel ACO algorithms for the minimum tardy task problem. Int. J. High Perform. Syst. Archit. **1**(1), 50–59 (2007)
2. Bullnheimer, B., Hartl, R.F., Strauß, C.: A new rank based version of the ant system: a computational study. Central Eur. J. Oper. Res. Econ. **7**(1), 25–38 (1999)
3. Bullnheimer, B., Kotsis, G., Strauß, C.: Parallelization strategies for the ant system. In: De Leone, R., Murli, A., Pardalos, P.M., Toraldo, G. (eds.) High Performance Algorithms and Software in Nonlinear Optimization. Applied Optimization, vol. 24, pp. 87–100. Springer, Boston (1998). doi:10.1007/978-1-4613-3279-4_6
4. Catala, A., Jaen, J., Mocholi, J.A.: Strategies for accelerating ant colony optimization algorithms on graphical processing units. In: IEEE Congress on Evolutionary Computation, CEC 2007, pp. 492–500. IEEE (2007)
5. Chu, D., Zomaya, A.: Parallel ant colony optimization for 3D protein structure prediction using the HP lattice model. In: Nedjah, N., de Macedo, M.L., Alba, E. (eds.) Parallel Evolutionary Computations. Studies in Computational Intelligence, vol. 22, pp. 177–198. Springer, Heidelberg (2006). doi:10.1007/3-540-32839-4_9
6. Dean, J., Ghemawat, S.: MapReduce: simplified data processing on large clusters. Commun. ACM **51**(1), 107–113 (2008)
7. Delévacq, A., Delisle, P., Gravel, M., Krajecki, M.: Parallel ant colony optimization on graphics processing units. J. Parallel Distrib. Comput. **73**(1), 52–61 (2013)
8. Delisle, P., Gravel, M., Krajecki, M., Gagné, C., Price, W.L.: A shared memory parallel implementation of ant colony optimization. In: Proceedings of the 6th Metaheuristics International Conference, pp. 257–264 (2005)
9. Doerner, K.F., Hartl, R.F., Benkner, S., Lucka, M.: Parallel cooperative savings based ant colony optimization - multiple search and decomposition approaches. Parallel Process. Lett. **16**(03), 351–369 (2006)
10. Dorigo, M.: Optimization, learning and natural algorithms. Ph.D. thesis, Politecnico di Milano, Italy (1992)
11. Dorigo, M., Gambardella, L.M.: Ant colonies for the travelling salesman problem. BioSystems **43**(2), 73–81 (1997)
12. Dorigo, M., Maniezzo, V., Colorni, A., Maniezzo, V.: Positive feedback as a search strategy. Technical report 91–016, Dipartimento di Elettronica, Politecnico di Milano, Italy (1991)
13. Gambardella, L.M., Dorigo, M.: Solving symmetric and asymmetric TSPs by ant colonies. In: International Conference on Evolutionary Computation, pp. 622–627 (1996)
14. Gambardella, L.M., Dorigo, M., Prieditis, A., Russell, S.: Ant-Q: a reinforcement learning approach to the traveling salesman problem. In: Proceedings of ML 1995, Twelfth International Conference on Machine Learning, pp. 252–260. Morgan Kaufmann (1995)

15. Goldberg, D.E., Holland, J.H.: Genetic algorithms and machine learning. Mach. Learn. **3**(2), 95–99 (1988)
16. Hoffman, K.L., Padberg, M., Rinaldi, G.: Traveling salesman problem. In: Gass, S.I., Fu, M.C. (eds.) Encyclopedia of Operations Research and Management Science, pp. 1573–1578. Springer, New York (2013)
17. Islam, M.T., Thulasiraman, P., Thulasiram, R.K.: A parallel ant colony optimization algorithm for all-pair routing in MANETs. In: Parallel and Distributed Processing Symposium, p. 8. IEEE (2003)
18. Jiening, W., Jiankang, D., Chunfeng, Z.: Implementation of ant colony algorithm based on GPU. In: 2009 Sixth International Conference on Computer Graphics, Imaging and Visualization, pp. 50–53. IEEE (2009)
19. Larrañaga, P., Kuijpers, C.M.H., Murga, R.H., Inza, I., Dizdarevic, S.: Genetic algorithms for the travelling salesman problem: a review of representations and operators. Artif. Intell. Rev. **13**(2), 129–170 (1999)
20. Lenstra, J.K., Kan, A.R., Lawler, E.L., Shmoys, D.: The Traveling Salesman Problem. A Guided Tour of Combinatorial Optimization. Wiley, New York (1985)
21. Manfrin, M., Birattari, M., Stützle, T., Dorigo, M.: Parallel ant colony optimization for the traveling salesman problem. In: Dorigo, M., Gambardella, L.M., Birattari, M., Martinoli, A., Poli, R., Stützle, T. (eds.) ANTS 2006. LNCS, vol. 4150, pp. 224–234. Springer, Heidelberg (2006). doi:10.1007/11839088_20
22. Middendorf, M., Reischle, F., Schmeck, H.: Multi colony ant algorithms. J. Heuristics **8**(3), 305–320 (2002)
23. Nvidia: Nvidia CUDA (2016). http://nvidia.com/cuda
24. Reinhelt, G.: TSPLIB: a library of sample instances for the TSP (and related problems) from various sources and of various types (2014). http://comopt.ifi.uniheidelberg.de/software/TSPLIB
25. Reynolds, R.G.: An introduction to cultural algorithms. In: Proceedings of the Third Annual Conference on Evolutionary Programming, pp. 131–139, Singapore (1994)
26. Scheuermann, B., So, K., Guntsch, M., Middendorf, M., Diessel, O., ElGindy, H., Schmeck, H.: Fpga implementation of population-based ant colony optimization. Appl. Soft Comput. **4**(3), 303–322 (2004)
27. Stützle, T.: Parallelization strategies for ant colony optimization. In: Eiben, A.E., Bäck, T., Schoenauer, M., Schwefel, H.-P. (eds.) PPSN 1998. LNCS, vol. 1498, pp. 722–731. Springer, Heidelberg (1998). doi:10.1007/BFb0056914
28. Stützle, T., Dorigo, M.: ACO algorithms for the traveling salesman problem. In: Evolutionary Algorithms in Engineering and Computer Science, pp. 163–183 (1999)
29. Stützle, T., Hoos, H.: MAX-MIN ant system and local search for the traveling salesman problem. In: 1997 IEEE International Conference on Evolutionary Computation, pp. 309–314. IEEE (1997)
30. Sun, X., Zhang, Y., Ren, X., Chen, K.: Optimization deployment of wireless sensor networks based on culture-ant colony algorithm. Appl. Math. Comput. **250**, 58–70 (2015)
31. Wang, P., Li, H., Zhang, B.: A GPU-based parallel ant colony algorithm for scientific workflow scheduling. Int. J. Grid Distrib. Comput. **8**(4), 37–46 (2015)
32. Wei, K.C., Wu, C.c., Wu, C.J.: Using CUDA GPU to accelerate the ant colony optimization algorithm. In: 2013 International Conference on Parallel and Distributed Computing, Applications and Technologies (PDCAT), pp. 90–95. IEEE (2013)
33. Wei, X., Han, L., Hong, L.: A modified ant colony algorithm for traveling salesman problem. Int. J. Comput. Commun. Control **9**(5), 633–643 (2014)

34. Xu, J., Zhang, M., Cai, Y.: Cultural ant algorithm for continuous optimization problems. Appl. Math. Inf. Sci **7**(2L), 705–710 (2013)
35. You, Y.S.: Parallel ant system for traveling salesman problem on GPUs. In: Eleventh Annual Conference on Genetic and Evolutionary Computation, pp. 1–2 (2009)
36. Yuan, S., Skinner, B., Huang, S., Liu, D.: A new crossover approach for solving the multiple travelling salesmen problem using genetic algorithms. Eur. J. Oper. Res. **228**(1), 72–82 (2013)

Combining Genetic Algorithm with the Multilevel Paradigm for the Maximum Constraint Satisfaction Problem

Noureddine Bouhmala(✉)

Department of Maritime Technology and Innovation, University of SouthEast, Horten, Norway
noureddine.bouhmala@hbv.no

Abstract. Genetic algorithms (GA) which belongs to the class of evolutionary algorithms are regarded as highly successful algorithms when applied to a broad range of discrete as well continuous optimization problems. This paper introduces a hybrid approach combining genetic algorithm with the multilevel paradigm for solving the maximum constraint satisfaction problem (Max-CSP). The multilevel paradigm refers to the process of dividing large and complex problems into smaller ones, which are hopefully much easier to solve, and then work backward towards the solution of the original problem, using the solution reached from a child level as a starting solution for the parent level. The promising performances achieved by the proposed approach are demonstrated by comparisons made to solve conventional random benchmark problems.

Keywords: Maximum constraint satisfaction problem · Genetic algorithms · Multilevel paradigm

1 Introduction

Many problems in the field of artificial intelligence can be modeled as constraint satisfaction problems (CSP). A constraint satisfaction problem (or CSP) is a tuple $\langle X, D, C \rangle$ where, $X = \{x_1, x_2, ...x_n\}$ is a finite set of variables, $D = \{D_{x_1}, D_{x_2}, ...D_{x_n}\}$ is a finite set of domains. Thus each variable $x \in X$ has a corresponding discrete domain D_x from which it can be instantiated, and $C = \{C_1, C_2, ...C_k\}$ is a finite set of constraints. Each k-ary constraint restricts a k-tuple of variables, $(x_1, x_2, ...x_k)$ and specifies a subset of $D_1 \times \ldots \times D_k$, each element of which are values that the variables can not take simultaneously. A solution to a CSP requires the assignment of values to each of the variables from their domains such that all the constraints on the variables are satisfied. The maximum constraint satisfaction problem (Max-CSP) aims at finding an assignment so as to maximize the number of satisfied constraints. Max-CSP can be regarded as the generalization of CSP; the solution maximizes the number of satisfied constraints. In this paper, attention is focused on binary CSPs, where all constraints are binary, i.e., they are based on the cartesian product

P.M. Pardalos et al. (Eds.): MOD 2016, LNCS 10122, pp. 330–340, 2016.
DOI: 10.1007/978-3-319-51469-7_28

of the domains of two variables. However, any non-binary CSP can theoretically be converted to a binary CSP [4]. Algorithms for solving CSPs apply the so-called 1-exchange neighborhood under which two solutions are direct neighbors if, and only if, they differ at most in the value assigned to one variable. Examples include the minimum conflict heuristic MCH [11], the break method for escaping from local minima [12], various enhanced MCH (e.g., randomized iterative improvement of MCH called WMCH [17], MCH with tabu search [6], evolutionary algorithms [19]. Weights-based algorithms attempt are techniques that work by introducing weights on variables or constraints in order to avoid local minima. Methods belonging to this category include genet [3], guided local search [16], the exponentiated sub-gradient [14], discrete Lagrangian search [15], the scaling and probabilistic smoothing [9], evolutionary algorithms combined with stepwise adaptation of weights [1], methods based on dynamically adapting weights on variables [13], or both (i.e., variables and constraints) [5]. Methods based on large neighborhood search have recently attracted several researchers for solving the CSP [10]. The central idea is to reduce the size of local search space relying on a continual relaxation (removing elements from the solution) and re-optimization (re-inserting the removed elements). Finally, the work introduced in [2] introduces a variable depth metaheuristic combing a greedy local search with a self adaptive weighting strategy on the constraints weights.

2 Algorithm

2.1 Multilevel Context

The multilevel paradigm is a simple technique which at its core involves recursive coarsening to produce smaller and smaller problems that are easier to solve than the original one. Multilevel techniques have been developed in the period after 1960 and are among the most efficient techniques used for solving large algebraic systems arising from the discretization of partial differential equations. In recent years it has been recognised that an effective way of enhancing metaheuristics is to use them in the multilevel context. The pseudo-code of the multilevel generic algorithm is shown in Algorithm 1. The multilevel paradigm consists of four phases: coarsening, initial solution, uncoarsening and refinement. The coarsening phase aims at merging the variables associated with the problem to form clusters. The clusters are used in a recursive manner to construct a hierarchy of problems each representing the original problem but with fewer degrees of freedom. The coarsest level can then be used to compute an initial solution. The solution found at the coarsest level is uncoarsened (extended to give an initial solution for the parent level) and then improved using a chosen optimization algorithm. A common feature that characterizes multilevel algorithms, is that any solution in any of the coarsened problems is a legitimate solution to the original one. Optimization algorithms using the multilevel paradigm draw their strength from coupling the refinement process across different levels.

```
input  : Problem P_0
output: Solution S_final(P_0)
begin
    level := 0;
    while Not reached the desired number of levels do
        P_level+1 := Coarsen (P_level);
        level := level + 1;
    end
    /* Initial Solution is computed at the lowest level */;
    S(P_level) =Initial Solution (P_level) ;
    while (level > 0) do
        S_start(P_level-1): = Uncoarsen (S_final(P_level));
        S_final(P_level-1) := Refine (S_start(P_level-1));
        level := level - 1;
    end
end
```

Algorithm 1. The Multilevel Generic Algorithm

2.2 Multilevel Genetic Algorithm

Genetic Algorithms (GAs) [8] are stochastic methods for global search and optimization and belong to the group of nature inspired metaheuristics leading to the so-called natural computing. It is a fast-growing interdisciplinary field in which a range of techniques and methods are studied for dealing with large, complex, and dynamic problems with various sources of potential uncertainties. GAs simultaneously examine and manipulate a set of possible solutions. A gene is part of a chromosome (solution), which is the smallest unit of genetic information. Every gene is able to assume different values called allele. All genes of an organism form a genome which affects the appearance of an organism called phenotype. The chromosomes are encoded using a chosen representation and each can be thought of as a point in the search space of candidate solutions. Each individual is assigned a score (fitness) value that allows assessing its quality. The members of the initial population may be randomly generated or by using sophisticated mechanisms by means of which an initial population of high quality chromosomes is produced. The reproduction operator selects (randomly or based on the individual's fitness) chromosomes from the population to be parents and enters them in a mating pool. Parent individuals are drawn from the mating pool and combined so that information is exchanged and passed to off-springs depending on the probability of the cross-over operator. The new population is then subjected to mutation and enters into an intermediate population. The mutation operator acts as an element of diversity into the population and is generally applied with a low probability to avoid disrupting cross-over results. Finally, a selection scheme is used to update the population giving rise to a new generation. The individuals from the set of solutions which is called population will evolve from generation to generation by repeated applications of an evaluation procedure that is based on genetic operators. Over many generations, the population becomes increasingly uniform until it ultimately converges to optimal or near-optimal solutions. The different steps of the multilevel weighted-genetic-algorithm are described as follows:

- **construction of levels:** Let $G_0 = (V_0, E_0)$ be an undirected graph of vertices V and edges E. The set V denotes variables and each edge $(x_i, x_j) \in E$ implies a constraint joining the variables x_i and x_j. Given the initial graph G_0, the graph is repeatedly transformed into smaller and smaller graphs $G_1, G_2, \ldots, G_m$ such that $|V_0| > |V_1| > \ldots > |V_m|$. To coarsen a graph from G_j to G_{j+1}, a number of different techniques may be used. In this paper, when combining a set of variables into clusters, the variables are visited in a random order. If a variable x_i has not been matched yet, then the algorithms randomly selects one of its neighboring unmatched variable x_j, and a new cluster consisting of these two variables is created. Its neighbours are the combined neighbors of the merged variables x_i and x_j. Unmatched variables are simply left unmatched and copied to the next level.
- **initial assignment:** The process of constructing a hierarchy of graphs ceases as soon as the size of the coarsest graphs reaches some desired threshold. A random initial population is generated at the lowest Level $G_k = (V_k, E_k)$. The chromosomes which are assignments of values to the variables are encoded as strings of bits, the length of which is the number of variables. At the lowest level, the length of the chromosome is equal to the number of clusters. The initial solution is simply constructed by assigning to all variable in a cluster, a random value v_i. In this work it is assumed that all variables have the same domain (i.e., same set of values), otherwise different random values should be assigned to each variable in the cluster. All the individuals of the initial population are evaluated and assigned a fitness expressed in Eq. 1 which counts the number of constraint violations where $< (x_i, s_i), (x_j, s_j) >$ denotes the constraint between the variables x_i and x_j where x_i is assigned the value s_i from D_{x_i} and x_j is assigned the value s_j from D_{x_j}.

$$Fitness = \sum_{i=1}^{n-1} \sum_{j=i+1}^{n} Violation(W_{i,j} <(x_i, s_i), (x_j, s_j)>) \qquad (1)$$

- **initial weights:** The next step of the algorithm assigns a fixed amount of weight equal to 1 across all the constraints. The distribution of weights to constraints aims at forcing hard constraints with large weights to be satisfied thereby preventing the algorithm at a later stage from getting stuck at a local optimum.
- **optimization:** Having computed an initial solution at the coarsest graph, GA starts the search process from the coarsest level $G_k = (V_k, E_k)$ and continues to move towards smaller levels. The motivation behind this strategy is that the order in which the levels are traversed offers a better mechanism for performing diversification and intensification. The coarsest level allows GA to view any cluster of variables as a single entity leading the search to become guided in far away regions of the solution space and restricted to only those configurations in the solution space in which the variables grouped within a cluster are assigned the same value. As the switch from one level to another implies a decrease in

the size of the neighborhood, the search is intensified around solutions from previous levels in order to reach better ones.

- **parent selection:** During the optimization, new solutions are created by combining pairs of individuals in the population and then applying a crossover operator to each chosen pair. Combining pairs of individuals can be viewed as a matching process. In the version of GA used in this work, the individuals are visited in random order. An unmatched individual i_k is matched randomly with an unmatched individual i_l.
- **genetic operators:** The task of the crossover operator is to reach regions of the search space with higher average quality. The two-point crossover operator is applied to each matched pair of individuals. The two-point crossover selects two randomly points within a chromosome and then interchanges the two parent chromosomes between these points to generate two new offspring.
- **survivor selection:** The selection acts on individuals in the current population. Based on each individual quality (fitness), it determines the next population. In the roulette method, the selection is stochastic and biased toward the best individuals. The first step is to calculate the cumulative fitness of the whole population through the sum of the fitness of all individuals. After that, the probability of selection is calculated for each individual as being $P_{Selection_i} = f_i / \sum_1^N f_i$, where f_i is the fitness of individual i.
- **updating weights:** The weights of each current violated constraint is then increased by one, whereas the newly satisfied constraints will have their weights decreased by one before the start of new generation.
- **termination condition:** The convergence of GA is supposed to be reached if the best individual remains suncharged during 5 consecutive generations.
- **projection:** Once GA has reached the convergence criterion with respect to a child level graph $G_k = (V_k, E_k)$, the assignment reached on that level must be projected on its parent graph $G_{k-1} = (V_{k-1}, E_{k-1})$. The projection algorithm is simple; if a cluster belongs to $G_k = (V_k, E_k)$ is assigned the value vl_i, the merged pair of clusters that it represents belonging to $G_{k-1} = (V_{k-1}, E_{k-1})$ are also assigned the value vl_i,

3 Experimental Results

3.1 Experimental Setup

The benchmark instances were generated using model A [18] as follows: Each instance is defined by the 4-tuple $\langle n, m, p_d, p_t \rangle$, where n is the number of variables; m is the size of each variable's domain; p_d, the constraint density, is the proportion of pairs of variables which have a constraint between them; and p_t, the constraint tightness, is the probability that a pair of values is inconsistent. From the $(n \times (n-1)/2)$ possible constraints, each one is independently chosen to be added in the constraint graph with the probability p_d. Given a constraint, we select with the probability p_t which value pairs become no-goods. The model A will on average have $p_d \times (n-1)/2$ constraints, each of which having on average $p_t \times m^2$ inconsistent pairs of values. For each pair of density tightness, we generate one soluble instance (i.e., at least one solution exists). Because of the stochastic nature of GA, we let each algorithm do 100 independent runs, each run with a different random seed. Many NP-complete or NP-hard problems show a phase transition point that marks the spot where we go from problems that are under-constrained and so relatively easy to solve, to problems that are over-constrained and so relatively easy to prove insoluble. Problems that are on average harder to solve occur between these two types of relatively easy problem. The values of p_d and p_t are chosen in such a way that the instances generated are within the phase transition. In order to predict the phase transition region, a formula for the constrainedness [7] of binary CSPs was defined by:

$$\kappa = \frac{n-1}{2} p_d log_m \left(\frac{1}{1-pt}\right). \tag{2}$$

The tests were carried out on a DELL machine with 800 MHz CPU and 2 GB of memory. The code was written in C and compiled with the GNU C compiler version 4.6. The following parameters have been fixed experimentally and are listed below:

- Population size = 50
- Stopping criteria for the coarsening phase: The reduction process stops as soon as the number of levels reaches 3. At this level, MLV-WGA generates an initial population.
- Convergence during the optimization phase: If there is no observable improvement of the fitness function of the best individual during 5 consecutive generations, MLV-WGA is assumed to have reached convergence and moves to a higher level.

3.2 Results

The plots in Figs. 1 and 2 compares the WGA with its multilevel variant MLV-WGA. The improvement in quality imparted by the multilevel context is

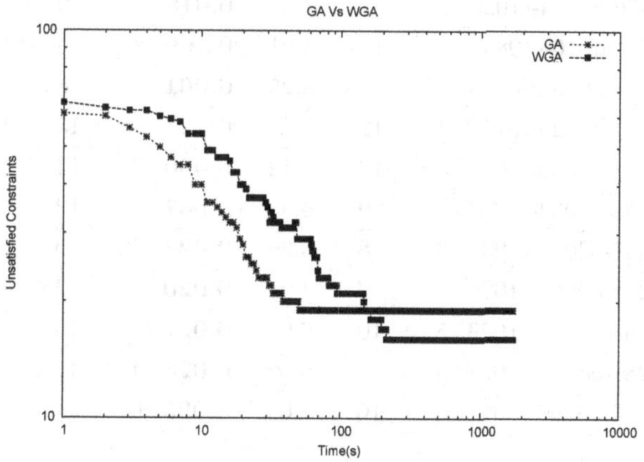

Fig. 1. MLV-GA Vs GA: evolution of the mean unsatisfied constraints as a function of time. csp-N30-DS40-C125-cd026ct063

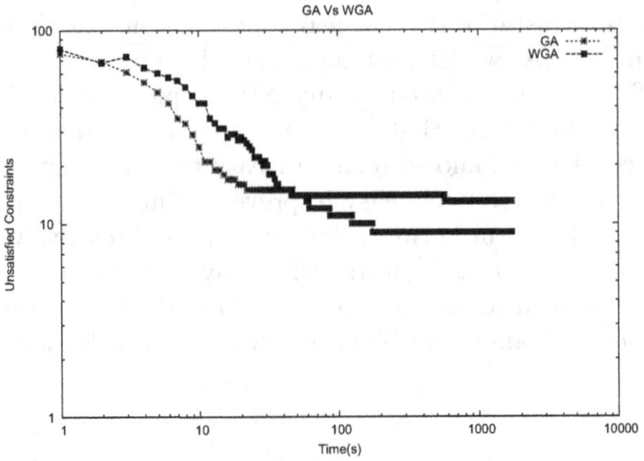

Fig. 2. MLV-GA Vs GA: evolution of the mean unsatisfied constraints as a function of time. csp-N35-DS20-C562-cd094-ct017

Table 1. MLV-WGA Vs WGA: number of variables: 25

Instance	MLV-WGA				WGA			
	Min	Max	Mean	RE_{av}	Min	Max	Mean	RE_{av}
N25-DS20-C36-cd-014-ct083	3	7	4.58	**0.128**	3	8	5.41	0.151
N25-DS20-C44-cd012-ct087	6	10	8.04	**0.183**	8	14	9.91	0.226
N25-DS20-C54-cd018-ct075	3	7	5.37	**0.100**	4	9	6.91	0.128
N25-DS20-C78-cd026-ct061	2	8	4.33	**0.056**	2	10	5.79	0.073
N25-DS20-C225-cd078-ct027	3	8	4.16	**0.019**	3	9	5.66	0.026
N25-DS20-C229-cd072-ct029	4	9	6.04	**0.014**	4	11	8.16	0.036
N25-DS20-C242-cd086-ct025	1	6	3.5	**0.015**	3	10	5.70	0.024
N25-DS20-C269-cd086-ct025	4	10	5.66	**0.022**	4	10	7.54	0.029
N25-DS20-C279-cd094-ct023	2	7	4.75	**0.018**	4	9	6.75	0.025
N25-DS40-C53-cd016-ct085	6	11	8.91	**0.169**	8	13	10.70	0.202
N25-DS40-C70-cd026-ct069	2	6	4.25	**0.061**	3	8	5.75	0.083
N25-DS40-C72-cd022-ct075	6	12	9	**0.125**	6	15	10.45	0.146
N25-DS40-C102-cd032-ct061	5	12	8.12	**0.080**	7	14	10.33	0.102
N25-DS40-C103-cd034-ct059	5	9	6.83	**0.067**	4	12	8.79	0.086
N25-DS40-C237-cd082-ct031	3	8	5.66	**0.024**	5	10	7.87	0.034
N25-DS40-C253-cd088-ct029	3	7	4.95	**0.020**	5	12	8.04	0.032
N25-DS40-C264-cd088-ct029	5	10	6.91	**0.027**	6	16	8.91	0.034
N25-DS40-C281-cd096-ct027	3	9	5.62	**0.020**	4	12	8.54	0.031
N25-DS40-C290-cd096-ct027	4	10	7.08	**0.025**	6	14	9	0.032

Table 2. MLV-WGA Vs WGA: number of variables: 30

Instance	MLV-WGA				WGA			
	Min	Max	Mean	RE_{av}	Min	Max	Mean	RE_{av}
N30-DS20-C41-cd012-ct083	2	6	3.70	**0.026**	3	7	5.08	0.124
N30-DS20-C71-cd018-ct069	1	7	3.37	**0.048**	3	10	5.66	0.080
N30-DS20-C85-cd020-ct065	3	9	6	**0.071**	5	12	8.37	0.099
N30-DS20-C119-cd028-ct053	3	10	5.70	**0.048**	6	12	8.83	0.075
N30-DS20-C334-cd074-ct025	6	13	8.16	**0.025**	6	14	9.87	0.030
N30-DS20-C387-cd090-ct021	3	9	6.66	**0.018**	5	13	8.70	0.033
N30-DS20-C389-cd090-ct021	2	9	6.08	**0.016**	4	14	8.95	0.024
N30-DS20-C392-cd090-ct021	3	10	7.08	**0.019**	5	15	9.16	0.024
N30-DS20-C399-cd090-ct021	5	13	7.70	**0.020**	6	14	9.79	0.025
N30-DS40-C85-cd020-ct073	5	11	7.75	**0.092**	7	14	10.87	0.152
N30-DS40-C96-cd020-ct073	8	12	16	0.167	11	19	14.58	0.015
N30-DS40-C121-cd026-ct063	8	14	10.5	**0.087**	9	19	14.33	0.152
N30-DS40-C125-cd026-ct063	8	18	12.20	**0.098**	10	19	15.58	0.125
N30-DS40-C173-cd044-ct045	4	10	6.41	**0.038**	6	14	9.20	0.054
N30-DS40-C312-cd070-ct031	7	14	10.5	0.033	7	19	13.33	0.025
N30-DS40-C328-cd076-ct029	6	13	10.37	**0.032**	10	18	13.45	0.042
N30-DS40-C333-cd076-ct029	7	13	10.25	**0.031**	9	18	12.62	0.038
N30-DS40-C389-cd090-ct025	6	13	9.33	**0.024**	9	17	12.20	0.032
N30-DS40-C390-cd090-ct025	6	14	9.29	**0.024**	10	17	13	0.031

immediately clear. Both WGA and MLV-WGA exhibits what is called a plateax region. A plateau region spans a region in the search space where cross-over and mutation operators leave the best solution or the mean solution unchanged. However, the length of this region is shorter with MLV-WGA compared to that of WGA. The multilevel context uses the projected solution obtained at $G_{m+1}(V_{m+1}, E_{m+1})$ as the initial solution for $G_m(V_m, E_m)$ for further refinement. Even though the solution at $G_{m+1}(V_{m+1}, E_{m+1})$ is at a local minimum, the projected solution may not be at a local optimum with respect to $G_m(V_m, E_m)$. The projected assignment is already a good solution leading WGA to converge quicker within few generations to a better solution. Tables 1, 2 and 3 show a comparison of the two algorithms. For each algorithm, the best (Min) and the worst (Max) results are given, while mean represents the average solution. MLV-WGA outperforms WGA in 53 cases out of 96, gives similar results in 20 cases, and was beaten in 23 cases. The performance of both algorithms differ significantly. The difference for the total performance is between 25% and 70% in the advantage of MLV-GA. Comparing the worst performances of both algorithms, MLV-WGA gave bad results in 15 cases, both algorithms give similar results in 8 cases, and

Table 3. MLV-WGA Vs WGA: number of variables 40.

Instance	MLV-WGA				WGA			
	Min	Max	Mean	RE_{av}	Min	Max	Mean	RE_{av}
N40-DS20-C78-cd010-ct079	6	12	8.91	**0.115**	5	13	9.04	0.116
N40-DS20-C80-cd010-ct079	7	13	9.62	**0.121**	7	13	10.04	0.153
N40-DS20-C82-cd012-ct073	4	9	6.25	**0.073**	4	11	6.95	0.085
N40-DS20-C95-cd014-ct067	2	8	4.45	0.047	2	7	4.12	**0.044**
N40-DS20-C653-cd084-ct017	2	14	9.37	**0.015**	6	16	10.62	0.018
N40-DS20-C660-cd084-ct017	6	14	9.12	**0.014**	7	6	9.75	0.015
N40-DS20-C751-cd096-ct015	6	13	9.91	0.014	5	13	9.83	0.014
N40-DS20-C752-cd096-ct015	5	17	9.29	0.013	3	13	9.20	0.013
N40-DS20-C756-cd096-ct015	6	15	9.95	0.014	5	16	8.75	**0.012**
N40-DS40-C106-cd014-ct075	7	14	11.08	**0.105**	7	16	11.5	0.109
N40-DS40-C115-cd014-ct075	12	20	15.5	0.135	11	20	15.5	0.135
N40-DS40-C181-cd024-ct055	6	17	12.04	0.067	7	17	11.75	**0.065**
N40-DS40-C196-cd024-ct055	11	12	16.58	0.085	12	20	15.54	**0.080**
N40-DS40-C226-cd030-ct047	7	14	10.91	0.051	7	16	11.16	**0.050**
N40-DS40-C647-cd082-ct021	11	23	15.66	0.025	11	20	15.20	**0.024**
N40-DS40-C658-cd082-ct021	11	22	16.33	**0.025**	13	21	16.70	0.026
N40-DS40-C703-cd092-ct019	9	21	13.41	0.020	8	20	13.58	0.020
N40-DS40-C711-cd092-ct019	12	23	15.75	0.023	8	20	14.87	0.021
N40-DS40-C719-cd092-ct019	8	21	16.54	0.024	10	20	15.16	**0.022**

MLV-WGA was able to perform better than WGA in 73 cases. Looking at the average results, MLV-WGA does between 16% and 41% better than WGA in 84 cases, while the differences are very marginal in the remaining cases where WGA beats MLV-WGA.

4 Conclusion

In this work, a multilevel weighted based-genetic algorithm is introduced for MAX-CSP. The results have shown that the multilevel genetic algorithm returns a better solution for the equivalent run-time for most cases compared to the standard genetic algorithm. The multilevel paradigm offers a better mechanism for performing diversification in intensification. This is achieved by allowing GA to view a cluster of variables as a single entity thereby leading the search becoming guided and restricted to only those assignments in the solution space in which the variables grouped within a cluster are assigned the same value. As the size of the clusters gets smaller from one level to another, the size of the neighborhood becomes adaptive and allows the possibility of exploring different regions in the search space while intensifying the search by exploiting the solutions from previous levels in order to reach better solutions.

References

1. Huerta-Amante, D.Á., Terashima-Marín, H.: Adaptive penalty weights when solving congress timetabling. In: Lemaître, C., Reyes, C.A., González, J.A. (eds.) IBERAMIA 2004. LNCS (LNAI), vol. 3315, pp. 144–153. Springer, Heidelberg (2004). doi:10.1007/978-3-540-30498-2_15
2. Bouhmala, N.: A variable depth search algorithm for binary constraint satisfaction problems. Math. Probl. Eng. **2015**, 10 (2015). Article ID 637809, doi:10.1155/2015/637809
3. Davenport, A., Tsang, E., Wang, C., Zhu, K.: Genet: a connectionist architecture for solving constraint satisfaction problems by iterative improvement. In: Proceedings of the Twelfth National Conference on Artificial Intelligence (1994)
4. Dechter, R., Pearl, J.: Tree clustering for constraint networks. Artif. Intell. **38**, 353–366 (1989)
5. Fang, Z., Chu, Y., Qiao, K., Feng, X., Xu, K.: Combining edge weight and vertex weight for minimum vertex cover problem. In: Chen, J., Hopcroft, J.E., Wang, J. (eds.) FAW 2014. LNCS, vol. 8497, pp. 71–81. Springer, Heidelberg (2014). doi:10.1007/978-3-319-08016-1_7
6. Galinier, P., Hao, J.-K.: Tabu search for maximal constraint satisfaction problems. In: Smolka, G. (ed.) CP 1997. LNCS, vol. 1330, pp. 196–208. Springer, Heidelberg (1997). doi:10.1007/BFb0017440
7. Gent, I.P., MacIntyre, E., Prosser, P., Walsh, T.: The constrainedness of search. In: Proceedings of the AAAI 1996, pp. 246–252 (1996)
8. Holland, J.: Adaptation in Natural and Artificial Systems. The University of Michigan Press, Ann Arbor (1975)
9. Hutter, F., Tompkins, D.A.D., Hoos, H.H.: Scaling and probabilistic smoothing: efficient dynamic local search for SAT. In: Hentenryck, P. (ed.) CP 2002. LNCS, vol. 2470, pp. 233–248. Springer, Heidelberg (2002). doi:10.1007/3-540-46135-3_16
10. Lee, H.-J., Cha, S.-J., Yu, Y.-H., Jo, G.-S.: Large neighborhood search using constraint satisfaction techniques in vehicle routing problem. In: Gao, Y., Japkowicz, N. (eds.) AI 2009. LNCS (LNAI), vol. 5549, pp. 229–232. Springer, Heidelberg (2009). doi:10.1007/978-3-642-01818-3_30
11. Minton, S., Johnson, M., Philips, A., Laird, P.: Minimizing conflicts: a heuristic repair method for constraint satisfaction and scheduling scheduling problems. Artif. Intell. **58**, 161–205 (1992)
12. Morris, P.: The breakout method for escaping from local minima. In: Proceeding AAAI 1993, Proceedings of the Eleventh National Conference on Artificial Intelligence, pp. 40–45 (1993)
13. Pullan, W., Mascia, F., Brunato, M.: Cooperating local search for the maximum clique problems. J. Heuristics **17**, 181–199 (2011)
14. Schuurmans, D., Southey, F., Holte, E.: The exponentiated subgradient algorithm for heuristic Boolean programming. In: 17th International Joint Conference on Artificial Intelligence, pp. 334–341. Morgan Kaufmann Publishers, San Francisco (2001)
15. Shang, E., Wah, B.: A discrete Lagrangian-based global-search method for solving satisfiability problems. J. Glob. Optim. **12**(1), 61–99 (1998)

16. Voudouris, C., Tsang, E.: Guided local search. In: Glover, F., Kochenberger, G.A. (eds.) Handbook of Metaheuristics. International Series in Operation Research and Management Science, vol. 57, pp. 185–218. Springer, Heidelberg (2003)
17. Wallace, R.J., Freuder, E.C.: Heuristic methods for over-constrained constraint satisfaction problems. In: Jampel, M., Freuder, E., Maher, M. (eds.) OCS 1995. LNCS, vol. 1106, pp. 207–216. Springer, Heidelberg (1996). doi:10.1007/3-540-61479-6_23
18. Xu, W.: Satisfiability transition and experiments on a random constraint satisfaction problem model. Int. J. Hybrid Inf. Technol. **7**(2), 191–202 (2014)
19. Zhou, Y., Zhou, G., Zhang, J.: A hybrid glowworm swarm optimization algorithm for constrained engineering design problems. Appl. Math. Inf. Sci. **7**(1), 379–388 (2013)

Implicit Location Sharing Detection in Social Media Turkish Text Messaging

Davut Deniz Yavuz and Osman Abul(✉)

Department of Computer Engineering,
TOBB University of Economics and Technology, Ankara, Turkey
{st111101036,osmanabul}@etu.edu.tr

Abstract. Social media have become a significant venue for information sharing of live updates. Users of social media are producing and sharing large amount of personal data as a part of the live updates. A significant share of this data contains location information that can be used by other people for many purposes. Some of the social media users deliberately share their own location information with other users. However, a large number of users blindly or implicitly share their own location without noticing it and its possible consequences. Implicit location sharing is investigated in the current paper.

We perform a large scale study on implicit location sharing detection for one of the most popular social media platform, namely Twitter. After a careful study, we prepared a training data set of Turkish tweets and manually labelled them. Using machine learning techniques we induced classifiers that are able to classify whether a given tweet contains implicit location sharing or not. The classifiers are shown to be very accurate and efficient as well. Moreover, the best classifier is employed in a browser add-on tool which warns the user whenever an implicit location sharing is predicted from just to be released tweet. The paper provides the followed methodology and the technical analysis as well. Furthermore, it discusses how these techniques can be extended to different social network services and also to different languages.

Keywords: Location privacy · Social media · Machine learning

1 Introduction

The majority of social media users are not aware of the potential risks when sharing their personal information in social media. A lot of people share large amount of personal data for various purposes, e.g. getting better service, entertaining and being recognized. People share photos, text messages, location, political and social expressions in social media venues like Twitter, Facebook, and Myspace. Whereas another group of people is eager to use such essential information for targeted advertising, marketing, and many others. For instance, managers may utilize the shared information to make promotion decisions within

This work has been supported by TUBITAK under grant number 114E132.

P.M. Pardalos et al. (Eds.): MOD 2016, LNCS 10122, pp. 341–352, 2016.
DOI: 10.1007/978-3-319-51469-7_29

employees by also taking into account the kind of their social media updates. Managers may also use this information while evaluating candidates when hiring. Society for Human Resource Management survey, dating October 2011, has elucidated that 18% of 500 recruiting members use social network searches to screen candidates [9]. Even worse, some criminals use the shared information for choosing their victims. For instance, thieves can exploit owner's location information to determine whether the house is vacant. Web sites pleaserobme.com and geosocialfootprint.com are examples of how easy to find people's exact location.

There is an extensive literature and several applications developed for providing location privacy in the context of explicit location sharing [1]. On the other hand, location privacy in the context of implicit location sharing is quite new and limited. Even such studies are available for English only and their semantic analysis is in English too. They mostly aim to raise awareness on how unsafe to share location implicitly or explicitly in social media. However, in this work, we propose a methodology and techniques on how to prevent or warn users about implicit location sharing while sending text messages.

In this study, we focus on Turkish text messages and extract location features based on Turkish language. We created a Turkish corpus of tweets from Twitter. The tweet data set and the extracted features are used to induce machine learning based classifiers which are able to detect implicit location sharing given a tweet content. Furthermore, we developed a Google Chrome extension tool leveraging the induced classifiers. The tool warns users about potential implicit location sharing.

The paper is structured as follows. The next section provides the related work. The methodology of the work is given in Sects. 3 and 4 presents the data set description and feature extraction. The classification procedure and results are provided in Sect. 5, and the tool featuring this classification is introduced in Sect. 6. Finally, Sect. 7 concludes.

1.1 Location Sharing on Social Media

Social media is becoming the first place that comes to people's mind to locate where a person is. In Facebook, Twitter and Foursquare users produce and share big amount of data, sizeable portion of which is personal data including status, text messages, photos, videos and many other kinds of personal updates. Every user is a data producer and at the same time enthusiastic on others' updates and hence constantly checking other's social updates.

With GPS enabled mobile devices people can share location with fine granularity and this potentially leads to location privacy leaks. Intruders may use personal location information in many ways, the most innocent of which is advertising and marketing. Consider the scenario that you check-in in a shopping mall by your social media profile, and some stores follow your profile and hence know the location information. Clearly, they may send you a targeted advertisement via the form of e-mail or SMS. This can be considered a location privacy violation with low risk. On the other hand, the main risk is that the disclosed location information can be used by attackers to track the user and make physical assaults. In Foursquare, one can add own home address as a place, and his friends can check-in at the home disclosing their current location. This may lead to friends of friends to know the person's home address. It is possible to find users' home location by 77.27% accuracy in a radius less than 20 km from Foursquare [10]. This can give rise to many location related risks. There are some precautions to prevent users from these kind of risks. In Foursquare, users can send an e-mail to `privacy@foursquare.com` about deleting their home address from the database [11]. Also, users can adjust their privacy settings so that only close friends can see check-ins.

In Facebook, Twitter and many mobile applications there are privacy settings aimed to protect users' location privacy. However, most of the people neglect these settings and are not aware of the risks of location sharing. This is because appropriately maintaining privacy setting in social media is not the primary concern for majority, i.e. they rather tend to use default settings. Even for the people making appropriate location privacy settings, the location sharing risks are still on the table. Clearly, they may type their location directly or indirectly and implicitly provide information that can be used to link to users' whereabouts. Hence, social media users need a preventive way that interactively check their text messages and detect whether the post shares private location information.

1.2 Location Sharing on Twitter

Twitter users can share their location with many different ways. For mobile phones, people can add a location label and with the help of built-in GPS can attach their exact locations to the label. One can also easily enable explicit location sharing while sending tweets. Another way to share location is to use other location-based social networking applications such as Foursquare. People can link their Twitter accounts to Foursquare in order to automatically tweet their location. In a preliminary study, we collected 537.125 Turkish tweets and observed that 20.903 of those tweets use Foursquare and 5.567 of them use Twitter's feature to share location. Figure 1 shows that approximately 5% of the tweets use explicit location sharing and 79% of them use Foursquare to share their location. We can assume that those people are aware of what they are doing as location sharing is on purpose. However, a group of people on Twitter are not aware of implicit location sharing. This happens when they do not use Foursquare or Twitter's location sharing option.

A lot of people send tweets somehow blindly and are not aware of resulting implicit location sharing. For example, a tweet in Turkish that says 'Çok sıkıldım, evde yapacak bir şey yok.' meaning 'I am very bored, nothing to do at home.' share location information implicitly that the user is at home at that time. But the user may not notice the fact that he is sharing his location. Yet another example is that the tweet in Turkish 'Armada'da gezecek çok mağaza var, gezerken çok yoruldum.' meaning 'There are a lot of shops in Armada, I feel tired of walking around.' shares implicit location too. Clearly, any follower reading this tweet understands that the user is currently in the renowned shopping

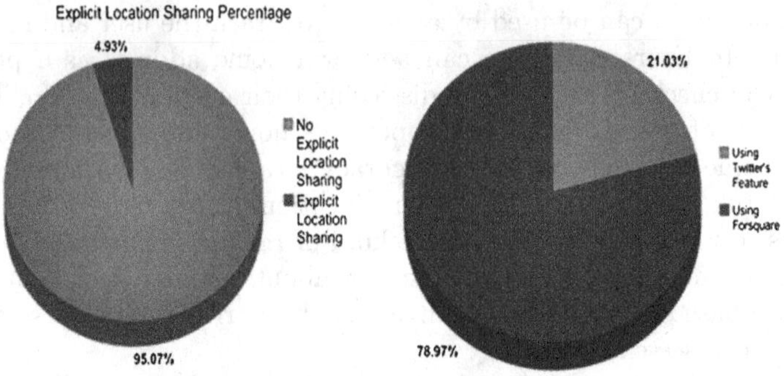

Fig. 1. Explicit location sharing ratio and source applications over all collected tweets

mall Armada in Ankara. Indeed a simple Google Maps search for Armada will reveal the exact user location on the map. These examples are hard to notice for users because the purpose of these tweets are to talk about how boring is the time and how tired the person is, respectively. These types of tweets are within the scope of implicit location sharing which is the main concern of our current study.

2 Related Work

Over the last several years, many researchers have developed ideas and done studies about social networks and location privacy on social media. Regarding location, there are two opposite directions of studies, raising awareness and techniques about location privacy and gathering precious information, including targeted people location, from social networks.

GeoSocial Footprint is an online tool that provides Twitter users with an opportunity to view their geosocial footprint [12]. Provided with a Twitter user name, this tool outputs a map of where the user was in the past. It also provides users with suggestions on how to decrease geosocial footprint. We have no doubt that this study will increase the awareness on location privacy. On the other hand, malicious attackers may use this service to harm their victims.

Weidemann [13] remarks that GIS science professionals are aware of the potential risks when using social network, but the general public usually does not know about these risk. It is also claimed that the real time record of the people's location is more precious information than credit card numbers and bank statements. The results of this study show that 0.8% of Twitter users share their current location via GPS or other technologies and 2.2% of all tweets provide ambient location data. We can conclude from these findings that Twitter users share enough data to followers who are interested in their location information for whatever purpose they desire.

Another notable example of location awareness is http://pleaserobme.com. This web site gives the opportunity to check your own timeline for check-ins.

The creators of this web site, Frank Groeneveld, Barry Borsboom and Boy van Amstel, mention that social networks have great search engines for users. By these functionalities people find their friend and the things they are interested. However, if you allow your messages to travel between different social networks it became more complicated to track your privacy and information you trust to your friends might end up somewhere else [14]. For instance, if you link your Foursquare account to your Twitter account your privacy settings in Foursquare will not work on your Twitter account and you can not protect your Foursquare location information from other public Twitter users. (An extract from Twitter privacy policy: "Our default is almost always to make the information you provide through the Twitter Services public for as long as you do not delete it, but we generally give you settings or features, like direct messages, to make the information more private if you want." [15].)

Large-scale tweet analysis is a hot research topic [2]. Sentiment analysis, classification of tweets as positive, negative or neutral on a specific issue, is more likely the most studied task on tweet data [3] as it has commercial implications, e.g. customer relation and campaign management. Sentiment analysis is successfully applied in many areas and even it reached to irony detection quality [7]. The sentiment analysis usually employs techniques from data mining, machine learning and natural language processing. Our methodology is very similar to supervised approaches for sentiment analysis like Taolu [7]. The methodology simply starts with collecting a relevant tweet data set and building classifiers. The classification problem is given the content of a given tweet, or in general any short text, whether it conveys implicit location sharing or not.

For the case of Twitter, there are a number of spatial indicators, including location mentions in tweets, friend network, location field, Website IP address, geotags, URL links and Time zone [4]. Indeed the location can be inferred by some means, for instance from content [5] and from social relationships [6]. Almost all of the content based methods are for English [4].

It is clear that, those tools, techniques and services increase the awareness of people about location sharing. However, very limited nation-wide research have been done to prevent social media users from implicit location sharing. In this study, we examined tweets in Turkish language and used data mining/machine learning algorithms to obtain classifiers preventing Twitter users from implicitly sharing location.

3 Methodology of the Study

Stefanidis presents three components of system architecture for collecting information from social media feeds, (i) extracting data from the data providers via APIs, (ii) parsing, integrating, and storing these data in a resident database, and (iii) analyzing these data to extract information of interest [8]. In this study, we also follow a similar approach to achieve what we intended to do.

The flow chart of the steps of our methodology is shown in Fig. 2. The first step is the data collection phase. In this step, we collected Turkish tweets via

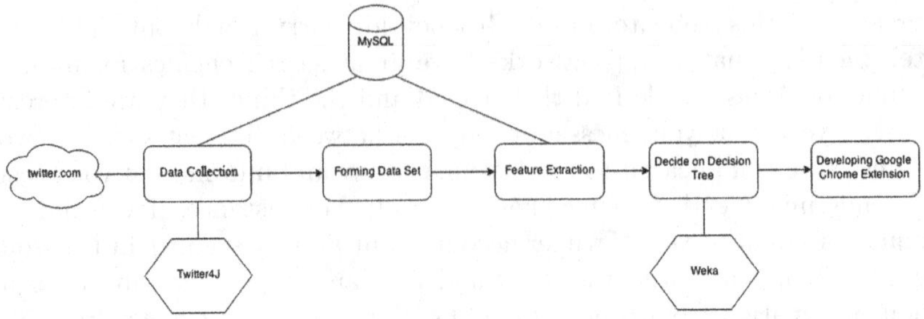

Fig. 2. The methodology followed in this study

Twitter4j from Twitter. The next step is constructing the training data set, a subset of all the tweets. The data set is created manually by looking at the collected tweets. After this step we came to the feature extraction phase. In this phase features are extracted by considering Turkish grammar properties related to location sharing. We employed Weka to find classifiers in the form of decision trees from the training data set. After evaluating a number of classifiers, we picked a few of good classifiers and employed the best performing decision tree classifier in our Google Chrome extension. The extension checks tweets very fast by first extracting feature values and then consulting the classifier.

4 Data Set and Feature Extraction

4.1 Data Collection

Twitter4j [16], a Java library for the Twitter API [17], is used to collect tweets from various random Twitter users living in Turkey, totaling to 537.127 tweets from 1813 distinct users. These tweets are in Turkish and there is no certain classification information to categorize the tweets to subjects and topics. So, our data set contains tweets almost every category one can think. MySQL [18] database management system is used to store the data set. Since the purpose of this study is to semantically analyze the tweets for location privacy, we exclude tweets which use Foursquare and similar explicit location sharing technologies.

4.2 Data Set

As our methodology follows supervised learning approach we need to have labelled data set for training. To do so, we picked 500 tweets from 537.127 in a few iteration so that 250 (positive set) of which implicitly share location but the other 250 (negative set) do not. Hence, the resulting manually labelled data set is balanced and appropriate for inducing supervised classifiers. This step was indeed labor intensive, as we need to sift through the negative set from seemingly positive ones. For instance, the tweets may mention a place with general sentences, e.g. a tweet meaning 'Armada's revenue this year is doubled' mentions Armada but there is no implication here that the user is sending the tweet from Armada. This also shows that mentioning a point of interest or a named entity in a message does not imply location whereabouts.

4.3 Feature Extraction

We considered six features that are relevant in expressing locations in Turkish language. Clearly a similar approach can be taken for other languages as well for extending this work to them. Although the class label, whether the tweet is implicitly sharing location, is marked manually in the previous, existence of the features are marked by SQL statements automatically. In other words, we are able to write SQL expressions, although some of which are too long, for each feature. Official TDK dictionary [19] is used for finding location indicative words and verbs. Index-Anatolicus [20] is also used for listing the city and well-known place names in Turkey.

Feature 1: Feature 1 consists of two Turkish suffixes which are 'deyim' and 'dayım'. These mean 'I am at ...' in English. For example, a tweet that says 'Sınavlar yüzünden bugün yine okuldayım' meaning 'Because of the exams I am at the school once again' shares location information. Another example is that 'Hastalığım yüzünden evdeyim, dşarı çıkamıyorum' meaning 'I am at home and I can't go outside because of my illness.' also shares location information. Checking this feature in the data set, four tweets have this feature and two of them share user's location. The other two are found to be non-location sharing suffixes.

Feature 2: Feature 2 is formed by two Turkish suffixes that are 'de' and 'da'. These two suffixes are searched at the end of the words. These words mean like 'at' and/or 'in' in English. As an example, a tweet like saying 'Okulda sınıf çok sıcak' meaning 'The class is too hot at the school' shares location information. Another tweet saying 'Terminalde arkadaşımı bekliyorum.' meaning 'I am waiting my friend at the terminal' also shares location. 158 tweets are marked for this feature, 97 of which are marked as location sharing statements.

Feature 3: Some special words specifying common locations are chosen to form feature 3. These include place names such as 'okul, ev, i, kafe, restoran' meaning 'school, home, work, cafe, restaurant'. Similar to them, 48 words are used in this feature. These words are significant words that have a great potential of location indicators when they are used in a sentence. One example is 'çu anda Avlu restorana gidiyorum' meaning 'I am going to Avlu restaurant right now' is sharing where you will be soon. Another example 'Ev çok dağınık' meaning 'The house is so messy' is location sharing too. 248 tweets contain feature 3, 191 of which are location sharing.

Feature 4: City names in Turkey are used for feature 4. There are 81 cities in Turkey, and all of them are included in feature 4 without case sensitivity. For example, 'Ankarada yapacak hiçbir çey yok.' meaning 'Nothing to do in Ankara', and 'Izmirde denize girmeyi çok özlemişim' meaning 'I just noticed I missed swimming in Izmir very much' share location information. On the other hand a tweet like 'I did not know that Ankara is the Turkey's capital' does not share location information. There are 178 tweets that have feature 4, 158 of which share location.

Feature 5: As discussed in the data set construction section, tweets that are due to Foursquare and other check-in featured social networks are not included in the data set. But these tweets are used to create feature 5. Indeed, we collected famous place names from such tweets automatically, e.g. a tweet meaning 'I'm at Marco Pascha in Ankara, Turkey https://t.co/HoJEeqXBhK'. We get the part 'Marco Pascha' and use it as a special place name in feature 5. Because all the special words in these tweets are renowned place names of cafes, restaurants, schools, universities, stadiums etc. This way we extracted 6560 distinct place names and used them for feature 5. For instance, a tweet like 'Armada'ya yemeğe geldik.' meaning 'We came to Armada to dine.' is an example in this category. 366 tweets are marked this way and 228 of which implicitly share location.

Feature 6: Feature 6 consists of 18 special Turkish verbs that use to describe where you are. For instance; 'geldim', 'geldik' and 'gitme' are the verbs that are used when you want to express where you are going or where you are. For example, 'Eve gidiyorum' meaning 'I am going to the home' conveys that the user is about to arrive home. Another example is 'Annemlere geldik' meaning 'We came to my mother's house' share location information too. 94 tweets have feature 6 and 89 of them are positive.

Figure 3 summarizes the individual relevancy of the features, i.e. their ability to distinguish positives from negatives on the 500 tweets data set.

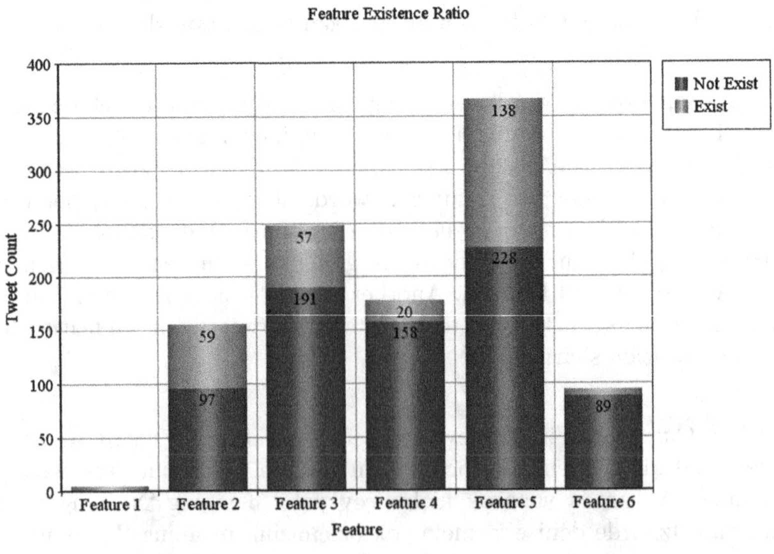

Fig. 3. Relevancy of features on the 500 tweets data set

5 Classification

5.1 Weka

Weka [21] is a free distribution software developed by Waikato University which provides tools and algorithms for data mining and predictive modelling. In this study, Weka 3.6 is used for choosing the most suitable classification algorithm among the available algorithms based on the feature classification and location privacy data.

5.2 Evaluating Data Mining Algorithms

Input Data Set: Input data given to classification algorithms is a collection of rows where the attributes are featurel, feature2, feature3, feature4, feature5, feature6 and class label. The data is exported from the tweet database by SQL statements. The full data set contains 500 tuples with 7 attributes (6 predictors and 1 class label). The class label value 1 indicates implicit location sharing and 0 indicates no location sharing.

Process: 73 classifiers and different test options are used in order to evaluate classifiers. First, 73 classifiers resulting from several classification algorithms are evaluated with the test option of 10-fold cross validation. After getting all the results, we rank them by looking at the ratio of correctly classified number of instances, i.e. accuracy. The top three classifiers are chosen. They are due to J48graft, J48 and END classification algorithms. In order to decide which one is the most suitable classifier, we try different test options with the same data for the three classifiers. As shown in the Fig. 4 J48graft algorithm is the most accurate one when classifying our instances.

Algorithm	Cross-validation	Percentage split	Correctly Classified Instances
J48graft		66	88,82
J48		66	88,23
END		66	88,23
J48graft	50		87,4
J48graft	10		87,6
J48		90	87,2
J48	10		87,4
J48	30		87,4
J48		50	87,2
J48	20		87,2
J48	70		87,2
END	50		87,2
END	10		87,4
J48graft		30	83,14

Fig. 4. Classifier accuracy comparison

Evaluation Results: J48graft produced the decision tree given bellow.

```
Decision Tree:
feature4 = 0
|   feature6 = 0
|   |   feature3 = 0: 0 (175.0/3.0)
|   |   feature3 = 1
|   |   |   feature2 = 0: 0 (55.0/17.0)
|   |   |   feature2 = 1: 1 (43.0/15.0)
|   feature6 = 1: 1 (49.0/5.0)
feature4 = 1
|   feature5 = 0: 0 (7.0/1.0)
|   feature5 = 1: 1 (171.0/14.0)
```

Since J48graft classifier gives the best result its decision tree is used in our Google Chrome extension. The J48 classifier also gives the same decision tree. Different decision trees resulting of different classifiers can be used for getting different behaviors. For instance, the decision tree of ADTree algorithm is shown below.

```
Decision Tree:
|  (1)feature4 = 0: -0.455
|  (1)feature4 = 1: 1.012
|  |    (3)feature5 = 0: -1.712
|  |    (3)feature5 = 1: 0.547
|  |  |        (7)feature1 = 0: 0.013
|  |  |        (7)feature1 = 1: 0.234
|  |    (6)feature1 = 0: -0.029
|  |    (6)feature1 = 1: 0.486
|  (2)feature3 = 0: -0.82
|  |    (4)feature5 = 0: -0.977
|  |    (4)feature5 = 1: -0.044
|  (2)feature3 = 1: 0.691
|  |     (5)feature5 = 0: -0.188
|  |    (5)feature5 = 1: 0.054
```

However, this decision tree is not used because of its poor prediction power. ADTree's classifier has the accuracy of 83% which is less than J48's results.

6 Google Chrome Extension

The Google Chrome extension is developed to prevent Twitter users from sharing their sensitive location information accidentally. After being installed, the extension is optional and it can be activated or deactivated from its popup menu. When it is activated the extension runs the classifier from J48graft. Extension checks the tweet before clicking the 'Tweetle' button. It first computes the values of six features from the text of the tweet and inputs it to the classifier. If the result is a positive prediction, it creates a warning message along with an icon. The warning text is 'Konum paylaşıyor olabilirsiniz!' meaning 'May be you are sharing your location!'. Figure 5 is a screenshot of the extension.

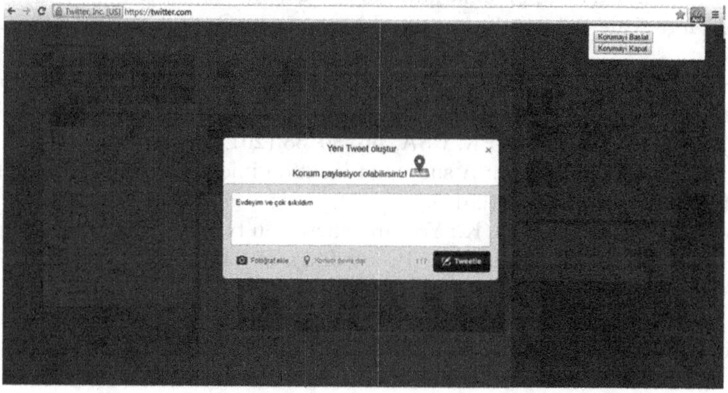

Fig. 5. A screenshot of our Google Chrome extension

The reason that we have embedded the classifier in a Google Chrome extension rather than a built-in Twitter functionality is due to the fact that we are unable to create a third-party Twitter service. If Twitter API provides such an option or collaboration, that will result in better look and feel warning.

7 Conclusion

This study has shown that implicit location sharing is a common privacy issue in the social networks era. Noticing this we have developed a system using machine learning techniques and for Turkish language tweets. The reason that Turkish is selected is two-fold, first we are native speakers and second we wanted to encourage other language specific studies.

The Google Chrome extension tool generated by this study could be very useful in various ways. It could help social media users to audit their location privacy sharing. It could be also useful to educate them on their location privacy awareness. Although the tool is implemented for Twitter, it can be easily deployed for other micro blogging social media platforms such as Facebook and MySpace.

Although the current accuracy is over 80%, future work may enhance this value by extracting new features mostly based on natural language studies and semantics. For instance, we can add more specific words to the available features but also adapt new features to help algorithm to work with more features and words based on the Turkish language understanding.

Following the same methodology, one can easily extend the work to other languages such as English and French. Keeping the methodology the same, it simply suffices to replace language specific features.

References

1. Shin, K.G., Ju, X., Chen, Z., Hu, X.: Privacy protection for users of location-based services. IEEE Wirel. Commun. **19**(1), 30–39 (2012)
2. Pandarachalil, R., Sendhilkumar, S., Mahalakshmi, G.S.: Twitter sentiment analysis for large-scale data: an unsupervised approach. Cogn. Comput. **7**(2), 254–262 (2015)
3. Agarwal, A., Xie, B., Vovsha, I., Rambow, O., Passonneau, R.: Sentiment analysis of Twitter data. In: Proceedings of the Workshop on Languages in Social Media (LSM 2011), Stroudsburg, PA, USA, pp. 30–38 (2011)
4. Ajao, O., Hong, J., Liu, W.: A survey of location inference techniques on Twitter. J. Inf. Sci. **41**(6), 855–864 (2015)
5. Cheng, Z., Caverlee, J., Lee, K.: You are where you tweet: a contentbased approach to geo-locating Twitter users. In: Proceedings of CIKM 2010, Toronto, Canada, pp. 759–768 (2010)
6. Jurgens, D.: That's what friends are for: inferring location in online social media platforms based on social relationships. In: Proceedings of ICWSM 2013, Boston, MA, pp. 273–282 (2013)
7. Taslioglu, H.: Irony detection on Turkish microblog texts. Master thesis, Middle East Technical University, Ankara (2014)
8. Stefanidis, A.: Harvesting ambient geospatial information from social media feeds (2012). http://www.academia.edu/1472285/Harvesting_Ambient_Geospatial_Information_from_Social_Media_Feeds
9. Kadaba, L.S.: What is privacy? As job-seekers are judged by their tweets and Facebook posts, uncertainty abounds (2012). http://articles.philly.com/2012-05-03/news/31539376_1_facebook-photos-facebook-passwords-employers
10. Pontes, T.: Beware of what you share inferring home location in social networks. In: IEEE 12th International Conference on Data Mining Workshops, Brussels, Belgium, pp. 571–578 (2012)
11. Foursquare: Privacy 101 (2016). https://foursquare.com/privacy/
12. Weidemann, C.: (2013). http://geosocialfootprint.com/
13. Weidemann, C.: Social Media Location Intelligence: The Next Privacy Battle - An ArcGIS add-in and Analysis of Geospatial Data Collected from Twitter.com (2013). http://journals.sfu.ca/ijg/index.php/journal/article/view/139
14. Groeveneld, F., Borsboom, B., Amstel, B.: Over-sharing and Location Awareness (2011). https://cdt.org/blog/over-sharing-and-location-awareness/
15. Twitter: Twitter Privacy Policy (2016). https://twitter.com/privacy
16. Twitter4J. http://twitter4j.org
17. Twitter developer. https://dev.twitter.com
18. MySQL. http://www.mysql.com/
19. Official Turkish Dictionary. http://www.tdk.gov.tr
20. Index Anatolicus. http://www.nisanyanmap.com/?lg=t
21. Weka. http://www.cs.waikato.ac.nz/ml/weka/

Fuzzy Decision-Making of a Process for Quality Management

Feyza Gürbüz[1(✉)] and Panos M. Pardalos[2]

[1] Department of Industrial Engineering, Erciyes University, Kayseri, Turkey
feyza@erciyes.edu.tr
[2] Center for Applied Optimization,
Department of Industrial and Systems Engineering,
Univesity of Florida, Gainesville, FL, USA
pardalos@ufl.edu

Abstract. Problem solving and decision-making are important skills for business and life. Good decision-making requires a mixture of skills: creative development and identification of options, clarity of judgement, firmness of decision, and effective implementation. SWOT analysis is a powerful tool that can help decision makers achieve their goals and objectives. In this study, data obtained through SWOT analysis from a quality department of a textile company were integrated by means of fuzzy multi criterian decision making. The aim of the study was to choose the policy most appropriate for the beneficial development of the quality department.

Keywords: Fuzzy decision making · SWOT analysis · Quality managament

1 Introduction

The globalized market conditions of our time and the unregulated operation of production factors have brought to the fore the question of quality. While only few decades ago the quality of a product or service was not thought to be important issue, now it has become a common factor for research in all aspects of daily life, such as the quality of life and/or the services provided by an organization.

Other factors that have also brought to the fore the whole concept of quality are the diversification of goods and services produced in the rapid process of change, the liberalization of international trade, the disappearance of commercial boundaries, technological progress and the new competitive conditions that these developments have brought about. For this reason, it is essential that companies determine and review their quality policies to achieve the highest possible standards and to endeavour to keep ahead of their rivals.

In many cases when making such an assessment, numerical values may prove ineffective in determining or accurately reflecting the complexities of real quality issues in life. Human thoughts and judgments inherently contain ambiguities and it may become impossible for individuals to express their preferences with simplistic dual analysis and conventional logic through the usage of such terminology as favorable/unfavourable, present/absent, yes/no. Every person experiences conditions in their daily

P.M. Pardalos et al. (Eds.): MOD 2016, LNCS 10122, pp. 353–378, 2016.
DOI: 10.1007/978-3-319-51469-7_30

lives which have many variables, and even if sometimes these are thought initially to be definite in the end they may turn out to be indefinite or inconclusive. In everyday real life the events people experience, almost without exception, contain ambiguity. Ambiguity is the most common of all causes to complicate the decision-making process. There are a multitude of methods that can be used to assess ambiguity. Fuzzy reasoning is one of the mathematical methods used to analyze ambiguity. It basically uses conventional reasoning expressions such as yes/no, right/wrong, black/white and also the values in between them. For example, fuzzy reasoning mathematically formulizes expressions like "somewhat warm" or "very cold". This in turn, facilitates the processing of such concepts by computers. Thus, it enables computers to be programmed in closer proximity to human thought. First developed by L.A. Zadeh (1965) of Azerbaijan, the Fuzzy set theory is basically concerned with the ambiguity of human thought and perception, and it tries to quantify this ambiguity.

One's knowledge and interpretations regarding physical occurrences generally manifest themselves as personal opinions. Therefore, there exists ambiguity in human thought, if not in quantity, and this is a source of knowledge.

The Fuzzy set theory enables such sources of knowledge to be used in the study of various events. Fuzzy set concepts function as a bridge between the verbal and the quantitative.

SWOT analysis is used to determine not only the strengths and weaknesses of the studied organizations technique, process or condition but also the opportunities and threats stemming from the external environment. The basic building blocks of SWOT analysis, i.e. the strengths of, weaknesses of, opportunities for and threats to the aforementioned entities were applied to the problem of quality analysis and policy choices under a fuzzy environment with the methodology of analytical hierarchy processing for the first time in this study.

Fuzzy AHP, fuzzy TOPSIS and fuzzy ELECTRE operations were performed with linguistic variables. All the assessments made were converted into triangular fuzzy numbers and quantified using a fuzzy weights matrix. Next, from a fuzzy decision matrix, a normalized fuzzy decision matrix was obtained. Then, subsequently, each of the other methods was used to apply, in each, case the requisite procedures of that particular method. In the second section of the study, the recent literature on multi-criteria decision-making under a fuzzy environment was reviewed. Latest, in the third section, methods of fuzzy multi-criteria decision making were studied.

The fourth section presents a study applying the above methodology within the quality department of a textile company. Finally, the last section is allocated for the assessment of the study's findings as well as to present some suggestions for further research studies in this area.

2 Literature Review

There are a large number of studies on the methods of multi-criteria decision making in existence, some of which are mentioned below.

In a study by Aytaç et al. [1], the method of fuzzy Electre was applied to the problems encountered by a textile operation in choosing a catering firm. In another study by Kaya and Kahraman [2] environmental effects were assessed using the fuzzy AHP and ELECTRE integrity methodology. Rouyendegh and Erol [3], in their study carried out a selection process for choosing the best project using the fuzzy ELECTRE method. Wang et al. [4], in their study, assessed the problem of choosing supplies by utilizing the TOPSIS method which they argue weighs up criteria more objectively and accurately in comparison with other methodologies. Using the Fuzzy ELECTRE method, the study by Marbini et al. [5] made an assessment of safety and health priority issues concerning the recycling of hazardous waste. Torlak et al. [6] carried out a comparative analysis on the subject of competition between operations using the TOPSIS method and then applied this system of methodology to make an assessment on Turkish Airlines. In a further study by Öztürk et al. [7] it was recommended that the methods of FAHP and fuzzy TOPSIS should be adopted to address ambiguities in verbal assessments made by decision takers.

To assist in the selection of the most appropriate policy decisions that should be made for the future maintenance of stability in the Turkish economy from an analysis of Turkey's last economic crisis, Yücenur et al. [8] devised, within their study, a hierarchical model based on the SWOT analysis approach. This model was then evaluated with the FAHP and the FANP methodology. In a study by Çinar [9], the problem of choosing the most suitable location for an institution to be founded was looked at and a decision support model was proposed to ensure the endurability of the most accurate choice. In the study by Erginel et al. [10] GSM operators' choice of criteria were determined by a working committee's review of the literature and also by its evaluation of customers' opinions by using linguistic variables to assist in the assessment. Cebeci [11] presented an approach by selecting a suitable ERP system for carrying out a study of the textile industry. FAHP, was used to compare these ERP system solutions, applied to a textile manufacturing company. Also, Gürbüz and Pardalos [12] used data mining and fuzzy cluster analysis to model the decision-making process of a ceramics production company. Triantaphyllou et al. [13] used fuzzy for the scales used in decision-making problems. Doumpos et al. [14] focused on the multi-criteria decision aid (MCDA) approach to review the main MCDA sorting techniques, and to present the multi-group hierarchical discrimination method.

Studies have also been conducted about multi-criteria decision making combined with SWOT analysis, as mentioned below.

Ekmekcioğlu et al. [15] used fuzzy TOPSIS integrated with fuzzy AHP to develop fuzzy multi-criteria SWOT analysis. Nuclear power plant site selection, which is a strategic and important issue for Turkey's energy policy making, is considered as an application case study that demonstrated the applicability of the developed fuzzy SWOT model. Prisenk and Borec [16], developed a combination of multi-criteria decision analysis (MCDA) and SWOT analysis by applying the DEX method for the identification of shortcomings in the production and marketing of local food products in Slovenia. Şevkli et al. [17] provided a quantitative basis to analytically determine the ranking of the factors in SWOT analysis via a conventional multi-criteria decision-making method, namely Analytic Network Process (ANP). Bas [18] proposed an integrated framework for the analysis of an electricity supply chain using an integrated SWOT-fuzzy TOPSIS methodology combined with Analytic Hierarchy

Process (AHP). Zavadskas et al. [19] proposed a methodology for determining management strategies in construction enterprises. SWOT analysis was used for formulating management strategies.

3 Methods of Multi-criteria Decision-Making Under Fuzzy Environment

Making a decision is the determination of existing or realizable choices and then ultimately it giving preference to one of them for the purpose of action (or indeed inaction). Fuzzy multi-criteria decision-making is known in the literature as the breaking down of a problem into its constituent parts. A wide range of scientific methods have been devised to solve fuzzy multi-criteria decision problems.

3.1 Fuzzy AHP

By its inherent nature it is not possible to express definitive criteria using Fuzzy logic, which is based on the fuzzy set theory, along the lines of whether "an object is a member of a set or not." Fuzzy logic therefore explains that the minimum degree of membership is equal to zero and the maximum degree of membership is equal to 1. Partial membership is defined as any intermediate value between zero and one.

The first study to apply fuzzy logic to AHP (FAHP) was performed in 1983 by Van Laarhoven and Pedryez [20]. In that study triangular membership functions were used for fuzzy relationships. Chang [21] suggested an expansion of the AHP method, which does not require an α-cut approach. FAHP has been applied in a variety of fields in its development process with success.

Sequencing can be done according to various characteristics, such as making the center of attraction the area below the function of membership degree cross-sectional points. Therefore, different sequencing methods can yield different sequencing results for the same data. There are a great many methods of FAHP. In this study the method of Liou and Wang [22] has been used.

3.2 Fuzzy TOPSIS

In real life, data are not deterministic. On the contrary they are defined as being fuzzy due to their being missing or some data being difficult to acquire. In general, preference-containing judgments are indefinite and preference, due to its inherent nature, cannot be expressed as a quantitative value. Therefore, the method of TOPSIS was designed in such a way so as to be able to use fuzzy data. This method facilitates the assessment and sequencing of alternatives by more than one decision maker according to multi-criteria under fuzzy conditions and this assists the decision - making process in making the correct choice.

In the first step of the TOPSIS methodology suggested by Chen [23], a committee of decision makers is formed. A caucus set consisting of N decision makers is expressed as $E = \{KV_1, KV_2, \ldots, KV_N\}$. Following the constitution of the committee, their existing alternatives $A = \{A_1, A_2, \ldots, A_m\}$ and the criteria to be used in the assessment of these alternatives, i.e. $K = \{K_1, K_2, \ldots, K_n\}$, are then determined. Then the linguistic variables to be used are chosen when assessing alternatives and when determining the weight significance of the criteria.

3.3 Fuzzy ELECTRE

To make a decision in a fuzzy ambience is a complicated task. This is because traditional methods are unable to formulate the subjectivity and ambiguity of the decision-making process. It is therefore necessary to use multi-criteria decision-making methods where the subjective criteria and weights of all criteria can be assessed with linguistic variables. Such a methodology has already been outlined in the aforementioned literature. Fuzzy Electre I is one of the fuzzy multi-criteria decision-making methods used by decision makers to solve the ambiguity of concepts that can arise during the course of their decisions making.

The fuzzy Electre suggested in this study is structured in Table 3 and the successfully implemented steps, which are an extension of Electre, are explained and summarized in Fig. 1 (Appendix 5).

4 Case Study

In this study, the quality policy most appropriate for the department was determined using the fuzzy multiple criteria decision-making method and also in accordance with the criteria derived from SWOT analysis obtained from within the quality department of a textile company. To assess the condition of this particular company and to selecti the policy most appropriate for this company fuzzy AHP, fuzzy TOPSIS, and fuzzy ELECTRE were essentially integrated utilizing SWOT analysis.

The objective was to choose the policy most appropriate for the development of the quality department. The assessment criteria used in achieving this objective were the basic building blocks of SWOT analysis, namely strength, weaknesses, opportunities and threats.

4.1 Criteria and Sub-criteria for the Choice of Quality Policy

The problem model suggested for analysis in the study was inherently hierarchical in its structure and consisted of different criteria, sub-criteria and choices. The aim was to choose the policy most appropriate for the quality department's development. As previously mentioned, the assessment criteria of SWOT analysis were used. The SWOT factors determined for the problem company model are presented in Table 4 [24] given in the Appendix 2. The hierarchical structure formed is presented in Fig. 2.

4.2 Fuzzy AHP

We realized the solution to our sample problem, for the first time, with the successful implementation and usage of fuzzy AHP methodology which was initiated using the data obtained from SWOT analysis. The linguistic expression in Table 5 was used in the assessment of the criteria and choice weights, shown in Appendix 3.

First the criteria were compared amongst themselves in the fuzzy triangular numbers as in Table 6 given in Appendix 3. Then, FAHP steps were taken. We calculated the synthetic ranking in the comparison matrix for the criteria above.

The decision maker's optimism was taken to be 0.5 and the total integral values were then calculated.

For the assessment of the different sub-criteria relating to each criterion the same procedure mentioned above was followed. In the last step of the fuzzy AHP methodology, however, priority weights were united as shown in Table 7 in Appendix 3.

At the end of the FAHP analysis of the hierarchical model, the alternative policies were arranged as A1-A2-A3. A1 was found to be the most important policy and was ranked the highest. This policy emphasized that the improvement of production quality was the most important factor necessary in meeting customers needs.

4.3 Fuzzy TOPSIS

In this model, the same criteria and alternative policies as in FAHP were used. First, 3 decision makers (DM1, DM2, DM3) made an assessment of the criteria and choices to be considered in selecting a policy. Then, these verbal variables were converted into fuzzy numbers. The criteria were assessed by the decision makers using the verbal variables in Table 2 (Appendix 1).

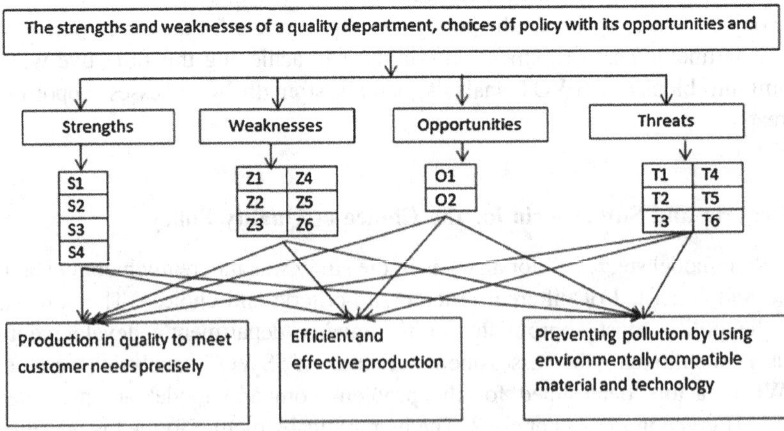

Fig. 2. Hierarchical structure

Then, using the verbal variables in Table 2 the decision makers assessed the three alternative policies as shown in Table 9 (Appendix 4).

Then, the verbal variables in Tables 8 and 9 (Appendix 4) were used to convert the triangle into fuzzy numbers as shown in Tables 10 and 11 (Appendix 4).

Following the assessment of the criteria and choices using the verbal variables assessment, the results of the 3 decision makers were reduced to a single value and significance weights relating to the criteria were determined. The significance weights are shown in Table 12 (Appendix 4).

After determining the criteria weights, the assessment of the alternatives in Table 12 was reduced to a single value and the fuzzy decision matrix was then constituted as shown in Table 13 (Appendix 4).

The fuzzy decision matrix was then constructed, together with the normalized fuzzy decision matrix shown in Table 14 (Appendix 4).

After the construction of the normalized fuzzy matrix, the positive ideal and fuzzy negative ideal solutions were constructed for all the criteria shown in Tables 16 and 17 (Appendix 4). Here the maximum and minimum values were found for each criterion on completion of the calculation.

After calculation of the normalized fuzzy matrix, the positive ideal and fuzzy negative ideal solutions were constructed for all the criteria. Here the maximum and minimum values were assessed and found for each criterion from this calculation.

Then, the differential of each choice for all the criteria between the fuzzy positive and negative ideal solutions was found. These differentials are presented in Tables 18 and 19.

After calculation of the differentials of all the criteria, calculations were made for both the fuzzy positive ideal and fuzzy negative ideal solutions, di^{*} and di^{-}, for the 3 policy choices. The results of these calculations are shown in Table 20 (Appendix 4).

Alternatives were arranged according to their varying differentials in comparison with the alternatives. Thus, the arrangement of the three alternative policies was expressed as being A1 > A2 > A3. This highlighted the most beneficial policy for the enhancement of quality most likely to satisfy the needs of customers.

4.4 Fuzzy ELECTRE

Since it employs the same first three steps as the fuzzy TOPSIS, the sections up to Table 15 should be followed in an identical fashion for this model and therefore have been skipped here. The explanation, therefore, continues from step 4. The prevalent choice in comparing the two alternatives was calculated by means of the weighted fuzzy decision matrix in Table 15. For this purpose, a comparison matrix was formed for each criterion. Compatibility and incompatibility sets were determined. In determining the compatibility and incompatibility sets, if a choices compared for the criterion ultimately became prevalent, $J_c = 0$, the differential distance for J_d could then be calculated. If it was not prevalent, then the difference between the two fuzzy numbers for J_c would have to be calculated. The differential computations for both criteria are illustrated in Table 21 (Appendix 5).

Based on the compatibility and incompatibility sets, the matrices C and D are then able to be constructed as in Tables 22 and 23 (Appendix 5). The weights of all the criteria within Matrix C (i.e. for $J_c = 0$) are then ascertained and recorded. Incompatibility matrix D is computed with the incompatibility sets (Table 23 in Appendix 5). Based on the compatibility matrix, dual matrix B can then be constructed as shown in Table 24 (Appendix 5). Dual matrix H is constructed according to the incompatibility matrix as shown in Table 25 (Appendix 5). Matrix Z was constructed by multiplying B by H as shown in Table 26 (Appendix 5). With the help of the general matrix in Table 27 (Appendix 5), the decision graph shown in Fig. 3 (Appendix 5), was then formulated and then sequencing was done in accordance with this graph.

The application of the fuzzy ELECTRE methodology to resolve the study task revealed that A1 established itself as being the most important policy in Table 27 (Appendix 5). In other words, the policy to be emphasized and focused on the most in developing the quality department was essentially the one 'tailor- made' for the production of quality to cater precisely for customer needs, namely the policy catgorized as A1.

5 Conclusion and Suggestions

SWOT analysis allows decision makers to bring transparency to the various factors that can affect the achievement of their business goals and objectives. Using the low-level details provided by SWOT analysis, businesses can clearly see what will or will not work for them towards their goals and objectives. This study established that it is first necessary to determine the criteria and sub-criteria to be considered for the quality department and these are determinable through SWOT analysis. Then the important policies required for the quality department have to be determined using fuzzy AHP, fuzzy TOPSIS and fuzzy ELECTRE methodology with multiple criteria.

Judging from the conclusions drawn from the study and presented in Table 28, we believe that the primary policy to be emphasized for implementation to successfully expand a quality department is the one that 'tailor-makes' production quality so as to meet all customers needs and requirements as closely and as thoroughly as possible. As a result of this process it can be deduced that productivity and efficiency would be more readily achievable.

Table 28. Comparisons of findings

	FAHP		FTOPSIS		FELECTRE
A1	**0.56**	**1**	**0.60**	**1**	**1**
A2	0.25	2	0.53	2	2
A3	0.19	3	0.26	3	3

This study is of importance in that it will help company workers shape their future and it can be used as a future precedent for other businesses. Future studies on the subject should focus on the assessment made with fuzzy, multiple-criteria decision making techniques arising out of the results of SWOT analysis and this should be utilized as part of the process within marketing and shipment departments, so that appropriate policies can be determined for each department.

As a result we can say that SWOT analysis can be a great tool to help us understand what is involved in making business decisions. To support their decision making process, a SWOT analysis was used in combination with multi-criteria decision making tehcniques.

Appendix 1: Linguistic Variables Used in the Study

See Table 1.

Table 1. Linguistic variables used in determining the weight of criteria

Very High (ÇY)	(0.8,1,1)
High (Y)	(0.7,0.8,0.9)
Medium High (BY)	(0.5,0.65,0.8)
Medium (O)	(0.4,0.5,0.6)
Medium Low (BD)	(0.2,0.35,0.5)
Low (D)	(0.1,0.2,0.3)
Very Low (ÇD)	(0,0,0.2)

Table 2. Linguistic variables used in assessing alternatives

Very Good (ÇI)	(8,10,10)
Good (I)	(7,8,9)
Medium Good (BI)	(5,6.5,8)
Fair (F)	(4,5,6)
Medium Poor (MK)	(2,3.5,5)
Poor (M)	(1,2,3)
Very Poor (ÇK)	(0,0,2)

Table 3. Proximity coefficient and level of acceptability

Proximity coefficient (CC_i)	Interpretation
$CC_i \in [0,0.2)$	Not recommended
$CC_i \in [0.2,0.4)$	Recommended with high risk
$CC_i \in [0.4,0.6)$	Recommended with low risk
$CC_i \in [0.6,0.8)$	Recommended
$CC_i \in [0.8,1.0)$	Acceptable and preferable

Appendix 2

Table 4. Swot factors

Swot group	Swot factor name	Swot factors
Strengths (S)	S1	Having a young and strong minded team
	S2	Having enough technological equipment to simplify the process (forklift, pallet, shelf, palm etc.)
	S3	The ability to rapidly load goods that are ready for shipping according to company policy
	S4	Taking photos after loading and keeping them for assessing & understanding problems that may be faced in the future
Weaknesses (W)	W1	Lack of ability to use the addressing system efficiently, having periodical problems with the system such as having apparent fabrics in the system where subsequently no such fabrics were found to exist
	W2	Sending unwanted fabrics, arising from adaptation problems to defective tool plans and defective systematically hipping problems
	W3	Mixed fabrics in the pallets
	W4	Problems occurring due to lack of knowledge about the priority firms that should have been shipped to with priority status and, as a result of this, the lack of ability to organize or reorganize
	W5	Lack of communication while programming shipment on weekdays
	W6	Shipping warehouses that are too small for the fabrics and as a result of this delayed work flows
	W7	Caring only about loading fabrics on time, especially for urgent fabrics, and ignoring the damages to goods during the loading process
Opportunities (O)	O1	Increasing warehouse arrangement and stock cycling speeds resulting from an increase in shipments
	O2	Ability to manage more successful shipping organization resulting from the systematical infrastructure used
Threats (T)	T1	Negative feedback from customers resulting from loading mistakes or missing loads, and the resultant loss of prestige in the eyes of customers
	T2	During the process of loading, workers suffering from cold in the trucks during the winter, as a result of cold draughts from loading tunnels outside and cold air

(*continued*)

Table 4. (*continued*)

Swot group	Swot factor name	Swot factors
		circulation in front of the doors, with the resultant decrease in work performance and increased health problems and ill health resulting from extremely hot weather in summer
	T3	Workers psychological problems and lack of concentration on their work due to excessive overtime work, loss of motivation and careless loading resulting from tiredness of workers, and the unwillingness and dissatisfaction of workers
	T4	Physical ailments of workers leading to poor and decreased work performance and loss of staff
	T5	Excessive waiting time in the trucks
	T6	Hazardous and insecure movement of forklift trucks in access ways and in the shipping area causing a higher incidence of industrial accidents, such as the risk of forklift crashes, fabrics being dropped, tumbling pallets etc
	T7	Intensive loading traffic and intensive work resulting in disease occurrence such as loin disease

Appendix 3: Tables of Fuzzy AHP

Table 5. FAHP assessment scale

Linguistic scale for importance and degrees of success	Fuzzy scale	Corresponding scale
Equally important/successful	(1,1,1)	(1/1, 1/1, 1/1)
Weakly more important/successful	(1,3,5)	(1/5, 1/3, 1/1)
Strongly more important/successful	(3,5,7)	(1/7, 1/5, 1/3)
Very strongly more important/successful	(5,7,9)	(1/9, 1/7, 1/5)
Absolutely more important/successful	(7,9,9)	(1/9, 1/9, 1/7)

Table 6. Comparison of criteria with one another

	O			T			S			W		
O	1	1	1	1	3	5	1	3	5	1	3	5
T	0.2	0.333	1	1	1	1	3	5	7	1	1	1
S	0.2	0.333	1	0.143	0.2	0.333	1	1	1	1	3	5
W	0.2	0.333	1	1	1	1	0.2	0.333	1	1	1	1

Table 7. Combination of choices with criteria and the priority order of the choices

The weight vector of main criteria	0.41	0.29	0.19	0.11	
Alternative policies	O	T	S	W	Total weight
A1	0.47	0.65	0.46	0.80	**0.56**
A2	0.28	0.20	0.31	0.13	0.25
A3	0.24	0.14	0.22	0.06	0.19

Appendix 4: Tables of Fuzzy TOPSIS

Table 8. Assessment of criteria by the decision maker

	DM1	DM2	DM3
S1	ÇY	Y	Y
S2	Y	BY	Y
S3	BY	O	BY
S4	O	O	ÇD
W1	O	Y	Y
W2	Y	ÇY	Y
W3	ÇD	ÇD	ÇD
W4	Y	Y	ÇY
W5	BY	D	O
W6	ÇD	D	BD
W7	O	BD	ÇD
Ol	O	O	O
O2	Y	Y	Y
T1	ÇY	ÇY	ÇY
T2	BY	Y	D
T3	Y	Y	Y
T4	O	O	O
T5	Y	O	Y
T6	Y	Y	BY
T7.	O	BY	BD

Table 9. Assessment of the alternatives under criteria

		DM1	DM2	DM3
S1	A1	ÇI	ÇI	ÇI
	A2	BI	BI	BI
	A3	F	F	
S2	A1	I	I	I
	A2	I	ÇI	I
	A3	F	BI	BI
S3	A1	I	ÇI	I
	A2	F	MK	BI
	A3	K	K	MK
S4	A1	I	BI	BI
	A2	ÇK	F	K
	A3	ÇK	ÇK	K
W1	A1	BI	BI	F
	A2	I	ÇI	I
	A3	MK	K	MK
W2	A1	ÇI	I	I
	A2	F	F	F
	A3	BI	MK	BI
W3	A1	BI	F	F
	A2	I	BI	I
	A3	F	F	MK
W4	A1	I	I	I
	A2	BI	BI	I
	A3	F	MK	BI
W5	A1	BI	I	BI
	A2	I	ÇI	BI
	A3	K	MK	K
W6	A1	BI	MK	F
	A2	F	BI	BI
	A3	K	K	K
W7	A1	I	I	F
	A2	BI	I	I
	A3	BI	F	I

(*continued*)

Table 9. (*continued*)

		DM1	DM2	DM3
O1	A1	I	BI	BI
	A2	BI	F	BI
	A3	MK	K	M
O2	A1	I	BI	I
	A2	BI	I	BI
	A3	MK	MK	MK
T1	A1	ÇI	ÇI	ÇI
	A2	I	ÇI	I
	A3	MK	BI	BI
T2	A1	F	I	I
	A2	I	F	F
	A3	K	K	MK
T3	A1	I	I	F
	A2	F	BI	BI
	A3	K	F	MK
T4	A1	F	BI	BI
	A2	I	BI	BI
	A3	K	K	MK
T5	A1	ÇK	K	ÇK
	A2	BI	BI	BI
	A3	K	ÇK	K
T6	A1	F	BI	F
	A2	MK	F	MK
	A3	MK	K	MK
T7	A1	BI	BI	BI
	A2	F	MK	MK
	A3	K	ÇK	K

Table 10. Assessment of crieria in the form of triangular fuzzy numbers

	DM1			DM2			DM3		
	l	m	u	l	m	u	l	m	u
S1	0,80	1,00	1,00	0,70	0,80	0,90	0,70	0,80	0,90
S2	0,70	0,80	0,90	0,50	0,65	0,80	0,70	0,80	0,90
S3	0,50	0,65	0,80	0,40	0,50	0,60	0,50	0,65	0,80
S4	0,40	0,50	0,60	0,40	0,50	0,60	0,00	0,00	0,20
W1	0,40	0,50	0,60	0,70	0,80	0,90	0,70	0,80	0,90
W2	0,70	0,80	0,90	0,80	1,00	1,00	0,70	0,80	0,90
W3	0,00	0,00	0,20	0,00	0,00	0,20	0,00	0,00	0,20

(*continued*)

Table 10. (*continued*)

	DM1			DM2			DM3		
	l	m	u	l	m	u	l	m	u
W4	0,70	0,80	0,90	0,70	0,80	0,90	0,80	1,00	1,00
W5	0,50	0,65	0,80	0,10	0,20	0,30	0,40	0,50	0,60
W6	0,00	0,00	0,20	0,10	0,20	0,30	0,20	0,35	0,50
W7	0,40	0,50	0,60	0,20	0,35	0,50	0,00	0,00	0,20
O1	0,40	0,50	0,60	0,40	0,50	0,60	0,40	0,50	0,60
O2	0,70	0,80	0,90	0,70	0,80	0,90	0,70	0,80	0,90
T1	0,80	1,00	1,00	0,80	1,00	1,00	0,80	1,00	1,00
T2	0,50	0,65	0,80	0,70	0,80	0,90	0,10	0,20	0,30
T3	0,70	0,80	0,90	0,70	0,80	0,90	0,70	0,80	0,90
T4	0,40	0,50	0,60	0,40	0,50	0,60	0,40	0,50	0,60
T5	0,70	0,80	0,90	0,40	0,50	0,60	0,70	0,80	0,90
T6	0,70	0,80	0,90	0,70	0,80	0,90	0,50	0,65	0,80
T7	0,40	0,50	0,60	0,50	0,65	0,80	0,20	0,35	0,50

Table 11. Presentation in triangular fuzzy numbers of the alternatives assessment

		DM1			DM2			DM3		
		l	m	u	l	m	u	l	m	u
S1	A1	8	10	10	8	10	10	8	10	10
	A2	5	6	8	5	6.5	8	5	6.5	8
	A3	4	5	6	4	5	6	2	3.5	5
S2	A1	7	8	9	7	8	9	7	8	9
	A2	7	8	9	8	10	10	7	8	9
	A3	4	5	6	5	6.5	8	5	6.5	8
S3	A1	7	8	9	8	10	10	7	8	9
	A2	4	5	6	2	3.5	5	5	6.5	8
	A3	1	2	3	1	2	3	2	3.5	5
S4	A1	7	8	9	5	6.5	8	5	6.5	8
	A2	0	0	2	4	5	6	1	2	3
	A3	0	0	2	0	0	2	1	2	3
W1	A1	5	6.5	8	5	6.5	8	5	6.5	8
	A2	7	8	9	8	10	10	7	8	9
	A3	2	3.5	5	1	2	3	2	3.5	5
W2	A1	8	10	10	7	8	9	7	8	9
	A2	4	5	6	4	5	6	4	5	6
	A3	5	6.5	8	2	3.5	5	5	6.5	8
W3	A1	5	6.5	8	4	5	6	4	5	6
	A2	7	8	9	5	6.5	8	7	8	9
	A3	4	5	6	4	5	6	2	3.5	5

(*continued*)

Table 11. (*continued*)

		DM1			DM2			DM3		
		l	m	u	l	m	u	l	m	u
W4	A1	7	8	9	7	8	9	7	8	9
	A2	5	6.5	8	5	6.5	8	7	8	9
	A3	4	5	6	2	3.5	5	5	6.5	8
W5	A1	5	6.5	8	7	8	9	5	6.5	8
	A2	7	8	9	8	10	10	5	6.5	8
	A3	1	2	3	2	3.5	5	1	2	3
W6	A1	5	6.5	8	2	3.5	5	4	5	6
	A2	4	5	6	5	6.5	8	5	6.5	8
	A3	1	2	3	1	2	3	1	2	3
W7	A1	7	8	9	7	8	9	4	5	6
	A2	5	6.5	8	7	8	9	7	8	9
	A3	5	6.5	8	4	5	6	7	8	9
O1	A1	7	8	9	5	6.5	8	5	6.5	8
	A2	5	6.5	8	4	5	6	5	6.5	8
	A3	2	3.5	5	1	2	3	1	2	3
O2	A1	7	8	9	5	6.5	8	7	8	9
	A2	5	6.5	8	7	8	9	5	6.5	8
	A3	2	3.5	5	2	3.5	5	2	3.5	5
T1	A1	8	10	10	8	10	10	8	10	10
	A2	7	8	9	8	10	10	7	8	9
	A3	2	3.5	5	5	6.5	8	5	6.5	8
T2	A1	4	5	6	7	8	9	7	8	9
	A2	7	8	9	4	5	6	4	5	6
	A3	1	2	3	1	2	3	2	3.5	5
T3	A1	7	8	9	7	8	9	4	5	6
	A2	4	5	6	5	6.5	8	5	6.5	8
	A3	1	2	3	4	5	6	2	3.5	5
T4	A1	4	5	6	5	6.5	8	5	6.5	8
	A2	7	8	9	5	6.5	8	5	6.5	8
	A3	1	2	3	1	2	3	2	3.5	5
T5	A1	0	0	2	2	3.5	5	0	0	2
	A2	5	5.5	8	5	6.5	8	5	6.5	8
	A3	1	2	3	0	0	2	1	2	3
T6	A1	4	5	6	5	6.5	8	4	5	6
	A2	2	3.5	5	4	5	6	2	3.5	5
	A3	2	3.5	5	1	2	3	2	3.5	5
T7	A1	5	6.5	8	5	6.5	8	5	6.5	8
	A2	4	5	6	2	3.5	5	2	3.5	5
	A3	1	2	3	0	0	2	1	2	3

Table 12. Significance weights criteria

	l	m	u
S1	0,73	0,87	0,93
S2	0,63	0,75	0,37
S3	0,47	0,60	0,73
S4	0,27	0,33	0,47
W1	0,60	0,70	0,80
W2	0,73	0,87	0,93
W3	0,00	0,00	0,20
W4	0,73	0,87	0,93
W5	0,33	0,45	0,57
W6	0,10	0,18	0,33
W7	0,20	0,28	0,43
O1	0,40	0,50	0,60
O2	0,70	0,80	0,90
T1	0,80	1,00	1,00
T2	0,43	0,55	0,67
T3	0,70	0,80	0,90
T4	0,40	0,50	0,60
T5	0,60	0,70	0,80
T6	0,63	0,75	0,87
T7	0,37	0,50	0,63

Table 13. Fuzzy decision matrix

	A1			A2			A3		
	l	m	u	l	m	u	l	m	u
S1	8,00	10,00	10,00	5,00	6,50	8,0O	3,33	4,50	5,67
S2	7,00	8,00	9,00	7,33	8,67	9,33	4,67	6,00	7,33
S3	7,33	8,67	9,33	3,67	5,00	6,33	1,33	2,50	3,67
S4	5,67	7,00	8,33	1,67	2,33	3,67	0,33	0,67	2,33
W1	4,67	6,00	7,33	7,33	8,67	9,33	1,67	3,00	4,33
W2	7,33	8,67	9,33	4,00	5,00	6,00	4,00	5,50	7,00
W3	4,33	5,50	6,67	6,33	7,50	8,67	3,33	4,50	5,67
W4	7,00	8,00	9,00	5,67	7,00	8,33	3,67	5,00	6,33
W5	5,67	7,00	8,33	6,67	8,17	9,00	1,33	2,50	3,67
W6	3,67	5,00	6,33	4,67	6,00	7,33	1,00	2,00	3,00
W7	6,00	7,00	8,00	6,33	7,50	8,67	5,33	6,50	7,67
Ol	5,67	7,00	8,33	4,67	6,00	7,33	1,33	2,50	3,67
O2	6,33	7,50	8,67	5,67	7,00	8,33	2,00	3,50	5,00
T1	8,00	10,00	10,00	7,33	8,67	9,33	4,00	5,50	7,00
T2	6,00	7,00	8,00	5,00	6,00	7,00	1,33	2,50	3,67

(*continued*)

Table 13. (*continued*)

	A1			A2			A3		
	l	m	u	l	m	u	l	m	u
T3	6,00	7,00	8,00	5,67	5,50	6,83	2,33	3,50	4,67
T4	4,67	6,00	7,33	5,67	7,00	8,33	1,33	2,50	3,67
T5	0,67	1,17	3,00	5,00	6,50	8,00	0,67	1,33	2,67
T6	4,33	5,50	6,67	2,67	4,00	5,33	1,67	3,00	4,33
T7	5,00	6,50	8,00	2,67	4,00	5,33	0,67	1,33	2,67

Table 14. Normalized fuzzy decision matrix

	A1			A2			A3		
	l	m	u	l	m	u	l	m	u
S1	0.80	1.00	1.00	0.50	0.65	0.80	0.33	0.45	0.57
S2	0.75	0.86	0.96	0.79	0.93	1.00	0.50	0.64	0.79
S3	0.79	0.93	1.00	0.39	0.54	0.68	0.14	0.27	0.39
S4	0.68	0.84	1.00	0.20	0.28	0.44	0.04	0.08	0.28
W1	0.50	0.64	0.79	0.79	0.93	1.00	0.18	0.32	0.46
W2	0.79	0.93	1.00	0.43	0.54	0.64	0.43	0.59	0.75
W3	0.50	0.63	0.77	0.73	0.87	1.00	0.38	0.52	0.65
W4	0.78	0.89	1.00	0.63	0.78	0.93	0.41	0.56	0.70
W5	0.63	0.78	0.93	0.74	0.91	1.00	0.15	0.28	0.41
W6	0.50	0.68	0.86	0.64	0.82	1.00	0.14	0.27	0.41
W7	0.69	0.81	0.92	0.73	0.87	1.00	0.62	0.75	0.88
O1	0.68	0.84	1.00	0.56	0.72	0.88	0.16	0.30	0.44
O2	0.73	0.87	1.00	0.65	0.81	0.96	0.23	0.40	0.58
T1	0.80	1.00	1.00	0.73	0.87	0.93	0.40	0.55	0.70
T2	0.75	0.88	1.00	0.63	0.75	0.88	0.17	0.31	0.46
T3	0.75	0.88	1.00	0.71	0.69	0.85	0.29	0.44	0.58
T4	0.56	0.72	0.88	0.68	0.84	1.00	0.16	0.30	0.44
T5	0.08	0.15	0.38	0.63	0.81	1.00	0.08	0.17	0.33
T6	0.65	0.83	1.00	0.40	0.60	0.80	0.25	0.45	0.65
T7	0.63	0.81	1.00	0.33	0.50	0.67	0.08	0.17	0.33

Table 15. Weighted normalized fuzzy decision matrix

	A1			A2			A3		
	l	m	u	l	m	u	l	m	u
S1	0.59	0.87	0.93	0.37	0.56	0.75	0.24	0.39	0.53
S2	0.48	0.64	0.84	0.50	0.70	0.87	0.32	0.48	0.68
S3	0.37	0.56	0.73	0.18	0.32	0.50	0.07	0.16	0.29

(*continued*)

Table 15. (*continued*)

	A1			A2			A3		
	l	m	u	l	m	u	l	m	u
S4	0.18	0.28	0.47	0.05	0.09	0.21	0.01	0.03	0.13
W1	0.30	0.45	0.63	0.47	0.65	0.80	0.11	0.23	0.37
W2	0.58	0.80	0.93	0.31	0.46	0.60	0.31	0.51	0.70
W3	0.00	0.00	0.15	0.00	0.00	0.20	0.00	0.00	0.13
W4	0.57	0.77	0.93	0.46	0.67	0.86	0.30	0.48	0.66
W5	0.21	0.35	0.52	0.25	0.41	0.57	0.05	0.13	0.23
W6	0.05	0.13	0.29	0.06	0.15	0.33	0.01	0.05	0.14
W7	0.14	0.23	0.40	0.15	0.25	0.43	0.12	0.21	0.38
O1	0.27	0.42	0.60	0.22	0.36	0.53	0.06	0.15	0.26
O2	0.51	0.69	0.90	0.46	0.65	0.87	0.16	0.32	0.52
T1	0.64	1.00	1.00	0.59	0.87	0.93	0.32	0.55	0.70
T2	0.33	0.48	0.67	0.27	0.41	0.58	0.07	0.17	0.31
T3	0.53	0.70	0.90	0.50	0.55	0.77	0.20	0.35	0.53
T4	0.22	0.36	0.53	0.27	0.42	0.60	0.06	0.15	0.26
T5	0.05	0.10	0.30	0.38	0.57	0.80	0.05	0.12	0.27
T6	0.41	0.62	0.87	0.25	0.45	0.69	0.16	0.34	0.56
T7	0.23	0.41	0.63	0.12	0.25	0.42	0.03	0.08	0.21

Table 16. Fuzzy positive ideal solutions

	L	M	U
S1	0,93	0,93	0,93
S2	0,87	0,87	0,87
S3	0,73	0,73	0,73
S4	0,47	0,47	0,47
W1	0,80	0,80	0,80
W2	0,93	0,93	0,93
W3	0,20	0,20	0,20
W4	0,93	0,93	0,93
W5	0,57	0,57	0,57
w6	0,33	0,33	0,33
w7	0,43	0,43	0,43
O1	0,60	0,60	0,60
O2	0,90	0,90	0,90
T1	1,00	1,00	1,00

(*continued*)

Table 16. (*continued*)

	L	M	U
T2	0,67	0,67	0,67
T3	0,90	0,90	0,90
T4	0,60	0,60	0,60
T5	0,80	0,80	0,80
T6	0,87	0,87	0,87
T7	0,63	0,63	0,63

Table 17. Fuzzy negative ideal solutions

	L	M	U
S1	0,24	0,24	0,24
S2	0,32	0,32	0,32
S3	0,07	0,07	0,07
S4	0,01	0,01	0,01
W1	0,11	0,11	0,11
W2	0,31	0,31	0,31
W3	0,00	0,00	0,00
W4	0,30	0,30	0,30
W5	0,05	0,05	0,05
W6	0,01	0,01	0,01
W7	0,12	0,12	0,12
O1	0,06	0,06	0,06
O2	0,16	0,16	0,16
T1	0,32	0,32	0,32
T2	0,07	0,07	0,07
T3	0,20	0,20	0,20
T4	0,06	0,06	0,06
T5	0,05	0,05	0,05
T6	0,16	0,16	0,16
T7	0,03	0,03	0, 03

Table 18. The differentials between AI and A* for each criterion

	d(Al, A*}	d(A2, A*)	d(A3, A*]
S1	0,20	0,40	O,55
S2	0,26	0,23	0,40
S3	0,23	0,42	0,57
S4	0,20	0,36	0,41
W1	0,37	0,21	0,58
W2	0,22	0,49	0,45
W3	0,17	0,16	0,17
W4	0,23	0,31	0,48
W5	0,24	0,21	0,44
W6	0,20	0,19	0,27
W7	0,21	0,20	0,22
O1	0,22	0,26	0,45
O2	0,25	0,30	0,58
T1	0,21	0,25	0,50
T2	0,22	0,28	0,49
T3	0,25	0,32	0,56
T4	0,26	0,22	0,45
T5	0,66	0,28	0,66
T6	0,30	0,44	0,54
T7	O,27	0,39	0,53

Table 19. The calculation of the differentials between A_I and A for each criterion

	d(Al, A-}	d(A2, A)	d(A3, A-]
S1	0,57	0,35	0,18
S2	0,37	0,40	0,23
S3	0,51	0,30	0,14
S4	0,32	0,12	0,07
W1	0,38	0,55	0,17
W2	0,48	0,19	0,25
W3	0,09	0,12	0,08
W4	0,48	0,40	0,23
W5	0,34	0,38	0,11
W6	0,17	0,20	0,07
W7	0,17	0,19	0,16
O1	0,39	0,33	0,13

(continued)

Table 19. *(continued)*

	d(Al, A-}	d(A2, A)	d(A3, A-]
O2	0,56	0,52	0,23
T1	0,59	0,50	0,26
T2	0,44	0,37	0,15
T3	0,53	0,42	0,20
T4	0,33	0,39	0,13
T5	0,15	0,56	0,13
T6	0,51	0,36	0,26
T7	0,43	0,26	0,11

Table 20. D_I^*, D_I^- and Cc_I values

	A1	A2	A3
DI*	5.16	5.90	9.31
DI−	7.80	6.92	3.27
DI* + DI−	12.96	12.82	12.58
CCI	**0.601**	0.539	0.260

Appendix 5: Tables of Fuzzy ELECTRE

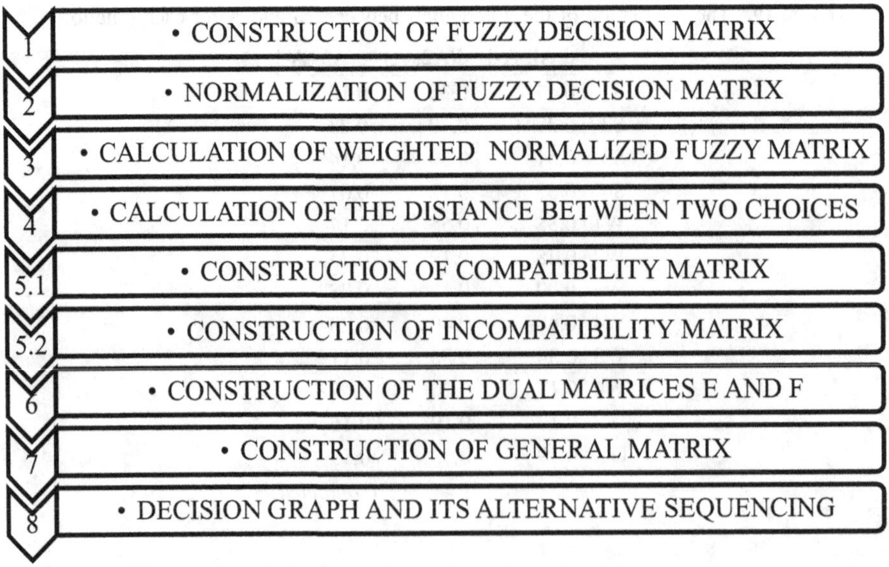
1
• CONSTRUCTION OF FUZZY DECISION MATRIX
2
• NORMALIZATION OF FUZZY DECISION MATRIX
3
• CALCULATION OF WEIGHTED NORMALIZED FUZZY MATRIX
4
• CALCULATION OF THE DISTANCE BETWEEN TWO CHOICES
5.1
• CONSTRUCTION OF COMPATIBILITY MATRIX
5.2
• CONSTRUCTION OF INCOMPATIBILITY MATRIX
6
• CONSTRUCTION OF THE DUAL MATRICES E AND F
7
• CONSTRUCTION OF GENERAL MATRIX
8
• DECISION GRAPH AND ITS ALTERNATIVE SEQUENCING

Fig. 1. Steps of the Fuzzy ELECTRE application

Table 21. Computation of the differentials between two alternatives for each criterion

	x11		x21		x31	
	JC	JD	JC	JD	JC	JD
x11	*	*	0,00	0,24	0,00	0,411462
x21	*	*	*	*	0,00	0,18
x32	*	*	*	*	*	*

	x12		x22		x32	
	JC	JD	JC	JD	JC	JD
x12	*	*	0,04	0,00	0,00	0,16
x22	*	*	*	*	0,00	0,19
x32	*	*	*	*	*	*

	x13		x23		x33	
	JC	JD	JC	JD	JC	JD
x13	*	*	0,00	0,22	0,00	0,39
x23	*	*	*	*	0,00	0,17
x33	*	*	*	*	*	*

	x14		x24		x34	
	JC	JD	JC	JD	JC	JD
x14	*	*	0,00	0,20	0,00	0,23
x24	*	*	*	*	0,00	0,06
x34	*	*	*	*	*	*

5,00	x15		x25		x35	
	JC	JD	JC	JD	JC	JD
x15	*	*	0,18	0,00	0,00	0,23
x25	*	*	*	*	0,00	0,41
x35	*	*	*	*	*	*

6,00	x16		x26		x36	
	JC	JD	JC	JD	JC	JD
x16	*	*	0,00	0,31	0,00	0,26
x26	*	*	*	*	0,06	0,00
x36	*	*	*	*	*	*

7,00	x17		x27		x28	
	JC	JD	JC	JD	JC	JD
x17	*	*	0,03	0,00	0,00	0,01
x27	*	*	*	*	0,00	0,04
x37	*	*	*	*	*	*

8,00	x18		x28		x38	
	JC	JD	JC	JD	JC	JD
x18	*	*	0,00	0,09	0,00	0,28
x28	*	*	*	*	0,00	0,19
x38	*	*	*	*	*	*

9,00	x19		x29		x39	
	JC	JD	JC	JD	JC	JD
x19	*	*	0,05	0,00	0,00	0,23
x29	*	*	*	*	0,00	0,28
x39	*	*	*	*	*	*

10,00	x110		x210		x310	
	JC	JD	JC	JD	JC	JD
x110	*	*	0,03	0,00	0,00	0,10
x210	*	*	*	*	0,00	0,13
x310	*	*	*	*	*	*

11,00	x11		x21		x31	
	JC	JD	JC	JD	JC	JD
x11	*	*	0,02	0,00	0,00	0,02
x21	*	*	*	*	0,00	0,04
x32	*	*	*	*	*	*

12,00	x11		x21		x31	
	JC	JD	JC	JD	JC	JD
x11	*	*	0,00	0,06	0,00	0,28
x21	*	*	*	*	0,00	0,22
x32	*	*	*	*	*	*

13,00	x11		x21		x31	
	JC	JD	JC	JD	JC	JD
x11	*	*	0,00	0,05	0,00	0,37
x21	*	*	*	*	0,00	0,32
x32	*	*	*	*	*	*

14,00	x11		x21		x31	
	JC	JD	JC	JD	JC	JD
x11	*	*	0,00	0,09	0,00	0,36
x21	*	*	*	*	0,00	0,27
x32	*	*	*	*	*	*

15,00	x11		x21		x31	
	JC	JD	JC	JD	JC	JD
x11	*	*	0,00	0,07	0,00	0,31
x21	*	*	*	*	0,00	0,24
x32	*	*	*	*	*	*

16,00	x11		x21		x31	
	JC	JD	JC	JD	JC	JD
x11	*	*	0,00	0,12	0,00	0,35
x21	*	*	*	*	0,00	0,25
x32	*	*	*	*	*	*

17,00	x11		x21		x31	
	JC	JD	JC	JD	JC	JD
x11	*	*	0,06	0,00	0,00	0,22
x21	*	*	*	*	0,00	0,28
x32	*	*	*	*	*	*

18,00	x11		x21		x31	
	JC	JD	JC	JD	JC	JD
x11	*	*	0,44	0,00	0,00	0,02
x21	*	*	*	*	0,00	0,45
x32	*	*	*	*	*	*

19,00	x11		x21		x31	
	JC	JD	JC	JD	JC	JD
x11	*	*	0,00	0,17	0,00	0,28
x21	*	*	*	*	0,00	0,11
x32	*	*	*	*	*	*

20,00	x11		x21		x31	
	JC	JD	JC	JD	JC	JD
x120	*	*	0,00	0,16	0,00	0,33
x220	*	*	*	*	0,00	0,16
x320	*	*	*	*	*	*

Table 22. Compatibility matrix

	A1			A2			A3		
	l	m	u	l	m	u	l	m	u
A1	*	*	*	6,97	8,43	9,57	9,83	12,00	14,17
A2	2,87	3,57	4,60	*	*	*	9,83	12,00	14,17
A3	0,00	0,00	0,00	0,73	0,87	0,93	*	*	*

Table 23. Incompatibility matrix

	A1	A2	A3
A1	*	1,42	0,00
A2	0,70	*	0,13
A3	1,00	7,50	*

Table 24. Matrix B

	A1	A2	A3
A1	*	1,00	1,00
A2	0,00	*	1,00
A3	0,00	0,00	*

Table 25. Matrix H

	A1	A2	A3
A1	**	1,00	1,00
A2	1,00	*	1,00
A3	1,00	0,00	*

Table 26. General matrix Z

	A1	A2	A3
A1	*	1,00	1,00
A2	0,00	*	1,00
A3	0,00	0,00	*

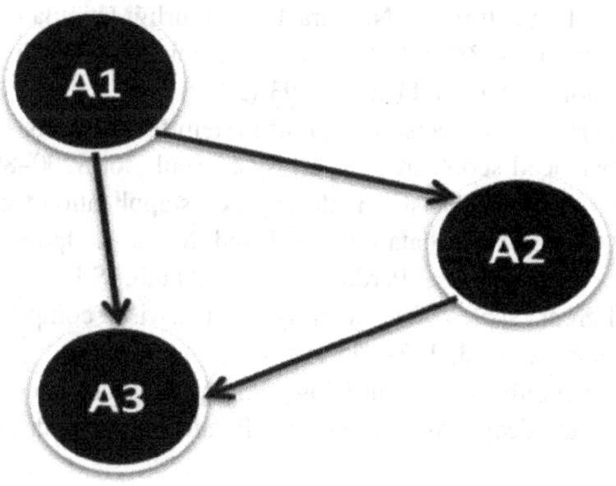

Fig. 3. Decision graph

Table 27. Final arrangement of criteria

		Arrangement
A1	A2, A3	1
A2	A3	2
A3	*	3

References

1. Aytaç, E., Tuşişik, A., Kundakci, N.: Fuzzy electre 1 method for evaluating firm alternatives. Ege Akademik Bakiş **11**, 125–134 (2011)
2. Kaya, T., Kahraman, C.: An integrated fuzzy AHP–ELECTRE methodology for environmental impact assessment. Expert Syst. Appl. **38**, 8553–8562 (2011)
3. Rouyendegh, B., Erol, S.: Selecting the best project using the fuzzy electre method. Math. Prob. Eng., Article ID 790142, 12 pages (2012). Hindawi Publishing Corporation
4. Wang, J., Cheng, C., Kun-Cheng, H.: Fuzzy hierarchical TOPSIS for supplier selection. Appl. Soft Comput. **9**, 377–386 (2009)
5. Marbini, A., Tavana, M., Moradi, M., Kangi, F.: Fuzzy group electre method for safety and health assessment in hazardous waste recycling facilities. Saf. Sci. **51**, 414–426 (2013)
6. Torlak, G., Şevkli, M., Sanal, M., Zaim, S.: Analyzing business competition by using fuzzy TOPSIS method: an example of turkish domestic airline industry. Expert Syst. Appl. **38**, 3396–3406 (2011)
7. Öztürk, A., Ertuğrul, İ., Karakaşoğlu, N.: Nakliye Firmasi Seçiminde Bulanik AHP ve Bulanik TOPSİS Yöntemlerinin Karşilaştirilmasi. Marmara üniversitesi İ.İ.B.F. Dergisi 25, 2 (2008)
8. Yücenur, G., Demirel, N., Demirel, T.: SWOT analysis and integrated fuzzy ahp/anp methodologies for strategic policy selection in turkish economy. J. Eng. Nat. Sci. **28**, 275–286 (2010)
9. Çinar, N.: Kuruluş Yeri Seçiminde Bulanik TOPSİS Yöntemi ve Bankacilik Sektöründe Bir Uygulama. KMÜ Sosyal ve Ekonomik Araştirmalar Dergisi **12**, 37–45 (2010)

10. Erginel, N., Çakmak, T., Şentürk, S.: Numara Taşinabilirliği Uygulamasi Sonrasi Türkiye'de Gsm Operatör Tercihlerinin Bulanik Topsis Yaklaşimi ile Belirlenmesi. Anadolu Üniversitesi Bilim Ve Teknoloji Dergisi **11**(2), 81–93 (2010)
11. Cebeci, U.: Fuzzy AHP-based decision support system for selecting ERP systems in textile industry by using balanced scorecard. Expert Syst. Appl. **36**, 8900–8909 (2009)
12. Gürbüz, F., Pardalos, M.P.: A decision making process application for the slurry production in ceramics via fuzzy cluster and data mining. J. Ind. Manage. Optim. **8**(2), 285–297 (2012)
13. Triantaphyllou, E., Lootsma, F., Pardalos, P.M., Mann, S.H.: On the evaluation and application of different scales for quantifying pairwise comparisons in fuzzy sets. J. Multi-Criteria Decis. Anal. **3**, 133–155 (1994)
14. Doumpos, M., Zopounidis, K., Pardalos, P.M.: Multicriteria sorting methodology: application to financial decision problems. J. Parallel Emergent Distrib. Syst. **15**(1–2), 113–129 (2000)
15. Ekmekçioglu, M., Kutlu, A.C., Kahraman, C.: A fuzzy multi-criteria SWOT analysis: an application to nuclear power plant site selection. Int. J. Comput. Intell. Syst. **4**(4), 583–595 (2011)
16. Prisenk, J., Borec, A.: A combination of the multi-criteria approach and SWOT analysis for the identification of shortcomings in the production and marketing of local food. Agric. Sci. J. **16**, 31–37 (2012)
17. Sevkli, M., Oztekin, A., Uysal, O., Torlak, G., Turkyilmaz, A., Delen, D.: Development of a fuzzy ANP based SWOT analysis for the airline industry in Turkey. Expert Syst. Appl. **39**, 14–24 (2012)
18. Bas, E.: The integrated framework for analysis of electricity supply chain using an integrated SWOT-fuzzy TOPSIS methodology combined with AHP: The case of Turkey. Electr. Power Energy Syst. **44**, 897–907 (2013)
19. Zavadskas, E.K., Turskis, Z., Tamosaitiene, J.: Selection of construction enterprises management strategy based on the SWOT and multi-criteria analysis. *Arch. Civ. Mech. Eng.* **XI**(4), 1063–1082 (2011)
20. Kahraman, C., Cebeci, U., Ruan, D.: Multi-attribute comparison of catering service companies using fuzzy AHP: the case of Turkey. Int. J. Prod. Econ. **87**, s.171–s.184 (2004)
21. Chang, D.Y.: Application of the extent analysis method on fuzzy AHP. Eur. J. Oper. Res. **95** (3), s.649–s.655 (1996)
22. Liou, T.S., Wang, M.J.: Ranking fuzzy numbers with integral value. Fuzzy Sets Syst. **50**, 247–255 (1992)
23. Chen, Chen-Tung: Extensions of the TOPSIS for group decision-making under fuzzy environment. Fuzzy Sets Syst. **114**, 1–9 (2000)
24. Gürbüz, F., Yalçin, N.: A swot-fahp application for a textile firm in Turkey. In: Business Modeling, Information Systems and Enterprise Optimization Conference, Tirana, Albania, 13 July–16 July (2011)

A Bayesian Network Profiler for Wildfire Arsonists

Rosario Delgado[1(✉)], José Luis González[2], Andrés Sotoca[3], and Xavier-Andoni Tibau[4]

[1] Departament de Matemàtiques, Universitat Autònoma de Barcelona, Edifici C- Campus de la UAB., Av. de l'Eix Central s/n, 08193 Cerdanyola del Vallès, Spain
delgado@mat.uab.cat

[2] Gabinete de Coordinación y Estudios, Secretaría de Estado de Seguridad, c/. D. Quijote, esquina con c/. de los Artistas, 28020 Madrid, Spain
jlga@interior.es

[3] Sección de Análisis del Comportamiento Delictivo, Unidad Técnica de Policía Judicial, Guardia Civil - c/. Príncipe de Vergara, 246, 28016 Madrid, Spain
asotoca@guardiacivil.es

[4] Facultat d'Economia i Empresa, Institut d'Economia de Barcelona, Universitat de Barcelona, Av. Diagonal, 690, 08034 Barcelona, Spain
xavitibau@gmail.com
http://gent.uab.cat/rosario_delgado

Abstract. Arson-caused wildfires have a rate of clarification that is extremely low compared to other criminal activities. This fact made evident the importance of developing methodologies to assist investigators in the criminal profiling. For that we introduce Bayesian Networks (BN), which have only recently be applied to criminal profiling and never to arsonists. We learn a BN from data and expert knowledge and, after validation, we use it to predict the profile (characteristics) of the offender from the information about a particular arson-caused wildfire, including confidence levels that represent expected probabilities.

Keywords: Bayesian network · Criminal profiling · Expert system · Wildfire arson

1 Introduction

Wildfire can be regarded as an environmental disaster which is triggered by either natural forces or anthropogenic activities, and is one of the most relevant threats to nature ecosystems and human societies according to the Food and Agriculture Organization of the United Nations (FAO) survey [10]. In this work we are

R. Delgado—This author is supported by Ministerio de Economía y Competitividad, Gobierno de España, project ref. MTM2015 67802-P.

P.M. Pardalos et al. (Eds.): MOD 2016, LNCS 10122, pp. 379–390, 2016.
DOI: 10.1007/978-3-319-51469-7_31

interested in the latter, the arson-caused wildfire, understood as "the large and destructive fire caused by humans that spreads quickly out control over woodland or brush, calcining forest fuels located in the mount and affecting vegetation in principle was not destined to burn", which is one of major environmental problems in Spain.

As it can be seen exploring the literature on the subject, wildfires have been studied mainly from the point of view of risk assessment. Just to mention some of these studies, in Thompson et al. [24] an integrated and systematic risk assessment framework to better manage wildfires and to mitigate losses to highly valued resources and assets is presented, with application to an area in Montana, United States, while in Penman et al. [19] patterns of natural and arson ignitions were examined within a region of south-eastern Australia, to determine the extent to which management can alter the risk of ignition. Different fire risk indices have been considered in Adab et al. [1] to delineate fire risk in northeastern Iran, which is a region subjected to frequent wildfires.

Although arson is one potential cause of many fires, yet the rate of clarification of arson-caused wildfires is extremely low compared to other criminal activities. According to the interim report of the Ministry of Agriculture, Food and Environment [17], 11,928 wildfires were committed in 2015 in Spain, of which 429 offenders have been identified, representing a clarification rate of 3.6%, disregarding that not all the fires were arson-caused. This fact highlights the importance of developing methodologies that can assist investigators to hunt down the culprits by means of the implementation of the criminal profiling, consisting of inferring features (behavioral, criminological, socio-demographic and of personality) of the offender from the analysis of the evidences obtained from the place where fire started.

Focusing on the case of arson-caused wildfire, apart from some few descriptive studies (see [21]), the only statistical approach to wildifre arsonist profiling stems from the work of González, Sotoca and collaborators [22]. Their approach to this problem is through the application of different techniques of multivariate data analysis. In this paper, however, we introduce in a novel way the methodology of Bayesian Networks (from now on, BN), that has only recently been used for criminal profiling (see, for instance, Baumgartner et al. [3,4]) and never, as far as we know, applied for profiling of any kind of arsonist.

BN are an increasingly popular methodology for modeling uncertain and complex domains, and in the opinion of many Artificial Intelligence researchers, the most significant contribution in this area in the last years (see Korb and Nicholson [14]). BN were introduced in the 1920s as a tool that describes probabilistic understanding of cause and effect and other relations between variables, and are the soundest framework for reasoning under uncertainty. From then, this methodology has proved to be useful in assisting in many decision-making procedures in a variety of fields, and its use in risk analysis is gaining popularity and includes applications in areas as diverse as economy (Adusei-Poku [2]), public health (Spiegelhalter [23] and Cruz-Ramírez et al. [6]), environmental risk (Borsuk et al. [5] and Pollino et al. [20]), emerging diseases (Walshe and Burgman

[26]), ecological disasters (Ticehurst et al. [25]) or nuclear waste accidents (Lee and Lee [15]). And with respect to criminology, for example, BN have been introduced as a novel methodology in assessing the risk of recidivism of sex offenders in Delgado and Tibau [7].

Regarding wildfire, there are some recent works as Papakosta and Straub [18], in which the authors study a wildfire building damage consequences assessment system, constructed from a BN with Hugin Expert software, and apply it to spatial datasets from the Mediterranean island of Cyprus. Dlamini develops a BN model in [8] with Netica software, from satellite and geographic information systems (GIS), with variables of biotic, abiotic and human kind, in order to determine factors that influence wildfire activity in Swaziland (see also [9]). However, up to our knowledge, there are no previous studies on the use of BN for profiling of the offender of a wildfire.

We learn a BN model from available data and expert knowledge and, subsequently, predict the profile of the offender from the particular characteristics of an arson-caused wildfire. The inferred arsonist characteristics include confidence levels that represent their expected probability, which enable for investigators to know what are the features of the person who sparked the blaze. Roughly speaking, we construct the most probable BN given the observed cases (learning procedure), and this model provides the optimal prediction for future cases (inference procedure), but its effectiveness for prediction purposes essentially depends on the sufficiency of the training database. We introduce different performance metrics to address this issue, and conclude that our database is actually large enough for our purposes. Having done that, we learn (*train*) both BN structure and parameters, to subsequently validate them both by split-validation and by $k-$fold cross-validation. The validation process provides information on the accuracy of the predictions.

2 Materials and Methods

The cases used in this study come from a database of policing clarified arson-caused wildfires (for which the alleged offenders have been identified), fed since 2008 throughout the entire Spanish territory, under the leadership of the Prosecution Office of Environment and Urbanism of the Spanish state. The database contains 1,423 solved cases. According to the experts, $n = 25$ variables have been chosen of the total set of 32 variables, because of their usefulness and predictive relevance. These variables are relatives to crime ($C_1, \ldots, C_{10}$) and to the arsonist ($A_1, \ldots, A_{15}$), and are described in Table 3, where their possible outcomes are also shown.

BN are graphical structures for representing the probabilistic relationships among variables, and for performing probabilistic inference with them. Given a set of random variables $V = \{X_1, \ldots, X_n\}$, a BN is a model that represents the joint probability distribution P over those variables. The graphical representation of the BN consists of a *directed acyclic graph (DAG)*, whose n nodes represent the random variables and whose directed arcs among the nodes represent conditional dependencies.

Using a BN to compute a posteriori probability is called "(Bayesian) inference": we enter an evidence and use it to update the probabilities of the network. From an evidence of the form $E = \{X_{i_1} = x_{i_1}, \ldots, X_{i_\ell} = x_{i_\ell}\}$, where $\{X_{i_1}, \ldots, X_{i_\ell}\} \subset V$ are the *evidence* variables, an inference consists in the computation of probabilities of the form $P(X_{j_1} = x_{j_1}, \ldots, X_{j_s} = x_{j_s} \,/\, E)$ with $\{X_{j_1}, \ldots, X_{j_s}\} \subset V \backslash \{X_{i_1}, \ldots, X_{i_\ell}\}$ the set of *query* variables. Variables of the BN that do not appear either as query or evidence are treated as unobserved. The prediction of a query variable X given the evidence E is the instantiation of X with the largest posterior probability. That is, if $x_1, \ldots, x_r$ are the possible instantiations of X, then $x^* = arg\,max_{k=1,\ldots,r} P(X = x_k \,/\, E)$ is the prediction for X, and $P(X = x^* \,/\, E)$ is said to be the *confidence level (CL)* of the prediction.

We adopt the *score-based* structure learning method (*"Search-and-score"*), which is an algorithm that attempts to find the structure that maximizes the score function, with the BIC (**Bayesian Information Criterion**) as score function. This score function is intuitively appealing because it contains a term that shows how well the model predicts the observed data when the parameter set is equal to its MLE estimation (the log-likelihood function), and a term that punishes for model complexity. Minimum Description Length (MDL) scoring metric is equivalent to the BIC score for Bayesian networks, while the Akaike's Information Criterion (AIC) only differs from BIC on the penalty term, which is less than that of BIC, implying that AIC tends to favor more complex networks than BIC. Some works suggest that MDL/BIC consistently outperforms AIC (see [16]), reason by which we have chosen the first as score function for our work. The *greedy search-and-score algorithm* with the BIC score function has *local scoring updating*, that it, this algorithm only needs locally recompute a few scores in each step to determine the score of the model in the next step, and performs a search through the space of possible network structures by adding, removing or reversing an arc, given a current candidate, subject to the constraint that the resultant graph does not contain a cycle, and "greedily" choosing the one that maximizes the score function, stopping when no operation increases this score.

A problem with this algorithm is that it could halt at a candidate solution that locally maximizes the score function rather than globally maximizes it. One way for dealing with this problem is what is named "iterated hill-climbing" (hc) algorithm, in which local search is done until a local maximum is obtained. Then, the current structure is randomly perturbed, and the process is repeated. Finally, the maximum over local maxima is used.

3 Constructing the BN Profiler

In this section we construct our expert system, which is a BN profiler. This construction is based on a data set of solved cases, from which the BN will be learned. The BN allows characterizing the wildfire arson in terms of the relationships between different variables and their strengths. These (in)dependence relationships are expressed in a very simple way in the BN, through the absence/presence

Crime variables	Author variables
C_1 = season	A_1 = age
C_2 = risk level	A_2 = way of living
C_3 = start time	A_3 = kind of job
C_4 = starting point	A_4 = employment status
C_5 = main use of surface	A_5 = educational level
C_6 = number of seats	A_6 = income level
C_7 = related offense	A_7 = sociability
C_8 = pattern	A_8 = prior criminal record
C_9 = traces	A_9 = history of substance abuse
C_{10} = who denounces	A_{10} = history of psychological problems
	A_{11} = stays in the scene
	A_{12} = distance from home to the scene
	A_{13} = displacement mean
	A_{14} = residence type
	A_{15} = wildfire motive

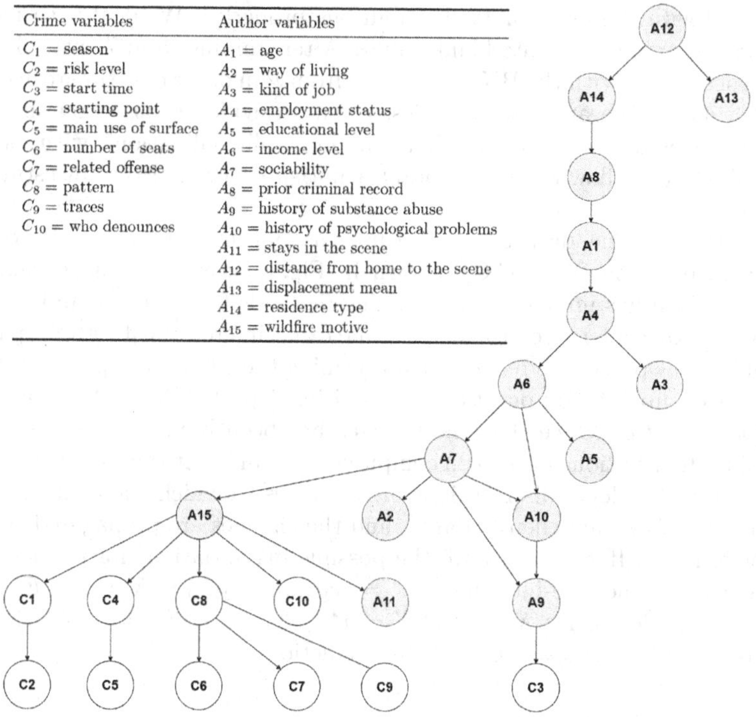

Fig. 1. Learned structure (DAG)

of arcs in its DAG (see Fig. 1). One of the most interesting features of BNs is their ability to compute posterior probabilities, that is, probabilities of some query variables given evidences. Once learned, the BN profiler will be used in the following way: given an evidence in terms of the crime variables for a given wildfire, we will predict the value of the query (arsonist) variables, which form the predicted profile of the arsonist.

With a view to "split-validation", the initial data set $\mathcal{D}$ with 1,423 cases is randomly partitioned into a training set $\mathcal{T}$ and a validation set $\mathcal{V}$ such that they are disjoint. Thus, none of the validation cases are used for training and can be considered to be new to the BN model. We use the common "80/20% rule" in order to do that, although there is an extensive literature on how to carry out this partition. Therefore, the training set $\mathcal{T}$ has M = 1,138 cases and the validation set $\mathcal{V}$ has N = 285 cases. The sufficiency of the set $\mathcal{T}$ in order to learn the BN is studied in Sect. 3.1. We use exclusively discrete (categorical) random variables, by discretizing the (few) continuous variables in the original database. By expert knowledge we choose a total number of 25 variables from the initial set of 32. The choice is the result of a balance between the benefits of having a high number of variables (more realistic model with higher accuracy) and the drawbacks arising from the corresponding increasing complexity (need for more data to estimate the probability distributions properly). We deal with missing values transforming them into blank values. After training from $\mathcal{T}$, we will apply an inference engine with the BN to the cases from $\mathcal{V}$ in order to estimate accuracy of predictions of the arsonist variables, given an evidence based on the crime variables, to predict new cases. We follow up "split validation" with "k–fold cross validation", allowing us to obtain a measure of the errors in estimating accuracy.

From the training data set $\mathcal{T}$, we use the bnlearn package of R to learn the structure of the BN (DAG), shown in Fig. 1, as well as the parameters. Package bnlearn has implemented sampling methods for inference and does not include any exact inference algorithm, while exact inference with gRain package is possible. This procedure relies on transforming the BN into a specially crafted tree to speed up the computation of conditional probabilities. Such a tree is called *junction tree*. Once the junction tree has been built and its conditional probability distributions have been computed, we can input the evidence into the junction tree. The local distributions of the nodes to which the evidence refers (evidence variables) are then updated, and the changes are propagated through the junction tree. If $x_1, \ldots, x_r$ are the possible instantiations for a variable X, with x_1 =blank, then we introduce $x^{**} = arg\,max_{k=2,\ldots,r} P(X = x_k \,/\, E)$ as the (corrected) prediction for X, and $P(X = x^{**} \,/\, E)/\big(1 - P(X = x_1 \,/\, E)\big)$ as the (corrected) *confidence level (CL)* of the prediction.

3.1 Performance Metrics

The performance of the BN learning algorithms (both for structure and parameters learning) always depends on the sufficiency of the training database, which is determined by the number of nodes, the size of their domain (set of possible instantiations), and the underlying probability distributions. While the nodes and their domain are known from the problem formulation, the underlying probability distributions are typically unknown a priori.

To determine whether the training database $\mathcal{T}$ with $M = 1{,}138$ cases is sufficient to estimate a BN model of criminal behavior of wildfire arsonists, subset samples from $\mathcal{T}$ of size ranging from 25 to 1,138 in increments of 5 cases, have been randomly generated. For every sample, we learn a BN model and the BIC score function is computed and plotted on a learning curve, as a function of the number of training cases, as Fig. 2 shows. We see that a saturation point is reached before attaining $M = 1{,}138$ (approximately at 900), from which the score does not improve even if we increase the number of training cases. Therefore, it does seem to have enough training data in $\mathcal{T}$ to learn the network. In addition, first graph in Fig. 2 shows clearly that if we impose no restriction at all (in black color and continuous line), the BIC score function takes higher values practically in all the range of the number of training cases than if we inhibited arcs connecting crime variables among them (in red color and dashed line). Moreover, inhibiting arcs from any crime variable to any arsonist variable gives the same model structure than with no restriction, reason by which we finally decide no consider restrictions on the learning structure procedure and obtain

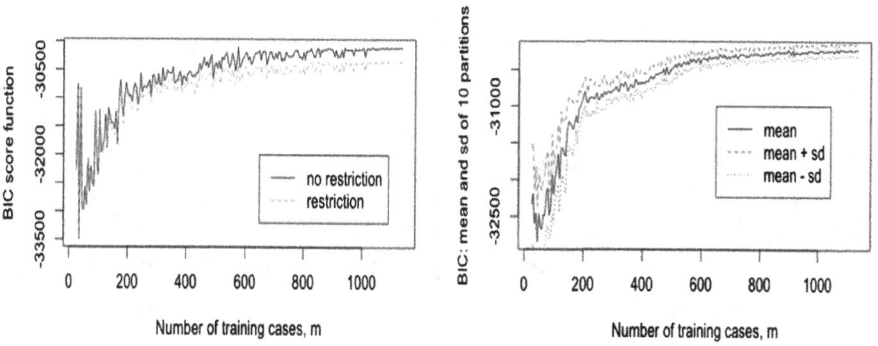

Fig. 2. BIC learning curve as function of m: with partition $\mathcal{T}-\mathcal{V}$, and with 10 partitions (mean and mean$\pm$ standard deviation). Increment $\Delta m = 5$ (Color figure online)

the DAG in Fig. 1. Second graph in Fig. 2 shows the smoother learning curve corresponding to the mean of the BIC values for 10 different random partitions (in black color and continuous line).

3.2 Internal Consistency

Before proceeding to consider validation of the constructed BN, we study if the training data set is internally consistent. To this end, we compute some measures (metrics) of discordance between the BN learned at each of the previous steps, that is, learned from each training subsample of $\mathcal{T}$ of size $m = 25, \ldots, M =$ 1,138, that we denote by BN_m, with respect to the final BN, which is that learned from the whole training database, denoted by BN_M. We also consider two measures of concordance. For that, we compare the binary matrix indicating directed arcs of network BN_m with that of the final network BN_M. We use notations: $a =$ number of matches 1 in the two matrices, $b =$ number of 1 in the matrix of BN_m being 0 in the matrix of BN_M, $c =$ number of 0 in the matrix of BN_m being 1 in the matrix of BN_M, $d =$number of matches 0 in the two matrices. The measures of concordance we consider are the *sensitivity*, which is $a/(a+c)$, that is, the proportion of arcs of BN_M that are already in BN_m, and the *accuracy*, which is $a/(a+b)$, that is, the proportion of arcs in BN_m that are in BN_M. Both increase to 1 as $m \to M$ as can be seen in Fig. 3.

The first measure of discordance we consider is the *Structural Hamming Distance (SHD)* (see [13]) that describes the number of changes that have to be made to BN_m for it to turn into BN_M (lower is better), is $b+c+e$ being e the number of directed arcs in BN_m that are incorrectly oriented comparing with BN_M. We see in Fig. 3 that, indeed, SDH tends to zero as $m \to M$. The second one is the *Jaccard dissimilarity* [12], defined by $1 - a/(a+b+c)$, and finally, we can also compute the dissimilarity index of Sokal-Michener (see [11]), which is $1-(a+d)/(a+b+c+d)$. Note that both Jaccard and Sokal-Michener dissimilarity measures take values in [0, 1], the first one taking into account the concordances

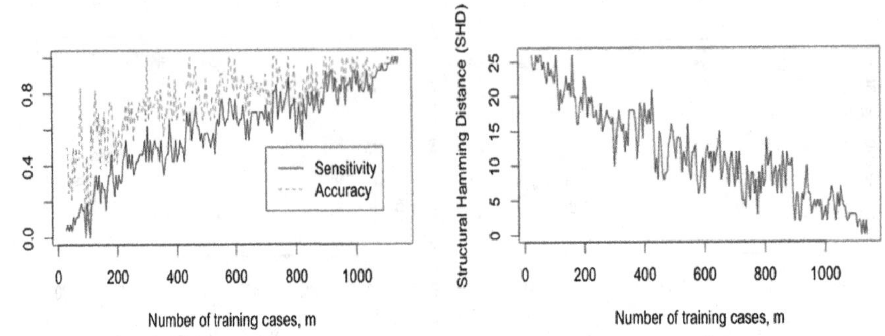

Fig. 3. Sensitivity and Accuracy, and Structural Hamming Distance metrics, as functions of number of training cases. Increment $\Delta m = 5$

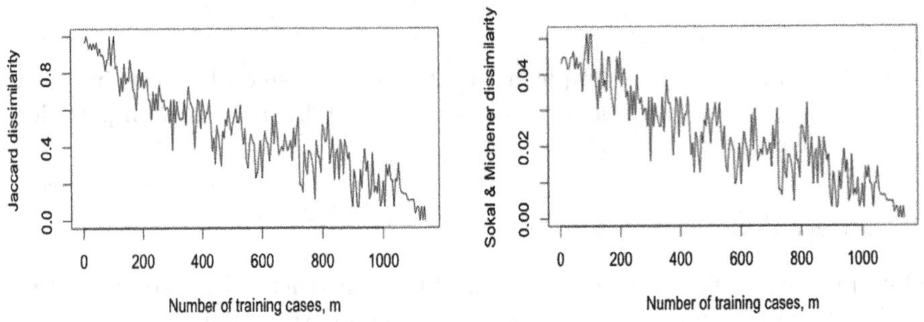

Fig. 4. Jaccard and Sokal-Michener metrics. Increment $\Delta m = 5$

only in the sense of present arcs in the networks, and the second one, also considering matches in absences, whilst SHD index takes values in $\mathbb{Z}^+$ and does not takes into account concordances. In Fig. 4 it can be seen the evolution of these two dissimilarity measures and how they converge to zero.

Graphs in Figs. 3 and 4 show the internal consistency of the dataset. Indeed, concordance measures increase to 1 while discordance metrics go to 0, as the subset of the training set increases to the whole set, at the same time that variability in the measures decreases.

4 Validation and Assessment of Predictive Accuracy

First we use "split-validation", from which the BN model is tested using the validation data set $\mathcal{V}$ composed of $N = 285$ solved cases, each reporting the offender variables. For each of the test cases in $\mathcal{V}$, we use the values of the crime variables as evidence in performing inference on the arsonist variables. Then, compare the predicted values for the arsonist variables (which are the highest-probability predictions and always different from blank, as explained above) to non-blank known case outcomes in $\mathcal{V}$, and take note of matches. If for a fixed case of $\mathcal{V}$ and

Table 1. Individual Predictive Accuracy (IPA) and Overall Predictive Accuracy (OPA). "Mean" and "Deviation" columns correspond to $k-$fold cross validation.

Arsonist variable IPA (%)	Split-validation	Mean	Deviation
A_1 = age	27.57	30.30	3.06
A_2 = way of living	53.21	53.46	3.01
A_3 = kind of job	76.44	74.62	3.02
A_4 = employment status	34.68	37.74	3.34
A_5 = educational level	46.63	52.24	5.44
A_6 = income level	44.27	44.05	5.04
A_7 = sociability	69.19	66.70	4.45
A_8 = prior criminal record	80.07	79.41	4.45
A_9 = history of substance abuse	78.50	82.07	2.82
A_{10} = history of psychological problems	79.70	82.71	4.83
A_{11} = stays in the scene	49.10	49.35	4.36
A_{12} = distance from home to the scene	41.91	42.46	3.02
A_{13} = displacement means to commit arson	44.53	42.79	4.66
A_{14} = residence type	40.52	45.12	5.28
A_{15} = wildfire motive	61.05	62.94	4.12
Total OPA (%)	54.04	55.19	0.05

a fixed arsonist variable, say A_i, the outcome of the variable is a blank, then the prediction of variable A_i for this case is not considered in order to compute the success rate of the BN in predicting. The success rate on predicting each arsonist variable from the evidence given by the crime variables for the N cases of $\mathcal{V}$ is called "IPA" (*Individual Predictive Accuracy*) (see column "Split-validation" in Table 1). It is obtained by dividing the number of correct predictions for each variable by the total number of predictions (excluding blanks).

From this table we see the wildfire arsonist characteristics that are typically correctly predicted (IPA $\geq$ 60%): A_3, A_7, A_8, A_9, A_{10} and A_{15}, and then can be used to narrow the list of suspects. Specially important are A_8, A_9 and A_{10}. Indeed, they are operative variables that greatly help the researcher to identify the offender, which is precisely the aim of profiling. Also of major importance is A_{15}. Its key role is evidenced since all crime variables except C_3 are descendants of A_{15}. It should also be borne in mind that for any variable we choose as prediction the outcome that maximizes the probability, causing failures in prediction when the second most likely outcome has a close probability. Table 1 also shows the average percentage of global correct accuracy, for all the arsonist variables, named "OPA" (*Overall Predictive Accuracy*). It is obtained by dividing the total number of matches (1,914) on all predictions by the total number of predictions (excluding blanks), which is 3,542.

Secondly, we also use "$k-$fold cross validation" with $k = 10$. That is, we reuse the data set $\mathcal{D}$ generating k splits of $\mathcal{D}$ into non-overlapping training and validate sets with (approx.) proportions $(k - 1)/k$ and $1/k$, respectively. For each split we obtain the IPA value as we did for the partition $\mathcal{T} - \mathcal{V}$. From these $k = 10$ values we compute the mean, as well as the standard deviation (see columns "Mean" and "Deviation" in Table 1).

Finally, we computed the "DIPA" (*Disincorporate Individual Predictive Accuracy*), which is the percentage of correct predictions, for each arsonist variable, from the validation set $\mathcal{V}$, according to the prediction that we made for it from the evidence given by the crime variables. For example, if the prediction for A_{15} was "gross negligence", then the DIPA tells us that the accuracy rate would be 79%, as can be seen in Table 2, while if the prediction for A_{15} was "revenge", this rate plummets to 16.67%.

Table 2. Disincorporate Individual Predictive Accuracy (DIPA) for A_{15}.

If prediction was	% accuracy (DIPA)
Profit	46.67
Gross negligence	79.00
Slight negligence	51.65
Impulsive	54.79
Revenge	16.67

Table 3. Outcomes of the variables in the database.

Variables	Outcomes
C_1 = season	Spring/winter/summer/autum
C_2 = risk level	High/medium/low
C_3 = wildfire start time	Morning/afternoon/evening
C_4 = starting point	Pathway/road/houses/crops/interior/forest track/others
C_5 = main use of surface	Agricultural/forestry/ livestock /interface/recreational
C_6 = number of seats	One/more
C_7 = related offense	Yes/no
C_8 = pattern	Yes/no
C_9 = traces	Yes/no
C_{10} = who denounces	Guard/particular/vigilance
A_1 = age	Up to 34 / 35 to 45 / 46 to 60 / more than 60
A_2 = way of living	Parents/in couple/single/others
A_3 = kind of job	Handwork/qualified
A_4 = employment status	Employee/unemployed/sporadic/retired
A_5 = educational level	Illiterate/elementary/middle/upper
A_6 = income level	High/medium/low/without incomes
A_7 = sociability	Yes/no
A_8 = prior criminal record	Yes/no
A_9 = history of substance abuse	Yes/no
A_{10} = history of psychological problems	Yes/no
A_{11} = stays in the scene	No/remains there/gives aid
A_{12} = distance from home to the scene	Short/medium/long/very long
A_{13} = displacement means to commit arson	On foot/ by car/ all terrain / others
A_{14} = residence type	Village/house/city/town
A_{15} = wildfire motive	Profit/gross negligence/slight negligence/impulsive/revenge

5 Conclusion

We present BN as a novel methodology for profiling of arsonists in wildfires. We learn the BN from the database and validate it, by estimating the accuracy of predictions for each arsonist variable individually (IPA) and all together (OPA). We also check internal consistency of the database. Therefore, we can use learned BN to predict the profile of the offender from the information about a particular arson-caused wildfire, and obtain confidence levels for the predictions of the arsonist variables. Globally, the estimation of network's accuracy is 55.19% (but varies between 30.30 and 82.71% according to the variable; take into account the number of possible values for each variable, which is high for some of them). By comparing with partial results obtained from experts (approx. 40% of accuracy), this value seems high enough for a profile model.

We think this approach is really innovative and helpful. Our purpose in the future is going deeper in the study of the obtained model and its performance as profiler, as well as its behavior as new solved cases are incorporated into the database, and also apply this methodology to other crimes (e.g. against persons).

Acknowledgments. The authors wish to thank the anonymous referees for careful reading and helpful comments that resulted in an overall improvement of the paper. They also would express their acknowledgment to the Prosecution Office of Environment and Urbanism of the Spanish state for providing data and promote research.

References

1. Adab, H., Kanniah, K.D., Solaimani, K.: Modeling forest fire risk in the northeast of Iran using remote sensing and GIS techniques. Nat. Hazards **65**, 1723–1743 (2013)
2. Adusei-Poku, K.: Operational risk management - implementing a BN for foreign exchange and money market settlement. Unpublished Ph.D. thesis, Göttinger University (2005)
3. Baumgartner, K., Ferrari, S., Palermo, G.: Constructing Bayesian networks for criminal profiling from limited data. Knowl.-Based Syst. **21**, 563–572 (2008)
4. Baumgartner, K., Ferrari, S., Salfati, C.: Bayesian network modeling of offender behavior for criminal profiling. In: Proceedings of the 44th IEEE Conference on Decision and Control, and the European Control Conference 2005, pp. 2702–2709 (2005)
5. Borsuk, M.E., Stow, C.A., Reckhow, K.H.: A BN of eutrophication models for synthesis, prediction, and uncertainty analysis. Ecol. Model. **173**, 219–239 (2004)
6. Cruz-Ramírez, N., Acosta-Mesa, H.G., Carrillo-Calvet, H., Nava-Fernández, L.A., Barrientos-Martínez, R.E.: Diagnosis of breast cancer using BN: a case study. Comput. Biol. Med. **37**, 1553–1564 (2007)
7. Delgado, R., Tibau, X.-A.: Las Redes Bayesianas como herramienta para la evaluación del riesgo de reincidencia: Un estudio sobre agresores sexuales, Revista Española de Investigación Criminológica, Artículo 1, Número 13 (2015). (in Spanish)
8. Dlamini, W.M.: A Bayesian belief network analysis of factors influencing wildfire occurrence in Swaziland. Environ. Model. Softw. **25**, 199–208 (2010)

9. Dlamini, W.M.: Application of Bayesian networks for fire risk mapping using GIS and remote sensing data. GeoJournal **76**, 283–296 (2011)
10. FAO, Fire Management - Global Assessment 2006. A Thematic Study Prepared in the Framework of the Global Forest Resources Assessment 2005, FAO Forestry Paper 151, Rome (2007)
11. Gower, J.C., Legendre, P.: Metric and Euclidean properties of dissimilarity coefficients. J. Classif. **3**, 5–48 (1986)
12. Jaccard, P.: Étude comparative de la distribution florale dans une portion des Alpes et des Jura. Bulletin de la Société Vaudoise des Sciences Naturelles **37**, 547–579 (1901). (in French)
13. Jongh, M., Druzdzel, M.J.: A comparison of structural distance measures for causal bayesian network models. In: Klopotek, M.A., Przepiórkowski, A., Wierzchon, S.T., Trojanowski, K. (eds.) Recent Advances in Intelligent Information Systems. EXIT (2009). ISBN:9788360434598
14. Korb, K.B., Nicholson, A.E.: Bayesian Artificial Intelligence, 2nd edn. CRC Press (Taylor & Francis Group), Boca Raton (2011)
15. Lee, C., Lee, K.J.: Application of BN to the probabilistic risk assessment of nuclear waste disposal. Reliab. Eng. Syst. Saf. **91**(5), 515–532 (2006)
16. Liu, Z., Malone, B., Yuan, C.: Empirical evalutation of scoring functions for Bayesian network model selection. BMC Bioinform. **12**(Suppl. 15), S14 (2012). http://www.biomedcentral.com/1471-2105/13/S15/S14
17. MAGRAMA - Ministerio de Agricultura, Alimentación y Medio Ambiente, Los Incendios Forestales en España: Avance informativo. 1 de enero al 31 de diciembre de 2015 (2016). (in Spanish). http://www.magrama.gob.es/es/desarrollo-rural/estadisticas/iiff_2015_def_tcm7-416547.pdf
18. Papakosta, P., Straub, D.: A Bayesian network approach to assessing wildfire consequences. In: Proceedings ICOSSAR, New York (2013)
19. Penman, T.D., Bradstock, R.A., Price, O.: Modelling the determinants of ignition in the Sydney Basin, Australia: implication for future management. Int. J. Wildland Fire **22**, 469–478 (2013)
20. Pollino, C.A., Woodberry, O., Nicholson, A., Korb, K., Hart, B.T.: Parameterisation and evaluation of a BN for use in an ecological risk assessment. Environ. Model Softw. **22**, 1140–1152 (2007)
21. Soeiro, C., Guerra, R.: Forest arsonists: criminal profiling and its implications for intervention and prevention. Eur. Police Sci. Res. Bull. (11), 34–40, Winter 2014/15. https://www.cepol.europa.eu/sites/default/files/science-research-bulletin-11.pdf
22. Sotoca, A., González, J.L., Fernández, S., Kessel, D., Montesinos, O., Ruíz, M.A.: Perfil del incendiario forestal español: aplicación del perfilamiento criminal inductivo. Anuario de Psicología Jurídica **2013**(23), 31–38 (2013). (in Spanish)
23. Spiegelhalter, D.J.: Incorporating Bayesian ideas into healthcare evaluation. Stat. Sci. **19**, 156–174 (2004)
24. Thompson, M.P., Scott, J., Helmbrecht, D., Calvin, D.E.: Integrated wildfire risk assessment: framework development and application on the Lewis and Clark National Forest in Montana USA. Integr. Environ. Assess. Manag. **9**(2), 329–342 (2012)
25. Ticehurst, J.L., Newham, L.T.H., Rissik, D., Letcher, R.A., Jakeman, A.J.: A BN approach for assessing the sustainability of coastal lakes in New South Wales, Australia. Environ. Model. Softw. **22**(8), 1129–1139 (2007)
26. Walshe, T., Burgman, M.: A framework for assessing and managing risks posed by emerging diseases. Risk Anal. **30**(2), 236–249 (2010)

Learning Optimal Decision Lists as a Metaheuristic Search for Diagnosis of Parkinson's Disease

Fernando de Carvalho Gomes(✉) and José Gilvan Rodrigues Maia

Computer Science Department and Virtual University Institute,
Federal University of Ceará, Campus do Pici, Bloco 952,
Fortaleza, CE 60455, Brazil
carvalho@ufc.br, gilvanmaia@virtual.ufc.br

Abstract. Decision Lists are a very general model representation. In learning decision structures from medical datasets one needs a simple understandable model. Such a model should correctly classify unknown cases. One must search for the most general decision structure using the training dataset as input, taking into account both complexity and goodness-of-fit of the underlying model. In this paper, we propose to search the Decision List state space as an optimization problem using a metaheuristic. We implemented the method and carried out experimentation over a well-known Parkinson's Disease training set. Our results are encouraging when compared to other machine learning references on the same dataset.

Keywords: Stochastic search · Learning decision lists · Parkinson's Disease classification

1 Introduction

Machine learning is becoming a research field of utmost interest for real-world applications such as automated security surveillance systems, spam detection, natural language processing and computer vision, among many others [1].

Current trends are to adopt nonlinear and overly complex models such as deep learning models consisting of many neural network layers [2]. Such a model cannot be interpreted by human beings without resorting to sophisticated visualization techniques. Therefore one could question whether knowledge can be obtained from such a model in an effective way.

In this paper, we are concerned to learn simple, interpretable knowledge representations. More specifically, we strive to obtain Decision Lists (DL) with reduced size from a dataset using a metaheuristic procedure. Decision lists' configuration space is a very interesting representation since it contains other decision models. Rivest and Blum, for instance, prove that decision lists exploratory space is more general than decision trees' state space [3,4].

P.M. Pardalos et al. (Eds.): MOD 2016, LNCS 10122, pp. 391–401, 2016.
DOI: 10.1007/978-3-319-51469-7_32

Bayesian Networks [5], Neural Networks [6], and Evolutionary/Genetic coding approach [7] representation spaces have been applied to heuristic search with better results than exact algorithms. Besides, DLs have been used for learning with deterministic heuristic algorithms [8,9]. On the other hand, it is not sufficient to search for a DL that matches most of the cases residing in the training set for a given application. A near-optimal decision list must correctly classify new cases that arise to the learnt model.

Nevertheless, one can count only with the training set in order to find the best DL. Vapnik [11] proposes to look for a compromise between model complexity and its goodness-of-fit. Here we consider that complexity is represented by the number of descriptors present in the learnt model which is proportional to its size. More recently, Nock and Gascuel [12] show that the problem of inducing the shortest consistent DL is NP-hard.

On the algorithmic point of view, it is often difficult (NP-hard) to find the classification procedure that represents the lower error frequency. One is therefore led to study approximation procedures. All the more that the observed frequency error on the training set, may not be a very good estimate of the actual error probability (that we really try to minimize). Finding the optimum at all costs therefore makes little sense, lower size complexity is also very important in order to obtain a good generalization power.

In this paper, we investigate how a DL can model the diagnosis of Parkinson's Disease based on measures of dysphonia. A combinatorial optimization based on multi-start Greedy Randomized Adaptive Search Procedure (GRASP) [13] is performed in order to derive DLs models that are understandable by human experts. Experimentation was carried out with cross-validation in order to demonstrate our approach can yield simple and effective models endowed with generalization capabilities.

The remainder of this paper is organized as follows. We present the DL search space problem formulation and the metaheuristic search algorithm in Sect. 2. Experimental evaluation and results are discussed in Sect. 3 for representative toy problems and in Sect. 4 for a problem from the medical domain. Finally, we present conclusions from this work and future research directions in Sect. 5.

2 Decision List Search Space

Decision Lists are classification procedures and they may be seen as a linearly ordered set of decision *rules*. Each rule is formed by conjunctive rules [3]. They have for shape:

$$
\begin{array}{lll}
\textbf{if} & \text{condition}_1 & \textbf{then return } c_1 \\
\textbf{elseif} & \text{condition}_2 & \textbf{then return } c_2 \\
\ldots & \ldots & \ldots \\
\textbf{elseif} & \text{true} & \textbf{then return } c_r
\end{array}
$$

where each of the r conditions $condition_i$ is a conjunction of literals associated with boolean attributes, and c_i is one of the c classes of the partition. An incoming feature vectors is classified using a given rule if and only if all its conditions hold for that sample. The last rule has always true as condition and is called default rule.

Decision trees, in their turn, are unordered rule sets in which each leaf l corresponds to a single conjunctive rule r. In this case, each predicate belonging to r is determined by the path from the root of the tree to l. Therefore all rules from a decision tree can be enumerated by any traversal algorithm in order to build an equivalent decision list in polynomial time on the decision list's depth p.

It also is worth mentioning that Boström [10] and Rivest [3] showed that decision lists limited to a maximum of p predicates per rule are strictly more expressive than decision trees with depth p. The next subsections are devoted to detail definitions about boolean attributes, conditions and rules within the context of this paper.

2.1 Deriving Binary Attributes

The optimization looks for a solution considering only binary attributes, therefore any non-binary attributes must be preprocessed in order to fit this representation. Alternatively, it is also possible to adopt any simple binary classification procedure given that such a "procedural" representation is supported by both diversification and intensification searches, more specifically for random model generation and when a neighbor model is derived, respectively.

Binarization can be executed without loss of generality. For example, a non-binary attribute may yield disjunctions of literals representing values of the original attribute or value intervals in the case of continuous attributes.

Each attribute is converted into a set of pairs of literals $(x, \neg x)$ stored in a "bag" of predicates. How to produce these pairs of literals from a given attribute depends on the nature of this attribute. There are two major classes of attributes: continuous attributes taking their values in $\Re$; and *qualitative* attributes assuming discrete values.

A continuous attribute can be binarized by simple comparison against a random threshold t taken from the corresponding range of values observed in the training data for that attribute. Different predicates can be generated at optimization time by selecting a specific attribute-threshold pair.

To illustrate the binarization for qualitative attributes, let us take the nominal attributes $Color < red, green, blue >$. In that case, binarization is carried out by the combination of all attribute modalities, for the *Color* attributes:

$$\begin{array}{cc} x & \neg x \\ < red > & < green, blue > \\ < green > & < red, blue > \\ < blue > & < red, green > \end{array}$$

2.2 Initial State

Conditions in a rule contain only boolean or continuous attributes, so that the latter require binarization. All boolean and continuous attributes are placed in a "bag" of predicates. The initial configuration is composed by a fixed number of empty rules. Empty rules are always false.

In order to start the GRASP search procedure we put in the rules n_r predicates randomly chosen from the "bag" of predicates. The maximum number of predicates in a single rule p is fixed. A predicate will not be replicated in a rule.

2.3 Operators

We used two operators to move from a state to another: (a) take out a literal from the condition part of the rule and replace it by one literal in the bag of predicates, both at random; and (b) randomly replace a class from the implication side of the rule. Both operators are carried out alternately with the same probability.

2.4 Cost Function

By assuming a combinatorial optimization standpoint, we admit there exists a myriad of solutions defined as a combination of rules composed by conditions associating incoming feature vectors with classes. Moreover, we must embrace an objective function describing the cost associated with each solution. A common procedure is to use the overall accuracy to define an objective function. However, in this work, we adopt a cost function that considers the product between accuracies obtained by the model S for each class c_i, so $f(S) = 1 - \prod accuracy(S, c_i)$. Therefore, $f(S)$ accounts for more general models that suit all classes.

2.5 Stopping Criterion

The multi-start GRASP algorithm (see below) is divided into two phases: diversification and intensification of the search. The central idea to stop searching for the best DL in the intensification phase is to stress the chosen region such that a large number of iterations has passed without improving the cost function. The diversification phase stops after a maximum number of iterations i is reached or a target accuracy is obtained by the best model.

2.6 Search Strategy: GRASP-DL (GDL)

Multi-start GRASP [13], intensify and diversify the search. Algorithm 1 takes a good solution S and looks for a better solution S^+ in its neighborhood and stops after performing a max number of iterations without improving the cost function f^+. Therefore, it executes a near greedy strategy. Algorithm 2 diversifies the search, *i.e.* it looks for a random feasible region S and pass it to Algorithm 1 in order to obtain a globally near-optimal solution S^* with minimal cost f^*.

Algorithm 1: Intensification Search

procedure IntensificationSearch(S)
    $S^+ \leftarrow S$;
    $f^+ \leftarrow f(S)$;
    while improves within max iterations **do**
        $S_{local} \leftarrow$**NextState**(S^+);
        if $f(S_{local}) < f^+$ **then**
            $S^+ \leftarrow S_{local}$;
            $f^+ \leftarrow f(S_{local})$;
        end if
    end while
    return S^+;
end procedure

Algorithm 2: GRASP algorithm

procedure Grasp(α)
    $f^* \leftarrow \infty$
    while stopping criterion not satisfied **do**
        $S \leftarrow$**RandomFeasibleSolution**(α);
        $S \leftarrow$**IntensificationSearch**(S);
        if $f(S) < f^*$ **then**
            $S^* \leftarrow S$;
            $f^* \leftarrow f(S)$;
        end if
    end while
    return S^*
end procedure

2.7 Algorithmic Complexity

Algorithmic complexity is defined in terms of the following variables:

- r is the maximum number of rules allowed per decision list;
- p is the maximum number of predicates allowed per rule;
- n denotes the number of samples in the training dataset;
- i is the number of iterations in the outer loop performing diversification search;
- k is the number of iterations for intensification search.

Model creation and modification perform in $O(rp)$ assuming that random predicate selection takes $O(1)$ time. In its turn, computing the cost function $f(S)$ requires a model S to be evaluated against the entire training dataset, therefore it takes $O(nrp)$ time considering the worst case when S uses the maximum number of allowed rules and predicates.

Finally, by assuming that a model that perfectly fits to all the training samples is never found, brute-force GDL requires an exhaustive search, which runs in time $O(iknrp)$. It is worth noting that, despite its polynomial time complexity, this algorithm may provide poor performance given large i and k values because intensification search poses as the bottleneck.

2.8 Speeding Up the Search

The issue of evaluating $f(S)$ can be addressed by incremental computation, in the following manner:

- First, we can resort to a data structures to store the training samples effectively classified by each rule found in the model S given as the input of the intensification step. This can be performed efficiently using two arrays per rule:
    - The first is an n-dimensional histogram telling which training samples were evaluated by the predicates of that rule.
    - The other is a $n+1$-dimensional array containing a permutation that sorts the samples in the histogram and puts a sentinel index for termination, thus allowing for efficient iteration over the samples. A radix sort algorithm can be used for sorting this permutation, thus performing in $O(n)$.

- Given the j-th rule is modified by the intensification search, we simply need to re-evaluate those training samples from that rule on, skipping any computation for the training samples falling into the prior $(n - j)$ rules. This can be done using the aforementioned arrays.

Moreover, implementations may resort to automatic parallelization features found in modern programming languages. This allows to dramatically speed up GDL by using all processing cores available at runtime.

It is important to observe that performance actually depends on how many training samples were handled by $(n - j)$ rules. Another optimization strategy can be derived by putting "strong", good-performing rules first at the decision list and thus preventing the intensification search from modifying those rules. However, such a greed approach may lead to poor diversification.

3 Validation Using Toy Domains

We developed an implementation of GDL using the C++ language for experimental evaluation. We start this validation by addressing some toy problems in order to show that models can actually handle simple non-linear binary separation and a more complex configuration inspired by a checkerboard. Moreover, an example of multi-class classification is also shown. Binary features are derived by thresholding over the sample points' coordinates.

Since this toy domain is two-dimensional, we can resort to a real-time visualization developed using OpenGL® [22] of how each model looks like. Due to its simplicity, we do not account for the default rule throughout the experiments.

3.1 Non-linear Curves

Our very first experiment already deals with non-linear classification. Two symmetric curves are described using 10 for each class. These curves are placed next to each other, so we cannot resort to linear separation. This case is handled gracefully by a model composed by 4 rules and 12 predicates, as shown below. The shape of the classifier can be seen in Fig. 1.

```
if                 (x1 < 0.510171) and (x1 < -0.448179) and (x0 >= 0.712665)             then return 1
elseif             (x1 >= 0.024501) and (x1 < 0.753306) and (x0 >= 0.965451)             then return 0
elseif (x0 < 1.18985) and (x1 < 0.651636) and (x0 >= 0.392175) and (x0 < 0.802772) then return 1
elseif                        (x1 < 0.520433) and (x0 < 0.768559)                         then return 1
elseif                                          true                                      then return 0
```

3.2 Checkerboard

We built a 3×3 checkerboard using 50 samples for each square. This dataset thus contains 450 samples that cannot be correcly classified by a linear model. Moreover this example considerably more difficult than the non-linear curves shown previously. For correctlyonly 4 rules and 11 predicates, which is simple and yet powerful enough to handle this situation correctly. This is illustrated by Fig. 2 for the model described as follows.

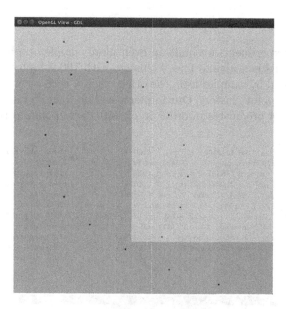

Fig. 1. A non-linear toy problem described by 20 symmetric points is handled correctly by a decision list with 4 rules and 12 predicates.

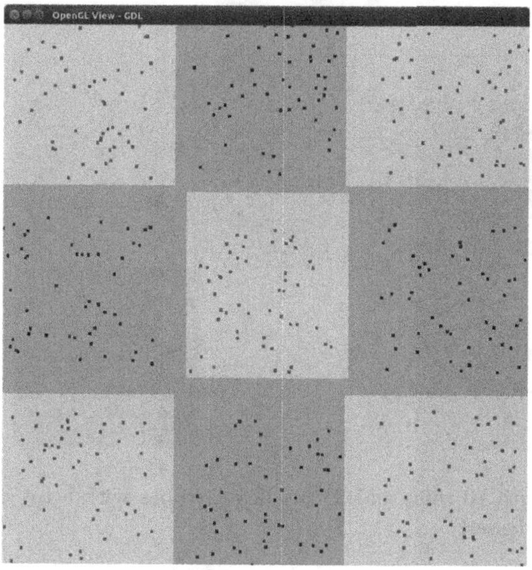

Fig. 2. A more complex, non-linear checkers toy problem described by 450 points. The model uses only 4 rules and 11 predicates. This illustrates how decision lists are powerful enough to cope with complex problems.

if	$(x_0 < 10.6085)$ and $(x_0 >= 7.40913)$ and $(x_1 < 10.9856)$ and $(x_1 >= 7.38758)$	**then return** 0
elseif	$(x_0 < 10.5403)$ and $(x_0 >= 7.18121)$	**then return** 1
elseif	$(x_1 < 11.1256)$ and $(x_1 >= 7.05512)$	**then return** 1
elseif	$(x_1 >= 8.59338)$ and $(x_1 < 12.2016)$ and $(x_1 < 9.05468)$	**then return** 1
elseif	true	**then return** 0

3.3 Multi-class

In the last toy experiment, we built an even more complex problem by placing samples from 5 distinct classes into 9 clusters with 100 samples each. The 5-th class is surrounded by samples from the other classes, each of these other classes is formed by two point clusters. Our implementation found a model that resorts to 10 rules and 24 predicates in order to classify these points correctly, as seen below in Fig. 3.

if $(x_0 >= 11.2558)$ and $(x_1 < 6.603)$ **then return** 1
elseif $(x_0 < 7.02799)$ and $(x_1 >= 6.75643)$ and $(x_1 < 10.9826)$ **then return** 1
elseif $(x_0 < 6.75632)$ and $(x_1 < 7.49381)$ **then return** 0
elseif $(x_0 >= 7.388)$ and $(x_1 < 7.46167)$ **then return** 3
elseif $(x_1 < 10.4178)$ and $(x_0 >= 10.6828)$ **then return** 2
elseif $(x_1 >= 6.08271)$ and $(x_0 >= 5.30955)$ and $(x_0 >= 10.7415)$ **then return** 3
elseif $(x_1 < 8.17738)$ and $(x_1 >= 12.2018)$ and $(x_0 >= 6.5438)$ **then return** 0
elseif $(x_0 >= 7.51555)$ and $(x_1 >= 4.26344)$ and $(x_1 < 10.6396)$ **then return** 4
elseif $(x_0 < 6.88642)$ and $(x_1 >= 4.32449)$ **then return** 2
elseif $(x_0 >= 6.35341)$ and $(x_1 >= 7.88943)$ **then return** 0
elseif true **then return** 3

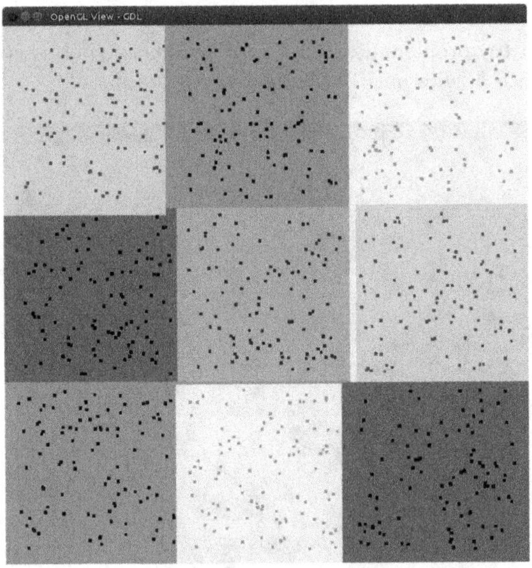

Fig. 3. A model with 10 rules and 35 predicates copes with a toy example with 900 samples from five classes.

4 Medical Domain Experimental Evaluation

Parkinson's Disease (PD) is the most common neurodegenerative disorder. The diagnosis of PD is challenging and currently none of the biochemical tests have proven to help in diagnosis [14]. Clinical symptoms and signals are the only option in order to correctly diagnose the disease.

4.1 Parkinson's Disease Dataset

The dataset presented by Little et al. [15,16] is composed of a range of biomedical voice measurements from 31 people, 23 with Parkinson's Disease (PD). Each column in the table is a particular voice measure, and each row corresponds one of 195 voice recording from these individuals. The main aim is to discriminate healthy people from those with PD, according to the class column which is set to 0 for healthy and 1 for PD. Vocal impairment is one important possible indicators at the PD's earliest stages, which is evidenced by dysphonia that, in its turn, can be measured in a non-invasive fashion. This is a key since PD symptoms undermines both effectiveness and patients' comfort when they undergo invasive tests.

4.2 Experiments

First, we test our hypothesis that choosing simpler models produces a fair compromise balance between generalization capabilities and accuracy. During the training phase we limited our search to consider models with a maximum of 10 rules with up to 4 conditions in order to keep models understandable. Results presented in Fig. 4 have been obtained by leave-one-out 10-fold cross validation by deriving models with an increasing number of rules. As expected, a model with 4 rules handles the problem with accuracy comparable to a model consisting of 10 rules and, furthermore, such a simple model can be easily understood by a medical specialist.

It is important noticing that our $C++$ implementation is CPU-intensive and requires less than 500 KB of RAM for executing on a 64-bit Ubuntu Linux machine when compiled with GCC 4.8.4.

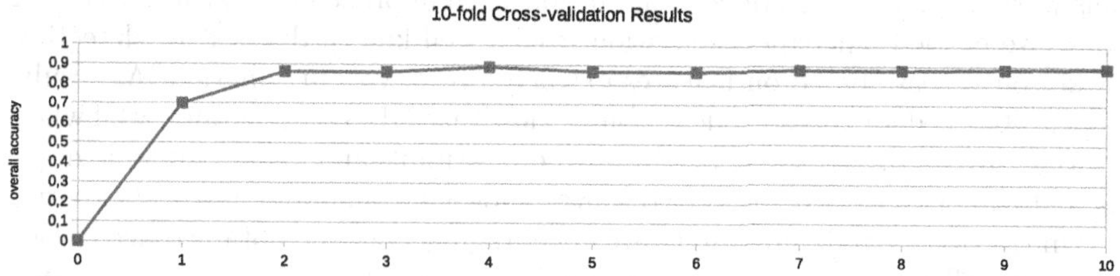

Fig. 4. Overall accuracy of 10-fold cross-validation when using variable number of rules for GRASP ran over $100k$ iterations with $10k$ intensification steps. A maximum of 4 conditions was allowed per rule in this experiment. Using 4 rules gives 88.94% accuracy. This shows our approach can produce good-quality models with reduced complexity that generalize well.

Different authors have used the same dataset applied to Neural Nets, Decision Trees, and Regression Models [17–19]. We compare our GDL approach results with theirs. Three outstanding simple models were obtained:

- A model with 3 rules and 5 conditions altogether, with overall accuracy of 94.44%, 100% for non-PD and 92.51% for PD class.

- A 4-rule model with 12 conditions altogether, with overall accuracy of 94.41%, 97.91% for non-PD and 93.19% for PD.
- A model composed of 5 rules and a total of 10 conditions, with overall accuracy of 95.45%, 100% for non-PD and 93.87% for PD.

We also obtained complex models that were very expensive to obtain and displayed even better results. For instance, models with 12 and 24 rules with overall accuracies of 97.43% and 98.46%, respectively. Table 1 summarizes how well our 5-rule model results compares to other classification strategies found in literature for the same problem. Our best model is simple and compares well against complex models such as GA-WK-ELM, SVM, and MLP/KNN.

Table 1. An overall comparison of our method (GDL) against the best results found in other works from literature on the same PD dataset.

Algorithm	Best run on test	Reference
GA-WK-ELM	96.81%	[21]
GDL	95.45% 5 rules	This paper
Parallel ANN	92.80%	[19]
Kernel SVM	91.40%	[16]
SVM	85.29%	[20]
MLP/KNN	82.35%	[20]

5 Conclusions

This work describes a combinatorial optimization approach for learning Decision Lists. We carried out experimentation over a well-known dataset for detecting Parkinson Disease based on features extracted from vocal phonation. As result, models obtained for this specific problem show that DLs can generalize well when few rules are used. Moreover, we obtained quite simple and understandable models that can effectively be used to discover knowledge alongside medical experts.

On the other hand, our results are encouraging since the obtained accuracies compare well against more complex models found on literature for the same problem. Future work include testing our approach to other medical domains and other PD datasets. One can also assume that our GDL approach may be helpful as a feature selection tool since the features pointed out by [16] are usually found as conditions in the best obtained models. However, this matter requires further investigations embracing a different methodology.

References

1. Jordan, M.I., Mitchell, T.M.: Machine learning: trends, perspectives, and prospects. Science **349**(6245), 255–260 (2015)
2. Goodfellow, I., Bengio, Y., Courville, A.: Deep Learning. MIT Press, Cambridge (2016). http://www.deeplearningbook.org

3. Rivest, R.L.: Learning Decision Lists (2001). http://people.csail.mit.edu/rivest/Rivest-DecisionLists.pdf
4. Blum, A.: Rank-r decision trees are a subclass of r-decision lists. Inf. Process. Lett. **42**(4), 183–185 (1992)
5. de Campos, L.M., Fernandez-Luna, J.M., Gamez, J.A., Puerta, J.M.: Ant colony optmization for learning Bayesian networks. Int. J. Approx. Reason. **31**, 291–311 (2002). Elsevier
6. Khan, K., Sahai, A.: A comparison of BA, GA, PSO, BP, and LM for training feed forward neural networks in e-Learning context. Int. J. Intell. Syst. Appl. **7**, 23–29 (2012). MECS Press
7. Bosman, P.A.N., LaPutré, H.: Learning and anticipation in online dynamic optimization with evolutionary algorithms: the stochastic case. In: Genetic and Evolutionary Computation Conference, London (2007)
8. Marchand, M., Sokolova, M.: Learning with decision lists of data-dependent features. J. Mach. Learn. Res. **6**, 427–451 (2005)
9. Franco, M.A., Krasnogor, N., Bacardit, J.: Post-processing operators for decision lists. In: GECCO 2012, Philadelphia (2012)
10. Boström, H.: Covering vs. divide-and-conquer for top-down induction of logic programs. In: Proceedings of the 14th International Joint Conference on Artificial Intelligence (IJCAI-1995), Montreal, Canada, pp. 1194–1200 (1995)
11. Vapnik, V.N.: The Nature of Statistical Learning Theory. Springer, New York (2000)
12. Nock, R., Gascuel, O.: On learning decision committes. In: Proceedings of the Twelfth International Conference on Machine Learning (2016)
13. Martí, R., Resende, M.G.C., Ribeiro, C.C.: Multi-start methods for combinatorial optimization. Eur. J. Oper. Res. **226**(1), 1–8 (2013)
14. Shiek, A., Winkins, S., Fatouros, D.: Metallomic profiling and linkage map analysis of early Parkinson's disease: a new insight to aluminum marker for the possible diagnosis. PLoS ONE **5**, 6 (2010)
15. Little, M.A., McSharry, P.E., Roberts, S.J., Costello, D.A.E., Moroz, I.M.: Exploiting nonlinear recurrence and fractal scaling properties for voice disorder detection. BioMed. Eng. OnLine, 6–23 (2007)
16. Little, M.A., McSharry, P.E., Hunter, E.J., Spielman, J.: Suitability of dysphonia measurements for telemonitoring of Parkinson's disease. IEEE Trans. Biomed. Eng. **56**(4), 1015–1022 (2008)
17. Das, R.: A comparison of multiple classification methods for diagnosis of Parkinson's Disease. Expert Syst. Appl. **37**, 1568–1572 (2010)
18. Singh, N., Pillay, V., Choonara, Y.E.: Advances in the treatment of Parkinson's Disease. Prog. Neurobiol. **81**, 29–44 (2007)
19. Astrom, F., Koker, R.: A parallel neural network approach to prediction of Parkinson's Disease. ESWA **38**(10), 12470–12474 (2011)
20. Sharma, A., Giri, R.N.: Automatic recognition of Parkinson's Disease via artificial neural network and support vector machine. Int. J. Innov. Technol. Explor. Eng. **4**(3), 35–41 (2014)
21. Avci, D., Dogantekin, A.: An expert diagnosis system for parkinson disease based on genetic algorithm-wavelet kernel-extreme learning machine. Parkinson's Dis. (2016). Article ID 5264743
22. Shreiner, D., Sellers, G., Kessenich, J.M., Licea-Kane, B.: OpenGL Programming Guide: The Official Guide to Learning OpenGL, Version 4.3. Addison-Wesley, Boston (2013)

Hermes: A Distributed-Messaging Tool for NLP

Ilaria Bordino, Andrea Ferretti, Marco Firrincieli, Francesco Gullo(✉), Marcello Paris, and Gianluca Sabena

R&D Department, UniCredit, Milan, Italy
{ilaria.bordino,andrea.ferretti2,marco.firrincieli,francesco.gullo, marcello.paris,gianluca.sabena}@unicredit.eu

Abstract. In this paper we present Hermes, a novel tool for natural language processing. By employing an efficient and extendable distributed-message architecture, Hermes is able to fullfil the requirements of large-scale processing, completeness, and versatility that are currently missed by existing NLP tools.

1 Introduction

Text is everywhere. It fills up our social feeds, clutters our inboxes, and commands our attention like nothing else. Unstructured content, which is almost always text or at least has a text component, makes up a vast majority of the data we encounter. *Natural Language Processing* (*NLP*) is nowadays one of the predominant technologies to handle and manipulate unstructured text. NLP tools can tackle disparate tasks, from marking up syntactic and semantic elements to language modeling or sentiment analysis.

The open-source world provides several high-quality NLP tools and libraries [2–5,9]. These solutions however are typically stand-alone components aimed at solving specific micro-tasks, rather than complete systems capable of taking care of the whole process of extracting useful content from text data and making it available to the user via proper exploration tools. Moreover, most of them are designed to be executed on a single machine and cannot handle distributed large-scale data processing.

In this paper we present Hermes,[1] a novel tool for NLP that advances existing work thanks to three main features. (*i*) *Capability of large-scale processing*: Our tool is able to work in a distributed environment, deal with huge amounts of text and arbitrarily large resources usually required by NLP tasks, and satisfy both real-time and batch demands; (*ii*) *Completeness*: We design an integrated, self-contained toolkit, which handles all phases of a typical NLP text-processing application, from fetching of different data sources to producing proper annotations, storing and indexing the content, and making it available for smart exploration and search; (*iii*) *Versatility*: While being complete, the

[1] In Greek mythology Hermes is the messenger of the gods. This is an allusion to our distributed-messaging architecture.

P.M. Pardalos et al. (Eds.): MOD 2016, LNCS 10122, pp. 402–406, 2016.
DOI: 10.1007/978-3-319-51469-7_33

tool is extremely flexible, being designed as a set of independent components that are fully decoupled from each other and can easily be replaced or extended.

To accomplish the above features, we design an efficient and extendable architecture, consisting of independent modules that interact asynchronously through a message-passing communication infrastructure. Messages are exchanged via distributed queues to handle arbitrarily large message-exchanging rates without compromising efficiency. The persistent storage system is also distributed, for similar reasons of scalability in managing large amounts of data and making them available to the end user.

2 Framework

The architecture of Hermes is based on using persisted message queues to decouple the actors that produce information from those responsible for storing or analyzing data. This choice is aimed to achieve an efficient and highly modular architecture, allowing easy replacement of modules and simplifying partial replays in case of errors.

Queues. Message queues are implemented as topics on Kafka[2] (chosen for easy horizontal scalability and minimal setup). There are at present three queues: news, clean-news and tagged-news. Producers push news in a fairly raw form on the news queue, leaving elaborate processing to dedicated workers down the line. Whilst all present components are written in Scala, we leave the road open for modules written in different languages by choosing a very simple format of interchange: all messages pushed and pulled from the queues are simply encoded as JSON strings.
The main modules of the system are depicted in Fig. 1 and described below.

Producers retrieve the text sources to be analyzed, and feed them into the system. Hermes currently implements producers for the following sources. (*i*)

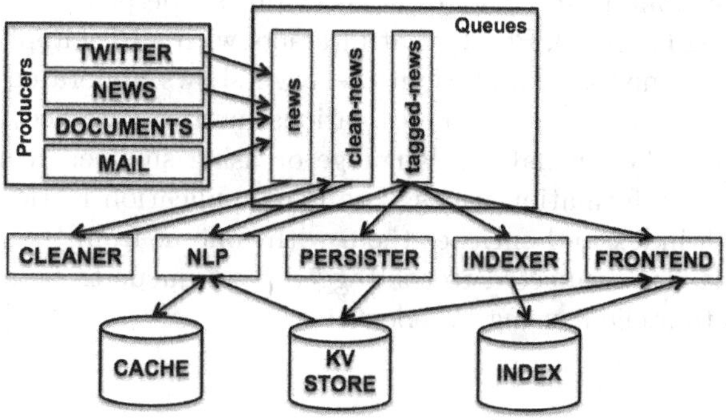

Fig. 1. Depiction of the architecture of Hermes

[2] http://kafka.apache.org.

Twitter: a long-running producer listening to a Twitter stream and annotating the relative tweets. (*ii*) *News articles*: this is a generic article fetcher that downloads news from the web following a list of input RSS feeds and can be scheduled periodically. (*iii*) *Documents*: this producer fetches a generic collection of documents from some known local or public directory. It can handle various formats such as Word and PDF. (*iv*) *Mail messages*: this producer listens to a given email account, and annotates the received email messages. We use Apache Camel's mail component[3] to process email messages. Each producer performs a minimal pre-processing and pushes the fetched information on the news queue. In the following we call news a generic information item pushed to this queue.

Cleaner. This module consumes the raw news pushed on the news queue, performs the processing needed to extract the pure textual content, and then pushes it onto the clean-news queue for further analysis. The module is a fork of Goose,[4] an article extractor written in Scala.

NLP. This module consists of a client and a service. The client listens for incoming news on the clean-news queue, asks for NLP annotations to the service, and places the result on the tagged-news queue. The service is an Akka[5] application providing APIs to many NLP tasks, from the simplest such as tokenization or sentence splitting, to complex operations such as parsing HTML or Creole (the markup language of Wikipedia), entity linking, topic detection, clustering of related news and sentiment analysis. All APIs can be consumed by Akka applications, using remote actors, or via HTTP by other applications. The algorithms we designed for NLP tasks are detailed in Sect. 3.

Persister and Indexer. In order to retrieve articles, we need two pieces of infrastructure: an index service and a key-value store. For horizontal scalability we respectively choose ElasticSearch and HBase.[6] Two long-running Akka applications listen to the clean-news and tagged-news queues, and respectively index and persist (on a table on HBase) news as soon as they arrive, in their raw or decorated form.

Frontend. The frontend consists of a JavaScript single-page client (written in CoffeeScript using Facebook React) that interacts with a Play application exposing the content of news. The client home page shows annotated news ranked by a relevance function that combines various metrics (see Sect. 3), but users can also search (either in natural language or using smarter syntax) for news related to specific information needs. The Play application retrieves news from the ElasticSearch index and enriches them with content from the persisted key-value store. It is also connected to the tagged-news queue to be able to stream incoming news to a client using WebSockets.

---

[3] http://camel.apache.org/mail.html.

[4] https://github.com/GravityLabs/goose/wiki.

[5] http://akka.io.

[6] https://www.elastic.co/, https://hbase.apache.org/.

Monitoring and Statistics. A few long-running jobs using Spark Streaming[7] connect to the queues and keep updated statistics on the queue sizes (for monitoring purposes) and the hottest topics and news per author, journal, or topic. Statistics are persisted for further analysis and can be displayed in the front end for summary purposes.

3 Algorithms

In this section we describe the algorithms implemented by the Hermes NLP module.

Entity Linking. The core NLP task in our Hermes is *Entity Linking*, which consists in automatically extracting relevant entities from an unstructured document.

Our entity-linking algorithm implements the TagMe approach [3]. TagMe employs Wikipedia as a knowledge base. Each Wikipedia article corresponds to an *entity*, and the anchor texts of the hyperlinks pointing to that article are the *mentions* of that entity. Given an input text, we first generate all the n-grams occurring in the text and look them up in a table mapping Wikipedia anchor-texts to their possible entities. The n-grams mapping to existing anchortexts are retained as mentions. Next, the algorithm solves the disambiguation step for mentions with multiple candidate entities, by computing a disambiguation score for each possible mention-sense mapping and then associating each mention with the sense yielding maximum score. The scoring function attempts to maximize the collective agreement between each candidate sense of a mention and all the other mentions in the text. The relatedness between two senses is computed as Milne and Witten's measure [7], which depends on the number of common in-neighbors.

Relevance of an Article to a Target. We use entities extracted from an article to evaluate the relevance of the article to a given target (represented as a Wikipedia entity too). To this end, we compute semantic relatedness between each entity in the text and the target, then we assign the text a relevance score given by the weighted average of the relatedness of all entities. We use a p-average, with $p = 1.8$, so that the contribution of relevant topics gives a boost, but non-relevant topics do not decrease the score much.

Related News. Every few hours (time window customizable), a clustering algorithm is run on the recent news to determine groups of related news. We use the K-means implementation provided by Apache Spark MLlib.[8] The feature vector of an article is given by the first ten entities extracted, weighted by their disambiguation scores.

Document Categorization. We categorize news based on the IPTC-SRS ontology.[9] Given a document, we compute the semantic relatedness between

[7] http://spark.apache.org/streaming/.
[8] http://spark.apache.org/mllib/.
[9] https://iptc.org/metadata/.

each extracted entity and each category (by mapping each category onto a set of editorially chosen Wikipedia entities that best represent them), and assign each entity the category achieving highest similarity. The ultimate category assigned to the article is chosen by majority voting.

Document Summarization. Following [6], we build a graph of sentences, where edges between two sentences are weighted by the number of co-occurring words, and rank sentences by Pagerank centrality. The highest ranked sentence is elected as a summary.

Sentiment Analysis. We follow the general idea [1,8] to have a list of positive and negative words (typically adjectives), and then average the score of the words in the text. The novelty here is that the list of positive and negative words is not fixed. Instead, a list of adjectives is extracted from Wiktionary, and a graph is formed among them, where edges are given by synonyms. Two opposite adjectives are then chosen as polar forms –say good and bad– and for each other adjective a measure of positivity is derived from the distance of an adjective from the two polars.

References

1. Chen, L., Wang, W., Nagarajan, M., Wang, S., Sheth, A.: Extracting diverse sentiment expressions with target-dependent polarity from Twitter. In: ICWSM (2012)
2. Clarke, J., Srikumar, V., Sammons, M., Roth, D.: An NLP Curator (or: how I learned to stop worrying and love NLP pipelines). In: LREC (2012)
3. Ferragina, P., Scaiella, U.: TAGME: on-the-fly annotation of short text fragments (by wikipedia entities). In: CIKM (2010)
4. Loper, E., Bird, S.: NLTK: the natural language toolkit. In: ACL ETMTNLP (2002)
5. Manning, C.D., Surdeanu, M., Bauer, J., Finkel, J., Bethard, S.J., McClosky, D.: The stanford CoreNLP natural language processing toolkit. In: ACL (2014)
6. Mihalcea, R., Tarau, P., Textrank: bringing order into text. In: EMNLP (2004)
7. Milne, D., Witten, I.H.: Learning to link with Wikipedia. In: CIKM (2008)
8. Thelwall, M., Buckley, K., Paltoglou, G.: Sentiment strength detection for the social web. JASIS&T **63**(1), 163–173 (2012)
9. Yosef, M.A., Hoffart, J., Bordino, I., Spaniol, M., Weikum, G.: AIDA: an online tool for accurate disambiguation of named entities in text and tables. PVLDB **4**(12), 1450–1453 (2011)

Deep Learning for Classification of Dental Plaque Images

Sultan Imangaliyev[1,2(✉)], Monique H. van der Veen[2], Catherine M.C. Volgenant[2], Bart J.F. Keijser[1,2], Wim Crielaard[2], and Evgeni Levin[1]

[1] Netherlands Organisation for Applied Scientific Research, TNO Earth, Life and Social Sciences, 3704HE Zeist, The Netherlands
sultan.imangaliyev@tno.nl

[2] Academic Centre for Dentistry Amsterdam, University of Amsterdam and Free University Amsterdam, 1008AA Amsterdam, The Netherlands

Abstract. Dental diseases such as caries or gum disease are caused by prolonged exposure to pathogenic plaque. Assessment of such plaque accumulation can be used to identify individuals at risk. In this work we present an automated dental red autofluorescence plaque image classification model based on application of Convolutional Neural Networks (CNN) on Quantitative Light-induced Fluorescence (QLF) images. CNN model outperforms other state of the art classification models providing a 0.75 ± 0.05 F1-score on test dataset. The model directly benefits from multi-channel representation of the images resulting in improved performance when all three colour channels were used.

Keywords: Deep learning · Convolutional neural networks · Computer vision · Bioinformatics · Computational biology · Quantitative light-induced fluorescence · Dentistry

1 Introduction

Oral diseases like caries and periodontitis are caused by the presence of pathogenic dental plaque [6]. A novel way to look at this plaque is the use of a Quantitative Light-induced Fluorescence (QLF) camera which uses an excitation wavelength of 405 nm with dedicated filters to inspect the teeth and make images. When using this system some dental plaque fluoresces red, which is suggested to be an indication for the pathogenicity of the dental plaque [5,10]. Since images have a special two-dimensional structure, a group of Deep Learning methods called Convolutional Neural Networks (CNN) explicitly uses the advantages of such a representation [3,4]. Although the intra- and inter-examiner agreement of QLF analysis is shown to be high [8,12], manual assessment of QLF-images might be expensive and laborious if the amount of images is large. Therefore, there is a need to automate this procedure by implementing CNN-based system for classification of QLF-images depending on the amount of red fluorescent

P.M. Pardalos et al. (Eds.): MOD 2016, LNCS 10122, pp. 407–410, 2016.
DOI: 10.1007/978-3-319-51469-7_34

dental plaque. The main aim of this study is to build an image classification system with predictive performance as high as possible using an appropriate modelling approach.

2 Convolutional Neural Networks

In this section we present short mathematical description of CNN following formulations [2,4,9]. For simplicity we assume a grayscale image that can be represented by an array of size $n_1 \times n_2$. Then, a colour image can be represented by an array of size $n_1 \times n_2 \times 3$ assuming three colour channels, for example Red-Green-Blue (RGB) representation. For simplicity, from now on, all definitions will be provided for grayscale images only. Given the filter $K \in \mathbb{R}^{2h_1+1 \times 2h_2+1}$, the discrete convolution of the image I with filter K is given by

$$(I * K)_{r,s} := \sum_{u=-h_1}^{h_1} \sum_{v=-h_2}^{h_2} K_{u,v} I_{r+u,s+v}. \tag{1}$$

Let layer l be a convolutional layer. The output of layer l consists of $m_1^{(l)}$ feature maps of size $m_2^{(l)} \times m_3^{(l)}$. The ith feature map in layer l, denoted $Y_i^{(l)}$, is computed as

$$Y_i^{(l)} = B_i^{(l)} + \sum_{j=1}^{m_1^{(l-1)}} K_{i,j}^{(l)} * Y_j^{(l-1)}, \tag{2}$$

where $B_i^{(l)}$ is a bias matrix and $K_{i,j}^{(l)}$ is the filter of size $2h_1^{(l)} + 1 \times 2h_2^{(l)} + 1$ connecting the jth feature map in layer $(l-1)$ with the ith feature map in layer l [4].

Convolutional layer is followed by a max-pooling layer. The architecture of the network is built based on stacking two convolutional layers with non-linearities each of them followed by a pooling layer. The feature maps of the final pooling layer are then fed into the actual classifier, consisting of one fully connected layer followed by a multinomial logistic regression function.

3 Dataset and Experimental Setup

The dataset was obtained during a clinical intervention study which was conducted at the Department of Preventive Dentistry of Academic Centre for Dentistry Amsterdam, studying the dynamic changes in red plaque fluorescence during an experimental gingivitis protocol [11]. During this intervention 427 QLF-images were collected and we translated them into the dataset of images with reduced resolution of 216×324 raw pixel intensities in three Red-Green-Blue channels. The images then were assessed by a panel of two trained dental experts using a modified form of a clinical plaque scoring system [12]. Throughout the experiment individuals were observed and split into classes depending on degree of red fluorescent plaque accumulation [11]. The datasets and Python scripts supporting the conclusions of this article are available on the http://www.learning-machines.com/ website.

The CNN model was implemented on Graphics Processing Unit (GPU) via Theano Python package [1]. The model was trained using stratified shuffled split when 80% of data were used as training, 10% of data as validation, and 10% as testing datasets. Model parameters such as the number of filters, filter shape, max-pooling shape and others were selected via an exhaustive grid search procedure combined with an early stopping to prevent overfitting. To compare CNN performance with one of the from other models we used various classification models implemented in the scikit-learn package [7]. Hyperparameters of those models were selected via a five-fold stratified shuffled cross-validation procedure. The reported final test and train F1-scores were obtained by averaging the results of ten random shuffles with fixed splits across all models.

4 Results

Results of simulations are provided in Fig. 1. As it is seen from Fig. 1 most of the models suffer from overfitting resulting in poor testing performance despite of a good training performance. Using only the Red channel results in a relatively good and comparable performance between both SVM models and Logistic Regression. Adding the Green and especially the Blue channels improves the performance of CNN compared to other models. As a result, the best model (CNN)

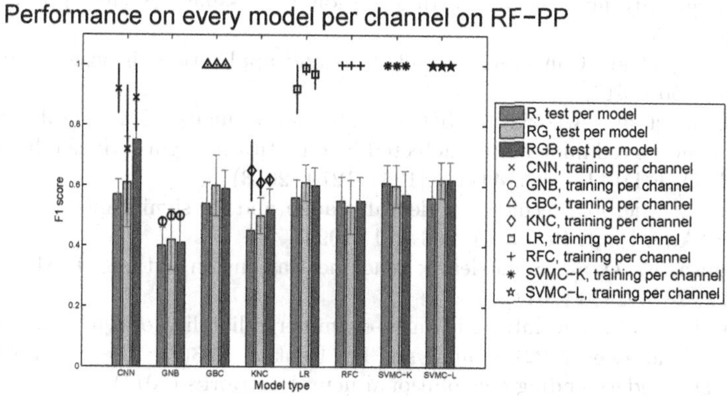

Fig. 1. Test and training performance of models on QLF-images using red fluorescent plaque groups as labels. Bars indicate the testing F1-score, symbols over bars the training F1-score. Labels on the x-axis refer to the following models: Convolutional Neural Network (CNN), Logistic Regression (LR), Support Vector Classifier with Gaussian Kernel (SVC-K), Support Vector Classifier with Linear Kernel (SVC-L), Gaussian Naïve Bayes Classifier (GNB), Gradient Boosting Classifier (GBC), K-Neighbors Classifier (KNC), and Random Forest Classifier (RFC). (Color figure online)

provided provided on average a 0.75 ± 0.05 testing F1-score and a 0.89 ± 0.11 training F1-score. The differences in full RGB representation between CNN and other models are statistically significant.

5 Conclusions

Based on the results we obtained, we conclude that CNN provided superior performance over other classification models provided that all three colour channels were used. It has been shown that the blue light illuminating the teeth excites green intrinsic fluorophores at the dentine-enamel junction, and induces red fluorescence at areas with prolonged bacterial activity such as old plaque accumulation [11]. Blue backscattered light is expected to produce sharper defined edges in images with little red fluorescent plaque, in comparison to images with a thick layer of red fluorescent plaque. The results suggest that hierarchical layer-wise feature training of CNN benefits from multi-channel representation of the images where each colour channel contains important information relevant to the classification task.

References

1. Bergstra, J., et al.: Theano: deep learning on GPUs with Python. In: NIPS 2011, BigLearning Workshop, Granada, Spain (2011)
2. Jarrett, K., et al.: What is the best multi-stage architecture for object recognition? In: 2009 IEEE 12th International Conference on Computer Vision, pp. 2146–2153. IEEE (2009)
3. Krizhevsky, A., et al.: Imagenet classification with deep convolutional neural networks. In: Advances in Neural Information Processing Systems, pp. 1097–1105 (2012)
4. LeCun, Y., et al.: Convolutional networks and applications in vision. In: ISCAS, pp. 253–256 (2010)
5. Lee, E.S., et al.: Association between the cariogenicity of a dental microcosm biofilm and its red fluorescence detected by quantitative light-induced fluorescence-digital (QLF-D). J. Dent. **41**(12), 1264–1270 (2013)
6. Marsh, P.: Microbial ecology of dental plaque and its significance in health and disease. Adv. Dent. Res. **8**(2), 263–271 (1994)
7. Pedregosa, F., et al.: Scikit-learn: machine learning in python. J. Mach. Learn. Res. **12**, 2825–2830 (2011)
8. Pretty, I., et al.: The intra-and inter-examiner reliability of Quantitative Light-induced Fluorescence (QLF) analyses. Br. Dent. J. **193**(2), 105–109 (2002)
9. Stutz, D.: Understanding convolutional neural networks (2014)
10. Van der Veen, M., et al.: Red autofluorescence of dental plaque bacteria. Caries Res. **40**(6), 542–545 (2006)
11. van der Veen, M.H., Volgenant, C.M., Keijser, B., Jacob Bob, M., Crielaard, W.: Dynamics of red fluorescent dental plaque during experimental gingivitis - a cohort study. J. Dent. **48**, 71–76 (2016)
12. Volgenant, C.M., Mostajo, M.F., Rosema, N.A., Weijden, F.A., Cate, J.M., Veen, M.H.: Comparison of red autofluorescing plaque and disclosed plaque-a cross-sectional study. Clin. Oral Invest. **20**, 1–8 (2016)

Multiscale Integration for Pattern Recognition in Neuroimaging

Margarita Zaleshina and Alexander Zaleshin(✉)

Moscow Institute of Physics and Technology, Moscow, Russia
zaleshin@gmail.com

Abstract. Multiscale, multilevel integration is a valuable method for the recognition and analysis of combined spatial-temporal characteristics of specific brain regions. Using this method, primary experimental data are decomposed both into sets of spatially independent images and into sets of time series. The results of this decomposition are then integrated into a single space using a coordinate system that contains metadata regarding the data sources. The following modules can be used as tools to optimize data processing: (a) the selection of reference points; (b) the identification of regions of interest; and (c) classification and generalization. Multiscale integration methods are applicable for achieving pattern recognition in computed tomography and magnetic resonance imaging, thereby allowing for comparative analyses of data processing results.

Keywords: Multiscale integration · Pattern recognition

1 Introduction

Modern-day radiologists typically must view hundreds or even thousands of images on a regular basis. For example, advanced multidetector computed tomography can produce several thousand images during a single examination [1].

Neuroimaging techniques are used to visualize the structural, functional, and/or biochemical characteristics of the brain. Structural neuroimaging usually involves the diagnosis of a tumor or injury using data obtained in the form of electronic medical records, including brain images. Brain mapping is the next analytic step for presenting the results of structural and functional neuroimaging in the form of sets of maps that contain additional thematic analytical layers. This paper describes the spatial data processing tools that are currently available for handling heterogeneous 2D slice images that are obtained using various brain scanning techniques.

Multilevel data mapping tools already exist in the field of geoinformatics. Applications of Geographic Information Systems (GIS) support the transformation of data, the integration of independent layers, and the conversion of raster images into vectors. Converting raster images into vector form allows the researcher to determine the precise position of observed phenomena and to calculate numerical characteristics of the areas being studied.

Here we present the possibilities regarding the use of GIS applications for the pattern recognition and comparative analysis of electronic medical records, with the ultimate goal of determining tumor area within the brain. The resulting data can be used

P.M. Pardalos et al. (Eds.): MOD 2016, LNCS 10122, pp. 411–418, 2016.
DOI: 10.1007/978-3-319-51469-7_35

to compare tumor size on the basis of aggregate data obtained over the course of several years using a variety of tomographic techniques. We also performed neuroimaging data classification using methods based on decision trees [2, 3].

2 Related Work

A current question in the field of neuroimaging is how cortical "activation" at the macro level (viewed, for example, using fMRI or EEG) is linked to micro level phenomena such as the activity of a single neuron in the spinal cord of awake animals and/or humans [4]. The selection of a suitable scale is generally based on several characteristics, including parameters related to individual neurons, ensembles of neurons, and brain regions. Outstanding reductions can be beneficial for developing clustering ensemble algorithms [5]. Time can also be regarded as a variable for revealing the constructive rules that underlie the spatial structure of task-related brain activity at a variety of levels [6].

The study of brain function often requires the accurate matching of cortical surface data in order to compare brain activity. In this context, several tasks are critical, including surface inflation for cortical visualization and measurements, surface matching, and the alignment of functional data for group-level analyses [7]. Rubinov and Sporns [8] describe measures that can be used to identify functional integration and segregation, quantify the centrality of individual brain regions and/or pathways, characterize patterns of local anatomical circuitry, and test the resilience of networks. The resulting data integration can help build a generalized model of the brain [9] as a spatially unbiased, high-resolution atlas (i.e., template) of the human brain and brainstem using MNI[1] coordinates.

Reference points are used to define coordinates (e.g., Talairach coordinates[2] or MNI coordinates), and these reference points are generally divided into the following two types: (a) regular points (e.g., grid cells), and (b) nodes or reference points, which are defined by convolutions. By connecting multiple spatial scales in order to decode the population activity of grid cells, Stemmler et al. [10] predicted the existence of neural and behavioral correlates of grid cell readout that transcend the known link between grid cells in the medial entorhinal cortex and place cells in the hippocampus.

3 Tools for Pattern Recognition in Neuroimaging

Neuroimaging data are usually decomposed into sets of independent spatial components for subsequent analysis [11]. In neuroimaging, multilevel separation is predicated either by the biological characteristics of the sources or by the properties of the primary

---

[1] The MNI Coordinate System originated at the Montreal Neurological Institute and Hospital and is used to normalize anatomical 3D datasets.

[2] Talairach coordinates is a 3D coordinate system of the human brain, which is used to map the location of brain structures independent from individual differences in the size and overall shape of the brain.

images. The resolution of the image slices is used to select the most suitable scale. Template sets for recognition also should contain information regarding the range of application scales.

3.1 Neuroimaging Software

A large number of neuroimaging programs have already been applied to pattern recognition in both computed tomography (CT) and magnetic resonance imaging (MRI), including BrainVISA (http://www.brainvisa.info), the FMRIB Software Library (FSL) (http://fsl.fmrib.ox.ac.uk/fsl/fslwiki/FSL), SPM (http://www.fil.ion.ucl.ac.uk/spm), and neuroVIISAS (http://139.30.176.116/neuroviisas.shtml).

These programs can perform complex analyses within specific scale ranges. However, researchers must develop additional tools for comparing mixed multiscale datasets, particularly when processing large numbers of radiographic images.

3.2 Adapting Existing Geoinformation Technologies to the Field of Neuroimaging

Tools and technologies that are already being used in other fields can be adapted for use in the healthcare field in order to improve the analysis, visualization, and navigation of large sets of medical images. New methods are needed in order to improve post-processing tasks, including the integration of 3D and multimodal image with non-image data [1].

The application of GIS tools to neuroimaging is relevant for the following reasons:

1. GIS technology has been used extensively to collect and analyze heterogeneous spatial data, as well as to organize and publish maps. Moreover, GIS tools are currently used for tasks that involve large numbers of images.
2. The tasks used in GIS technology have many similarities with neuroimaging and brain mapping, including data recognition, processing of spatial data, and calculating routes. In addition, the use of integrated vector-and-raster models — which are implemented in GIS — enables researchers to create a representation of both the surface and interior of anatomical structures, thereby allowing the researcher to identify anatomical structures and perform spatial analyses [12].
3. Recent studies have shown that the brain contains biological structures that are directly responsible for navigation and recognition, including "place cells" and "grid cells" [10, 13].

4 Materials and Methods

4.1 Materials

Electronic medical records were obtained from the period 2012 through 2015, including images obtained using various CT and MRI equipment and processed in accordance with the Digital Imaging and Communications in Medicine Standard (DICOM) (http://dicom.nema.org). Control primary image series were grouped by date and type of observation in MicroDicom (http://www.microdicom.com).

Main DICOM Tags for the primary image series (s1–s5) are summarized in Table 1. Bold numbers indicate the number of slides in the observations.

Table 1. Primary image series

DICOM Tags	s1	s2	s3	s4	s5
StudyDate	2012-02-07	2012-09-06	2013-02-07	2014-05-16	2015-08-06
Modality	MR	PR	MR	CT	MR
Manufacturer	Marconi Medical Systems	Philips Medical Systems	GE MEDICAL SYSTEMS	TOSHIBA	SIEMENS
Manufacturer ModelName	Polaris 1.0T	EWS	DISCOVERY MR750w	Aqui lion	Avanto
ImpVersion Name	Merge COM 3_330	Makhaon Networks	ArhiMed	DICOM Viewer	MR_2004V_VB11A
Software Versions	VIA 2.0E.004	R2.6.3.2	DV23.0 V01_1142.a	V1.8.1	Syngo MR 2004V 4VB11D
Number of slices by tags (*Abbreviated names of tags: TmpR = Temporal Resolution, STh = Slice Thickness, PxS = Pixel Spacing*)	TmpR = 131660: 45 slices TmpR = 9348: 3 slices TmpR = 83910: 14 slices	STh = 5, PxS = 0.8984 \0.8984: 230 slices	STh = 10, PxS = 0.5859 \0.5859: 10 slices STh=4, PxS = 0.4296 \0.4296, 0.4688\0.4688: 118 slices STh = 4, PxS= 0.8594\0.8594: 805 slices STh = 1.2, PxS = 0.4688 \0.4688: 542 slices	STh = 2, PxS = 0.384 \0.384: 76 slices	STh = 10, PxS = 0.5468 \0.5468: 3 slices STh = 5, PxS = 0.8437 \0.8437: 35 slices STh = 4, PxS = 0.5273 \0.5273,0.4882 \0.4882,0.875 \0.875,0.5078 \0.5078: 127 slices STh = 1, PxS = 0.5 \0.5: 124 slices
Total slices number = 2129	62 slices	230 slices	1475 slices	76 slices	286 slices

4.2 Methods

The data were processed using the open source software program QGIS (http://qgis.org), including additional analysis modules.

The following GIS methods were applied in the data analysis (Table 2): (a) the selection of reference points; (b) the identification of regions of interest (ROIs); (c) classification and generalization; and (d) multilayer comparison.

Table 2. Data processing methods

	Methods	Description	Tools
(a)	Contouring	Contouring allows converting raster data to vector. Isoline calculation with a given tolerance is performed using Gdal_contour plugin	**Gdal_contour** generates a vector contour file from the input raster: http://www.gdal.org/gdal_contour.html
	Selection of reference points	Plugin allows extracting nodes from isolines and polygon layers and then outputting extracted nodes as reference points. In the work reference points were identified by the most detailed data set (s3 in Table 1)	**Extract nodes** is a tool for nodes extraction: http://docs.qgis.org/2.6/en/docs/user_manual/processing_algs/qgis/vector_geometry_tools/extractnodes.html
	Georeferencing of image series	To georeference an image, one first needs to establish reference points, input the known local coordinates of these points, choose the coordinate system and other projection parameters and then minimize residuals. Residuals are the difference between the actual coordinates of the reference points and the coordinates predicted by the spatial model	**Georeferencer Plugin** is a tool for snapping rasters to single coordinate system with the help of reference points: http://docs.qgis.org/2.0/en/docs/user_manual/plugins/plugins_georeferencer.html
(b)	ROIs identification	Subsets of samples of the tumor area are selected as regions of interest (ROIs). Vector ROIs are the basis for next template creation and classification	**Semi-Automatic Classification** is a plugin for the semi-automatic supervised classification of images: [14] (in the work modified version was used)
(c)	Data Classification and Template Creation	DTclassifier helps to allocate data on the image with the same characteristics. How it works: (1) Selecting training datasets, (2) Selecting data to classify, (3) Refining templates	**DTclassifier** is a plugin that allows classification of data in QGIS: [15]. It uses a particular classification algorithm - "decision trees" [2, 3]
(d)	Multilayers Comparison	To compare parameter data from different layers, used the analytical tools of fTools Plugin. It provides a growing suite of spatial data management and analysis functions that are both fast and functional	**fTools Plugin** for analysis functions: http://docs.qgis.org/2.8/en/docs/user_manual/plugins/plugins_ftools.html

5 Results

Here we performed the following steps:

1. Neuroimaging slides were integrated into one project that contains series of independent layers with the appropriate metadata.
2. ROIs were vectorized, and coordinates for reference points were defined.
3. Templates were created in order to define tumor areas.
4. Tumor size was calculated and compared between various times.

The primary datasets were grouped by layers in accordance with the date of the original study and then entered into QGIS (Fig. 1A). All of the image series were then

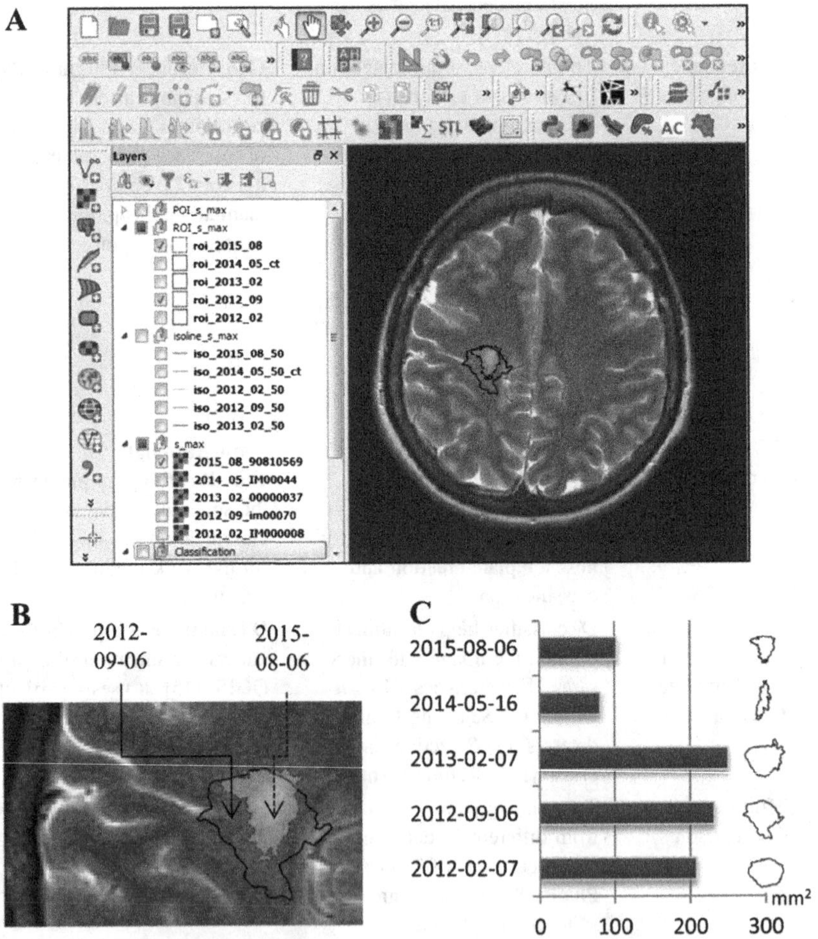

Fig. 1. Analysis of electronic medical records: A. QGIS project. B. Example of selected ROIs (tumor areas). C. Comparison of tumor sizes at different times (tumor contours with maximum section square are shown in the right column).

combined into a common coordinate system, and layers taken at the same approximate location were then overlaid with each other.

For each set of layers, contouring coefficients were identified and used to create isolines, after which ROIs were highlighted (Fig. 1B). The parameters of the regions that were filled with tumor were then saved in templates.

Using the data classification, tumor area locations were determined for different points in time. Lastly, a comparative analysis of tumor size was performed for each dataset using the precise coordinates for tumor borders (Fig. 1C).

The above comparison revealed that tumor size decreased in this patient by approximately 50% from 2012 through 2015 (Table 3).

Table 3. Maximum tumor size at the indicated dates

	s1	s2	s3	s4	s5
Study Date	2012-02-07	2012-09-06	2013-02-07	2014-05-16	2015-08-06
STumor_max (mm^2)	208.16	231.90	250.60	80.63	102.86
Percentage (normalized to s1)	100.0%	111.4%	120.4%	38.7%	49.4%

6 Conclusions and Future Work

Here we examined the feasibility of using GIS technology to process data obtained from electronic medical records. Our results demonstrate that GIS applications can be applied to neuroimaging, thereby providing added value such as the ability to work with vectors, integrate multilayers, identify regions of interest, and classify and compare data.

Future research will include the use of spatial operations for neuroimaging and brain mapping, including questions regarding the identification of reference points and the selection of characteristic scales. Practical significance of such works is determined by the necessity of extracting meaningful and relevant information from large volume of neuroimaging data.

References

1. Andriole, K.P., Wolfe, J.M., Khorasani, R., Treves, S.T., Getty, D.J., Jacobson, F.L., Steigner, M.L., Pan, J.J., Sitek, A., Seltzer, S.E.: Optimizing analysis, visualization, and navigation of large image data sets: one 5000-section CT scan can ruin your whole day. Radiology **259**, 346–362 (2011)
2. Murthy, S.K.: Automatic construction of decision trees from data: a multi-disciplinary survey. Data Min. Knowl. Discov. **2**, 345–389 (1998)
3. Michie, D., Spiegelhalter, D.J., Taylor, C.C., Michie, E.D., Spiegelhalter, D.J., Taylor, C.C.: Machine learning, neural and statistical classification. Ellis Horwood Ser. Artif. Intell. **37** (xiv), 289s (1994)

4. Nazarova, M., Blagovechtchenski, E.: Modern brain mapping - what do we map nowadays? Front. Psychiatry. **6**, 1–4 (2015)
5. Vega-Pons, S., Avesani, P.: On pruning the search space for clustering ensemble problems. Neurocomputing **150**, 481–489 (2015)
6. Papo, D.: Time scales in cognitive neuroscience (2013). http://www.frontiersin.org/Journal/Abstract.aspx?s=454&name=fractal_physiology&ART_DOI=10.3389/fphys.2013.00086
7. Lombaert, H., Arcaro, M., Ayache, N.: Brain transfer: spectral analysis of cortical surfaces and functional maps. In: Ourselin, S., Alexander, D.C., Westin, C.-F., Cardoso, M.J. (eds.) IPMI 2015. LNCS, vol. 9123, pp. 474–487. Springer, Heidelberg (2015). doi:10.1007/978-3-319-19992-4_37
8. Rubinov, M., Sporns, O.: Complex network measures of brain connectivity: uses and interpretations. NeuroImage **52**, 1059–1069 (2010)
9. Diedrichsen, J.: A spatially unbiased atlas template of the human cerebellum. NeuroImage **33**, 127–138 (2006)
10. Stemmler, M., Mathis, A., Herz, A.: Connecting multiple spatial scales to decode the population activity of grid cells. Sci. Adv. **1**, e1500816 (2015)
11. McKeown, M.J., Makeig, S., Brown, G.G., Jung, T., Kindermann, S.S., Bell, A.J., Sejnowski, T.J.: Analysis of fMRI data by blind separation into independant spatial components. Hum. Brain Mapp. **6**, 160–188 (1998)
12. Barbeito, A., Painho, M., Cabral, P., O'neill, J.: A topological multilayer model of the human body. Geospat. Health **10**, 199–204 (2015)
13. Moser, E.I., Roudi, Y., Witter, M.P., Kentros, C., Bonhoeffer, T., Moser, M.-B.: Grid cells and cortical representation. Nat. Rev. Neurosci. **15**, 466–481 (2014)
14. Congedo, L.: Semi-automatic classification plugin for QGIS. http://fromgistors.blogspot.com/p/semi-automatic-classification-plugin.html
15. Brui, A., Dubinin, M.: Raster classification with DTclassifier for QGIS. http://gis-lab.info/qa/dtclassifier-eng.html

Game Theoretical Tools for Wing Design

Lina Mallozzi[1(✉)], Giovanni Paolo Reina[2], Serena Russo[1], and Carlo de Nicola[1]

[1] University of Naples Federico II, Naples, Italy
{mallozzi,denicola}@unina.it, serusso@unisa.it
[2] Second University of Naples, Aversa, Italy
giovannipaolo.reina@unina2.it

Abstract. In the general setting of modeling for system design it is assumed that all decision-makers cooperate in order to choose the optimal set of the variable design. Sometimes there are conflicts between the different tasks, so that the design process is studied as a multi-player game. In this work we deal with a preliminary application of the design of a wing by studying its optimal configuration by considering some of the standard parameters of the plant design. The choice of the parameters value is done by optimizing two tasks: the weight and the drag. This two-objective optimization problem is approached by a cooperative model, just minimizing the sum of the weight and the drag, as well by a non-cooperative model by means of the Nash equilibrium concept. Both situations will be studied and some test cases will be presented.

Keywords: Strategic games · Noncooperative equilibrium solutions · Experimental design

1 Introduction

In the experimental design framework ([1]), a decision-maker is investigating a typical engineering system related to structures or mechanisms. We assume that this system is unambiguous and described by finite-dimensional variable vector (design variables vector), say $\mathbf{x} = (x_1, ..., x_n)$ with $\mathbf{x} \in X$, being X the admissible set, and by some parameters. If the designer wishes to optimize the design with respect to several objective functions such as $f_1(\mathbf{x}), ..., f_m(\mathbf{x})$, then the problem reduces to finding a design variable vector $\mathbf{x}$ that solves the following standard multiobjective formulation:

$$\min_{\mathbf{x} \in X} \{f_1(\mathbf{x}), ..., f_m(\mathbf{x})\}.$$

Usually, the parameters are fixed quantities over which the designer has no control, while some structural and physical constraints on the design variables must be considered. The tuple $< X; f_1, ..., f_m; x_1, ..., x_n >$ is called experimental design situation.

As done in several situations in the context of scalarization procedures for multicriteria optimization problems, one can optimize just the sum of the different criteria. In this case the solution has a cooperative nature and it is called in a

P.M. Pardalos et al. (Eds.): MOD 2016, LNCS 10122, pp. 419–426, 2016.
DOI: 10.1007/978-3-319-51469-7_36

game theoretical setting *cooperative solution.* On the other hand, it may happen that the minimization of one criterium can hurt another criterium: a conflictual situation between the different requirements with respect to the different design variables may appear. In this case a *noncooperative solution* concept fits to better describe the model ([2,3]).

In this paper we consider the optimization problem of the wing design typical of the aerospace field of engineering: find out the optimal setting for the strategical variables important to built a wing. The principal tasks in this case are the drag and the weight, both to be minimized. So that we model the design problem as a two criteria optimization problem and we investigate the cooperative approach as well as the noncooperative one using the well known solution concept of Nash equilibrium. We present a numerical procedure based on a genetic type algorithm in line with previous papers ([4–9]).

In Sect. 2 some technical concepts are recalled, in Sect. 3 the wing design model is described and the results of our procedure are shown. Some final discussions are contained in Sect. 4.

2 Preliminaries

In this section we point out a general description of the optimization problem as applied to a real case of wing design. In this context, a designer has to decide the values of three different variables in a set given by fluid dynamic and structural constraints: the main aim is the minimization of drag and weight given by two functions depending on the design variables, according to well known laws (details of the wing model will be given in the next Section). So the ingredients of the model are two real valued functions depending on three real design variables varying in a bounded and closed set of $\mathbb{R}$ under some constraint limitations. These ingredients give us the possibility to translate the problem into a two-player normal form game.

A two-player normal form game, that consists of a tuple $< 2; X_1, X_2; f_1, f_2 >$ where X_i is the set of player i's admissible choices and $f_i : X_1 \times X_2 \to \mathbb{R}$ is player i's payoff function ($\mathbb{R}$ is the set of real numbers), for $i = 1, 2$. We suppose here that players are cost minimizing, so that player i has a cost $f_i(x_1, x_2)$ when player 1 chooses $x_1 \in X_1$ and player 2 chooses $x_2 \in X_2$.

Definition 1. *A Nash equilibrium [10, 11] for the game* $< 2; X_1, X_2; f_1, f_2 >$ *is a strategy profile* $\hat{\mathbf{x}} = (\hat{x}_1, \hat{x}_2) \in X = X_1 \times X_2$ *suchthat for any* $i = 1, 2$, $\hat{x}_i$ *is solution to the optimization problem*

$$\min_{x_i \in X_i} f_i(x_i, \hat{x}_{-i})$$

where $x_{-i} = x_j$ *for* $j = 1, 2$ *and* $j \neq i$.

The notion of Nash equilibrium strategy captures the idea that each player optimizes his payoff and no single player can individually improve his welfare by deviating from it. So, the Nash equilibrium strategy acquires the notion of a "stable solution". Hence, $\hat{\mathbf{x}}$ is a Nash equilibrium profile if no player has an incentive to unilaterally deviate from $\hat{\mathbf{x}}$ because his payoff will not decrease with such an action. In experimental design problems, the Nash equilibrium concept describes the idea of 2 (virtual) players competing in order to choose the optimal value of the assigned design variables minimizing his own objective in a noncooperative way.

On the other hand, the designer may also have a cooperative approach and looks for the cooperative solution.

Definition 2. *We call cooperative solution for the experimental design situation* $< X_1, X_2; f_1, f_2; x_1, x_2 >$ *a vector* $\hat{\mathbf{x}} = (\hat{x}_1, \hat{x}_2) \in X$ *that minimizes the function* $f_1 + f_2$, *i.e. a solution to*

$$\min_{(x_1,x_2)\in X_1\times X_2} f_1(x_1, x_2) + f_2(x_1, x_2).$$

It should be possible, of course, to use other different solution concepts to approach the experimental design situation. Some addressed issues are discussed in the concluding Section.

3 Wing Design

The aerodynamic characteristics of the wing can be described starting from its two representations: the planform and the airfoil. The wing planform is the shape of the wing as viewed from directly above looking down onto the wing: span (b), tip chord (c_t) and root chord (c_r) are geometric parameters that, in planform design, are very important to the overall aerodynamic characteristic of the wing (see Fig. 1).

The airfoil is obtained from the intersection of the wing surface with a plane perpendicular to the span b (for example plane AA of Fig. 1). Figure 2 shows an example of airfoil section: it consists of a leading edge (LE), a trailing edge (TE) and the line, joining these two, called chord (c). The distance between the upper and lower surface and perpendicular to the chord line represents the airfoil thickness, t, and its maximum t_{max} is also designated as a percent of chord length. From all these quantities, it is possible to derive some non-dimensional

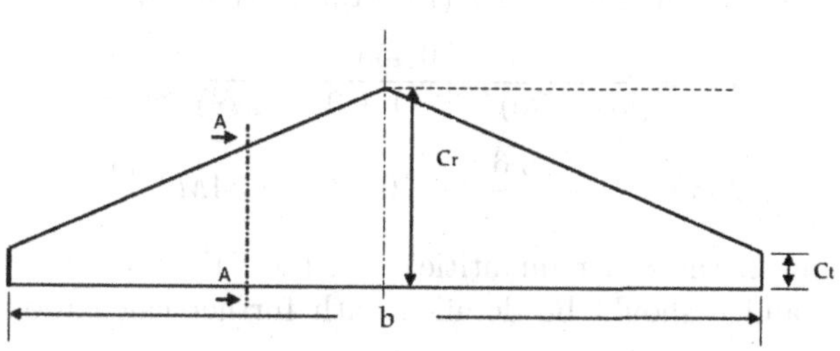

Fig. 1. Main characteristics of a tapered wing planform

parameters, that are generally used for the characterization of the wing. These are respectively the taper ratio ($\lambda = c_t/c_r$) and the thickness ratio ($\tau = t_{max}/c$).

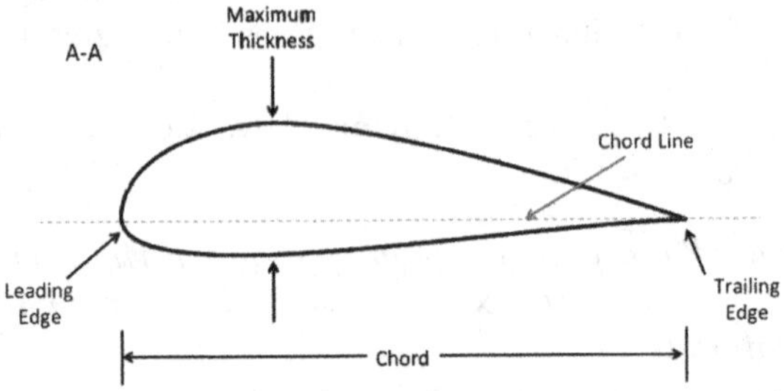

Fig. 2. Scheme of a wing section

In the preliminary wing design, two quantities are taken into account: the wing weight W_w and the aerodynamic drag force coefficient C_D. The triplette $\omega = (b, \lambda, \tau)$ is in a domain $\Omega = [\underline{b}, \overline{b}] \times [\underline{\lambda}, \overline{\lambda}] \times [\underline{\tau}, \overline{\tau}]$ of $\mathbb{R}^3$. Aim of the study is to decide the optimal values for the mentioned parameters trying to minimize both the drag coefficient and the wing weight. Two virtual player are minimizing the cost functions: we assume that one player chooses at the same time the pair b, λ in order to minimize the drag and the second virtual player chooses the third design variable τ minimizing the weight.

In this paper the value of the wing area S is assumed to be fixed (as detailed in the next section), so, if $AR(b) = b^2/S$ represents the aspect ratio, according to the Basic Aerodynamics Theory, the definition for C_D (see [12]) and a classical estimator for W_w (as derived from [13]) are respectively expressed as functions of the variable design set $\omega = (b, \lambda, \tau)$ as in the following:

$$C_D(b, \lambda, \tau) = C_{D0}(\tau) + \frac{C_L^2}{\pi \cdot AR(b) \cdot e}, \tag{1}$$

$$W_w(b, \lambda, \tau) = 0.0051 \cdot (W_{dg} \cdot N_z)^{0.557} \cdot S^{0.649} \cdot AR(b)^{0.5} \cdot \tau^{-0.4} \cdot (1+\lambda)^{0.1} \cdot S_{cs}^{0.1}, \tag{2}$$

where

$$C_{D0}(\tau) = c_f \cdot FF(\tau) \cdot (1.977 + 0.52\tau), \tag{3}$$

$$c_f = \frac{0.455}{(\log_{10} Re)^{2.58} \cdot (1 + 0.144M^2)^{0.65}}, \tag{4}$$

$$FF(\tau) = [1 + \frac{0.6}{0.25}\tau + 100\tau^4] \cdot \left[1.34M^{0.18}\right]. \tag{5}$$

In Eqs. 1 and 2, the other quantities (i.e. C_L, W_{dg}, N_z, S_{sc}, Re, M) are held constant[1] and e should be detailed with further calculations. Since the

[1] In our computation $C_L = 0.45, W_{dg} = 44000lb, N_z = 3.75, S_{sc} = 0.09 \cdot S, Re = 2 \cdot 10^7, M = 0.41$.

drag coefficient C_D is a non-dimensional quantity, W_w has been properly adimensionalized with a reference quantity W_{ref}, in order to be consistent with C_D to obtain a weight coefficient $C_W(b, \lambda, \tau) = W_w(b, \lambda, \tau)/W_{ref}$. Denoting $X_1 = [\underline{b}, \overline{b}] \times [\underline{\lambda}, \overline{\lambda}]$ and $X_2 = [\underline{\tau}, \overline{\tau}]$, we consider the two-player normal form game $< 2; X_1, X_2; C_D, C_W >$. By considering, as a reference, the wing of the regional transport airplane, the ATR-72 depicted in Fig. 3 ([14]), we present a game theoretical setting for this wing design modelling in which two players interact strategically, one of whom - choosing (b, λ) - wants to minimize the wing weight and the other - choosing τ - wishes to minimize the drag coefficient. By using a genetic type procedure, a Nash equilibrium solution for the game is computed and compared with the minimizer of the function $C_D + C_W$, i.e. the cooperative solution.

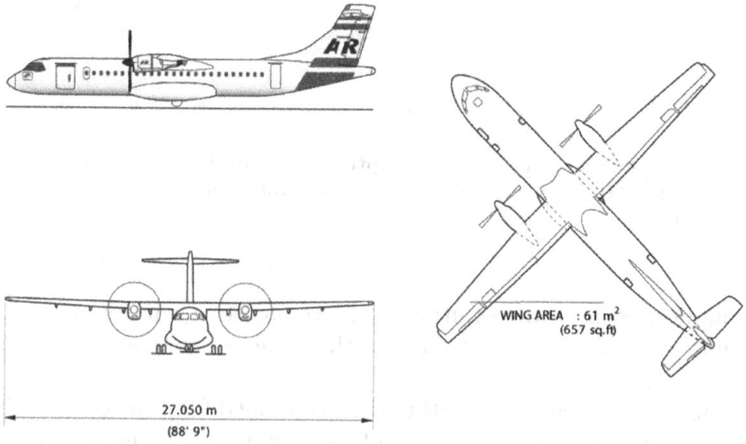

Fig. 3. ATR-72

3.1 Results

This section is devoted to the computational results; they will be compared with those of the real wing of ATR-72. In this work the wing area S is fixed to the ATR-72 value, that is equal to 650 sq.ft and, for design reasoning linked to the architecture of the airplane, the root chord cannot exceed 12 ft: these requirements induce indirect limitations on b and λ. For all these reasons the geometrical bounds of our problem are $\Omega' = \Omega \cap A \subset \mathbb{R}^3$ where

$$\Omega = [60, 90]ft \; \times [0.3, 1.0] \times [0.1, 0.3], \tag{6}$$

$$A = \Big\{(b, \lambda, \tau) \in \Omega : S = 650 \; sq. \; ft \;\&\; c_r \leq 12 \; ft\Big\}. \tag{7}$$

Table 1. Results of the wing design problem

	b [ft]	λ	τ
Real wing (ATR-72)	88.9	0.57	0.14
Cooperative solution	85.5	0.52	0.22
Nash solution	85.8	0.51	0.10

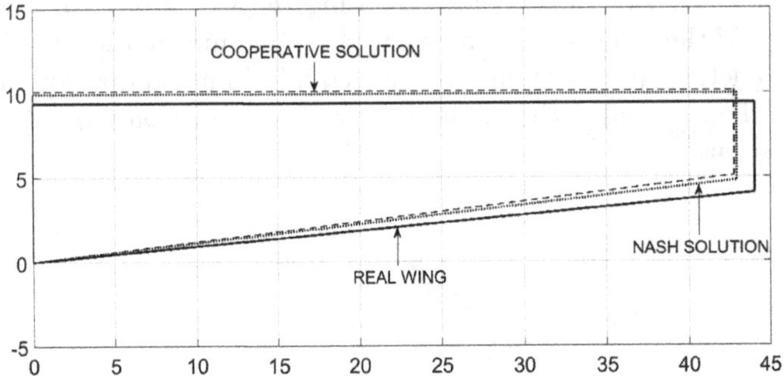

Fig. 4. Wing planforms as results of the optimization problem: black curve refers to the real wing of ATR-72, dots curve to the Nash solution and dashed curve to the cooperative solution.

The results derived from the solution of the cooperative solution and the Nash Equilibrium are reported in Table 1 and they can be compared with those of ATR-72.

According to variables b and τ, the results obtained from the cooperative and Nash solution are very similar to those of the real wing. The major discrepancy is observed in the τ parameter: this is probably due to the fact that the wing weight is an increasing function of τ so it has a minimum on the lower bound of the domain, while the real value is the result of further choices and calculations. This aspect is interesting because it should be possible for the future, to consider additional criteria for the wing design problem.

Figure 4 shows the comparison between the three different wing planforms: black line, dots line and dashed line refer respectively to real wing, the Nash and cooperative solutions.

Remark 1. An element $x^* \in X$ is called a feasible solution and the vector $z^* = f(x^*) \in \mathbb{R}^2$ is called its outcome. In multi-objective optimization, there does not typically exist a feasible solution that minimizes all objective functions simultaneously. A fundamental definition in this context concerns the *Pareto optimal solutions*, i.e. solutions that cannot be improved in any of the objectives without degrading at least one of the other objectives. In mathematical terms, a feasible solution $x^1 \in X$ is said to Pareto dominate another solution $x^2 \in X$, if

$$f_i(x^1) \leq f_i(x^2) \quad \forall i \in \{1, 2\}$$

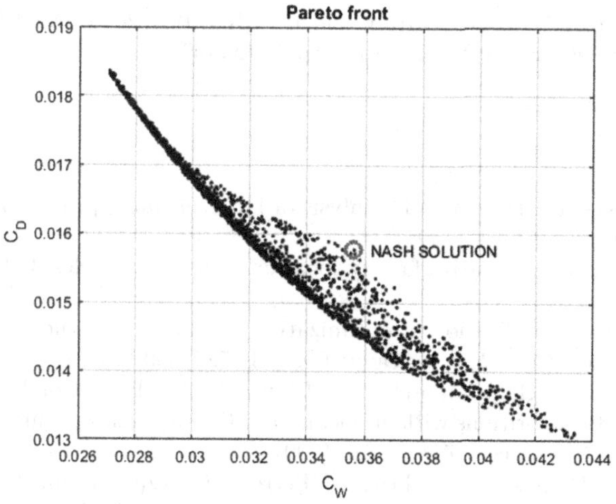

Fig. 5. Pareto front and Nash equilibrium outcome pair.

and the strict inequality holds for at least one index. A solution $x^1 \in X$ (and the corresponding outcome $f(x^*)$) is called Pareto optimal, if there does not exist another solution that dominates it. The set of Pareto optimal outcomes is often called the Pareto front or Pareto boundary. Solving the wing design problem by means of the Nash equilibrium solution, it turns out important to explore the Pareto dominance. As well known, it may happen that a Nash equilibrium is not Pareto optimal. In Fig. 5 the Pareto front of the wing design problem is depicted with respect to the Nash equilibrium payoff pair.

4 Conclusion

In this paper we studied two different approaches to model a preliminary wing design problem, namely the cooperative approach and the Nash equilibrium solution. In this case the noncooperative protocol identifies the most favorable solution. It remains the question of possible multiple Nash equilibria: this analysis will be performed in the future.

Future work will focus on the influence of more aerodynamic and structural parameters, by coupling these optimization tools with aerodynamic flow solver.

Another aspect worth to be investigated is modeling such a problem by mean of the Stackelberg solution concept, where one of the players has the leadership in playing the game and decides first his choice. This approach corresponds to a hierarchical situation where upstream decisor makes decisions that remain superior even though the requirements of the downstream decisor are yet unknown.

Acknowledgments. The work has been supported by STAR 2014 (linea 1) "Variational Analysis and Equilibrium Models in Physical and Social Economic Phenomena",

University of Naples Federico II, Italy and by GNAMPA 2016 "Analisi Variazionale per Modelli Competitivi con Incertezza e Applicazioni".

References

1. Dean, A., Voss, D.: Design and Analysis of Experiments. Springer Texts in Statistics. Springer, New York (1998)
2. Mallozzi, L.: Noncooperative facility location games. Oper. Res. Lett. **35**, 151–154 (2007)
3. Mallozzi, L.: An application of optimization theory to the study of equilibria for games: a survey. Cent. Eur. J. Oper. Res. **21**, 523–539 (2013)
4. Periaux, J., Chen, H.Q., Mantel, B., Sefrioui, M., Sui, H.T.: Combining game theory and genetic algorithms with application to DDM-nozzle optimization problems. Finite Elem. Anal. Des. **37**, 417–429 (2001)
5. D'Amato, E., Daniele, E., Mallozzi, L., Petrone, G.: Equilibrium strategies via GA to Stackelberg games under multiple follower's best reply. Int. J. Intel. Syst. **27**, 74–85 (2012)
6. Mallozzi, L., D'Amato, E., Daniele, E.: A game theoretical model for experiment design optimization. In: Rassias, T., Floudas, C.A., Christodoulos, A., Butenko, S. (eds.) Optimization in Science and Engineering. In Honor of the 60th Birthday of P.M. Pardalos, pp. 357–368. Springer, New York (2014)
7. Rao, J.R., Badhrinath, K., Pakala, R., Mistree, F.: A study of optimal design under conflict using models of multi-player games. Eng. Optim. **28**, 63–94 (1997)
8. Tang, Z., Desideri, J.-A., Periaux, J.: Multicriterion aerodynamic shape design optimization and inverse problem using control theory and nash games. J. Optim. Theor. Appl. **135**, 599–622 (2007)
9. Xiao, A., Zeng, S., Allen, J.K., Rosen, D.W., Mistree, F.: Collaborative multidisciplinary decision making using game theory and design capability indices. Res. Eng. Des. **16**, 57–72 (2005)
10. Başar, T., Olsder, G.J.: Dynamic Noncooperative Game Theory. SIAM, Philadelphia (1999). Reprint of the second edition (1995)
11. Mallozzi, L.L.: Basic concepts of game theory for aerospace engineering applications. In: Vasile, M., Becerra, V.M. (eds.) Computational Intelligence in the Aerospace Sciences, pp. 149–179. American Institute of Aeronautics and Astronautics AIAA (2014)
12. Anderson, J.D.: Fundamentals of Aerodynamics, 4th edn. McGraw-Hill, New York (2007)
13. Raymer, D.P.: Aircraft Design: A Conceptual Approach, 5th edn. AIAA, Boston (2012)
14. Jackson, P.: Jane's All the World's Aircraft. IHS Global Inc, Coulsdon (2013). 104th 2013–2014 edn.

Fastfood Elastic Net: Combining Variable Selection with Kernel Expansion Approximations

Sonia Kopel(✉), Kellan Fluette, Geena Glen, and Paul E. Anderson

College of Charleston, Charleston, USA
{kopels,fluetteka}@g.cofc.edu, gmglen6@gmail.com, andersonpe2@cofc.edu
http://anderson-lab.github.io/

Abstract. As the complexity of a prediction problem grows, simple linear approaches tend to fail which has led to the development of algorithms to make complicated, nonlinear problems solvable both quickly and inexpensively. Fastfood, one of such algorithms, has been shown to generate reliable models, but its current state does not offer feature selection that is useful in solving a wide array of complex real-world problems that spans from cancer prediction to financial analysis.

The aim of this research is to extend Fastfood with variable importance by integrating with Elastic net. Elastic net offers feature selection, but is only capable of producing linear models. We show that in combining the two, it is possible to retain the feature selection offered by the Elastic net and the nonlinearity produced by Fastfood. Models constructed with the Fastfood enhanced Elastic net are relatively quick and inexpensive to compute and are also quite powerful in their ability to make accurate predictions.

Keywords: Kernel methods · Data mining · Algorithms and programming techniques for big data processing

1 Introduction

The value of effective prediction methods is self-evident: being able to predict future outcomes can be applied to nearly any field. The methods and algorithms that are used to generate these predictive models continue to evolve to handle large, complicated real-world datasets that have high dimensionality and large sample sizes. The fields that these datasets come from range from stock market and financial analysis to disease screening to weather prediction. For such datasets, many of the machine learning techniques that are commonly used to generate models either fail or are too expensive in either their required storage space or their runtime, rendering them inefficient.

More data offers the ability to train better, more realistic models, but the cost of generating these models is often computationally intractable because of the scope of the mathematical operations that a computer must perform during

P.M. Pardalos et al. (Eds.): MOD 2016, LNCS 10122, pp. 427–432, 2016.
DOI: 10.1007/978-3-319-51469-7_37

training. As a result, these more sophisticated models often cannot be computed in real-time, rendering them useless in many disciplines. Storage space is another concern. While the computer running the algorithms may be able to hold terabytes of data, it cannot hold a square matrix of corresponding dimensions. Thus, even in the rare cases when time is plentiful, the required computing resources to generate models from large datasets is unavailable to many researchers.

Linear techniques, such as Elastic net [6], tend not to succumb to these pitfalls because of their relative simplicity. However, complex datasets with hundreds, if not thousands, of features are unlikely to have linear decision boundaries. As a result, the linear models produced by these techniques are quite often unreliable or inaccurate.

One of the simpler methods for finding nonlinear decision boundaries is the kernel trick. The kernel trick transforms the data by implicitly mapping the features into a higher, possibly infinite, dimension and from there calculating a linear decision boundary [6]. For instance, if a dataset is not linearly separable in two dimensions, it can be transformed into a higher dimensional space. Depending on the function used to transform the data, it may be possible to find a linear decision boundary in this new feature space. The trick to this technique lies in the fact that the mapping function (ϕ) need never be explicitly defined. Rather, the individual points can be transformed by taking their dot product with a known kernel function (k), like the sigmoid function or the radial basis function [3]. The relationship between ϕ and k is as follows:

$$\langle \phi(x), \phi(y) \rangle = k(x, y) \tag{1}$$

$$f(x) = \langle w, \phi(x) \rangle = \sum_{i=1}^{N} a_i k(x_i, x) \tag{2}$$

Unfortunately, the kernel trick may also become intractable to compute as the computation and storage requirements for the kernel matrix are exponentially proportional to the number of samples in the dataset [3]. However, the Random Kitchen Sinks (RKS) algorithm chooses to approximate the kernel function more effectively by randomizing features instead of optimizing them [5]. It does this by randomly selecting these features from a known distribution. The authors show that with this method shallow neural nets achieve comparable accuracy to AdaBoost, a popular ensemble method which adjusts weak learners in favor of instances misclassified by previous classifiers [2]. Overall, RKS has comparable predictive ability to the commonly used AdaBoost, but does not require the same level of rigor on the part of the computer during training. However, RKS makes use of dense Gaussian random matrix multiplications which are computationally expensive. Le and Smola mitigate this problem by replacing these multiplications by multiplying several diagonal matrices in their algorithm: Fastfood [3].

Like Fastfood, Elastic net can be combined with machine learning techniques. One implementation of Elastic net combines it with support vector machines (SVMs) yielding a substantial performance improvement without sacrificing accuracy [6]. Our work builds directly upon the Fastfood algorithm, which while capable of generating models quickly does not provide a variable importance measure or feature section. Herein, we describe our research into developing a computationally efficient variable importance measure for Fastfood by leveraging the built-in feature selection of Elastic net [6]. Our results indicate that models generated with our improved variation of Fastfood retains the benefits of the Fastfood while also providing variable importance, which is not available in the standard Fastfood algorithm.

2 Methods

In order to generate nonlinear models with feature selection, we combine two existing algorithms: Fastfood and Elastic net. Our method, Fastfood Elastic Net (FFEN) retains the nonlinearity of Fastfood while incorporating feature selection of Elastic net to provide a variable importance measurement.

To reduce the already quick run-time complexity of RKS, Fastfood combines Hadamard matrices with Gaussian scaling matrices [3]. The algorithm replaces the Gaussian random matrices in RKS with this combination. Because of the relative ease of these computations in contrast multiplying to Gaussian random matrices, the complexity of the runtime is reduced from $O(nd)$ to $O(n \log d)$ [3]. This allows Fastfood to approximate the kernel feature map quickly. The main takeaway from this is that unlike in the kernel trick where ϕ is never defined, in Fastfood (and RKS) it is approximated.

Another appeal of Fastfood is that it has previously been combined with other machine learning techniques, most notably neural nets. By implementing the algorithm at each layer of the neural net, additional nonlinearity is added to the training of the neural net and the training process is also sped up [1].

Despite the successes of Fastfood in making accurate predictions in loglinear time, it is unable to inherently measure variable importance and perform variable selection because it relies on kernel approximation and projection into a new feature space. Elastic net, however, implicitly provides variable selection, but is a linear technique characterized in Eq. 3 [6]. Thus, Elastic net in isolation is not capable of creating a predictive model for complicated datasets with nonlinear decision boundaries.

$$\hat{\beta} = \arg\min_{\beta}(||y - \mathbf{X}\beta||^2 + \lambda||\beta||^2 + \alpha||\beta||_1) \tag{3}$$

where α and λ are parameters to be specified by the user, X is the data matrix, y are the labels or dependent variable, and β is a vector of coefficients representing the model.

Our method takes the original feature matrix (X) of size ($n \times d$) and applies Fastfood. This transforms the original features to a new feature matrix (F) of size ($n \times p$), where $p \geq d$. In this step, Fastfood approximates ϕ to transform X into a higher dimensional space.

$$\underset{n\times p}{F} = \underset{n\times d}{X} \times \underset{d\times p}{\phi} \tag{4}$$

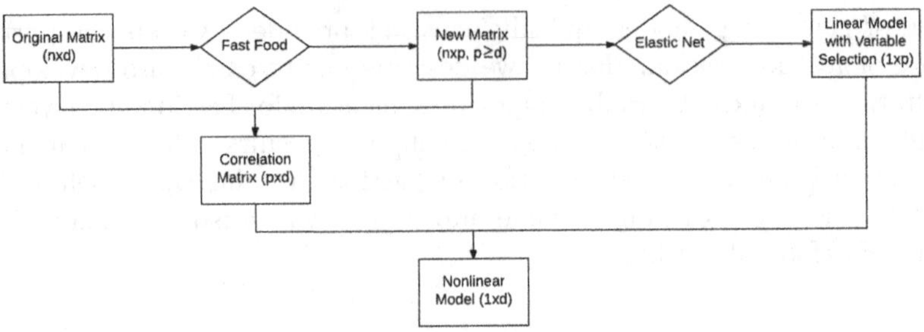

Fig. 1. Process

Elastic net is applied to this new feature matrix to generate a model with variable selection (L). Optimization of parameters λ and α within the Elastic Net algorithm are also optimized throughout the process. To reduce back to the original dimensionality, we calculate a correlation matrix (ρ) between the new features and the original features. Finally, these correlations are aggregated to reduce and relate the Elastic net coefficient vector of size $(1 \times p)$ to the original feature space of size $(1 \times d)$. The overall process of FFEN is shown in Fig. 1.

$$\underset{1\times d}{O} = \underset{1\times p}{L} \times \underset{p\times d}{\rho} \tag{5}$$

This method can then be run iteratively to remove extraneous variables or a single run can be used to integrate variable importance in the original dimensional space with the kernel approximation basis space.

3 Results and Discussion

For evaluation, tests have been performed on both simulated linear datasets as well as real datasets from the UCI machine learning repository [4]. The simulated datasets had random variables masked, providing a known gold standard for feature selection and variable importance. Our Fastfood enhanced Elastic net (FFEN) was compared to lasso: a common implementation of Elastic net. To keep the comparisons fair, parameters λ and α for lasso were optimized in the same way as they were for FFEN. For the simulated linear data, the true β values for each variable were known, so the primary metric that was used to compare the FFEN model to lasso was average mean absolute error (MAE) for all of the β values. Table 1 depicts the results for simulated datasets of differing dimensions.

The results indicate that FFEN improves as the number of samples increases and the number of dimensions decreases. To justify this, we repeated our simulation 20 times on datasets of varying size (see Table 1). The average MAE for FFEN was 0.03097 as opposed to 0.0473 for lasso. Given the small standard

Table 1. Simulated linear data

n	d	FFEN MAE	Lasso MAE
400	100	0.2103	0.0401
1000	80	0.1162	0.0301
4000	10	0.0471	0.0300
5000	10	0.0234	0.0400

deviations, 0.007572 and 0.00725 respectively, we can with more than 99% certainty conclude that FFEN outperforms lasso on simulated linear datasets with those dimensions which we confirmed by performing a paired T-Test.

To compare performance on real-world datasets, we generated 20 different training and test sets, and ran both FFEN and lasso on each set. We then compared R^2 values and ran a paired T-Test to test to see which model was better. Table 2 depicts the average R^2 values for the 20 runs as well as the standard deviations (σ) for these runs. FFEN performed slightly worse than lasso for the KEGG Metabolic Reaction Network Dataset though the performance dropped slightly from approximately 0.90 to 0.88. The high value of R^2 for lasso indicates that this is a linear dataset, and therefore, FFEN provides marginal benefits. However, FFEN significant outperforms lasso on the Physicochemical Properties of Protein Tertiary Structure Dataset (Protein) and also outperformed lasso on the UCI million song dataset. Because UCI provided a preferred training and test set partition for the UCI million song dataset, only one run was needed to generate the results. The R^2 values for FFEN and lasso from this run are also depicted in Table 2.

Table 2. Empirical data

Dataset	n	d	FFEN R^2	FFEN σ	Lasso R^2	Lasso σ
Protein	45729	9	0.3366	0.0268	0.2387	0.0038
KEGG	64608	27	0.8754	0.0120	0.8992	0.0023
Music	515345	90	0.2245		0.2098	

In conclusion, FFEN has been shown to offer promising results. The models generated by our algorithm are comparable or better than those of Elastic net on simulated linear datasets when measuring the accuracy of the resulting coefficients. Further, FFEN shows improved accuracy when appropriate complex nonlinear datasets while incorporating novel variable importance not available in standard Fastfood. The use of Fastfood allows us to generate these models both quickly and inexpensively.

In the future, we hope to adjust the algorithm more efficiently and accurately perform feature selection. In addition, we are working on testing a wider range of datasets with varying sample sizes and dimensionality. Further, we would like to compare the runtime of this algorithm to other runtime optimized machine learning algorithms, such as Elastic net enhanced SVMs (SVEN).

Acknowledgments. The authors would like to thank the College of Charleston for hosting the NSF Omics REU which is funded by the National Science Foundation DBI Award 1359301 as well as the UCI machine learning repository [4].

References

1. Bowick, M., Neiswanger, W.: Learning fastfood feature transforms for scalable neural networks. In: Proceedings of the International conference on..., pp. 1–15 (2013)
2. Freund, Y., Schapire, R.R.E.: Experiments with a new boosting algorithm. In: International Conference on Machine Learning, pp. 148–156 (1996)
3. Le, Q., Sarlós, T., Smola, A.J.: Fastfood – approximating kernel expansions in log-linear time. Int. Conf. Mach. Learn. **28**(1), 1–29 (2013)
4. Lichman, M.: UCI machine learning repository (2013)
5. Rahimi, A., Recht, B.: Weighted sums of random kitchen sinks: replacing minimization with randomization in learning. Adv. Neural Inf. Process. Syst. **1**(1), 1–8 (2009)
6. Zou, H., Hastie, T.: Journal of the Royal Statistical Society. Series B: Statistical Methodology **28** (2013)

Big Data Analytics in a Public General Hospital

Ricardo S. Santos[1](✉), Tiago A. Vaz[3], Rodrigo P. Santos[3], and José M. Parente de Oliveira[2]

[1] Universidade de Mogi das Cruzes, Mogi das Cruzes, SP, Brazil
ricardosantos@umc.br, rsantos@ita.br
[2] Aeronautics Institute of Technology, São José dos Campos, SP, Brazil
parente@ita.br
[3] Hospital de Clínicas de Porto Alegre, Porto Alegre, RS, Brazil
{tvaz, rpsantos}@hcpa.edu.br

Abstract. Obtaining information and knowledge from big data has become a common practice today, especially in health care. However, a number of challenges make the use of analytics in health care data difficult. The aim of this paper is to present the big data analytics framework defined and implemented at an important Brazilian public hospital, which decided to use this technology to provide insights that will help improve clinical practices. The framework was validated by a use case in which the goal is to discover the behavior patterns of nosocomial infections in the institution. The architecture was defined, evaluated, and implemented. The overall result was very positive, with a relatively simple process for use that was able to produce interesting analytical results.

Keywords: Big data · Data analysis · Medical information · Decision support system · Nosocomial infections

1 Introduction

Obtaining information and knowledge from big data has become a common practice today, especially in health care. Different types of health institutions, from regulatory agencies to hospitals, have applied analytical techniques of big data to solve different types of problems [1–6].

Health care is a data-rich industry. Administrative systems, electronic health records (EHR), and other medical software hold a tremendous number of transactions for each patient treated. Thus, the potential for big data analytics to improve quality of care and patient outcomes is very high. However, there are a number of challenges that make the use of analytics in health care data difficult.

Some studies [1, 2, 7–9] point to challenges such as: huge data segmentation by departments; protecting patients' privacy; reluctance of providers and physicians to share data; the amount of unstructured data; and especially, the adequate selection of relevant patients and subset of data for analysis to avoid bias. For instance, when certain group of patients systematically undergoes more complete medical evaluation,

P.M. Pardalos et al. (Eds.): MOD 2016, LNCS 10122, pp. 433–441, 2016.
DOI: 10.1007/978-3-319-51469-7_38

they tend to be over-represented in diagnostic incidence rates. This phenomenon is known as diagnostic sensitivity bias.

Beyond the challenges presented, we should consider that application of big data in health care is at a nascent stage, and the evidence demonstrating a useful big data analytics framework is limited [8]. Thus, implementation of this kind of solution in a hospital can be hard work and risky.

The aim of this paper is to present the big data analytics framework defined and implemented at an important Brazilian public hospital, Hospital de Clínicas de Porto Alegre (HCPA), which decided to use this technology to provide insights that will help improve clinical practices. The proposed architecture tries to overcome the main challenges of big data analytics for health care, cited above.

The framework was validated by a use case whose goal is to discover, from the large amount of existing data, the behavior patterns of nosocomial infections in the institution. The business purposes is identify action points that could help reduce or better manage the incidence of nosocomial infections.

Nosocomial infection was chosen by the hospital management for the use case due to problem relevance. Currently, it is one of the major public health problems. Despite the development of prevention and control mechanisms, incidences are still significant and the cost of treatment is very high. Developed countries have reported incidence between 5% and 15% of patients in regular wards, while a recent meta-analysis of 220 studies of developing countries found an average incidence of 7.4% [10, 11].

The next section presents the architecture of the solution, addressing its components or features to overcome the main challenges mentioned above. Section 3 presents the results obtained from the use case applied to validate the framework.

2 Methodology

The term big data is defined in different ways. One of the more accepted definitions is a set of data characterized by three V's (high-volume, high-velocity, and high-variety) which becomes large enough that it can not be processed using conventional methods [12]. The practice of analytics can be defined as the process of developing actionable insights through the application of statistical models and other computational techniques on a set of data [13].

A big general hospital can certainly qualify as a high-data volume organization, considering that a huge number of transactions are stored for each patient treated. The transactional data in the hospital changes very rapidly, patient data are inputted and updated into electronic health records multiple times a day, and the results of diagnostic tests are recorded electronically in real time. All of these attributes support the assertion that health care data meet the high-velocity criteria. Finally, general hospital data vary from discrete coded data elements to images of diagnostic tests to unstructured clinical notes, thus meeting the high-variety criteria.

The HCPA is a public general university hospital that belongs to the Brazilian National Health System (SUS). It has 6,133 employees, 845 hospital beds, and attends yearly approximately 35,000 hospital admissions, 600,000 medical consultation, 3,500,000 medical exams, 48,000 surgical procedures, 4,000 births, and 350,000 outpatient procedures.

The hospital has an excellent computerized system for recording patients' medical records, which includes all support and administrative activities. This system, called AGHU (Management Application of University Hospitals), has been in full operation for more than fifteen years. The amount of data currently available is approximately 100 TB, including data from AGHU, diagnostic images, and other software applications.

The ultimate target of this work is to define and implement a Big Data Framework for HCPA to provide clinical and managerial knowledge hidden in the huge volume of data to improve patient care and hospital management. Therefore, this framework should consider the following assumptions:

- (i) there are several sources of data;
- (ii) there are unstructured data;
- (iii) the patient's privacy must be protected;
- (iv) the necessity of sharing data between institutions;
- (v) a way to easily select the subset of data for analysis should exist.

2.1 Framework Architecture

Considering the framework premises, the solution architecture was defined as shown in Fig. 1. Flexibility and scalability were also considered in the design of architecture.

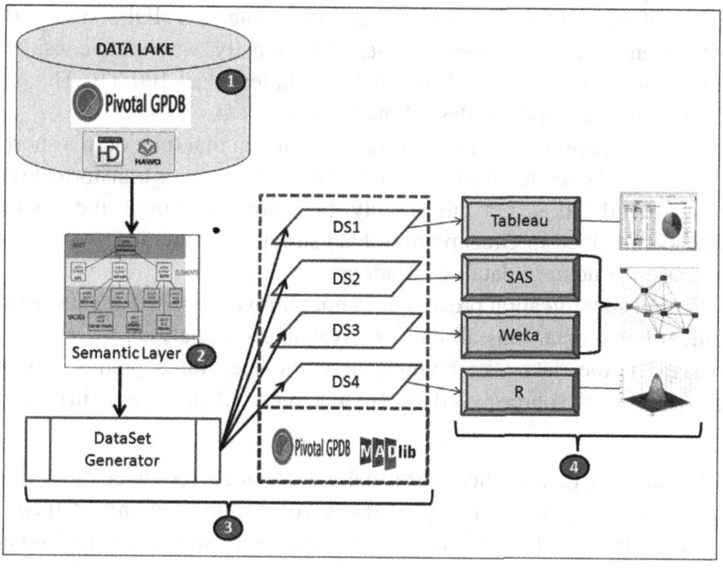

Fig. 1. Solution architecture

The proposed architecture consists of four basic components: (1) data lake, which stores the entire available hospital data; (2) a semantic representation of the date lake; (3) a mechanism to generate datasets for analysis; and (4) a set of analytical tools for data exploration and analysis.

Data Lake. According to Aasman [14], a data lake is a set of data stored in its original format. Unlike data warehouses, data lakes are "pre-processed" only when required for analysis. Thus, they are kept in their original format, and when a part is required for a particular analysis, only that part is processed to meet this specific demand.

The proposed architecture defines a Date Lake as storing all available data, in their original format (structured and unstructured data), from several hospital information systems, such as EHR, administrative systems, visitor control system, laboratory exams, etc., as well as external data from governmental and regulatory agencies.

Due to data variety, the data lake will be physically stored in two distinct and complementary technologies: Greenplum and Hadoop. The first one stores structured data and the second unstructured data.

Greenplum is a database manager for massively parallel processing (MPP) that is based on PostgreSQL and is an open-source technology. An MPP system is a high-performance system architecture that enables efficient load distribution of several terabytes and can use all the system resources in parallel to process a query, because each instance has its own memory and storage [15].

Pivotal HD (PHD) is an open source distribution powered by Apache Hadoop. Hadoop is an open source framework for large-scale data processing that enables the storage of a massive volume of data over a number of distributed servers [16].

The reason for the choice of an MPP database (Greenplum) to store part of the data lake is that the great portion of available data is structured and stored in a relational database; thus, it does not make sense to load these data in a HDFS system (Hadoop) whose range of analytic tool options is lower. Additionally, we should consider that the MPP databases work very well with volume of data less than 100 TB. The part of the data lake stored in Hadoop includes all unstructured data.

The data loading into the data lake should be done in real time or in a short time (at least once a day) and does not have any task of data cleaning or transformation, except the patient disidentification. For this activity, two tools were used: the loading utility called GPLoad contained in Greenplum to load structured data; and a script developed in Java to load unstructured data into Hadoop.

The patient disidentification process does not remove any attribute, just replaces the values of identifiable data, thus all data are available for analysis.

The design of the data lake component overcomes three premises of big data framework: (i) several sources of data; (ii) unstructured data, and (iii) the patient's privacy.

Semantic Layer. Big data applications usually deal with a very wide range of data in a data lake. Due to this variety, many works advocate the mapping of these data by means of a semantic layer [17–19]. Its purpose is to direct the data analyst to help select the data subset necessary for analysis, as well as prevent the selection of inadequate data that may produce spurious correlations.

The proposed framework adopted a semantic layer for mapping the data contained in the data lake based on SNOMED-CT (Systematized Nomenclature of Medicine - Clinical Terms).

The SNOMED-CT is a taxonomy of medical terms that provides codes, synonyms, and definitions which cover anatomy, diseases, procedures, microorganisms, substances, etc. This internationally accepted standard enables a consistent way to index, store, retrieve, and analyze medical data in all specialties and levels of care.

To implement the semantic layer, an ontology represented by the Web Ontology Language (OWL) was adopted, due to its greater power of representation.

Gruber defines ontology as an explicit specification of a conceptualization [20]. The specification can be represented in a declarative formalism that identifies a set of objects (classes) and relations between them. Several languages and tools can be used to organize and facilitate the construction and manipulation of ontologies, among them the OWL is a semantic markup language used to publish and share ontologies on the World Wide Web [21].

The semantic layer is an extremely important component of framework and is the raw material for Dataset Generator.

Mapping the data lake from SNOMED has two objectives: The first is to allow doctors and other health professionals to benefit from big data. Because they are familiar with such language they could, better than simple statisticians, select a suitable subset of data for their analysis, avoiding bias. This helps overcome the premise (v) of big data framework: easily select the subset of data for analysis.

The second goal, which is essentially important in the big data context, is the sharing of data with other institutions. Once all particularities of a specific EHR were mapped to univocal concepts, data from various health institutions can be loaded within the data lake without generating duplicate records or ambiguous understanding. In this way, the fourth premise (the necessity of sharing data between institutions) is overcome.

Dataset Generator. In order to meeting the premises (iv) and (v), beyond the semantic layer, a component was designed whose function is to generate a dataset from concepts selected by the user.

The Dataset Generator reads the semantic layer, which is stored in an OWL file, presents to the user (data analyst) the concepts in the SNOMED format, and allows the selection of those concepts that should be analyzed. Each SNOMED concept in the semantic layer is associated with a set of data contained in the data lake. Based on this equivalence, the Dataset Generator accesses the data lake and produces the dataset, which is stored as a table in Greenplum database, corresponding to selected concepts. Figure 2 shows the flowchart of the dataset generator.

No datasets are overlapping or excluded after its generation. This ensures the non-volatility of analyzes and reports produced from that datasets.

Another important feature of the dataset generator is to store the data as a relational table in a MPP database. This allows the use of a vast range of analytical tools available for relational databases with the execution performance of the MPP systems.

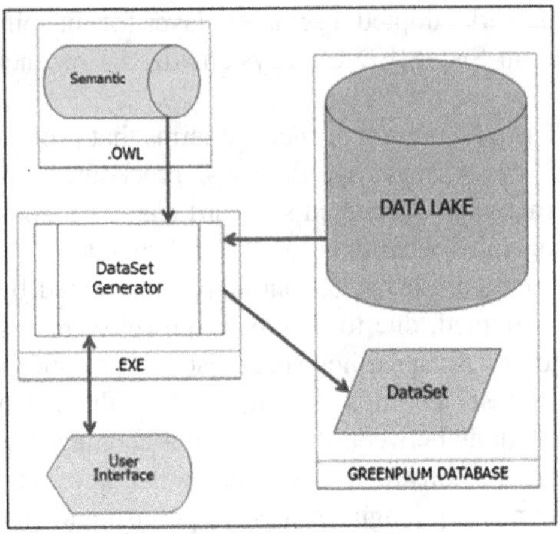

Fig. 2. Dataset generator flowchart

Analytical Tools. In the teaching hospital context, many health professionals already use some analytical tool to support their research activity. Therefore, one of the desirable features in big data for health care is that different analytical tools can be connected to the data lake and used for different data analyst needs. This feature is easily meet by the proposed architecture because datasets are physically stored in Greenplum tables. Thus, they are easily accessed and manipulated by the major analytical tools such as Tableau, R, SAS, Weka, etc.

3 Results

As mentioned in the introduction, the validation of the proposed big data framework was done by using of this framework in order to discover the behavior patterns of nosocomial infections in the HCPA.

For this study, the data lake was loaded with 15 years of complete medical records (all of the care provided to patients) from 2000 to 2014. This volume corresponds to approximately 500,000 hospitalizations and 2 TB of data. Diagnostic images and unstructured data from anamnesis were not considered because they are not related with nosocomial infections, which is an infection acquired inside of hospital.

All the data analysis process was done by one data scientist, specialized in health care, and a team from the committee for control of nosocomial infections (CCNI). The analytical tool used was the library of statistic and machine learning called MADLIB, which is a component of Greemplum.

The principal evaluation criteria are the discovery of interesting and unfamiliar patterns of nosocomial infections.

Initially, a data exploration task was done using the Tableau analytic tool in order to obtain an overview of the dataset. This step allowed validation of the consistency of the data loaded into the data lake.

After the exploratory phase, several analytical models from data mining techniques were created and submitted for evaluation by the CCNI team, which examined the relevance and coherence of the models.

A total of, 96 models were produced and submitted to the CCNI, which elected 12 of them as interesting behavior patterns of nosocomial infections, some of them are discussed below. These models were summarized and presented to the user in the form of a Dashboard (Fig. 3).

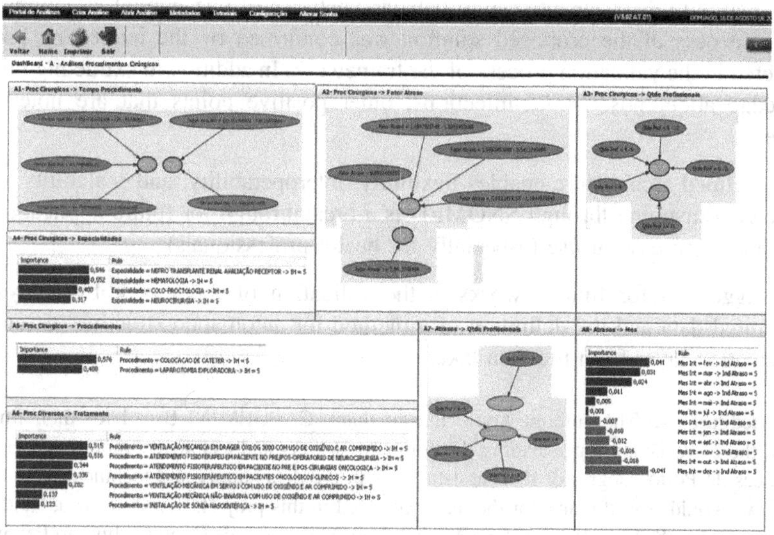

Fig. 3. User dashboard

Three data mining techniques were used to build the models: Association Rules, using Apriori method; Classification, using C 4.5 algorithm; and Clustering methods.

One of the interesting discovered patterns was the correlation between the duration of the surgical procedures and the incidence of nosocomial infection. Surgical procedures lasting more than 3.5 h are more likely to lead to a nosocomial infection. Such evidence coincides with the medical literature, which establishes long surgery as one of the factors that contribute to the occurrence of nosocomial infection. However, another not so familiar pattern was discovered; the correlation between infections and the extra time in the surgical procedures.

Another pattern that appeared is the correlation between the number of professionals attending the surgery and the incidence of nosocomial infection. The model found a high incidence of infection in the procedures carried out with 10 or more professionals in the operating room.

Considering that the length of surgical procedures is related to hospital infections, factors associated with long surgeries were examined. The analysis found an interesting correlation with the months that the procedure was performed: the months of February, March, and April are strongly related with extra time in surgical procedures. Although unusual, the presented results were analyzed and considered consistent by the medical staff, who explained that the new groups of residents begin in February.

4 Conclusion

This paper presents the definition and implementation of a data analytics framework at a public general hospital. The proposed framework tries to overcome some challenges of big data analytics for health care. The framework was validated by a study case whose goal is to discover interesting behavior patterns of nosocomial infections.

The success of the proposed solution was confirmed by the interesting patterns discovered in the validation process of the framework. In addition to extracting relevant knowledge about nosocomial infections, other positive points that are noteworthy include:

- The defined architecture enables flexibility, interoperability, and scalability.
- The data mapping through SNOMED is a breakthrough for future applications.
- Emphasis on ease of use (especially for health professionals).

A suggestion for futures works is the validation of framework by analysis of unstructured data and the definition of a method for automatic extraction of semantic representation layer from the data lake.

Acknowledgments. The authors would like to thank CAPGEMINI Brazil for their financial support; the EMC Brazil for providing the computer servers and technical support; and Hospital de Clínicas de Porto Alegre for their assistance with the data supply and technical information. Finally, we would like to thank for the team involved in this project: Ulisses Souza, Raul Hara, Margarita Bastos, Kefreen Batista, Jean Michel, Luis Macedo, Lais Cervoni, Juliano Pessini and the staff of CAPGEMINI.

References

1. White, S.E.: A review of big data in health care: challenges and opportunities. Open Access Bioinf. **6**, 13–18 (2014)
2. Raghupathi, W., Raghupathi, V.: Big data analytics in healthcare: promise and potential. Health Inf. Sci. Syst. **2**(3), 1–10 (2014)
3. Dao, T.K., Zabaneh, F., Holmes, J., Disrude, L., Price, M., Gentry, L.: A practical data mining method to link hospital microbiology and an infection control database. Am. J. Infect. Control. **36**(3), S18–S20 (2008)
4. Murdoch, T.B., Detsky, A.S.: The inevitable application of big data to health care. J. Am. Med. Assoc. **309**(13), 1351–1352 (2013)
5. Dhar, V.: Big data and predictive analytics in health care. Big Data **2**(3), 113–116 (2014)
6. Kamal, N., Wiebe, S., Engbers, J.D., Hill, M.D.: Big data and visual analytics in health and medicine: from pipe dream to reality. J. Health Med. Inf. **5**(5), 1–2 (2014)
7. Peek, N., Holmes, J.H., Sun, J.: Technical challenges for big data in biomedicine and health: data sources, infrastructure, and analytics. IMIA Yearb. Med. Inf. **9**(1), 42–47 (2014)

8. Rumsfeld, J.S., Joynt, K.E., Maddox, T.M.: Big data analytics to improve cardiovascular care: promise and challenges. Nat. Rev. Cardiol. **13**, 350–359 (2014)
9. Ellaway, R.H., Pusic, M.V., Galbraith, R.M., Cameron, T.: Developing the role of big data and analytics in health professional education. Med. Teach. **36**, 216–222 (2014)
10. Khan, H.A., Ahmad, A., Mehboob, R.: Nosocomial infections and their control strategies. Asian Pac. J. Trop. Biomed. **5**(7), 509–514 (2015)
11. Breathnach, A.S.: Nosocomial infections and infection control. Medicine **41**(11), 649–653 (2013)
12. Minelli, M., Chambers, M., Dhiraj, A.: Big data analytics: emerging business intelligence and analytic trends for today's business. Wiley, Hoboken (2013)
13. Cooper, A.: What is analytics? Definition and essential characteristics. CETIS Analytics Ser. **1**(5), 1–10 (2012). http://publications.cetis.ac.uk/wp-content/uploads/2012/11/What-is-Analytics-Vol1-No-5.pdf
14. Aasman, J.: Why data lakes require semantics. Information Management Online, 11 June 2015
15. GoPivotal: Greenplum Database: Getting Started, Version 4.2, Rev. A01. Product Documentation (2014)
16. GoPivotal: Pivotal Hadoop: Getting Started, Version 3.0. Product Documentation (2016)
17. Kuiler, E.W.: From big data to knowledge: an ontological approach to big data analytics. Rev. Policy Res. **31**(4), 311–318 (2014)
18. Kim, J., Wang, G., Bae, S.T.: A survey of big data technologies and how semantic computing can help. Int. J. Seman. Comput. **8**(1), 99–117 (2014)
19. Berman, J.: Principles of Big Data. Morgan Kaufmann, Waltham (2013)
20. Gruber, T.R.: Toward principles for the design of ontologies used for knowledge sharing. Int. J. Hum Comput Stud. **43**, 907–928 (1995)
21. Dean, M., Schreiber, G.: Web Ontology Language (OWL) Reference. Product Documentation, Version OWL2 (2009). https://www.w3.org/TR/owl-ref

Inference of Gene Regulatory Network Based on Radial Basis Function Neural Network

Sanrong Liu, Bin Yang(✉), and Haifeng Wang

School of Information Science and Engineering,
Zaozhuang University, Zaozhuang, China
batsi@126.com

Abstract. Inference of gene regulatory network (GRN) from gene expression data is still a challenging work. The supervised approaches perform better than unsupervised approaches. In this paper, we develop a new supervised approach based on radial basis function (RBF) neural network for inference of gene regulatory network. A new hybrid evolutionary method based on dissipative particle swarm optimization (DPSO) and firefly algorithm (FA) is proposed to optimize the parameters of RBF. The data from E.*coli* network is used to test our method and results reveal that our method performs than classical approaches.

Keywords: Gene regulatory network · Radial basis function neural network · Particle swarm optimization · Firefly algorithm

1 Introduction

Inference of gene regulatory network (GRN) is a central problem in the field of systems biology, which is essential to understand inherent law of life phenomenon and analyzing complex diseases [1–3]. Technologies like microarray and high-throughput sequencing could give the expression data of a number of genes. Reconstruction of GRN based on genome data is still a big challenge [4, 5].

Many computational methods have been proposed to infer gene regulatory network. These could be divided into two categories: unsupervised methods and supervised methods. The unsupervised methods mainly contain Boolean network [6], Bayesian network [7], differential equation [8] and information theory [9]. However Maetschke et al. performed an extensive evaluation of inference methods on simulated and experimental expression data and the results revealed that the supervised approaches perform better than unsupervised approaches [10]. The problem of inferring GRN could be split into many binary classification subproblems [11]. Each regulatory factor (TF) associates with one subproblem. TF-gene pairs reported are assigned to the positive class and other pairs are assigned to the negative class. Mordelet et al. proposed a new method (SIRENE) based on support vector machine (SVM) algorithm for the inference of gene regulatory networks from a compendium of expression data [12]. Gillani et al. developed a tool (CompareSVM) based on SVM to compare different kernel methods for inference of GRN [13].

P.M. Pardalos et al. (Eds.): MOD 2016, LNCS 10122, pp. 442–450, 2016.
DOI: 10.1007/978-3-319-51469-7_39

Neural networks (NN) have been widely applied in pattern recognition for the reason that neural-networks based classifiers can incorporate both statistical and structural information and achieve better performance than the simple minimum distance classifiers. In this paper, radial basis function (RBF) neural network is proposed as classifier to infer gene regulatory network. However it is difficult to determine the number, center and width of the function, and the weights of the connections of RBF neural network. Single evolutionary algorithm is not able to meet the need of the present search. In order to improve search ability and population diversity, many hybrid evolutionary algorithms have proposed, such as Particle Swarm Optimization (PSO) with Nelder–Mead (NM) search [14], PSO with artificial bee colony (ABC) [15], genetic algorithm (GA) with PSO [16]. In order to optimize the parameters of RBF neural network, a new hybrid evolutionary method based on dissipative particle swarm optimization (DPSO) and firefly algorithm (FA) is proposed. The sub network from E.coli network is used to validate our method.

The paper is organized as follows. In Sect. 2, we describe the details of our method, containing RBF neural network and hybrid evolutionary method based on DPSO and FA. In Sect. 3, the example is performed to validate the effectiveness and precision of the proposed method. Conclusions are reported in Sect. 4.

2 Method

2.1 RBF Neural Network

The RBF neural network is a feedforward propagation network, which consists of input layer, hidden layer and output layer [17]. As described in Fig. 1, the input layer

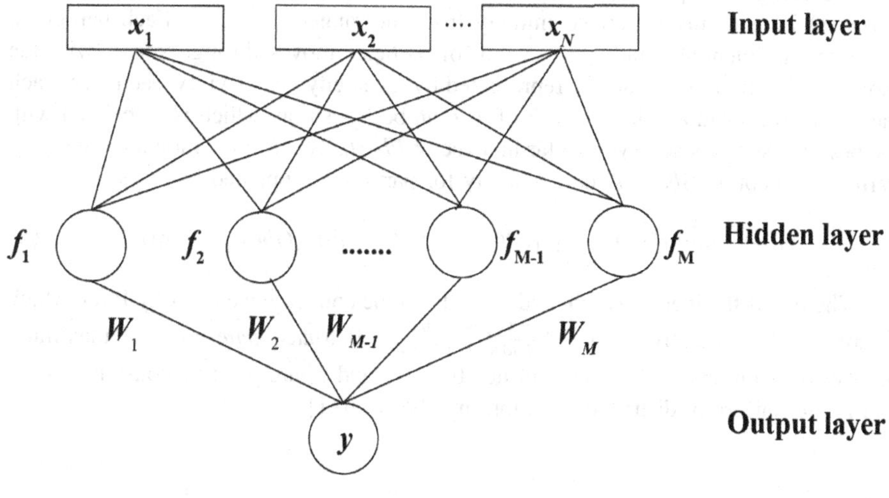

Fig. 1. The example of RBF network.

comprises N nodes, which represent the input vector $[x_1, x_2, \ldots, x_N]$. The RBF as "base" of the neuron in hidden layer is used to transform the input vector.

The output of the network is defined as followed.

$$y = \sum_{k=1}^{M} W_k \phi(\|x - c_k\|) \tag{1}$$

Where W_k denotes the connection weight from the k-th hidden unit to the output unit, c_k is prototype of center of the k-th hidden unit, and $\|\cdot\|$ indicates the Euclidean norm [17].

The RBF $\phi(\cdot)$ is usually Gaussian function. Equation is computed as

$$f_i = e^{-(x_1-\mu_1)^2/2\sigma_1^2} * e^{-(x_2-\mu_2)^2/2\sigma_2^2} \ldots * e^{-(x_N-\mu_N)^2/2\sigma_N^2} \tag{2}$$

The output functions are described as followed.

$$y = f_1 W_1 + f_2 W_2 + \ldots + f_{M-1} W_{M-1} + f_M W_M \tag{3}$$

The parameters (μ, σ, W) need be optimized. The number is $2NM + M$. In this study, hybrid evolutionary algorithm is used to train a RBF neural network.

2.2 Hybrid Evolutionary Method

Dissipative Particle Swarm Optimization. Dissipative particle swarm optimization was proposed to optimize parameters of neural network in order to prevent that the evolution process will be stagnated as time goes on [18].

According to the problem size, the particle vector $[x_1, x_2, \ldots, x_n]$ (n is the number of particles) is randomly generated initially from the range $[x_{\min}, x_{\max}]$. Each particle x_i represents a potential solution. A swarm of particles moves through space, with the moving velocity of each particle represented by a velocity vector vi. At each step, each particle is evaluated and keeps track of its own best position, which is associated with the best fitness it has achieved so far in a vector $Pbest_i$. The best position among all the particles is kept as *Gbest*. A new velocity for particle i is updated by

$$v_i(t+1) = w^* v_i(t) + c_1 r_1 (Pbest_i - x_i(t)) + c_2 r_2 (Gbest(t) - x_i(t)). \tag{4}$$

Where w is the inertia weight and impacts on the convergence rate of DPSO, which is computed adaptively as $w = \frac{\max\_generation - t}{2 \cdot \max\_generation} + 0.4$($max\_generation$ is maximum number of iterations, and t is current iteration), c_1 and c_2 are positive constant, and r_1 and r_2 are uniformly distributed random number in [0,1].

In this step, DPSO adds negative entropy through additional chaos for each particle using the following equations.

$$\text{IF}(rand() < c_v)\ \text{THEN}\ v_i(t+1) = rand()^{*} v_{max} \tag{5}$$

$$\text{IF}(rand() < c_l)\ \text{THEN}\ x_i(t+1) = random(x_{\min}, x_{max}) \tag{6}$$

Where c_v and c_l are chaotic factors randomly created in the range [0, 1], v_{max} is the maximum velocity, which specified by the user [18].

Based on the updated velocity v_i, particle x_i changes its position according to the following equation:

$$x_i(t+1) = x_i(t) + v_i(t+1). \tag{7}$$

Firefly Algorithm. Firefly algorithm (FA) is an efficient optimization algorithm, which was proposed by Xin-She Yang in 2009 [19]. It is very simple, has few parameters and easy to apply and implement, so this paper uses firefly algorithm to optimize the parameters of RBF neural network.

Firefly algorithm is the random optimization method of simulating luminescence behavior of firefly in the nature. The firefly could search the partners and move to the position of better firefly according to brightness property. A firefly represents a potential solution. In order to solve optimization problem, initialize a firefly vector $[x_1, x_2, \ldots, x_n]$ (n is the number of fireflies). As attractiveness is directly proportional to the brightness property of the fireflies, so always the less bright firefly will be attracted by the brightest firefly.

The brightness of firefly i is computed as

$$B_i = B_{i0} * e^{-\gamma r_{ij}}. \tag{8}$$

Where B_{i0} represents maximum brightness of firefly i by the fitness function as $B_{i0} = f(x_i) \cdot \gamma$ is coefficient of light absorption, and r_{ij} is the distance factor between the two corresponding fireflies i and j.

The movement of the less bright firefly toward the brighter firefly is computed by

$$x_i(t+1) = x_i(t) + \beta_i(x_j(t) - x_i(t)) + \alpha\varepsilon_i. \tag{9}$$

Where α is step size randomly created in the range [0, 1], and ε_i is Gaussian distribution random number.

The Flowchart of Hybrid Method. The flowchart of our hybrid evolutionary method, which is used to optimize the parameters of RBF neural network, is described in Fig. 2.

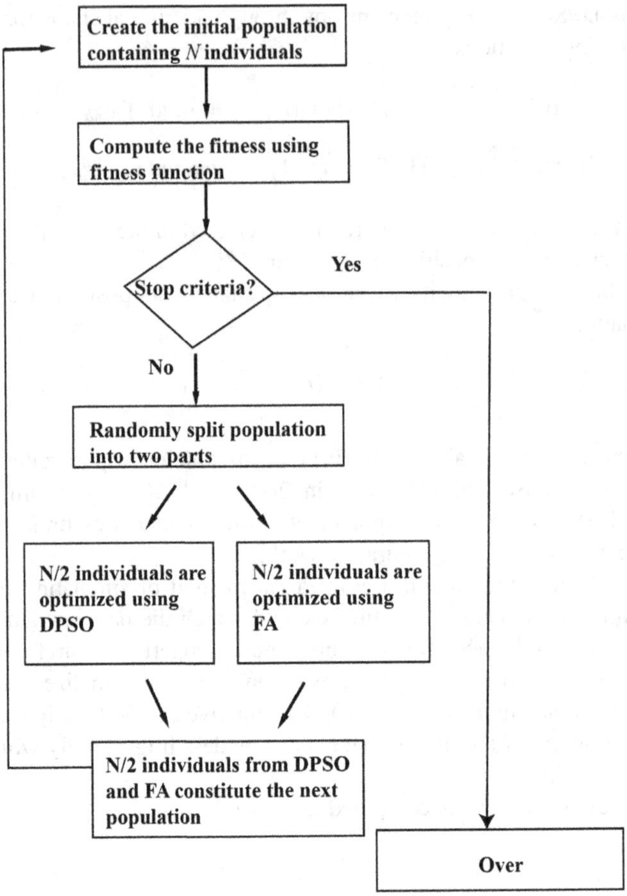

Fig. 2. The flowchart of our hybrid evolutionary method.

3 Experiments

In this part, the expression data generated from sub network from E.*coli* network using three different experimental conditions (knockout, knockdown and multifactorial) are used to test our method [13]. This network contains 150 genes and 202 true regulations. To evaluate the performance of our method, we compare it with CLR (context likelihood to relatedness) [20], back-propagation and SVM [12]. The parameters in CLR, BPNN and SVM are set by default.

Five criterions (sensitivity or true positive rate (TPR), false positive rate (FPR), positive predictive (PPV), accuracy (ACC) and F-score) are used to test the performance of the method. Firstly, we define four variables, i.e., TP, FP, TN and FN are the number of true positives, false positives, true negatives and false negatives, respectively. Five criterions are defined as followed.

$$
\begin{aligned}
TPR &= TP/(TP+FN),\\
FPR &= FP/(FP+TN),\\
PPV &= TP/(TP+FP),\\
ACC &= (TP+TN)/(TP+FP+TN+FN),\\
F-score &= 2PPV * TPR/(PPV+TPR)
\end{aligned} \tag{10}
$$

The parameters of hybrid evolutionary method are listed in Table 1. Through several runs, the results are listed in Table 2. From results, we can see that our method performs better than CLR except TPR with multifuactorial data. We also compare our method with classical supervised methods (SVM and BPNN) in Table 2. From the results, it can seen that our method has the higher F-score, which means that RBF could identify more true regulations and less false true regulations. In addition, to assess the effectiveness of our proposed criterion function, the ROC curves obtained by CLR and RBF on E.*coli* network with different experimental conditions are shown in Fig. 3. The results shows that RBF with two experimental conditions (knockout and knockdown) performs better than CLR and RBF with multifactorial experimental condition are similar with CLR.

Table 1. Parameters for experiment.

Parameters	Values
Population size	100
Generation	100
PSO $[V_{min}, V_{max}]$	[−2.0, 2.0]
PSO c_1, c_2	2
FA γ	0.6
FA α	0.02

Table 2. Comparison of four methods on E.*coli* network with different experimental conditions.

		TPR	FPR	PPV	ACC	F-score
Knockout data	CLR [20]	0.4356	0.3478	0.0114	0.6444	0.0222
	SVM [12]	0.4554	0.0076	0.3552	0.9786	0.3991
	BPNN	0.4703	0.0081	0.348	0.9783	0.4
	RBF	**0.485**	**0.0076**	**0.363**	**0.9787**	**0.4152**
Knockdown data	CLR [20]	0.4406	0.3602	0.0111	0.6323	0.0217
	SVM [12]	0.5198	**0.0073**	**0.3962**	0.9796	0.4497
	BPNN	0.5347	0.0087	0.36	0.9782	0.4303
	RBF	**0.5495**	0.0082	0.3827	**0.9796**	**0.4512**
Multifactorial data	CLR [20]	**0.8168**	0.3355	0.0219	0.6600	0.0427
	SVM [12]	0.5445	**0.0076**	0.3971	0.9795	0.4593
	BPNN	0.5495	0.0085	0.3725	0.9786	0.444
	RBF	0.5842	0.0078	**0.4069**	**0.9796**	**0.4797**

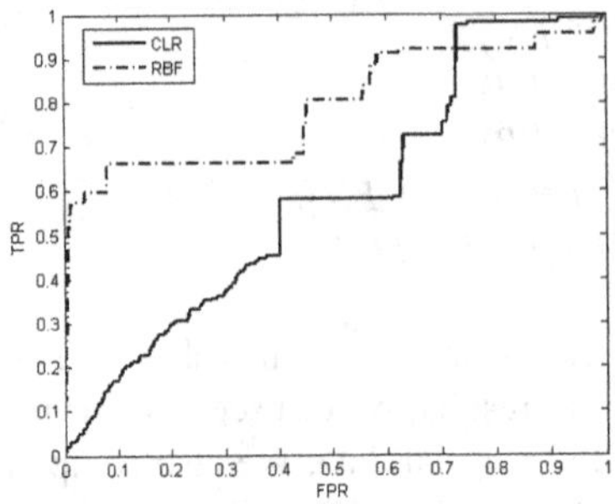

(a)

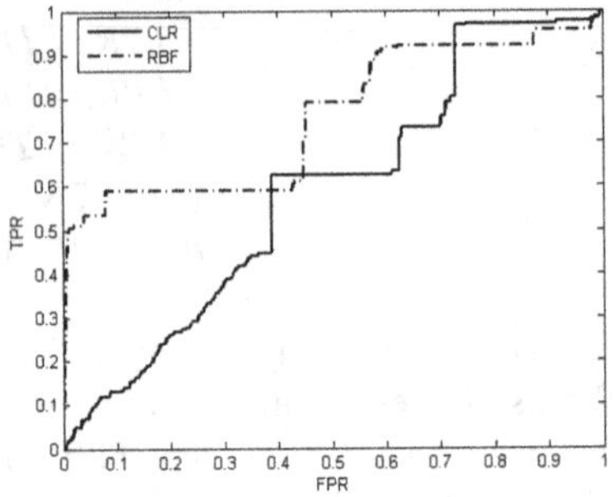

(b)

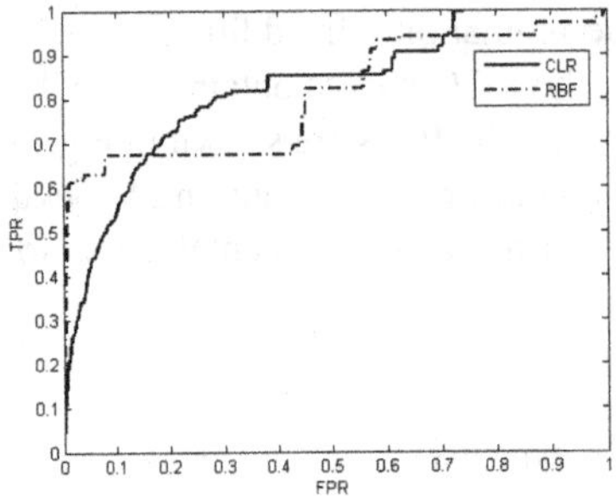

(c)

Fig. 3. ROC curves of two methods with knockdown data (a), knockouts data (b) and multifactorial data (c).

To test the effectiveness of our proposed hybrid evolutionary method (DPSO+FA), we make comparison with DPSO and FA. Through several runs, the results are described in Fig. 4, which shows that our proposed hybrid evolutionary method could gain more optimal solutions than DPSO and FA with the same generation.

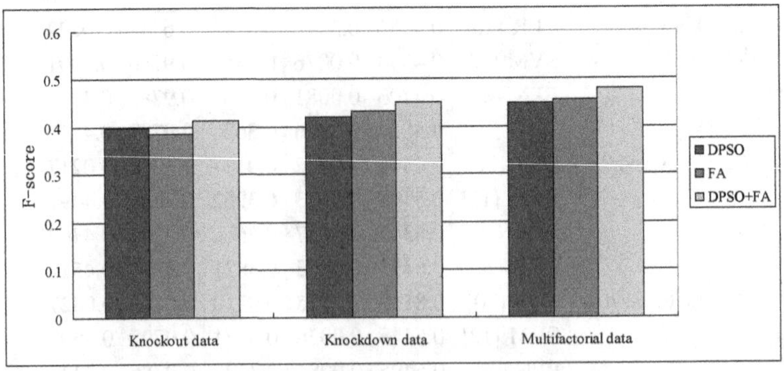

Fig. 4. F-score comparison among PSO, FA and hybrid evolutionary method (DPSO+FA).

4 Conclusions

To summarize, RBF neural network can be used to infer gene regulatory network from a compendium of gene expression data. The sub network with 150 genes from E.*coli* network is used to validate our method. TPR, FPR, PPV, ACC, F-score and ROC curves reveal that our method could gain higher accuracy for biological datasets (knockout, knockdown and multifactorial) than CLR, BPNN and SVM.

In the future, we will apply our method to more large-scale real gene regulatory network identification and develop the parallel program in order to improve the runtime.

Acknowledgements. This work was supported by the PhD research startup foundation of Zaozhuang University (No. 2014BS13), and Shandong Provincial Natural Science Foundation, China (No. ZR2015PF007).

References

1. Ellwanger, D.C., Leonhardt, J.F., Mewes, H.W.: Large-scale modeling of condition-specific gene regulatory networks by information integration and inference. Nucleic Acids Res. **42** (21), e166 (2014)
2. Vera-Licona, P., Jarrah, A., Garcia-Puente, L.D., McGee, J., Laubenbacher, R.: An algebra-based method for inferring gene regulatory networks. BMC Syst. Biol. **8**, 37 (2014)
3. Xie, Y., Wang, R., Zhu, J.: Construction of breast cancer gene regulatory networks and drug target optimization. Arch. Gynecol. Obstet. **290**(4), 749–755 (2014)
4. Penfold, C.A., Millar, J.B., Wild, D.L.: Inferring orthologous gene regulatory networks using interspecies data fusion. Bioinformatics **31**(12), i97–i105 (2015)
5. Baur, B., Bozdag, S.: A canonical correlation analysis-based dynamic bayesian network prior to infer gene regulatory networks from multiple types of biological data. J. Comput. Biol. **22**(4), 289–299 (2015)
6. Yang, M., Li, R., Chu, T.: Construction of a Boolean model of gene and protein regulatory network with memory. Neural Netw. **52**, 18–24 (2014)
7. Adabor, E.S., Acquaah-Mensah, G.K., Oduro, F.T.: SAGA: a hybrid search algorithm for Bayesian Network structure learning of transcriptional regulatory networks. J. Biomed. Inform. **53**, 27–35 (2015)
8. Sun, M., Cheng, X., Socolar, J.E.: Causal structure of oscillations in gene regulatory networks: Boolean analysis of ordinary differential equation attractors. Chaos **23**(2), 025104 (2013)
9. Wang, J., Chen, B., Wang, Y., Wang, N., Garbey, M., Tran-Son-Tay, R., Berceli, S.A., Wu, R.: Reconstructing regulatory networks from the dynamic plasticity of gene expression by mutual information. Nucleic Acids Res. **41**(8), e97 (2013)
10. Maetschke, S.R., Madhamshettiwar, P.B., Davis, M.J., Ragan, M.A.: Supervised, semi-supervised and unsupervised inference of gene regulatory networks. Brief Bioinform. **15**(2), 195–211 (2014)
11. Cerulo, L., Elkan, C., Ceccarelli, M.: Learning gene regulatory networks from only positive and unlabeled data. BMC Bioinform. **11**, 228 (2010)

12. Mordelet, F., Vert, J.P.: SIRENE: supervised inference of regulatory networks. Bioinformatics **24**(16), i76–i82 (2008)
13. Gillani, Z., Akash, M.S., Rahaman, M.D., Chen, M.: CompareSVM: supervised, Support Vector Machine (SVM) inference of gene regularity networks. BMC Bioinform. **15**, 395 (2014)
14. Taher, N., Ehsan, A., Jabbari, M.: A new hybrid evolutionary algorithm based on new fuzzy adaptive PSO and NM algorithms for distribution feeder reconfiguration. Energy Convers. Manage. **54**(1), 7–16 (2011)
15. Kuan-Cheng, L., Yi-Hsiu, H.: Classification of medical datasets using SVMs with hybrid evolutionary algorithms based on endocrine-based particle swarm optimization and artificial bee colony algorithms. J. Med. Syst. **39**, 119 (2015)
16. Reza, M., Fatemi, G., Farzad, Z.: A new hybrid evolutionary based RBF networks method for forecasting time series: a case study of forecasting emergency supply demand time series. Eng. Appl. Artif. Intell. **36**, 204–214 (2014)
17. Jia, W., Zhao, D., Shen, T., Su, C., Hu, C., Zhao, Y.: A new optimized GA-RBF neural network algorithm. Comput. Intell. Neurosci. **2014**, 6 (2014). 982045
18. Xie, F.X., Zhang, W.J., Yang, Z.L.: A dissipative particle swarm optimization. In: Congress on Evolutionary Computation (CEC), pp. 1456–1461 (2002)
19. Yang, X.-S.: Firefly algorithms for multimodal optimization. In: Watanabe, O., Zeugmann, T. (eds.) SAGA 2009. LNCS, vol. 5792, pp. 169–178. Springer, Heidelberg (2009). doi:10.1007/978-3-642-04944-6_14
20. Butte, A.J., Tamayo, P., Slonim, D., Golub, T.R., Kohane, I.S.: Discovering functional relationships between RNA expression and chemotherapeutic susceptibility using relevance networks. Proc. Natl. Acad. Sci. U.S.A. **97**(22), 12182–12186 (2000)

Establishment of Optimal Control Strategy of Building-Integrated Photovoltaic Blind Slat Angle by Considering Interior Illuminance and Electricity Generation

Taehoon Hong, Jeongyoon Oh, Kwangbok Jeong(✉), Jimin Kim, and Minhyun Lee

Department of Architectural Engineering, Yonsei University, Seoul 03722, Republic of Korea
{hong7, omk1500, kbjeong7, cookie6249, mignon}@yonsei.ac.kr

Abstract. A building-integrated photovoltaic blind (BIPB), in which blind and PV system is combined to generate energy in the building exterior and reduce the heating and cooling load in building by shading function. This study aimed to establish the optimal control strategy of BIPB slat angle by considering interior illuminance and electricity generation. First, in terms of interior illuminance considering overall light (i.e., daylight and artificial illumination) and electricity generation from BIPB, it was determined that the optimal blind slat angle is 80° at all time. Second, in terms of interior illuminance considering daylight and electricity generation from BIPB, it was determined that the optimal blind slat angle is 10° (9:00), 20° (10:00–11:00, 14:00–15:00) and 30° (12:00–13:00). Based on results of this study, effective use of BIPB can be induced by providing information for optimal blind slat angle to users that are considering BIPB implementation.

Keywords: Optimization · Building-integrated photovoltaic blind · Electricity generation · Interior illuminance

1 Introduction

To solve the issue of global pollution and increasing needs for energy, attention for new and renewable energy is increasing [1]. Especially, the photovoltaic (PV) system is easy to implement on buildings, and the potential of energy substitution is superior [2]. Meanwhile, the South Korea government amended *act on the promotion of green buildings* in May 2015, and decided that the exterior window in public building with total floor area larger than 3,000 m^2 should have shading device installed [3]. Accordingly, the research should be conducted on building-integrated photovoltaic blind (BIPB), in which blind and PV system is combined to generate electricity in the building exterior and reduce the heating and cooling load in building by shading function.

Meanwhile, the previous studies have been focused on technical-economic performance analysis according to BIPB's design variables, and there are not enough

P.M. Pardalos et al. (Eds.): MOD 2016, LNCS 10122, pp. 451–454, 2016.
DOI: 10.1007/978-3-319-51469-7_40

studies regarding both the interior illuminance and electricity generation from BIPB [4, 5]. Therefore, this study aimed to establish the optimal control strategy of BIPB slat angle by considering interior illuminance and electricity generation, which was conducted in three steps: (i) step 1: definition of design variables; (ii) step 2: estimation of interior illuminance and electricity generation from BIPB using energy simulation; and (iii) step 3: optimization of BIPB slat angle by considering interior illuminance and electricity generation.

2 Material and Method

2.1 Step 1: Definition of Design Variables

This study considered the design variables that affect the interior illuminance and electricity generation from BIPB in three aspects [5, 6]: (i) architectural design elements (i.e., region and orientation); (ii) window design elements (i.e., visible transmittance (VT) and exterior window area); and (iii) BIPB design elements (i.e., blind slat angle and efficiency of PV panel).

2.2 Step 2: Estimation of Interior Illuminance and Electricity Generation from BIPB Using Energy Simulation

To establish the optimal control strategy of BIPB blind slat angle, this study was conducted the estimation of the interior illuminance and electricity generation from BIPB by using the 'Autodesk Ecotect Analysis' and 'Radiance' software program. The analysis case has been set for one class from 'H' elementary school in South Korea, and the analysis time has been set as 9:00–15:00 (8 h), based on the elementary school's time for classes at the summer solstice [7].

The Table 1 shows the basic information of design variables. First, the region and orientation as an architectural design elements were set as Seoul and Southern, respectively. Second, the VT and exterior window area as a window design elements were set as 62.3% and 11.3 m^2, respectively. Third, the blind slat angle and efficiency of PV panel were set as 0–90° and 11.7% through the market research, respectively [5]. This study performed the estimation of interior illuminance and electricity generation based on the 10 discrete values of blind slat angle (e.g., 0°, 10°, and etc.).

Table 1. Basic information of design variables

Class	Architectural design elements		Window design elements		BIPB design elements	
	Region	Orientation	Visible transmittance	Exterior window area	Blind slat angle	Efficiency of PV panel
'H' elementary school	Seoul	Southern	62.3%	11.3 m^2	0–90°	11.7%

2.3 Step 3: Optimization of BIPB Slat Angle by Considering Interior Illuminance and Electricity Generation

This study considered two optimization objectives for establishment of optimal control strategy of BIPB slat angle: (i) interior illuminance considering overall light (i.e., daylight and artificial illumination) or daylight; and (ii) electricity generation from BIPB. First, in terms of interior illuminance considering overall light or daylight, the optimal blind slat angle has been defined to satisfy standard illuminance in classroom (i.e., 400 lx–600 lx) [8]. Second, in terms of electricity generation from BIPB, the optimal blind slat angle has been defined to have the maximum energy generation.

3 Results and Discussions

The optimal control strategy of BIPB slat angle considering interior illuminance and electricity generation was established as follows: (i) strategy 1 (interior illuminance considering overall light and electricity generation from BIPB); and (ii) strategy 2 (interior illuminance considering daylight and electricity generation from BIPB) (refer to Table 2).

- *Strategy 1(Interior illuminance considering overall light and electricity generation from BIPB)*: The interior illuminance considering overall light (i.e., daylight and artificial illumination) satisfies the standard illuminance in classroom at all times when the blind slat angle is 70–90°, and the electricity generation from BIPB is maximized when the setting is 60° (9:00 and 12:00–15:00) and 80° (10:00–11:00). Therefore, it was determined that the optimal blind slat angle is 80° to meet the standard illuminance in classroom and maximum electricity generation from BIPB.
- *Strategy 2 (Interior illuminance considering daylight and electricity generation from BIPB)*: The interior illuminance considering daylight satisfies the standard illuminance in classroom when the blind slat angle are 0° and 10° (9:00–15:00), 20° (10:00–15:00), and 30° (12:00–13:00). It means that the lighting energy consumption can be reduced by using only daylight. Therefore, it was determined that the optimal blind slat angle is 10° (9:00), 20° (10:00–11:00 and 14:00–15:00) and 30° (12:00–13:00) to meet the standard illuminance in classroom and maximum electricity generation from BIPB.

Table 2. Optimal BIPB slat angle by control strategy

Class	Time						
	9:00	10:00	11:00	12:00	13:00	14:00	15:00
Strategy 1	80°	80°	80°	80°	80°	80°	80°
Strategy 2	10°	20°	20°	30°	30°	20°	20°

4 Conclusion

This study aimed to establish the optimal control strategy of BIPB slat angle considering interior illuminance and electricity generation, which was conducted in three steps: (i) step 1: definition of design variables; (ii) step 2: estimation of interior illuminance and electricity generation from BIPB using energy simulation; and (iii) step 3: optimization of BIPB slat angle by considering interior illuminance and electricity generation. The main findings could be summarized as follows. First, in terms of interior illuminance considering overall light and electricity generation from BIPB, it was determined that the optimal blind slat angle is 80° to meet the standard illuminance in classroom and maximum electricity generation from BIPB. Second, in terms of interior illuminance considering daylight and electricity generation from BIPB, it was determined that the optimal blind slat angle is 10° (9:00), 20° (10:00–11:00 and 14:00–15:00) and 30° (12:00–13:00) to meet the standard illuminance in classroom and maximum electricity generation from BIPB. Based on results of this study, effective use of BIPB can be induced by providing information for optimal blind slat angle to users that are considering BIPB implementation.

Acknowledgements. This work was supported by the National Research Foundation of Korea (NRF) grant funded by the Korea government (MSIP; Ministry of Science, ICT & Future Planning) (NRF-2015R1A2A1A05001657).

References

1. Korea Energy Economics Institute (KEEI): Energy demand outlook (2015)
2. Korea Institute of Energy Research: New & renewable energy white paper (2014)
3. The Act on the Promotion of Green Buildings: Ministry of Land Infrastructure and Transport. MOLIT, Sejong (2015)
4. Kang, S., Hwang, T., Kim, J.T.: Theoretical analysis of the blinds integrated photovoltaic modules. Energ. Build. **46**, 86–91 (2012)
5. Hong, T., Koo, C., Oh, J., Jeong, K.: Nonlinearity analysis of the shading effect on the technical-economic performance of the building-integrated photovoltaic blind. Appl. Energy (2016, in press)
6. Tzempelikos, A.: The impact of venetian blind geometry and tilt angle on view, direct light transmission and interior illuminance. Sol. Energy **82**(12), 1172–1191 (2008)
7. Seoul Haengdang Elementary School. http://hd.es.kr/index/index.do. Accessed 06 Oct 16
8. Korean Agency for Technology and Standards. https://standard.go.kr/. Accessed 06 Oct 16

Author Index

Printed in the United States
by Bookmasters

Printed in the United States
By Bookmasters